Green Chemistry

Principles and Case Studies

Green Chemistry

Principles and Case Studies

By

Felicia A. Etzkorn

Virginia Tech, USA
Email: fetzkorn@vt.edu

ROYAL SOCIETY OF **CHEMISTRY**

Print ISBN: 978-1-78801-798-5
EPUB ISBN: 978-1-83916-016-5

A catalogue record for this book is available from the British Library

The Royal Society of Chemistry is a charity, registered in England and Wales, Number 207890, and a company incorporated in England by Royal Charter (Registered No. RC000524), registered office: Burlington House, Piccadilly, London W1J 0BA, UK, Telephone: +44 (0) 20 7437 8656.

Visit our website at www.rsc.org/books

Printed in the United Kingdom by CPI Group (UK) Ltd, Croydon, CR0 4YY, UK

Preface

Green Chemistry is a radically different approach to producing the products of the modern world in a sustainable, non-polluting, and non-hazardous way. Green Chemistry is efficient chemistry, wasting neither materials nor energy. Of course, the Platonic ideal of waste-free, completely benign chemical processes and products is a lofty goal and may not be entirely possible. But the harder we try, the more creative and ingenious we are, the better off humanity and the planet will be.

Green Chemistry, as a field of investigation, was unknown when I was a student. I first started thinking about chemistry as a senior in high school taking Environmental Biology with Mrs. Grote. When I asked her "If I want to do something about the environment, should I major in Biology?" She said no, I should become a chemist if I really wanted to change the world. Well, here I am, hoping to help my students change the world. Mrs. Grote was prescient. I thought she meant for me to become an environmental chemist. But even in college, I was discouraged by environmental chemistry because it meant cleaning up the messes people had already made, and still are making. Nevertheless, Environmental Chemistry is a critical field of study, a prerequisite for Green Chemistry, because we have to know and define what pollution is, what is harmful or toxic, before we can avoid it.

The Principles of Green Chemistry, developed by Paul C. Anastas and John Warner in 1995, are the tools by which we attempt to *anticipate* and avoid using and creating hazardous substances to create the wonderful variety of consumer products to make our lives

Green Chemistry: Principles and Case Studies
By Felicia A. Etzkorn

Published by the Royal Society of Chemistry, www.rsc.org

easier. Chemists make everything from carbon fiber for jet planes to new antibiotics that save lives. We cannot hope to cover every possible situation that a chemist might face, but I hope by giving students the tools of Green Chemistry, they will invent creative new ways to make useful materials that will have minimal negative consequences for people, animals, plants or the environment. This is what we mean by sustainability.

This textbook is intended to give instructors a tool to use in active-learning classrooms. The exact practices used in the classroom, and the ways the textbook could be used will vary as widely as the people using it. The book is arranged into chapters corresponding to each of the Principles of Green Chemistry. I have used case studies to illustrate various aspects of each principle within each chapter. Although I tried to match the case studies with particular Principles, real cases are complex. Most of the case studies embody more than one principle, so the student must be aware that such a linear pathway through the material is artificial. Referring often to the Principles and to the introductions to subsequent chapters may be helpful. Each chapter has problems at the end that are intended to help students actively engage with the material of that chapter. A set of very useful Appendixes contain data essential to solving chemical problems.

Because a deep understanding of chemical mechanisms allows us to design greener processes and products, I place a heavy emphasis on student engagement with a wide range of mechanisms, from nucleophilic substitution to organometallic catalytic cycles. A review of organic functional groups is given in Appendix A. Step-by-step instructions for drawing a mechanism or a molecular orbital diagram are given in Appendix B. Slides with spoken commentary on basic methods for drawing mechanisms and molecular orbital diagrams are available at https://pubs.rsc.org/en/content/ebook/978-1-78801-798-5 to allow the student to work step-by-step repeatedly if desired.

Other means of active learning that I have used in my courses are: (1) student literature oral presentations, and (2) group proposals. The literature presentation is a good way for students to develop the habit of reading current literature in the field, while developing critical thinking about the content of those articles. It does happen, there are mistakes in the literature, or maybe the experiments could be designed better. In addition, it is just as important to understand how the work was done, why the control experiments were done, what one can and cannot conclude from a particular set of results, and how useful those results might be in the real world. A set of guidelines to help students prepare a good oral presentation can be requested by email.

The group proposal serves as a launch pad for student creativity in Green Chemistry. A set of guidelines to help students identify and develop a proposal that is neither trivial nor impossible can be requested by email. Group proposals might even become the foundation for a new small business—this is my dream. I remain optimistic, both for our students, and for our future.

Felicia A. Etzkorn

Acknowledgements

I am forever grateful to all of the Virginia Tech students, too numerous to mention by name, who have taken Green Chemistry with me through the years, from its nascent 1 credit-hour form in 2003, through the thick and thin of the first 3 credit-hour form disrupted by the horrific event of April 16, 2007, to the introduction of my only half-completed textbook in spring of 2015, and onwards to the present. They engaged enthusiastically with me in learning what this field of Green Chemistry is all about. They found errors in the text. They taught me so much about what helps them learn, and what does not. I am grateful for the students' choices of truly important green chemistry papers to present orally in class; we learned so much together. I'd like to particularly thank the students whose chosen paper ended up as a case study in this book; I might not have come across the life cycle analysis of polylactic acid, or the use of the water hyacinth, *Eichhornia crassipes*, to recover Pd from wastewater. I am forever grateful to Virginia Tech graduate student, Paul Arcoria, for his close reading of the entire text, and for his diligent grading as the first teaching assistant I've had for this course. I always felt like Paul had my back, thanks.

I am ever so grateful to Professor Paul Anastas (Yale University), and Dr John C. Warner (Warner Babcock Institute for Green Chemistry) for leading the way in establishing Green Chemistry as a discipline. Their vision was powerful and it will truly make the world a better place. I am grateful to Professor Michael Cann (University of Scranton) who led the way with two volumes of *Real-World Cases in Green Chemistry* that I have used previously in my Green Chemistry courses. He was my guiding light in organizing, developing, and

Green Chemistry: Principles and Case Studies
By Felicia A. Etzkorn

Published by the Royal Society of Chemistry, www.rsc.org

writing this book. I thank Professor David Brown (Davidson College) and Dr Rachel Schmidt-Radde (Beindersheim, Germany) for editing Chapter 2 very carefully. I particularly want to thank Professor Jamie Ferguson (Emory & Henry College) for her friendship, her great suggestions, and her close reading of Chapters 3, 4 and 5. I am grateful to Professor Diego Troya (Virginia Tech) for his insistence on the precision of language, and for editing Chapter 6, and to Professor David Kingston (Virginia Tech) for his careful editing and excellent suggestions for Chapter 7. Thanks go to Professor Geoff Coates (Cornell University) and Professor Stephen Miller (University of Florida) for their comments and corrections on the case studies referring to their work in Chapter 10.

I thank all of my many fellow travelers who organized, spoke at, or attended the *ACS Green Chemistry & Engineering Conferences*. Thanks to all the green chemists around the world for your creativity and hard work in making the world a better place. I thank my editor at the Royal Society of Chemistry, Helen Armes, for her encouragement and prompt responses to all my niggling questions. I am grateful to Connor Sheppard at the Royal Society of Chemistry for his prompt responses to my almost weekly questions about copyright permissions. I hope the reader will forgive any errors—scientific, grammatical, attribution, or otherwise. I take full responsibility for the entire contents. All errors are my own.

Dedication

In memory of my dad, Larry Etzkorn, who exemplified an ethical life.

Green Chemistry: Principles and Case Studies
By Felicia A. Etzkorn

Published by the Royal Society of Chemistry, www.rsc.org

The 12 Principles of Green Chemistry

Adapted from P. T. Anastas and J. C. Warner, *Green Chemistry: Theory and Practice*, Oxford University Press, New York, 1998, 30, with permission from the authors. © 1998 Paul T. Anastas and John C. Warner.

1. **Prevent Waste.** It is better to prevent waste than to treat or clean up waste after it has been created.
2. **Synthetic Efficiency.** Synthetic methods should be designed to maximize the incorporation of all materials used in the process into the final product.
3. **Benign Syntheses.** Wherever practicable, synthetic methods should be designed to use and generate substances that possess little or no toxicity to human health and the environment.
4. **Benign Products.** Chemical products should be designed to effect their desired function while minimizing their toxicity.
5. **Avoid Auxiliaries.** The use of auxiliary substances (e.g., solvents, separation agents, etc.) should be made unnecessary wherever possible and innocuous when used.
6. **Energy Efficiency.** Energy requirements of chemical processes should be recognized for their environmental and economic impacts and should be minimized. If possible, synthetic methods should be conducted at ambient temperature and pressure.
7. **Renewable Feedstocks.** A raw material or feedstock should be renewable rather than depleting whenever technically and economically practicable.
8. **Avoid Protecting Groups.** Unnecessary protecting groups (or other temporary modification of physical/chemical processes)

Green Chemistry: Principles and Case Studies
By Felicia A. Etzkorn

Published by the Royal Society of Chemistry, www.rsc.org

should be minimized or avoided if possible, because such steps require additional reagents and can generate waste.

9. **Catalysis.** Catalytic reagents (as selective as possible) are superior to stoichiometric reagents.
10. **Degradation or Recovery.** Chemical products should be designed so that at the end of their function they break down into innocuous degradation products and do not persist in the environment. Products that do not degrade should be designed to fully recover the chemical materials from which they were made.
11. **Real-time Analysis.** Analytical methodologies need to be further developed to allow for real-time, in-process monitoring and control prior to the formation of hazardous substances.
12. **Prevent Accidents.** Substances and the form of a substance used in a chemical process should be chosen to minimize the potential for chemical accidents, including releases, explosions, and fires.

Contents

Green Chemistry: Principles and Case Studies
By Felicia A. Etzkorn

Published by the Royal Society of Chemistry, www.rsc.org

5 Avoid Auxiliaries 125

6 Energy Efficiency 169

1 Prevent Waste

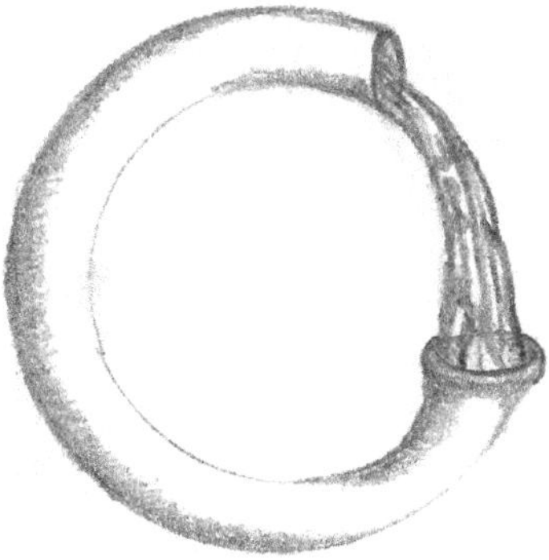

"Principle 1: It is better to prevent waste than to treat or clean up waste after it has been created."[1]

1.1 Green Chemistry: First, Do No Harm

Green chemistry aspires to give us all of the wonderful products we have come to expect in our lives, without the associated pollution of the past. The job of green chemists is to eliminate hazards completely from our processes and products, so that no accidental release or exposure would be possible. That is the ideal, the lofty goal we aspire to, but the reality is one of more gradual shift and compromise, tempered by economics and regulations. The Green Chemistry Institute was founded in 1995, the same year the US Presidential Green Chemistry Challenge (PGCC) Awards were established by the US Environmental Protection Agency (EPA) under President Bill Clinton. These awards

Green Chemistry: Principles and Case Studies
By Felicia A. Etzkorn

Published by the Royal Society of Chemistry, www.rsc.org

have now inspired industrial and academic chemists to invent new products and processes that are "benign by design," or sustainable without polluting the environment.

Chemists have serious responsibilities in our profession. Whether a chemist or a company chooses to ignore the consequences of their chosen chemicals and methods, or to take great pains to produce the best products in the least polluting way that one can imagine, chemistry has consequences. Green chemistry holds that each chemist, each company, is responsible for preventing accidents and pollution, and for using the least amount of resources necessary. Green chemistry has as its core value, a chemist's version of the Hippocratic Oath, "First, Do No Harm."[2] This injunction is particularly apt in the case of chemistry, where the products of our labor may last for years, or even centuries, for better or for worse.

1.1.1 Definition of Green Chemistry

> "Green chemistry is the design of chemical products and processes that reduce or eliminate the use and generation of hazardous substances."[†]

A course in Green Chemistry might just as well be called Industrial Chemistry, because green chemistry is about making commercial products by clean and efficient methods. In many ways, industry is leading the way in making greener products. Half of the PGCC awards have been to industry. PGCC awards also go to academic chemists with a strong desire that their greener process or product find application in society, so much so that they often start a company to commercialize their chemistry. Many of the chemicals described in the case studies of this book are commodity chemicals, such as starting materials for things like plastics or dyes. Commodity chemicals are defined by the large scale of their production, and the global market for their distribution. Others are fine chemicals, made on smaller scale, but just as crucial, like medicines. Green chemistry is *all* about the applications, thus this book will be grounded in case studies, primarily from the US Presidential Green Chemistry Challenge (PGCC) Awards.

1.1.2 Principles of Green Chemistry

In 1998, Paul Anastas and John C. Warner published a seminal book, *Green Chemistry: Theory and Practice*, which established the 12 Principles of Green Chemistry as the foundation for this dramatic shift in the way

[†]From ref. 3. Copyright © 2017 Yale University. All rights reserved.

we do chemistry.[1] The 12 Principles fall into three general categories: (1) use and produce no toxic chemicals, (2) minimize the use of chemical and energy resources, and (3) prevent accidents. The chapters in this book are organized by Principle, but many of the case studies use several of the principles to achieve their goals. These illustrative case studies are designed to inspire the student to find new ways to apply the principles in their future work.

1.1.3 Economic Driving Force

One of the cool things about green chemistry is that greener processes and products are also typically more economical, with renewable feedstocks, fewer steps, lower energy costs, and lower hazardous waste disposal costs.

In 2001, Shaw Industries was bought by Berkshire-Hathaway Inc. headed by Warren Buffett. Shaw developed EcoWorx carpet tiles that were completely recyclable back into carpeting.[4] A number of economic advantages propelled EcoWorx to a market share of 80% of the carpet sold in their product line. The feedstocks were old carpet tiles sent back for remanufacturing. The carpet tiles were made from lighter materials that required less energy in manufacturing and shipping. The EcoWorx product line is highly successful.

Carpet tiles originated with Interface, Inc., whose owner, Ray Anderson said, "Sustainability is proving to be incredibly good for business. What began as the right thing to do quickly became the smart thing to do. Sustainability doesn't cost, it pays."[5]

Unfortunately, we are still in a time of cheap fossil feedstocks—petroleum and coal, and this can doom a greener process. For example, Dupont developed a new method for breaking down polyethylene terephthalate (PET) into the two monomers that could then be repolymerized into virgin PET. They built a pilot plant in North Carolina in 1996 that produced 100 million pounds of PET per year, but a drop in oil prices led to the plant closure and dismantling in 2000. There may come a day when oil prices rise, and recycling PET could again become cost competitive.

1.2 Better Living Through Chemistry

There is no doubt that the modern world has benefited enormously from the efforts of chemists, without which we would not have most medicines, fuels, computers, in short, the stuff of modern life. In 1935, Walter Carothers invented a nearly indestructible fabric, nylon.

The same year, Dupont asserted in an advertising slogan, "Better Things and Better Living...through Chemistry," which was later co-opted and shortened to "Better Living Through Chemistry." One of the earliest organic synthesis industries, the German synthetic color dye industry, spawned the first medicinal chemistry program. The first antibiotic, sulfanilamide (first a red dye),[6] saved countless lives, and was forerunner to the whole class of antibiotic drugs. We are now on the cusp of individualized medicine, in which each patient's unique physiology and disease progression are considered when designing and prescribing medical treatment. The tricky part will be whether we can, at the same time, avoid side effects and the production of toxic pollutants in the process of converting to individualized medicine.

1.2.1 Predicting Harm: Evolution of Refrigerants

We have to know what the potential harm might be before we can know that we should avoid it. In the early 1900s, refrigerators that used ammonia as the coolant were used commercially, but the acute toxicity of ammonia gas prevented widespread use in homes. The solution came when chemists at General Motors developed Freon (Figure 1.1), patented in 1928 as a refrigerant, which was non-flammable and of very low toxicity. Perhaps this could be considered an early attempt at green chemistry. Fifty years later, chemists Frank S. Rowland and Mario J. Molina, discovered that Freon could destroy the ozone layer of the stratosphere that protects living things from UV radiation. Together with Paul Crutzen, they were awarded the Nobel Prize in Chemistry in 1995 for their work on ozone depletion. The chemists at General Motors who invented CFCs did not anticipate, nor could they even have been expected to anticipate, any larger problems with CFCs.

The next step in replacements for CFCs contained only fluorine, carbon, and hydrogen, primarily hydrofluorocarbon (HFC)-134a, 1,1,1,2-tetrafluoroethane, which does not destroy the ozone layer (Figure 1.1). Unfortunately, the saturated HFCs are potent greenhouse gases.[7] HFC-134a has a 100-year global warming potential (GWP) that is equivalent to 1430 times that of CO_2.[8] Currently, hydrofluoroolefin (HFO)-1234yf (2,3,3,3,-tetrafluoropropene) is considered an efficient

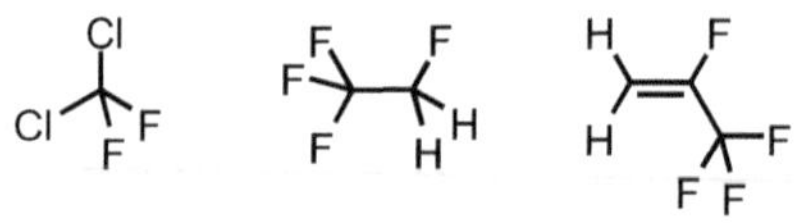

Figure 1.1 Refrigerants, L to R: Freon, (HFC)-134a, and (HFO)-1234yf.

and safe refrigerant with a 100-year GWP less than 1 (Figure 1.1). US and Japanese automakers are switching to HFO-1234yf, however Mercedes-Benz engineers in Germany found that it ignited on a hot engine when mixed with the compressor oil necessary for air conditioning.[9] Highly corrosive HF gas, given off in the fire, etched the windshield in their test.[9] Under ordinary release conditions, HFO-1234yf degrades by oxidation to trifluoroacetic acid (TFA),[10] a strong, toxic acid for which the environmental safety has not yet been fully assessed.[11]

The most benign refrigerant under development is CO_2.[12] CO_2 has a GWP of 1 by definition, well below the European Union requirement for refrigerants to be below a GWP of 150. CO_2 is cheap, abundant, non-toxic, and all natural; it is not subject to government regulation and can be released to the atmosphere requiring no special recovery methods. Any CO_2 used in refrigeration would either be withdrawn from the atmosphere or recovered from burning fossil fuel for energy. One problem with CO_2 is that it operates as a refrigerant best under transcritical conditions, above the critical pressure of 73 atm.[13] This is essentially an engineering design problem, requiring new systems with stronger containers and special valves. There are also engineering advantages. The volumetric refrigeration capacity is higher for CO_2, so units can be smaller.[13] The compression ratio is about half that of HFC compressors, increasing the efficiency.[13] The problems have been solved for stationary refrigeration, and German automakers are working on a solution for vehicles.[14]

We have yet to discover a truly benign, efficient substitute for HFCs. This illustrates the difficulty of finding truly green chemical solutions for significant societal needs. No one wants to go back to the days of smelly, moldy iceboxes, or worse yet, no refrigeration or air conditioning at all. Replacing HFCs remains a significant opportunity for green chemists.

The chemists who developed Freon, indeed all of us, were unaware of stratospheric ozone destruction by CFCs. Our collective ignorance leads one to wonder if there are other types of environmental or biological harmful effects that we do not yet know, cannot predict, or have missed in the literature. It is wise for chemists to pay attention to the work of toxicologists and environmental chemists. Whatever type of harm is known, we should take care to avoid in our work.

1.3 Environmental Pollution

Most people lack interest in, or feel incapable of comprehending, the chemistry used to make the products they use. We just want our things to function well. We want durable goods, like a refrigerator, a house, or a car, to last a long time and to still function. We want a product that is supposed to be temporary, like a plastic bag, to go away; we want to drop it in the trash and have it disappear. And this is where chemists have succeeded heroically, and also failed miserably. Our work is a work in progress.

What most people do care about is a clean environment, at a minimum in their own backyard—hence the term "not in my back yard" or NIMBY, that refers, often derogatively, to people fighting against contamination of their home or community environment. We want clean air to breathe, clean water to drink, clean food to eat, and clean places to live, play and work. The connection between the chemistry used to make products and a clean environment is obscure to many, though it has become more mainstream since 1970, when the US Clean Air Act was passed.[15] The US Federal Water Pollution Control Act Amendments of 1972, now known as the Clean Water Act of 1977,[16] did much the same for water pollution consciousness.

1.3.1 DDT and Silent Spring

The environmental consciousness of the modern era has been attributed to publication in 1962 of the book, *Silent Spring* by Rachel Carson.[17,18] In this moving and superbly written account of the dramatic deformities and stillbirths of birds caused by the pesticide 2,2-*p*,*p*′-dichlorodiphenyl-1,1,1-trichloroethane (DDT, Figure 1.2), Carson was first to observe endocrine disruption in birds. It was Carson who first pointed the finger at ourselves, "...no enemy action had silenced the rebirth of new life in this stricken world. The people had done it themselves."[18] This new attitude found its way into popular culture with Walt Kelly's *Pogo* comic strip on the newly founded Earth Day in 1971 (Figure 1.3).

1.3.2 Times Beach and Love Canal Super Fund Sites

The environmental movement had a big job cut out for it. Although toxicology is an ancient science, the twist was that chemical

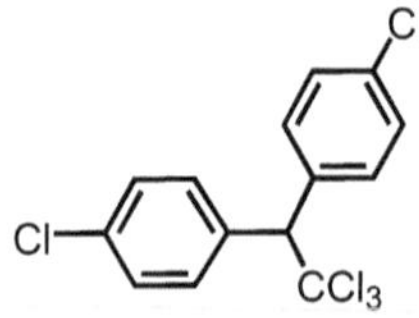

Figure 1.2 The structure of 2,2-*p*,*p*′-dichlorodiphenyl-1,1,1-trichloroethane (DDT).

Figure 1.3 We have met the enemy, and he is us. *Pogo* comic strip by Walt Kelly published on Earth Day, April 22, 1971. Copyright Okefenokee Glee & Perloo, Inc. Used by permission. Contact permissions@pogocomics.com.

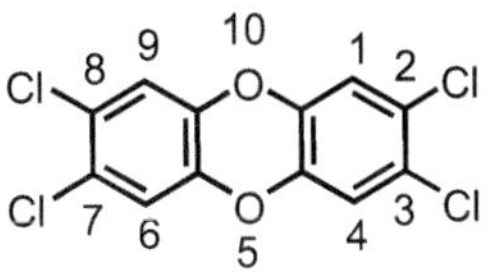

Figure 1.4 Structure of the most common dioxin: 2,3,7,8-tetrachlorodibenzo [b,e][1,4]dioxin.[19]

manufacturing was producing and selling novel synthetic or "man-made" substances for which there were no health effects data, and none were required by law. Two events galvanized a public now attuned to environmental issues. In 1972, Russell Martin Bliss sprayed waste oil contaminated with dioxin (2,3,7,8-tetrachlorodibenzodioxin, Figure 1.4) in Times Beach, Missouri. The dioxin and trichlorophenol caused many birds and horses to die immediately, and caused severe health effects to the people living in the town. Dioxins are among the most potent toxins known.

There is still much concern about the formation of dioxins in municipal waste incinerators and paper bleaching. In order to completely destroy dioxins during incineration, three conditions are necessary: (1) high combustion temperature, (2) adequate combustion time, and (3) high combustion turbulence.[19] Dioxins form as flue gases are cooling during emissions or flue gas filtration, so incinerator design is vital, especially when chlorine-containing compounds are being incinerated.[19]

In 1977, severe health effects due to toxins leaking from the Hooker Chemical Co. waste site at Love Canal, NY came to light by the efforts of two investigative newspaper reporters for the Niagara Falls Gazette, David Pollak and David Russell. The result was the Super Fund, or Comprehensive Environmental Response, Compensation and Liability Act of 1980, which is used to this day to clean up hazardous waste sites.

The sad lesson of Love Canal is that burial of toxic waste in landfills is no guarantee of safety on a human history time scale, or in the case of nuclear waste, on a geological time scale. Thus, the primary goal of green chemistry is neither to use nor produce toxic substances in the first place.

1.4 Risk Is a Function of Hazard and Exposure

Green chemistry differs from environmental chemistry in that we seek to avoid using and creating chemical products that are in any way hazardous. In my first job out of college, I worked with and synthesized very hazardous substances—radioactive carcinogens—that were used in environmental and biological studies of pollution. Our primary goal was, understandably, to protect ourselves with fume hoods, double-door entries, rubber shoe covers, white paper "zoot" suits worn over our clothes, double gloves, safety glasses, and weekly urine and lab monitoring. This approach epitomizes the exposure-avoidance model of safety that was in vogue in 20th century industry, which is still used in most chemistry laboratories. What we did not much consider was that the hazardous fumes were going out the hood into the community, or the barrels of radioactive, carcinogenic, and flammable waste that were sealed and put into landfills. (Explosives were taken out into a field and shot with a rifle to detonate and inactivate them.) Little consideration was given to what happens when those barrels corrode and begin to leak into the environment, such as what happened at Love

Canal in Niagara Falls, NY in 1953. After many other infamous chemical spills, releases and exposures later, including the horrific explosion in Bhopal, India, and the spraying of dirt roads and horse arenas with dioxin-contaminated oil in Times Beach, Missouri (unsettlingly close to my home town), we are finally wising up.

Anastas and Warner set out the principle of risk as an equation.[1] If *either* hazard or exposure is zero, then zero risk is involved. Zero risk, that is, in a perfect world.

$$\text{Risk} = f\,[\text{hazard, exposure}]^{1}$$

Their point was that we have historically aimed to eliminate risk by eliminating exposure, through the use of barriers, such as gloves, laboratory coats, fume hoods, blast shields, double-walled reactors, double-hulled petroleum tankers, *etc.* The sad truth is that barriers break, gloves tear, fume hoods emit toxins into the local air, ships are breached, and we are all eventually exposed to toxins that begin leaking from hazardous waste sites once a steel drum begins to rust, like at Love Canal.

Many chemists have a sense of pride in their ability to handle hazardous chemicals safely. I am one of these—I have used ^{3}H and ^{14}C (both radioactive), Hg, HF, and many other extremely hazardous chemicals in my work over the years, and I am proud of my skills in their safe handling. There are many chemicals I would rather not ever touch. Derek Lowe, a medicinal chemist well known for his blog "In the Pipeline," has an entertaining, long, and growing list of "Things I Won't Work With."[20] A better way, the ideal of green chemistry, is to use non-toxic, non-flammable, and non-explosive chemicals to eliminate the hazard part of the risk equation entirely, so that exposure is no longer an issue, and risk ideally becomes zero.

We will see that green chemists sometimes compromise. In a case study on the BHC synthesis of ibuprofen in Chapter 2, extremely dangerous HF is used as a Lewis acid to replace $AlCl_3$ in a Friedel–Crafts reaction, which eliminates tons of hazardous waste ($AlCl_3 \cdot xH_2O$). The HF is easily recycled because of its low boiling point. This is a compromise because the reaction is more atom economical. Yet HF was the first chemical Derek Lowe posted on his "Things I Won't Work With" blog, a good starting list of chemicals to avoid for prevention.

1.4.1 Hazards: Union Carbide Explosion in Bhopal, India

Toxicity is not the only hazard that green chemistry seeks to avoid. Highly reactive, compressed gas, explosive, flammable, corrosive, strongly acidic, basic, oxidizing or reducing chemicals are also hazard concerns. On December 3, 1984, a terrible explosion at a Union Carbide India plant resulted in the death of 5200 people, and thousands of others were permanently or partially injured in Bhopal, India.[21] A storage unit that contained 42 tons of methyl isocyanate (MIC) had water leak into it through a faulty valve, initiating an exothermic reaction. The pressure built up and the tank exploded, releasing MIC along with other reaction products. MIC is flammable, reactive with water, and very toxic to the respiratory system.[22] MIC has a boiling point of just 39 °C, so that it becomes a gas with minimal temperature elevation. MIC is extremely hazardous—the cause of most of these deaths was pulmonary edema, or excess fluid in the lungs.[23] One of the lessons from Bhopal is that highly reactive chemicals, such as MIC, should be avoided in chemical synthesis if at all possible.

1.5 Toxicology and Environmental Chemistry

In order to design better processes and products, chemists need to know what is hazardous, and why, in order to avoid it. Toxicology is the discipline that studies the biological effects of toxic chemicals, both the qualitative (severity), and quantitative (potency) aspects of toxicity, as well as the biological mechanism of action. We will see that there are ways to avoid toxicity by design in Chapter 4.

Environmental chemistry is quite distinct from green chemistry, yet environmental chemistry is a critical partner in ensuring that chemistry becomes sustainable, that all chemistry becomes green chemistry. Wikipedia says, "Environmental chemistry can be defined as the study of the sources, reactions, transport, effects, and fates of *chemical* species in the air, soil, and water environments."[24] Since all matter is chemical, we should be more specific and say that environmental chemistry is the study of synthetic or mined pollutants in the environment. There are other types of pollution, such as soil loss due to agriculture or construction, but this is typically not the domain of chemists.

Environmental chemists have been analyzing ever-lower concentrations of toxic and hazardous substances in a wide array of environmental and biological samples. For example, three years after the initial dioxin spraying, and after 18 inches of topsoil were removed, soil samples at the Shenandoah Stables in Times Beach were found to contain 30 ppm of dioxin, which is a very high level of this potent toxin.[25] Now, environmental chemists routinely find parts-per-trillion (ppt), equivalent to picomolar (10^{-12} M), concentrations of dioxin.[26] This illustrates that there is no such thing as a zero concentration of a pollutant in the environment. It is a question of the detection level of an instrument for a specific substance, by a particular method, in a particular matrix.

Once the type and amount of a substance is found, environmental chemists are charged with figuring out how to clean up the mess, whether to attempt to destroy it, or to gather it up for disposal in a landfill, which of course could leak later. Environmental chemistry has the unenviable job of detecting, analyzing, and cleaning up manmade pollution.

1.6 Life Cycle Analysis

The best tool that has emerged for pollution prevention with green chemistry is the life cycle analysis (LCA), the study of the production of a product from cradle to grave, or better yet, as McDonough and Braungart titled their book, *cradle-to-cradle*.[27] LCA is an extremely useful method for chemists and engineers to analyze the entire life cycle of a product.

LCA is a 21st century method for analyzing the sustainability of a product or process from the extraction, mining, or growing of a resource to the ultimate fate of the product, whether landfilled, recycled, or composted. LCA is typically both qualitative and quantitative. The new greener product is compared with products currently on the market. Sustainability is measured by the overall efficiency of the process in the use of raw materials or feedstocks, energy, greenhouse gases, water, and the output of pollution. Life cycle thus refers to the entire process to make a product, use it, and dispose of it throughout the product's useful lifetime. Greener products or processes are compared with traditional products or processes that perform the same or similar functions. Each LCA establishes the criteria to be used to assess sustainability.

The first step in an LCA is to decide what will be measured. These are frequent candidates: materials in, materials out, energy used, energy produced, water used, water released. This may sound formidable, yet it is already a necessary part of the economics of designing a chemical production plant, or the process chemistry to synthesize a medicine.

A complete LCA can sometimes deliver surprising results. Hexamethylene diamine (HMDA) is one of the two key monomers used to make nylon-6,6, which is the nylon used in our backpacks, clothing,

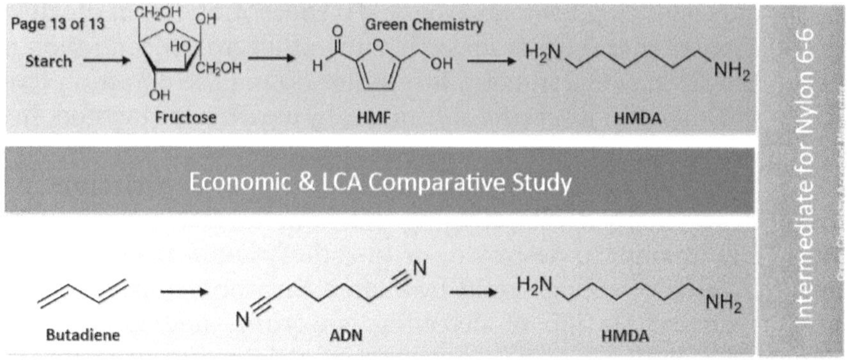

Figure 1.5 Life cycle analysis (LCA) of the production of HMDA from food crop starch *vs.* petroleum. Reproduced from ref. 28 with permission from the Royal Society of Chemistry.

carpet, car parts, and many others. Solvay (China) Co. discovered that production of HMDA from starch is more polluting and energy-intensive than production from petroleum (Figure 1.5).[28] Although the bio-based route from sugars had greenhouse gas emission advantages over the petroleum-based route, energy-intensive drying of the hydroxymethylfurfural (HMF) intermediate increased marine and freshwater eutrophication. Growing corn and potatoes for the starch feedstock was also found to have negative environmental consequences.[28] These challenges are not insurmountable, however. Improvements in drying technology and agricultural practices could result in flipping the LCA outcome.

1.7 Case Study: Polylactic Acid (PLA) (NatureWorks®)

An early example of a complete LCA was performed by NatureWorks®, a subsidiary of Cargill Dow LLC, for their polymer, polylactic acid (PLA).[29] Cargill Dow has defined sustainability by a "triple bottom line approach" of economic, social and environmental sustainability. We will only be concerned with the environmental sustainability, although the social sustainability exemplifies some green chemistry principles as well. For example, under the social sustainability criterion, Cargill Dow does not allow the endocrine disrupter (Chapter 3) bis-phenol A (BPA) to be used as a plasticizer in products made with their PLA, and they do not allow PLA to be used for packaging tobacco products.[29]

PLA is a biodegradable polymer derived entirely from the renewable resource cornstarch. The carbon source is thus entirely carbon dioxide and water, which plants take up to synthesize polysaccharides using energy from the sun by photosynthesis.[29] The cornstarch is enzymatically converted into natural D-glucose, also called dextrose (Figure 1.6). Dextrose is then fermented at neutral pH to produce lactic acid. A pre-polymerization step is necessary to remove water before the lactic acid can be cyclized to produce the key intermediate, lactide. The lactide is then distilled in one of the higher energy-intensive steps (Figure 1.6). The final step is a solvent-free ring-opening polymerization (ROP) to produce the high molecular weight PLA product with the desired properties.

Cargill Dow chose three indicators for the LCA of PLA: (1) the use of fossil energy in production, (2) net greenhouse gases produced, and (3) water use.[29] The indicators for PLA were compared with nine petrochemical-based polymer products, and cellophane, made from an older renewable feedstock, cellulose. LCA methodology is very complex, with many decisions to be made about inclusion or exclusion of data, weighting of various factors, and approximations.[29] First, the use of petroleum, both as feedstock and as the energy source in the production of plastics, was evaluated (Figure 1.7). Since PLA is derived from a bio-feedstock instead of petroleum, the total amount of fossil energy used in production is less than all of the other plastics. A net

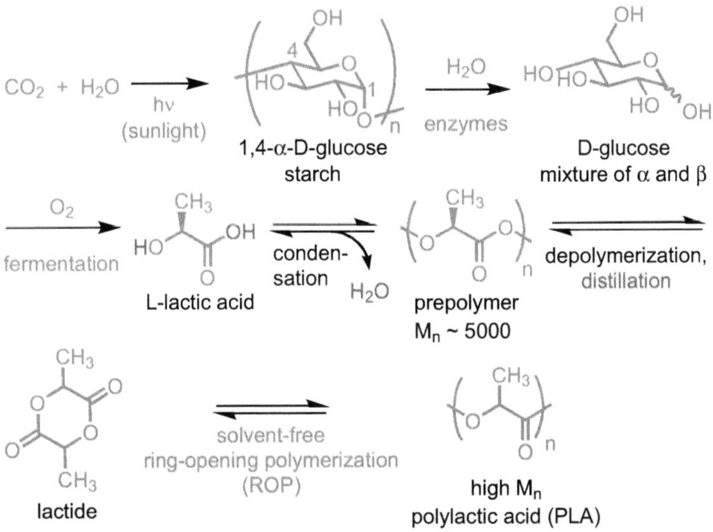

Figure 1.6 NatureWorks® process for manufacturing polylactic acid (PLA) from cornstarch.[29]

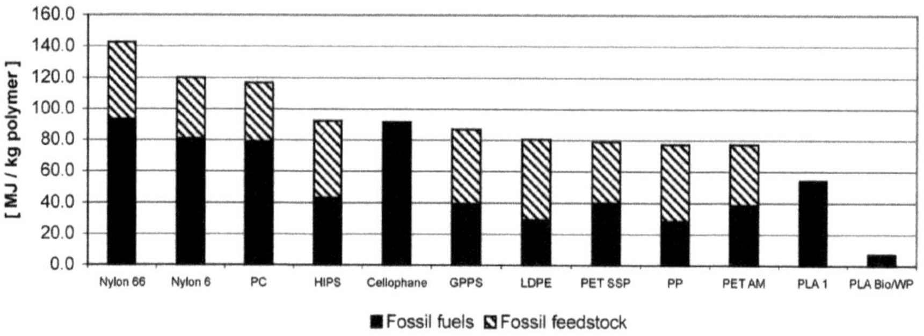

Figure 1.7 Fossil energy requirement for some petroleum-based polymers and polylactide. The cross-hashed part of the bars represents the fossil energy used as chemical feedstock (the fossil resource to build the polymer chain). The solid part of each bar represents the gross fossil energy use for the fuels and operations supplies used to drive the production processes. PC = polycarbonate; HIPS = high impact polystyrene; GPPS = general purpose polystyrene; LDPE = low density polyethylene; PET SSP = polyethylene terephthalate, solid state polymerization (bottle grade); PP = polypropylene; PET AM = polyethylene terephthalate, amorphous (fibers and film grade); PLA1 = polylactide (first generation); PLA B/WP (polylactide, biomass/wind power scenario). Reprinted and reproduced from ref. 29 with permission from Elsevier. Copyright 2003.

use of 54 MJ kg^{-1} of fossil energy was for agricultural purposes in growing the corn for feedstock, for transportation, and for chemical processing, minus the embodied solar energy in the corn feedstock. With the inclusion of lactide from agricultural waste (stalks, husks and leaves, called "corn stover") as the biomass feedstock (Bio) and wind power (WP), fossil energy use was decreased to only about 9 MJ kg^{-1} (Figure 1.7). Improvements in process energy would also lead to a more cost competitive product. Another advantage of using corn stover is that the process would not compete with the food supply. Thus, energy efficiency improves all three components of the triple bottom line: environment, economics, and society.[29]

Solar energy warms objects on earth, think of a rock in the sun, or a parking lot. Greenhouse gases trap heat radiated from these objects in the atmosphere. The mechanism is simply absorption of electromagnetic radiation, and emission in the infrared region back into the atmosphere. Of course, some heat is radiated out to space, but the effect of higher concentrations of greenhouse gases is net global warming. Not all geographical locations will experience the same warming; northern latitudes are warming at the greatest rate of all, and the eastern US has experienced cooler temperatures, even as the western US is burning, literally. This is why we now call it climate change. The three major greenhouse gases are rated in terms of CO_2 equivalents (C-eq): CO_2 is designated 1 C-eq; CH_4 has 21 C-eq; and N_2O has 310 C-eq.[30]

Cargill Dow included these three greenhouse gases in the LCA of PLA. In this analysis, the first generation PLA1 from feed corn did not fare much better than low density polyethylene or polypropylene, but it was much better than Nylon-6,6 or Nylon-6 (Figure 1.8). The production of nylon is very energy intensive as well, which illustrates the connection of greenhouse gas production to energy use (Figure 1.7). Using corn stover as feedstock and wind energy in the production of PLA B/WP actually removes greenhouse gases from the atmosphere because growing corn uses CO_2 as the most basic feedstock (Figure 1.8).

The final impact assessed in the LCA was the use of water in plastic production. Three uses of water were analyzed: irrigation water, process water, and cooling water (Figure 1.9). Even though the corn feedstock requires irrigation with water to grow it, PLA is competitive with all the major commercial plastics studied, and it beats the nylons and cellophane by a substantial margin.

Overall, Cargill Dow's LCA of PLA is a model for the use of LCA for other products. They found that the current production of PLA from feed corn already has a significant environmental advantage over other polymers, and that switching to corn stover feedstock and wind power represents a dramatic improvement even over the current process. Notably, Cargill Dow did not include an analysis of the pollution prevention impact of PLA, which is a compostable product, keeping more permanent polymers, such as polyethylene terephthalate (PET) and nylons out of landfills and the oceans.

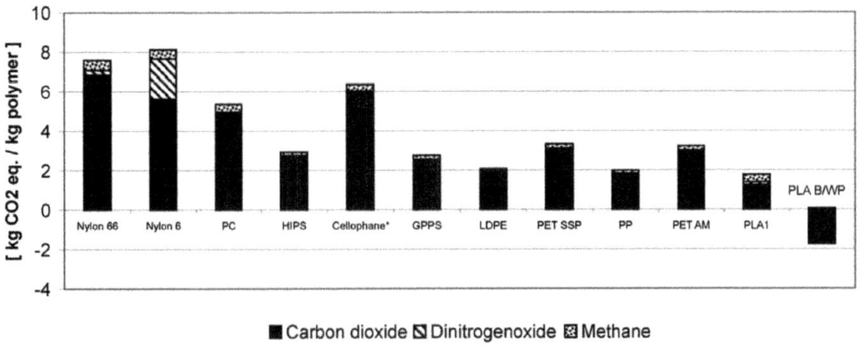

Figure 1.8 Contributions to global climate change for some petrochemical polymers and the two polylactide polymers. Reprinted and reproduced from ref. 29 with permission from Elsevier. Copyright 2003.

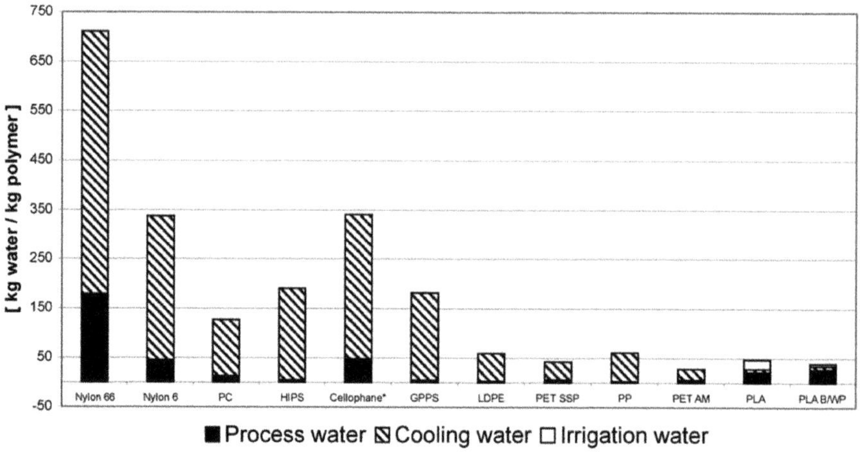

Figure 1.9 Gross water use by petrochemical polymers and the two PLA cases. Reprinted and reproduced from ref. 29 with permission from Elsevier. Copyright 2003.

1.8 Resources: The Scientific Literature

As you work through this textbook, many ideas and problems will require the use of the scientific literature. Several databases and journal resources are particularly useful and complete.

1.8.1 Databases

The most complete database for chemical research is *Scifinder*. An important tool in searching databases is the use of applicable key words. Sometimes search words will need to be made more specific to narrow the range of hits, other times search words need to be made more general. For example, in 2015, a search in *Scifinder* using "green chemistry" as search words turned up thousands of hits; it was too broad. On one hand, "sustainable (or green) synthesis of adderall" turned up no relevant hits; it was too narrow. On the other hand, "synthesis of adderall" garnered 669 hits; it would be extremely difficult to read and assess the green merits of each synthesis. One way to get around this is to select a recent comprehensive review article, and find the most efficient synthesis described therein. To get the best result from a search—a search result of 10 to 100 hits is useful and manageable—one must try different search words, combinations of search words, chemical structures, and reactions. Searching the literature is like a video game in that persistence and experience help you find your way, gather tools, and reach your goals.

Scifinder, chemical literature, by library subscription.
PubMed, biomedical literature, free to the public.
Web of Science, multidisciplinary, by library subscription.

1.8.2 Journals

Green chemistry articles are increasingly published in traditional chemical journals. Specific society collections of journals, such as the American Chemical Society (ACS: pubs.acs.org) and the Royal Society of Chemistry (RSC: pubs.rsc.org), have useful search routines that are specific for the chemical literature. The "Advanced" search links are especially useful, because you can narrow your search by years, author, or journal. Specific journals devoted to green chemistry may be useful in identifying current areas of research, and finding solutions to particular problems. You can keep up to date by browsing the table of contents of these on a regular basis.

Green Chemistry, Royal Society of Chemistry, since 1999 http://pubs.rsc.org/en/journals/journalissues/gc#!recentarticles&adv, accessed July 19, 2019.
ACS Sustainable Chemistry and Engineering, American Chemical Society, since 2013 http://pubs.acs.org/journal/ascecg, accessed July 7, 2019.

1.9 Implementation of Green Chemistry

John Warner and Paul Anastas established the Principles of Green Chemistry 25 years ago. Where are we now? In 2010, Scientific American published an article that said hazardous chemicals had been reduced by about 0.5 billion kg over 15 years, but that is a drop in the bucket compared with the 33.5 billion kg per day that the US produces or imports.[31] There is still a lot of disagreement about how to make *all* chemistry *green* chemistry, that is to make the "green" in green chemistry superfluous.[31]

Edward Woodhouse, a political scientist at Rensselaer Polytechnic Institute said, "One way to think about it is to ask yourself: 'What is the purpose of government? Why isn't everything done by voluntary exchange among willing buyers and sellers?' The answer is, of course, that a lot of important things that need doing won't be done voluntarily."[31]

"[John] Warner favors the 'build a better mousetrap' philosophy: Do green chemistry by making alternatives that are not only safer but effective and economical, and chemical companies will eagerly adopt them."[31]

As you learn about this new way of doing chemistry, think about whether you agree with Woodhouse, or with Warner, about the best way to achieve these goals, to make it happen.

1.10 Summary

Green chemistry is a conscious change in the way we do chemistry—to do no harm. Chemistry is a powerful tool to improve the quality of our lives. We all reap the benefits of chemists who have improved the quality of modern life. Tragic experiences, such as *Silent Spring*, Times Beach, Love Canal, and Bhopal, have proven that it is nearly impossible to eliminate exposure to hazardous chemicals that have already been used in the production of these beneficial modern products. So green chemistry seeks to eliminate risk by avoiding the use and production of hazardous chemicals in the first place. Of great use to green chemistry are the tools of toxicology, environmental chemistry, LCA, and of course, the scientific literature that encompasses all of these areas. Cargill Dow exemplified the use of LCA in the production of NatureWorks® polylactic acid from corn. Implementing green chemistry is a question of public policy: should government attempt to force the change, or will market forces make it happen?

1.11 Problems: Prevent Waste

Problem 1.1

(a) Three general categories were given for the 12 Principles of Green Chemistry. List which principles fit into each of the three general categories, using the short title of each principle. Some principles may apply to more than one category.

(b) Explain your choices with as few words as possible. (Ask your instructor whether or not to use complete sentences.)

Problem 1.2

(a) Terephthalic acid (1,4-benzenedicarboxylic acid) is one of the monomers for PET. Is terephthalic acid considered a fine chemical, or a commodity chemical?

(b) Ibuprofen is an anti-inflammatory over-the-counter drug. Is Ibuprofen a fine chemical, or a commodity chemical?

Problem 1.3

DDT is used as a potent insecticide against the malaria mosquito in Africa. An article in *Scientific American* (http://www.scientificamerican.com/article/ddt-use-to-combat-malaria/) lays out the cases for and against using DDT to control malaria.

(a) Read the article, and perhaps find others. Then choose one side or the other and write a brief argument in support.
(b) What can chemists do to prevent malaria *and* the use of DDT?
(c) Find one scientific paper in the chemical literature (Scifinder or Pubmed) describing a new mosquito insecticide, and assess if it would be more environmentally degradable, less toxic, or less susceptible to insect resistance. Give the full reference (Authors, Title, Journal, Year, Volume, page numbers).

Problem 1.4

DDT is considered an endocrine disrupter, a substance that interferes with the hormone chemical messenger system, often a mimic of estrogen.

(a) Draw the structure of 17β-estradiol.
(b) Next to it, redraw the structure of DDT.
(c) Circle and label the parts of the two molecules that are of similar type (polar, hydrophobic, charged), and similar arrangement in space.

Problem 1.5

(a) What are the disadvantages of choosing CO_2 as a refrigerant?
(b) What are the disadvantages of choosing N_2O as a refrigerant?

Problem 1.6

(a) Draw the balanced reaction of methyl isocyanate (MIC) with excess water.
(b) What are the physical forms of the products?
(c) Why did the MIC tank in Bhopal explode when water leaked into it?

Problem 1.7

(a) Dioxin can be destroyed by incineration. What are the three criteria for achieving complete combustion of dioxins?
(b) Is this environmental or green chemistry? Explain briefly.
(c) Terrence Collins invented a catalyst that can be used to accelerate paper bleaching by hydrogen peroxide and avoids the formation of dioxins. Is this environmental or green chemistry? Explain briefly.

Problem 1.8

Use PubMed and Scifinder to search for "paper bleaching with hydrogen peroxide" and "health effects of dioxin."

(a) How many hits do you get for each search term in each of the two databases?
(b) Is one database better than the other for certain topics?

Problem 1.9

(a) Draw a curved arrow mechanism for the ring-opening polymerization (ROP) of lactide into PLA initiated by H_2O.
(b) Is this a step-growth or a chain-growth polymerization? (Hint: Look up the criteria for step-growth and chain-growth polymerization.)
(c) Why do you think the lactide intermediate must be distilled before the ROP?

References

1. P. T. Anastas and J. C. Warner, *Green Chemistry: Theory and Practice*, Oxford University Press, New York, 1998.
2. Hippocrates, *Hippocratic Oath*, Wikipedia, 5–3 BCE, https://en.wikipedia.org/wiki/Hippocratic_Oath, accessed July 19, 2019.
3. Center for Green Chemistry and Engineering at Yale, *Principles of Green Chemistry*, https://greenchemistry.yale.edu/about/principles-green-chemistry accessed January 18, 2019.
4. A. Larson, A. Anderson and K. O'Brien, *Shaw Industries: Sustainable Business, Entrepreneurial Innovation, and Green Chemistry*, 2009, http://www.acs.org/content/dam/acsorg/greenchemistry/industriainnovation/Shaw-Industries-business-case-study.pdf, accessed July 23, 2019.
5. R. Anderson, *Qual. Prog.*, 2004, **37**, 32–37.
6. T. Hager, *The Demon Under the Microscope*, Three Rivers Press, Crown Publishing Group, Random House, Inc., New York, 2006.

7. G. J. M. Velders, A. R. Ravishankara, M. J. Molina, M. K. Miller, J. Alcamo, J. S. Daniel, D. W. Fahey, S. A. Montzka and S. Reimann, *Science*, 2012, **335**, 922–923.
8. *Climate Change 2007: The Physical Science Basis*, Contribution of Working Group I to the Fourth Assessment Report of the IPCC, ed. S. Solomon, D. Qin, M. Manning, Z. Chen, M. Marquis, K. Averyt, M. M. B. Tignor and H. L. Miller Jr., Intergovernmental Panel on Climate Change (IPCC), Cambridge UK and New York, NY USA, 2007, p. 996.
9. C. Hetzner, *Coolant safety row puts the heat on Europe's carmakers*, Reuters, https://uk.reuters.com/article/uk-europe-cars-refrigerant/coolant-safety-row-puts-the-heat-on-europes-carmakers-idUKBRE8BB0HE20121212, accessed July 23, 2019.
10. M. D. Hurley, T. J. Wallington, M. S. Javadi and O. J. Nielsen, *Chem. Phys. Lett.*, 2008, **450**, 263–267.
11. M. H. Russell, G. Hoogeweg, E. M. Webster, D. A. Ellis, R. L. Waterland and R. A. Hoke, *Environ. Toxicol. Chem.*, 2012, **31**, 1957–1965.
12. I. Urieli, *Carbon Dioxide (R744) - The New Refrigerant*, 2013, https://www.ohio.edu/mechanical/thermo/Applied/Chapt.7_11/Chapter9.html, accessed July 23, 2019.
13. J. Staub, B. D. Rasmussen and M. Robinson, *CO_2 As Refrigerant: The Transcritical Cycle*, The NEWS Media, BNP Media, 2004, https://www.achrnews.com/articles/94092-co2-as-refrigerant-the-transcritical-cycle, accessed June 25, 2018.
14. J. L. Hardcastle, *Automakers Develop CO_2-Based Air Conditioning*, Environmental Leader, 2013, https://www.environmentalleader.com/2013/03/automakers-develop-co2-based-air-conditioning/, accessed July 19, 2019.
15. U. S. Congress, *The Clean Air Act (Air pollution prevention and control)*, 1970, https://www.govinfo.gov/content/pkg/USCODE-2008-title42/pdf/USCODE-2008-title42-chap85.pdf, accessed July 23, 2019.
16. *Federal Water Pollution Control Act*, 1977, https://legcounsel.house.gov/Comps/Federal%20Water%20Pollution%20Control%20Act.pdf, accessed July 23, 2019.
17. E. Griswold, The Wild Life of 'Silent Spring', *The New York Times Magazine*, Sept. 23, 2012, http://www.nytimes.com/2012/09/23/magazine/how-silent-spring-ignited-the-environmental-movement.html?_r=0, accessed July 23, 2019.
18. R. Carson, *Silent Spring*, Fawcett Publications, Inc., Greewich, CT, 1962.
19. P. Van der Avert and B. M. Weckhuysen, *Angew. Chem., Int. Ed.*, 2002, **41**, 4730–4732.
20. D. Lowe, *Things I Won't Work With, in In the Pipeline*, © 2017 American Association for the Advancement of Science, 2017, vol. 2019. http://blogs.sciencemag.org/pipeline/archives/category/things-i-wont-work-with.
21. Union Carbide Corporation, *Bhopal Gas Tragedy Information*, 2001–2015, http://www.bhopal.com/ accessed September 17, 2015.
22. E. E. McConnell, J. R. Bucher, B. A. Schwetz, B. N. Gupta, M. D. Shelby, M. I. Luster, A. R. Brody, G. A. Boorman, C. Richter, M. A. Stevens and B. Adkins Jr., *Environ. Sci. Technol.*, 1987, **21**, 188–193.
23. U.S. Environmental Protection Agency, *Health and Environmental Effects Profile for Methyl Isocyanate*, Environmental Criteria and Assessment Office, Office of Health and Environmental Assessment and Office of Research and Development, Washington, DC, 1986, accessed Sept. 2015.
24. *Environmental Chemistry*, Wikimedia, 2015, https://simple.wikipedia.org/wiki/Environmental_chemistry, accessed July 23, 2019.
25. R. Allen, *The Dioxin War: Truth and Lies About a Perfect Poison*, Pluto Press, London, 2004.
26. T. Zacharewski, *Environ. Sci. Technol.*, 1997, **31**, 613–623.
27. W. McDonough and M. Braungart, *cradle to cradle*, North Point Press, New York, NY, 2002.
28. A. B. Dros, O. Larue, A. Reimond, F. De Campo and M. Pera-Titus, *Green Chem.*, 2015, **17**, 4760–4772.

29. E. T. H. Vink, K. R. Rábago, D. A. Glassner and P. R. Gruber, *Polym. Degrad. Stab.*, 2003, **80**, 403–419.
30. D. Albritton, R. Derwent, I. Isaksen, M. Lal and D. Wuebbles, Radiative Forcing of Climate Change, *Climate Change 1995—the Science of Climate Change*, IPCC Intergovernmental Panel on Climate Change, Cambridge University Press, Cambridge (UK), 1995.
31. E. Laber-Warren, *Green Chemistry: Scientists Devise New "Benign by Design" Drugs, Paints, Pesticides and More*, Scientific American, May 28, 2010, http://www.scientificamerican.com/article/green-chemistry-benign-by-design/, accessed July 23, 2019.

2 Synthetic Efficiency

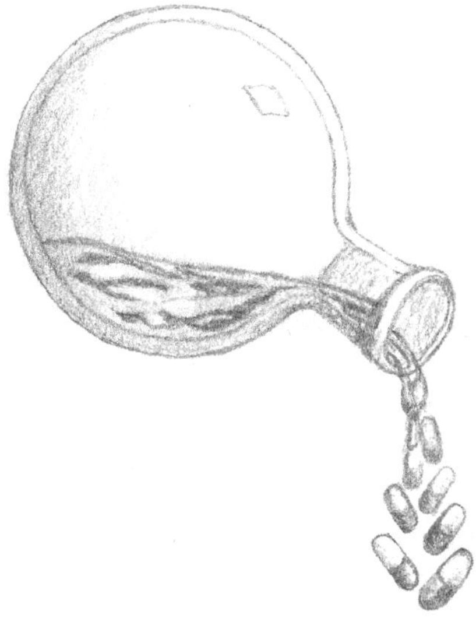

"Principle 2: Synthetic methods should be designed to maximize the incorporation of all materials used in the process into the final product."[1]

To prevent waste, all materials used in a synthesis should be incorporated efficiently into the desired product. Once again, this is our aspiration; the reality is often a series of compromises. Using materials efficiently makes for an economical process as well as a greener process. The materials used include reactants (feedstocks or starting materials), reagents, solvents, purification solutions and chromatographic supports. Catalysts are exempt from this type of analysis because they are

Green Chemistry: Principles and Case Studies
By Felicia A. Etzkorn

Published by the Royal Society of Chemistry, www.rsc.org

not consumed in a reaction. Ideally, catalysts can be recycled many times. In this chapter, we will focus on the efficient incorporation of reactants into products, through single- and multiple-step syntheses. Solvents and auxiliaries, such as chromatographic supports, will be examined in depth in Chapter 5, and catalysts in Chapter 9. Efficient incorporation of materials often takes the form of fewer reaction steps in a synthesis, and of combining steps in one-pot reactions.

2.1 Intermediates and Reagents

A challenge for green chemistry is the design of synthetic routes that do not include hazardous intermediates and reagents. Most chemicals are synthesized in multistep processes through synthetic intermediates that are typically isolated and purified in between reactions. Intermediates are defined in part as "the substances generated by one step and used for the succeeding step."[2] High-energy intermediates, as distinguished from synthetic intermediates, are formed during the course of a single reaction. Such intermediates are hypothesized as part of a mechanism, and evidence for their existence is deduced from kinetic and spectroscopic data.

Reagents can be confused with reactants; they are both used to make a reaction proceed. Sometimes atoms of the reagent end up in the product of a reaction or intermediate, so they could be considered reactants too. A working definition of a reagent for the bench chemist is typically whatever is commercially available.

In the Presidential Green Chemistry Challenge (PGCC) Award-winning BHC synthesis of the analgesic ibuprofen, carbon monoxide (CO) is used in the final step and incorporated into the final product. Is it a reagent or a reactant? It does not really matter, as long as we incorporate the atoms into the product as efficiently as possible.

Green chemistry also seeks to avoid hazardous reagents and processes. A well-known example of a hazardous intermediate is methyl isocyanate (MIC), the reactive intermediate that caused the explosion at Bhopal, India. MIC is an intermediate in the synthesis of carbaryl, which is the common name of the pesticide sold under the brand name Sevin™. The starting material or feedstock is phosgene (Cl–(C=O)–Cl), an even more toxic and reactive compound. Chemical hazards to be avoided will be considered in depth in Chapter 3. To avoid such toxic or hazardous intermediates, green chemists must redesign the synthetic route. In Chapter 12, we will see an example of a redesigned route to carbaryl.

A green synthesis is resource efficient. By minimizing the number of synthetic steps, we eliminate waste and achieve a more efficient and economical process. Hazardous intermediates and reagents can often be eliminated as well. When catalysts are used, the activation barrier for a reaction is lowered and less reactive chemicals can serve as intermediates or reagents. Catalysts also decrease the energy requirements of a process as we will see in Chapter 6, Energy Efficiency.

2.2 Calculating Efficiency

Three ways to look at the efficiency of a synthesis will be described: (1) % atom economy, (2) overall % yield, and (3) Sheldon's E-factor or Process Mass Intensity (PMI). Atom economy is a very useful measure for quickly comparing the efficiency of possible alternative synthetic routes to a product. Overall yield gives a realistic assessment of the efficiency of a synthetic route in practice. The E-factor and the PMI include both of these accountings, and they also account for other materials used in the *actual* production, including aqueous workup and purification materials.

2.2.1 Atom Economy

Barry Trost coined the term "atom economy" in 1991 to reflect the "maximum number of atoms of reactants appearing in the products".[3,4] Atom economy is defined as the percent mass of all reactant atoms incorporated into the final mass of product; 100% is optimal. The atom economy relies solely on the theoretical maximum yield of each reaction. It does not account for losses due to side products, unreacted starting materials, solvents and other auxiliaries, or catalysts. The atom economy concept is most useful for comparing several different synthetic routes to give the same desired product.

Calculating the atom economy is fairly simple. First, determine the masses by calculating the molecular weight of each reactant and product. Next, determine the stoichiometric ratio of each reactant to product. Rarely is this other than 1 : 1 in organic chemistry, but balance the reaction first, and be aware of the possibility. The atom economy is the product mass as a percent of the mass of all reactants (eqn (2.1)).

$$\%\text{ Atom Economy} = \frac{\text{mass desired product}}{\sum \text{masses of reactants}} \times 100\% \tag{2.1}$$

Simple reactions fall into four categories: addition, elimination, substitution, and rearrangement. Classic examples of each reaction type follow. If it has been a while since you studied organic chemistry, you may wish to review the basic functional groups of organic chemistry found in Appendix A.

2.2.2 Addition Reactions

Examples of addition reactions are hydrogenation of a ketone (Figure 2.1), or addition of HCl to an alkene. Addition reactions display maximum atom economy because all atoms are incorporated into the product. In the example that follows, both cyclopentanone and hydrogen are incorporated fully into the product to give 100% atom economy. The catalyst is heterogeneous 5% Pd/C, and the solvent is methanol. The convention used in this book (which is not always followed in the literature) is that reagents or reactants are shown above the arrow, and catalysts and solvents are beneath the arrow.

2.2.3 Elimination Reactions

Elimination reactions are by definition wasteful, unless all products are ultimately used. For example, the elimination of HCl from a chloroalkane to give an alkene produces the waste by-products *t*-butanol and KCl (Figure 2.2). A strong, bulky base, such as potassium *t*-butoxide, must be used to obtain useful yields of the alkene product instead of the substitution product. By-products are typically not shown in organic reactions. In green chemistry, unlike typical organic reactions, all reactions should be balanced to show any waste products explicitly in the reaction.

Elimination can produce several alkene structural isomers; in the case above, 2-hexene would also be produced. Therefore, selectivity is

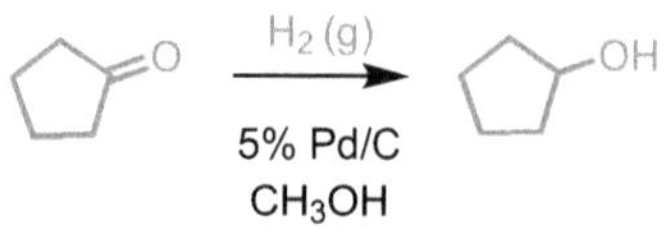

Figure 2.1 Hydrogenation of a ketone.

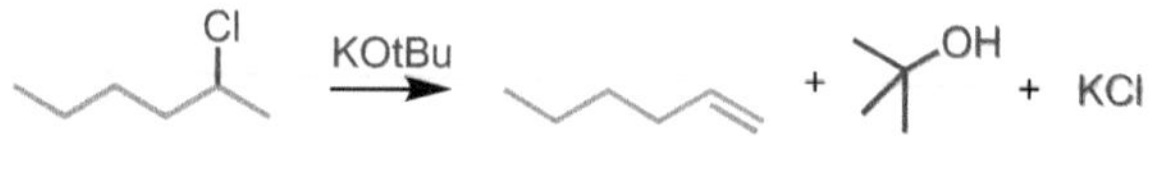

Figure 2.2 Elimination of HCl to give 1-hexene.

an issue, and the yield is unlikely to be above 90% for the desired alkene. Separation of the isomeric alkene products is another issue, which could require energy-intensive distillation, or the use of a chromatographic support, such as silica gel, and chromatography solvents. Chromatography materials will be examined in Chapter 5.

2.2.4 Substitution Reactions

Substitution reactions lie on a continuum between addition and elimination in terms of atom economy. The defining characteristic of a substitution reaction for atom economy is the mass of the leaving group. If iodide ion is the leaving group in an S_N2 reaction, the atom economy will be lower than if chloride ion is the leaving group.

The classic example of a substitution reaction is the S_N2 reaction, substitution nucleophilic bimolecular. The reaction in Figure 2.3 is left for you to balance. What is the waste product?

A less obvious substitution reaction is the Wittig reaction, discovered by Georg Wittig. Many variants of the Wittig reaction have been invented, mainly to improve the selectivity. Note that the Wittig reaction is invariably a three-step process, although these steps are usually done sequentially in one pot (Figure 2.4). First, an alkyl halide is treated with a phosphorus reagent, triphenylphosphine (PPh_3) in this case, to make the phosphonium salt. Second, deprotonation of the phosphonium salt with a strong base, such as butyllithium (*n*-BuLi), gives the phosphorus ylide (with a C=P bond). Third, the phosphorus ylide reacts with a carbonyl compound to give the alkene. By-products are created in the second and third steps for a fairly inefficient process. The beauty of the

Figure 2.3 Substitution of bromide with methyl sulfide.

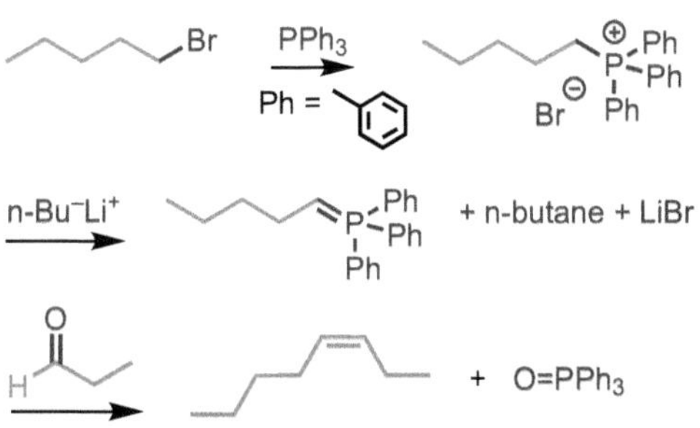

Figure 2.4 Wittig reaction to produce an alkene creates wasteful by-products.

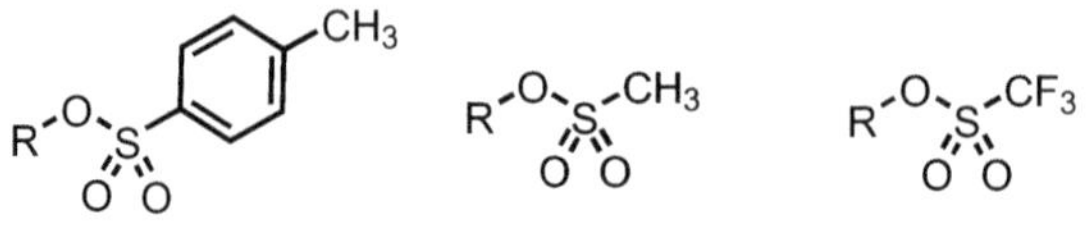

Figure 2.5 Sulfonate esters are used to convert the hydroxy groups in alcohols into sulfonate leaving groups.

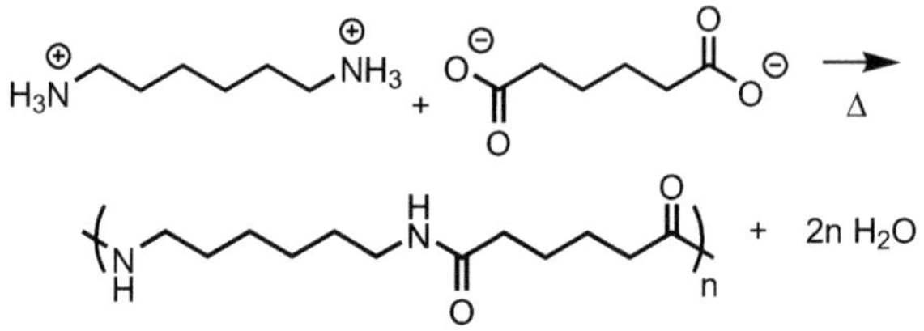

Figure 2.6 Synthesis of nylon-6,6 by a substitution reaction.

Wittig reaction is complete control over the location of the double bond that is hard to achieve by other methods.

Since alcohols are somewhat acidic ($pK_a \approx 16$), substitution using basic nucleophiles does not work. The acid–base reaction is faster and gives the more stable salt product. For substitution reactions with alcohols (R–OH), the hydroxy functional group (–OH) is turned into a good leaving group such as the sulfonate esters: tosylate, mesylate, or triflate (Figure 2.5).

Other important types of substitution reactions are substitution at a carboxylic acid derivative by addition–elimination, electrophilic aromatic substitution, and nucleophilic aromatic substitution.

The synthesis of Nylon-6,6, a polymer connected by amide bonds, is an example of carbonyl substitution by addition–elimination (Figure 2.6). First, a 1 : 1 salt of 1,6-hexanediamine and hexanedioic acid (common name: adipic acid) is prepared. Heating the purified salt produces the polyamide, Nylon-6,6. In this case, the atom economy is high, but the energy cost of heating to high temperatures so that the water can be driven out makes the process less green. Dupont has several patents for a lower temperature process at about 130 °C.[5]

2.2.5 Rearrangement Reactions

Rearrangement reactions can be the most atom economical of all, unless derivatives or reagents must be used to initiate the reaction.

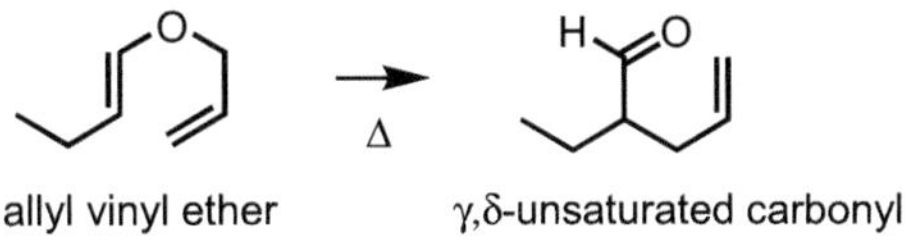

Figure 2.7 A Claisen [3,3]-sigmatropic rearrangement with 100% atom economy.

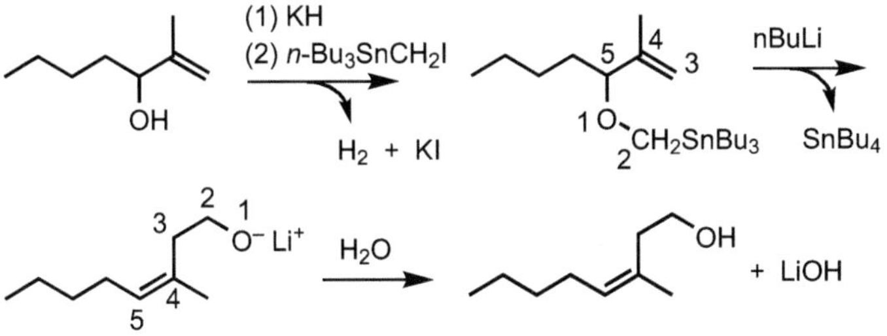

Figure 2.8 A Still-Wittig [2,3]-sigmatropic rearrangement.[6]

An example is the Claisen rearrangement, a [3,3]-sigmatropic rearrangement (Figure 2.7). Sigmatropic means that one sigma bond is broken and another formed in an intramolecular rearrangement; the π-bonds are rearranged as well. The numbers in brackets refer to the number of heavy atoms (C, N, O) on each side of the breaking and forming bonds; the atom economy is 100%.

As a counter example, a Wittig [2,3]-sigmatropic rearrangement with quite poor atom economy was invented by W. Clark Still (Figure 2.8).[6] In its favor, the reaction is highly selective for the trisubstituted (*Z*)-alkene isomer, which can be difficult to obtain otherwise. The tributyltin derivative is necessary to obtain the carbanion intermediate that undergoes the rearrangement. One by-product is tetrabutyl tin ($SnBu_4$), which is toxic.[7] (What are the other by-products?) In organic reactions, aqueous workup is usually not shown explicitly. In the Still–Wittig reaction, water is necessary to make the final alcohol from the lithium alkoxide (RO^-Li^+) (Figure 2.8).

A more atom economical alternative is the sila-[2,3]-Wittig rearrangement.[8] A trimethylsilyl (TMS) methyl ether is prepared from the alcohol precursor (Figure 2.9). The TMS group is removed with a fluoride reagent, in this example n-Bu_4NF. The carbanion created can then undergo the [2,3]-Wittig rearrangement with stereochemical control to give primarily the exocyclic alkene isomer (Figure 2.9).

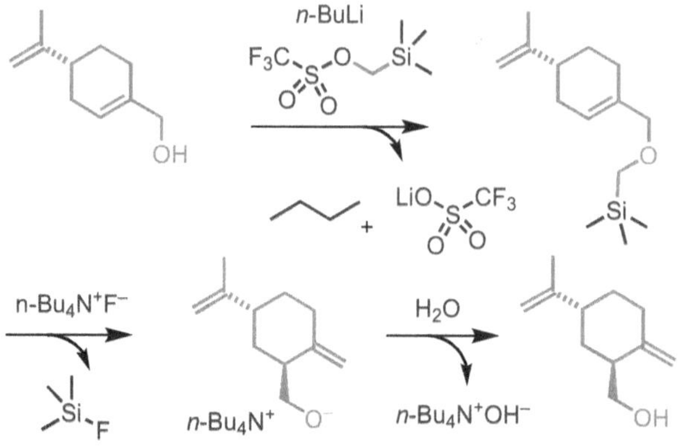

Figure 2.9 A more atom economical sila-[2,3]-Wittig rearrangement.[8]

2.3 Case Study: Atom Economical Ibuprofen Process (BHC Company)

Ibuprofen is a popular anti-inflammatory analgesic drug. Since 1983, it has been sold over the counter because of its relative pharmacological safety. The original production-scale synthesis of ibuprofen by the Boots Co. required six steps from (2-methylpropyl)benzene (Figure 2.10).[9] The key step in the synthesis was the Friedel–Crafts acylation using $AlCl_3$ in stoichiometric amounts. This first step alone produced many tons of $AlCl_3{\cdot}H_2O$ per year, which was disposed of in landfills.[9]

To calculate the overall atom economy of the reaction, it is only necessary to calculate the molecular weights of each reactant and the final ibuprofen product, not the intermediates. Careful mechanistic analysis of the last step, the hydrolysis of a nitrile, shows that two molar equivalents of water must be used to generate the two carboxylic acid oxygens; only one of the hydrogens ends up in the product.

Many of you may be astute enough to figure out that the total mass of unused atoms (shown in brown in Figure 2.10) can be added to the mass of the final product, ibuprofen, to obtain the denominator of the equation, which may be quicker in some cases. It is certainly one way to check your calculation (see Table 2.1 for data).

The modern synthesis of ibuprofen was developed by a joint venture of Boots Co. and Hoechst Celanese Co. called BHC (Figure 2.11). The synthesis won a PGCC award in 1997 for Greener Synthetic Pathways.[9] The BHC process is used currently by BASF in Bishop, Texas for large-scale production of ibuprofen.

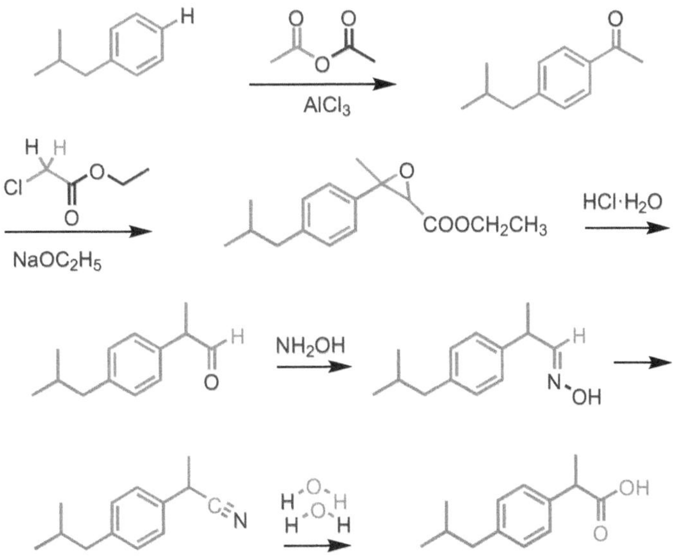

Figure 2.10 Boots Co. synthesis of ibuprofen. Atoms used in the final product are shown in green, and unused atoms are shown in brown. Adapted with permission from ref. 34, Copyright 2000 American Chemical Society. See also ref. 9.

Table 2.1 Atom economy of the Boots synthesis of ibuprofen: molecular formulas and weights of all reactants and the product.

Compound	Molecular formula	Molecular weight (g mole^{-1})
2-Methylpropylbenzene	$C_{10}H_{14}$	134.2
Acetic anhydride	$C_4H_6O_3$	102.1
Aluminum chloride	$AlCl_3$	133.3
Ethyl 2-chloroacetate	$C_4H_7O_2Cl$	122.6
Sodium ethoxide	C_2H_5ONa	68.1
Hydrochloric acid·hydrate	H_3OCl	54.5
Hydroxylamine	H_3NO	33.0
2 water	$2(H_2O)$	36.0
Ibuprofen	$C_{13}H_{18}O_2$	206.3

Boots process ibuprofen atom economy calculation.

$$\%\ AE = \frac{206.3}{134.2 + 102.1 + 133.3 + 122.6 + 68.1 + 54.5 + 33.0 + 36.0} \times 100\%$$

$$= \frac{206.3}{683.8} \times 100\% = 30\%$$

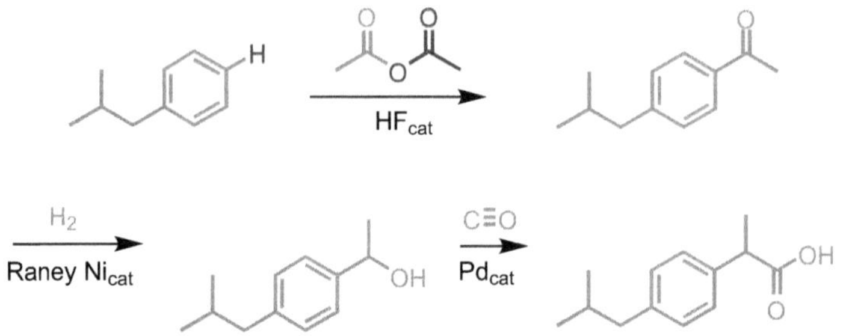

Figure 2.11 PGCC Award-winning BHC synthesis of ibuprofen. Unused atoms are shown in brown, the catalysts are shown in black under the arrows, and the atoms incorporated into the final product are shown in green. Adapted with permission from ref. 34, Copyright 2000 American Chemical Society. See also ref. 9.

In this synthesis, hydrofluoric acid is used as both solvent and Friedel–Crafts acylation catalyst in a closed-loop recycled system (Figure 2.11). HF is a liquid with a low boiling point of 19.5 °C, which makes it very efficient to recycle. HF is extremely hazardous, so this step is a compromise between efficiency and hazard. On the plus side, only three steps are used in contrast to the six in the Boots process, and all three steps use catalysts, which are neutral in terms of atom economy because they are recycled. In addition, the acetic acid left over from the first step is recycled to produce more acetic anhydride, which can be reused in the process. This is a classic real-world case of dramatically improved atom economy through green chemistry.

2.3.1 Overall Yield: No Such Thing as 100%

The atom economy calculation does not account for losses due to less than 100% yield for each synthetic step. So now, let us take up the concept of overall yield. Organic chemists are suspicious of reports of 100% yield, their own or anyone else's, and rightly so. At the end of a reaction the unpurified product typically contains small amounts of remaining starting material and possibly, some form of by-product or structural or stereochemical isomers. Mostly we are quite happy with yields greater than 90%. Now, imagine a sequence of 10 reactions with each giving a 90% yield, *i.e.* an overall yield of $0.90^{10} \times 100\% = 35\%$. The overall yield is based on the moles of product obtained from the moles of each limiting reactant. An easy way to calculate the overall yield from a published synthesis is by multiplying the yields as mole fractions.

2.3.2 Sheldon's E-factor and Process Mass Intensity (PMI)

The third method for determining waste in a synthesis was devised by Roger Sheldon in 1992.[10,11] A phloroglucinol (1,3,5-trihydroxybenzene) production plant closed "because the cost of disposing of the waste was rapidly approaching the selling price of the product."[11] Sheldon's survey revealed that, in a typical fine-chemical synthesis of 1 kg of product, tens of kg of waste are generated. Thus, he proposed a measure of the *actual* efficiency of a chemical synthesis. Because oil refining and bulk (commodity) chemical production are already quite efficient, Sheldon focused on the fine-chemical and pharmaceutical segments of the chemical industry (Table 2.2).[11]

The E-factor is a waste metric, calculated as mass of waste divided by mass of product (eqn (2.2)). In practice, the amount of waste is calculated by taking the difference between the masses of inputs used and the desired product. The PMI was developed by the ACS Green Chemistry Round Table in February 2011 for pharmaceutical processes.[12] The emphasis in PMI is on the *efficiency* of input use in the process rather than the waste.[12] PMI is very similar to the E-factor, using only mass inputs instead of waste in the numerator (eqn (2.3)). A perfect E-factor is zero, while a perfect PMI is one.

$$E = \frac{\text{mass inputs} - \text{mass product}}{\text{mass product}} \tag{2.2}$$

$$\text{PMI} = \frac{\text{mass inputs}}{\text{mass product}} \tag{2.3}$$

Originally, the E-factor included water used in the process, and the PMI still does. In this book, the water used will be included in calculations of E-factor or PMI.

The calculation seems quite simple, but finding the actual amounts used in a synthesis can be quite tricky from the outside. An

Table 2.2 Sheldon's E-Factor calculated for different chemical industry segments. Reproduced from ref. 11 with permission from the Royal Society of Chemistry.

Industry segment	Product tonnage	E-factor (kg waste kg per product)
Oil refining	10^6–10^8	<0.1
Bulk chemicals	10^4–10^6	<1 to 5
Fine chemicals	10^2–10^4	5 to >50
Pharmaceuticals	10–10^3	25 to >100

experimental procedure published in a journal or patent will typically state "...was purified by chromatography to give..." without mentioning the amount of silica gel or solvents used in the chromatography. Aqueous workup reagents are typically given as a volume "...washed with 5% $NaHCO_3$ (3×50 mL)..." which is unhelpful for calculating the *mass* used, unless you know the density of 5% $NaHCO_3$ and all of the other solutions used. Least helpful are statements such as "...was worked up in the usual way."

For example, in Winkler's synthesis of Ritalin™, the key catalytic step is described in the Supporting Information as follows (Figure 2.12).[13] The first two steps of the synthesis to prepare the phenyldiazoacetate were reported by Moritani *et al.*[14]

"An oven-dried 5 mL flask was fitted with a stir bar and septa and flame-dried under an Ar purge. *N*-Boc-piperidine (0.35 mL) was added and degassed, and then freshly distilled cyclohexane (0.35 mL) was added. Rh_2(5*R*-MEPY)$_4$ (2.3 mg, 0.00251 mmol) was added, and the reaction was heated at 50 °C for 20 min. The methyl phenyldiazoacetate (70.4 mg, 0.40 mmol) was injected over 4 hours into the blue reaction mixture at 50 °C. The reaction mixture was allowed to cool to room

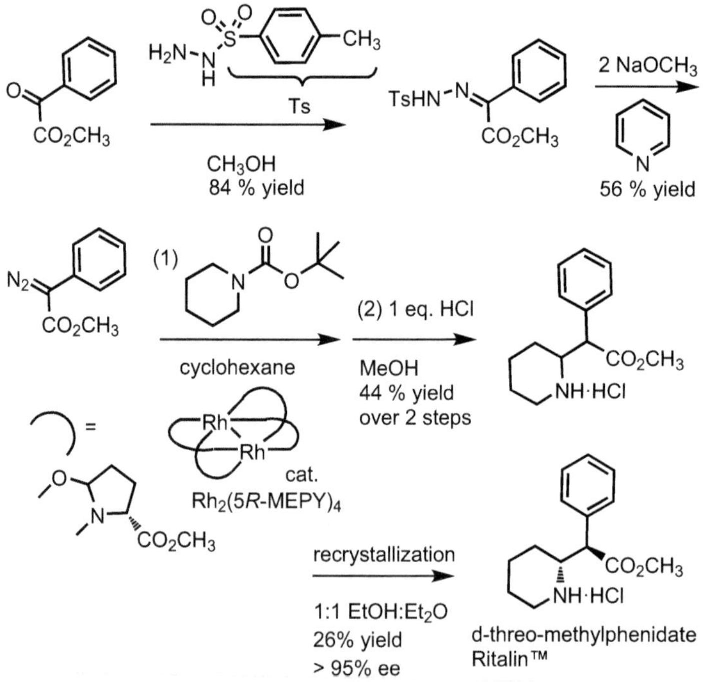

Figure 2.12 Winkler's synthesis of Ritalin™,[13] from Moritani's methyl phenyldiazoacetate.[14]

temperature and then filtered through a pipette column of silica with ether to remove the catalyst. The solution was then concentrated *in vacuo* to give a yellow oil which was purified *via* flash chromatography on silica gel (20×250 mm, 10% Et_2O/pet ether), giving the *N*-Boc-methylphenidate (0.086 g, 64.5%) as a green oil (threo isomer)."[13]

In this case, no aqueous workup was done. There appears to be no attempt to recover the catalyst, which is quite typical for synthetic reactions conducted on a small scale. Surprisingly, the dimensions of the column are given, which leads to a fairly good estimate of the amount of silica gel used in the chromatography. But the amount of solvent used, the only unknown, must be guessed based on the typical method of chromatography.[15] In the future, perhaps the journals and patents will insist on including the amounts of *all* materials used in grams, much like a French cookbook!

The chemical industries have a much easier handle on the E-factor or PMI, since the cost of raw materials purchased for a particular process is a critical expenditure, and the mass output of the desired product is essential to the bottom line of the production plant. E-factor and PMI differ from the concept of atom economy in that they are highly useful measures of the actual material efficiency of a process, but not as useful at comparing the *potential* efficiency of different multistep synthetic routes to a final product.

2.4 Chirality

2.4.1 Enantiomers: Non-superimposable Mirror Images

In living systems, nearly every molecule is chiral and asymmetric, that is, biomolecules typically exist as a single stereoisomer, either an enantiomer or a diastereomer. Synthesis of biological molecules is achieved in living systems using chiral catalysts, which are mainly proteins called enzymes. Enzymes effortlessly catalyze reactions by binding and stabilizing the transition state structure, lowering the activation barrier of the reaction.[16] The transition state of a reaction is exceedingly short-lived, on the order of a bond vibration, about 10^{-14} second. We will explore more about enzymes in Chapter 9 on catalysts because they are so very useful in green chemistry.

Medicinal chemists have long recognized that the drug they design works better if its three-dimensional shape, including the stereochemistry, complements the biological target that it binds. Chiral drugs have been classified into three groups: eutomers with one bioactive enantiomer, racemic drugs with equally bioactive enantiomers, and racemic drugs that undergo chiral inversion.[17] For example, the (*S*)-ibuprofen isomer is 100 times more biologically active than the inactive (*R*)-isomer. The (*R*)-isomer is converted to the active (*S*)-isomer *in vivo* by liver enzymes, but not the reverse.[17]

The challenge for green chemistry lies in designing the processes to achieve each asymmetric synthesis cheaply and cleanly. In 1992, the US Food and Drug Administration (FDA) established a goal that encouraged pharmaceutical companies to develop single-enantiomer drugs.[18] This is an important step on the way towards making medicinal chemistry greener.

2.4.2 Biological Effectiveness

The chiral drug market exploded in the 1990s because the FDA required biological evaluation of both enantiomers of new drugs. By 2000, chiral drugs accounted for \$115 billion share of the market for pharmaceuticals.[19] The development of single-enantiomer drugs was also seen as a way of extending the patent on older, racemic drugs. Some of the enthusiasm wore off as chiral drugs were often found to have comparable effectiveness as the racemic versions.[20] Yet medicinal chemists remain undeterred; both synthetic methods and the market for chiral drugs continue to grow.

In 1956, thalidomide was approved as a sedative in Europe and Canada, and in 1957 it was marketed to pregnant women to prevent morning sickness (Figure 2.13). Approval in the US was denied six times thanks to Frances Oldham Kelsey, a pharmacologist with the FDA, because the applications did not include adequate test results.[21] Babies born to women who had taken the drug during the first three months of pregnancy had many limb abnormalities (flipper arms), and it became clear that thalidomide was a teratogen. Thalidomide is currently used to treat certain cancers and leprosy, of course with restrictions on the use in pregnant women.

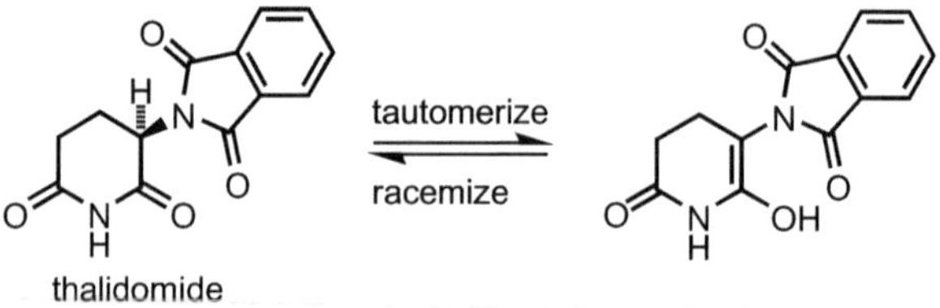

Figure 2.13 Tautomerization-racemization of single-enantiomer drug, thalidomide.

The stereochemistry behind thalidomide toxicity was important. The levorotatory isomer is highly teratogenic, but the dextrorotatory isomer is also mildly teratogenic. This was difficult to sort out because the enantiomers racemize *in vivo*. The racemization is likely to occur by deprotonation at the chiral center; this proton is acidic because of the adjacent electron-withdrawing C=O and $N(C{=}O)_2$ groups (Figure 2.13).

In several cases, one enantiomer of a drug has greater toxicity than the other.[17] In other cases, stereochemistry does not have such serious health consequences. Typically, the undesired isomer is either neutral or detrimental to health. Thus, production of a racemic drug is wasteful if one enantiomer is biologically inactive in the best case, and harmful in the worst case.

2.4.3 Stereochemistry Designations

Let us step back and review some basic stereochemistry. We often call stereochemistry "handedness." If you look at your two hands, they are more or less mirror images of each other. More significantly, they are *non-superimposable* mirror images of each other, and this is the definition of enantiomers. It is a bit harder to visualize this with molecules, unless you build 3-D models. Organic chemists have developed a system for designating enantiomers as *R* (rectus, Latin for right) or *S* (sinister, Latin for left). In biological chemistry, sugars and amino acids are designated with a one-font-size smaller capital D or L (Latin for dexter and laevus, confusingly also right and left). The D or L is based on their stereochemical relationship to D- or L-glyceraldehyde. Yet a third designation is the lowercase italicized letters *d* or *l*, based on whether the stereoisomer rotates plane-polarized light to the right (clockwise) or left (counterclockwise), with no particular relationship to the *R*/*S* or D/L systems. The *d* or *l* designation is now often supplanted with the direction or sign of the rotation, as in (−)-shikimic acid, which we will see later in this chapter.

As an example, alanine is the simplest chiral amino acid, with a methyl group on the side chain (Figure 2.14). To determine the *R*/*S* designation of an enantiomer with one chiral center, number the groups attached to the chiral center starting with the atom having the highest atomic mass. If two attached atoms are the same, such as the carbons of the methyl group and the carboxylic acid, go out to the next atom and find the tiebreaker; in this case oxygen makes the acid rank higher. Direct the lowest ranked group, in this case hydrogen, so that it points back away from the view. Now look at the procession of

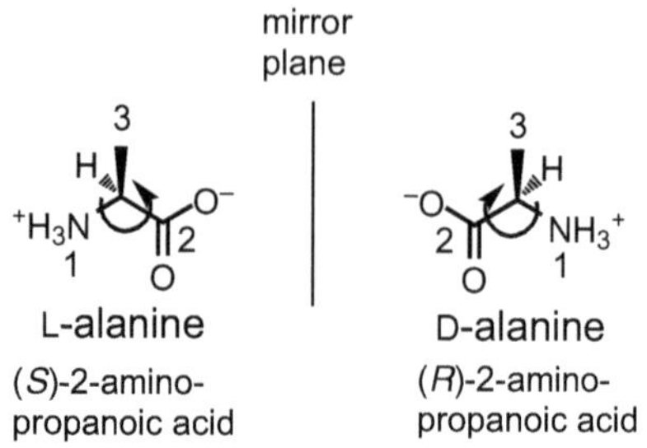

Figure 2.14 Method to determine the *R/S* stereochemical designation of alanine.

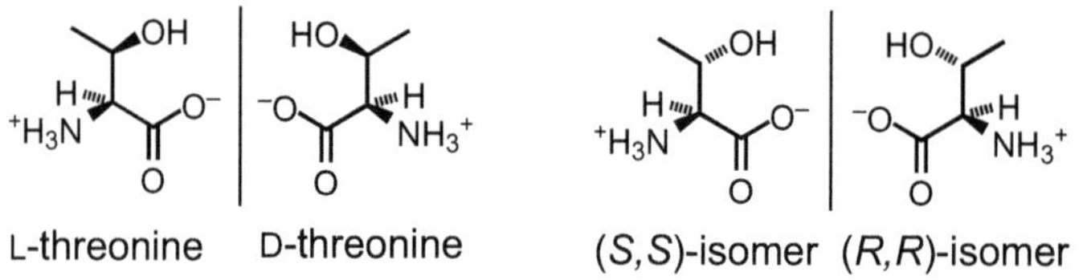

Figure 2.15 Stereochemical designation of diastereomers of threonine.

the groups in order of rank. If the groups are arranged clockwise, the molecule is named (*R*), and if counterclockwise, the molecule is named (*S*) in the IUPAC system.

For alanine, the nitrogen is ranked 1 because it has the highest atomic mass, then the carboxy group is ranked 2, then the methyl group is ranked 3, and the hydrogen is ranked 4, placed going back behind the plane of the paper (Figure 2.14). L-Alanine is counted 1-2-3 counterclockwise, so it is designated as the (*S*)-enantiomer. D-Alanine is then the (*R*)-enantiomer of 2-aminopropanoic acid.

2.4.4 Enantiomers and Diastereomers

Matters become a little more complicated when there are two or more stereocenters in a molecule. The method for designating overall stereochemistry is the same as for enantiomers; just add more *R*/*S* designations. But the shape is what matters, not the name, "...a rose by any other name would smell as sweet" (W. Shakespeare). Now with n stereocenters, there are 2^n possible stereoisomers; if $n = 2$, there are four stereoisomers, and two pairs of enantiomers, as in threonine (Figure 2.15). L- and D-Threonine are mirror images of each other, so they are still enantiomers. Notice that the left two structures are not mirror images of either structure to the right, so they are diastereomers.

2.4.5 Physical, Chemical, and Biological Properties

In synthesizing asymmetric bioactive molecules, several factors must be considered. Enantiomers have identical physical and chemical properties, except that they rotate plane-polarized light to an equal degree, but in opposite directions. This means that enantiomers cannot be easily separated from one another by ordinary means. In some way, enantiomers must be made into diastereomers, which *do* have different physical (*e.g.* boiling point or crystal form) and chemical (*e.g.* reactivity) properties. Then we can separate them by physical methods, such as distillation, crystallization, or chromatography. Chemists work tirelessly to avoid having to separate enantiomers by clever use of chiral starting materials or chiral catalysts. And these, too, are tools of green chemistry.

2.4.6 Methods for Synthesis of Single Enantiomers

If a drug is synthesized without attention to stereochemical control, labor, materials, and energy are wasted. In a racemic synthesis of a drug that has only one active enantiomer, half of the reagents, solvents, auxiliaries, energy, and labor are wasted. At the end of the synthesis half of the product is waste. And when the inactive or toxic enantiomer is metabolized or passes through a human's system it is also wasted, or worse it sickens the patient with side effects. Pharmaceuticals and their metabolites are increasingly found in our surface waters and thus in our water supplies, because they often pass through wastewater treatment plants unchanged. Limiting the pharmaceuticals entering the system through the design and production of single-enantiomer drugs is one way to decrease the contamination of our water supplies.

Keep in mind the caveat that a stereocontrolled synthesis could be much longer or less efficient than the synthesis of the same molecule without regard to stereochemistry. The tools of calculating synthetic efficiency and life cycle analysis can give definitive guidance for choosing whether to synthesize a single enantiomer, a diastereomer, or a racemic mixture.

2.5 Greener Solutions for Single Stereoisomers

The tools for producing single-stereoisomer drugs have been described as "...chiral pool technology, resolution, biological asymmetric methods, and chemical asymmetric techniques...."[22]

Organic chemists have been working to control stereochemistry from the beginning. Emil Fischer published a famous series of papers in which he logically and methodically determined the stereochemistry of all the natural sugar molecules, in part by synthesizing them. We stand on the shoulders of giants as we design new drugs, and new asymmetric syntheses of drugs, to eliminate the waste and potential toxicity associated with the wrong stereoisomer. All of the methods described below also apply to the design and synthesis of pesticides, since pesticides are also biologically active species.

2.5.1 Design Achiral Drugs

Aspirin (Figure 2.16) remains one of the most effective and safe drugs ever invented. Currently aspirin is used most often as a blood thinner to prevent heart disease, but it is still a highly effective painkiller. It can be derived from salicylic acid found in willow bark and other natural sources, by acetylation of the phenol, hence the name, acetylsalicylic acid. But it is mass-produced from petroleum feedstock now. There are no chiral centers, so controlling the stereochemistry during synthesis is not a factor. If it is possible to design an effective drug without any stereocenters, an achiral drug, this is the simplest way to achieve the goals of the FDA policy.

2.5.2 Asymmetric Catalysts

Asymmetric catalysts can be either biological, as in an enzyme, or chemical. Catalysts accelerate the reaction rate by lowering the energy barrier for the reaction. Since the original form of a catalyst is regenerated over and over during the reaction, very little catalyst is needed, typically 0.1% up to 10% by weight of the limiting reagent. Efficient recycling of a stable catalyst makes a process even greener.

Enzymes have been used to catalyze many organic reactions. Enzymes are proteins, which can be prepared using molecular biology techniques in *E. coli*. Because enzymes are composed of L-amino acids, most are naturally stereospecific—they take a single stereoisomer of substrate and convert it to a single stereoisomer of product.

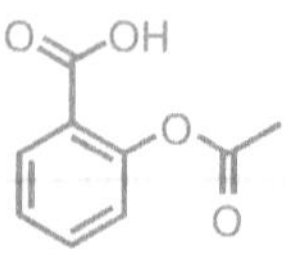

Figure 2.16 Aspirin (acetylsalicylic acid) is an achiral drug.

Sometimes an enzyme is used in a reaction with the desired isomer to carry on to product. Other times an enzyme is used in a reaction with the undesired isomer so it can be isolated and discarded.

Heterogeneous catalysts, which are solids and can be filtered out of a reaction solution for easy recovery, have a distinct advantage over homogeneous catalysts, which are soluble. Catalysts often confer a great deal of chemical specificity on the outcome of the reaction, improving yield by guiding the reaction along the desired pathway. Best of all, chiral catalysts can be designed to give a high yield of a desired stereoisomer. For example, Barry Sharpless designed an asymmetric epoxidation catalyst, for which he shared the Nobel Prize in 2001 (Figure 2.17).[23] The starting material is an allylic alcohol, without any chirality. The catalyst is built *in situ* from a cheap, readily available enantiomer of diethyl tartrate as the chiral ligand, and a titanium compound as a Lewis acid catalyst. *t*-Butyl hydroperoxide is the oxidant. The stereoisomer of the tartrate controls the stereochemical outcome of the Sharpless epoxidation. This asymmetric catalyst, and many others that have followed, is a very effective way of introducing an asymmetric center into a molecule.

One enantiomer of the catalyst will give a very high yield of one stereoisomer of the product, while the opposite enantiomer of the catalyst will give the opposite stereoisomer. Catalysts are usually designed based on a working knowledge of the mechanism of the reaction, *e.g.* whether it is acid or base catalyzed. To find the best specific catalyst for a particular reaction, a large number of different metals and ligands are usually combined and tested on a benchmark

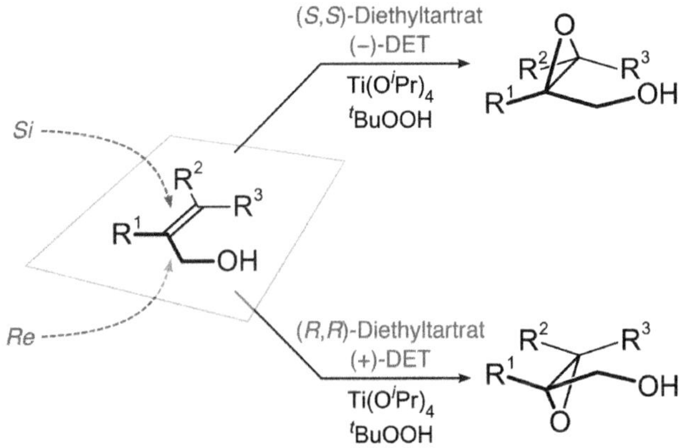

Figure 2.17 The Sharpless Epoxidation. By Sponk reproduced from https://en.wikipedia.org/wiki/Sharpless_epoxidation, accessed July 26, 2019.

substrate. Conditions of temperature, solvent, and pressure may be varied as well.

2.5.3 Single-enantiomer Bio-feedstocks (Chiral Pool Technology)

The origin of all stereoisomers can eventually be traced back to living things. Almost all of the building blocks of life are asymmetric, such as sugars and amino acids. Compounds with a stereogenic center can be readily obtained from abundant natural sources, so these are great feedstocks for green chemical synthesis. Once the asymmetric center has been introduced, it is necessary to avoid processes that might racemize or epimerize at the chiral center. (Epimerize is the word used to describe equalizing isomers at one stereocenter when there are two or more stereocenters in the molecule.) The asymmetric center is carried through the rest of the synthesis, and may perhaps even direct the formation of new stereocenters along the way. Glucose is particularly abundant because cellulose and amylose, the principle structural materials of all plants, are polymers of glucose monomers (Figure 2.18). The amino acid serine is often used as a chiral feedstock because it is highly functionalized for many different reaction possibilities (Figure 2.18). If the unnatural isomer of a sugar or amino acid is desired, they are also available, but more expensive because they cannot be obtained directly from nature.

2.6 Case Study: Synthesis of Tamiflu from Shikimic Acid (Nie *et al.*)

Tamiflu (oseltamivir phosphate) is the only approved drug proven to shorten the duration of influenza illness. In 2009, Nie, Shi, Ko and Lu synthesized Tamiflu from (−)-shikimic acid in eight steps (Figure 2.19).[24] At that time, the commercial production route used by Roche had 15 steps with only 18% overall yield from shikimic acid.[25]

In 2009, John Andraos compared six industrial and nine academic syntheses of oseltamivir using the reaction mass efficiency (RME), a

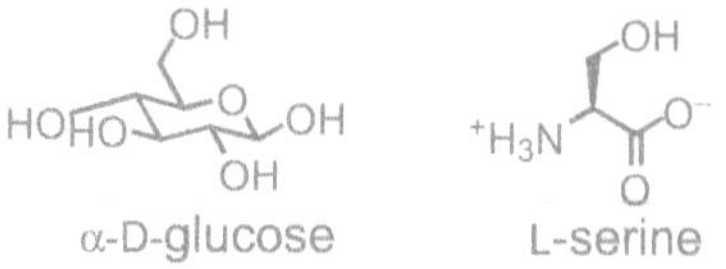

Figure 2.18 Example single-enantiomer bio-feedstocks.

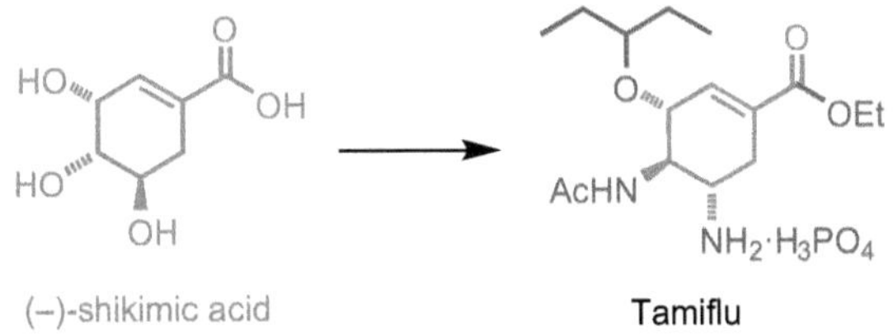

Figure 2.19 Tamiflu is synthesized from the asymmetric bio-feedstock (−)-shikimic acid.

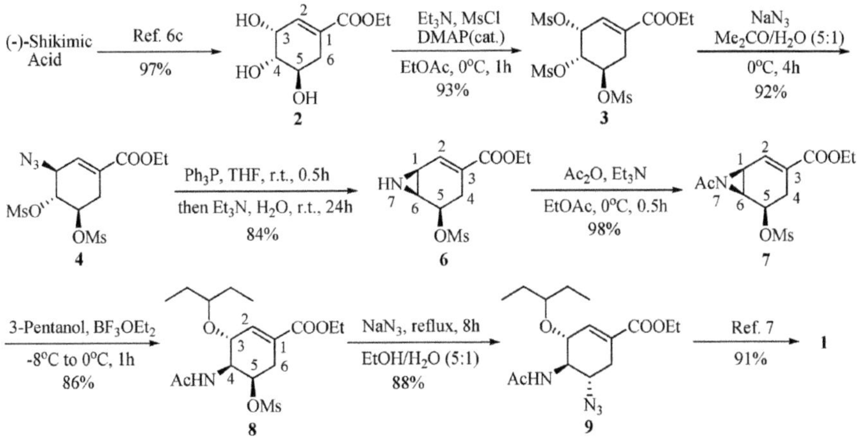

Figure 2.20 The Nie *et al.* synthesis of Tamiflu from shikimic acid. Reprinted with permission of American Chemical Society from ref. 24, Copyright © 2009.

method he developed for the analysis.[26] The Roche synthesis from shikimic acid was found to be the most efficient. Since the Nie synthesis was published the same year, it was not included in the analysis.

In the Nie synthesis, each of the three stereogenic centers of shikimic acid is used in the final product, two by inversion, and the third by a double inversion (Figure 2.20).[24] Most of the synthesis is fairly atom economical; small reagents such as NaN_3 are used to install each of the two nitrogen atoms. The most inefficient step in terms of atom economy is reduction of the azide to an amine using PPh_3, a reagent also used in the Wittig reaction. Hydrogenation with a Pd or Pt catalyst is an efficient, but hazardous, way to reduce azides to amines.

The Nie synthesis makes elegant use of an aziridine (three-membered nitrogen-containing) ring to transfer the first installed nitrogen to the 6-position with inversion of configuration (Figure 2.20).

Next, chiral resolution is organized into three categories in order of the potential for making a process greener.

2.6.1 Enzymatic Resolution

The advantage of enzymatic resolution is that the chiral catalyst reacts selectively with only a single enantiomer of a racemic compound (intermediate or product), converting it to a single-enantiomer compound that has different physical properties from the starting compound (regardless of the chirality). Immobilized enzymes (solids) are most effective because they can be filtered from the reaction for recovery and reuse, as in the following case study.

Stiripentol is an anti-epileptic drug for which the (+)-isomer is 2.4 times as active as the (−)-isomer.[27] Interestingly, the less active isomer is found at a 5-6 fold excess in the brain.[27] This can result in tolerance, a situation where the more active isomer is ineffective. Lipases are enzymes that hydrolyze or transesterify (switch) esters. Lipase A from *Candida antarctica* was used to isolate the more active (+)-isomer of stiripentol in a two-step process. First, racemic (*rac*)-stiripentol was transesterified with vinyl butanoate using lipase A and the ester product was predominantly the (3*R*) isomer (Figure 2.21). In the second step, the ester was hydrolyzed, the stereochemistry was maintained, and the desired (+)-isomer was formed (Figure 2.21).[28]

Chiral feedstocks and intermediates may also be resolved enzymatically. For example, the enzyme lipase is often used to hydrolyze

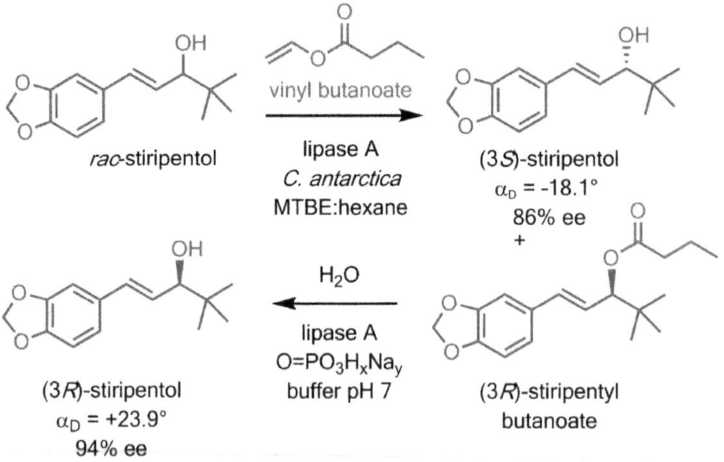

Figure 2.21 Two-step enzymatic resolution of the anti-epileptic drug, stiripentol, using lipase A.[28]

one enantiomer of a methyl ester to give a single enantiomer of a carboxylic acid, and the opposite enantiomer remains as the methyl ester. Either the acid or the ester may be used as the single-enantiomer feedstock.

2.6.2 Resolution of Diastereomeric Salts

Making and separating diastereomeric salts is the next most efficient way to resolve a single enantiomer during the synthesis of a drug molecule. For example, the amine ephedrine is readily available as a single enantiomer. To isolate a single enantiomer of a carboxylic acid, the racemic mixture of acids can be mixed in a 1 : 1 ratio with ephedrine, and one of the diastereomeric salts can be crystallized (Figure 2.22). Recall that diastereomers have different physical properties, so they crystallize differently. After separation, the desired carboxylic acid enantiomer can be freed by acidification and extraction into an organic solution. On a large scale, it is feasible to recover the chiral amine base and reuse it for the same or a different chiral resolution.

For example, $AcSCH_2CH(CH_2Ph)CO_2H$ is a key intermediate in the synthesis of inhibitors of encephalin and angiotensin converting enzyme (ACE), which are used as analgesics and cardiovascular agents. Resolution of the intermediate acid using ephedrine was described in a European patent application (Figure 2.22).[29]

If the resolution of the desired stereoisomer is performed late in the synthesis, all of the materials and energy in steps that use a racemic mixture are half wasted. So the earlier in the synthesis that the resolution can be performed, the better, but only if the chiral center is stable to the conditions of the remaining reactions. Resolution of a

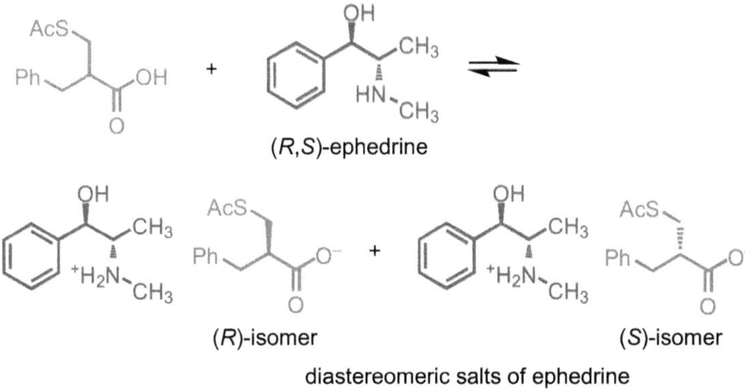

Figure 2.22 Resolution of a drug intermediate with diastereomeric ephedrine salts.[29]

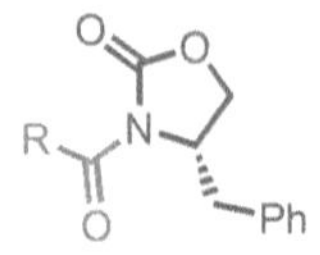

Figure 2.23 A covalent Evans chiral derivative.

racemic mixture of products is the least efficient, least economical, and least green method of obtaining a chiral drug.

2.6.3 Resolution of Diastereomeric Covalent Derivatives

In a similar way, the two enantiomers in a racemic mixture can be made into diastereomers by forming a covalent bond with a single-enantiomer auxiliary. David Evans developed a chiral auxiliary based on an oxazolidinone (Figure 2.23). The resulting diastereomeric derivatives can be separated by crystallization or chromatography, and a second reaction must be performed to remove the auxiliary. Chiral auxiliaries can be recycled, but this process adds two steps to the synthesis. Covalent derivatives are typically resolved by chromatography, which adds to the auxiliary burden of the synthesis. A racemic acid, RCOOH, is reacted with the chiral auxiliary to form an amide bond, the diastereomeric amides are separated, and the chiral auxiliary is removed.

The FDA has encouraged pharmaceutical companies to synthesize single-enantiomer drugs because they are the most biologically active, with fewer side effects, and they are associated with less waste than racemic drugs. The tools to achieve this goal in order of preference are: (1) design achiral drugs, (2) use asymmetric catalysts, (3) use single-enantiomer bio-feedstocks, (4) use enzymatic resolution, (5) resolve diastereomeric salts, and (6) resolve covalent diastereomer derivatives. In every case, it is best to separate stereoisomers as early in the synthesis as possible. The longer the stereoisomeric mixture is carried through the synthesis, the more material is wasted.

2.7 Multistep Synthetic Efficiency

2.7.1 Retrosynthetic Analysis

E. J. Corey, who won the 1990 Nobel Prize in Chemistry, published a book with Xue-Min Cheng called *The Logic of Chemical Synthesis*.[30]

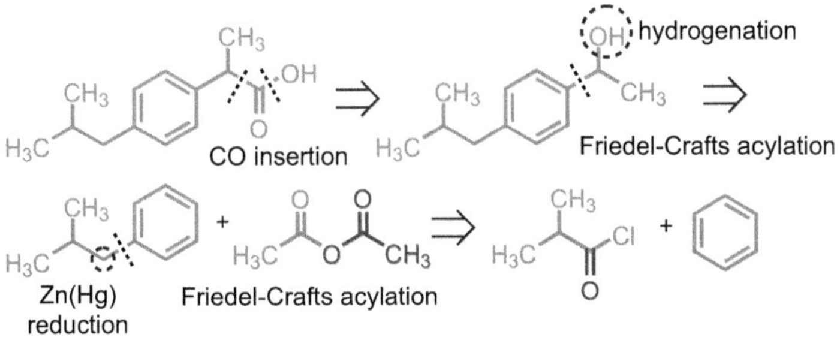

Figure 2.24 A retrosynthetic analysis of the BHC process for ibuprofen.

The breakthrough idea of this book was called retrosynthetic analysis. The organic chemist uses this logic in deciding how to approach the total synthesis of a target molecule. As we saw in the case study on Tamiflu, there is as much art as analysis in this method. One has to have a vast knowledge of organic reactions and then be able to recognize the most efficient way to use those reactions to put fragments of a molecule back together. In a retrosynthetic analysis, bonds are "broken" stepwise in hypothetical disconnections until commercially available pieces are found. Each disconnection requires a forward reaction that will be high yielding and highly selective for the desired intermediate. Not every single reaction needs to be outlined in a retrosynthetic analysis. Usually the disconnections focus on key carbon–carbon bond forming reactions.

A retrosynthetic analysis of the BHC synthesis of ibuprofen is shown in Figure 2.24. The words in the figure are not usually included in a retrosynthetic analysis. Notice that the retrosynthetic arrows have a double stem, to distinguish them from forward reaction arrows (Appendix B). Disconnections are marked with dashed lines, and the exact reagents to be used are not usually given. For example, CO insertion is done with carbon monoxide, but it requires a catalyst, as does the Friedel–Crafts acylation. An additional retrosynthetic step not shown in the BHC forward synthesis for the atom economy calculation (Figure 2.24) has been included. Even though isobutyl benzene is commercially available, it is fairly expensive, and it has to be synthesized somehow. Whether it is fair to include the synthetic efficiency of commercially available compounds used in a process is debatable.

2.7.2 Convergent Synthesis

One of the defining features of E. J. Corey's method for determining the most efficient pathway to a target molecule is the idea of a convergent synthesis. In a retrosynthetic analysis, breaking the molecule into evenly sized pieces with similar complexity results in a convergent forward synthesis. For example, in the first disconnection of Corey's retrosynthetic analysis of prostaglandin, a natural inflammatory molecule, three pieces, one with five carbons, and two with seven carbons of the final target are obtained (Figure 2.25).[31] The central ring is more complex than the two alkyl chains, but this is considered a classic convergent synthesis.

2.7.3 Biomimetic Synthesis

Students are typically exposed to carbocation rearrangements in second-year Organic Chemistry. Natural steroid hormones, such as cholesterol, are synthesized by carbocation rearrangements *in vivo*.[32] We now know that sterol biosynthesis proceeds *via* cationic cyclization of 1,5-dienes. William S. Johnson used this understanding to synthesize 11α-hydroxyprogesterone by a biomimetic pathway (Figure 2.26).[33] This is a key intermediate in the commercial production of the common pharmaceutical cream hydrocortisone acetate. This is an astonishing stereospecific transformation, in which three carbon–carbon bonds are formed in a single cascade reaction. Cascade reactions will be described in more detail in Chapter 8 because they are frequently successful without protecting groups. Notice that the larger groups, the connections to other rings, are all equatorial in the product giving the more thermodynamically stable diastereomer (Figure 2.26).

Biomimetic synthesis can be much greener than traditional methods by using asymmetric bio-feedstocks, enzymes to catalyze key stereochemical steps, and no derivitization with protecting groups or asymmetric auxiliaries. An example of such a biomimetic synthesis will be discussed in Chapter 8, Avoid Protecting Groups.

2.8 Summary

The efficiency of a synthesis can be described in three different ways: (1) atom economy, (2) overall yield, and (3) the E-factor or PMI. The atom economy is most useful for comparing the efficiency of different proposed synthetic routes. Overall yield is used widely in the organic literature to measure the production of product based on the limiting reagent, but yield does not include the atom economy or the use of auxiliaries. Sheldon's E-factor and the PMI are quantitative ways to

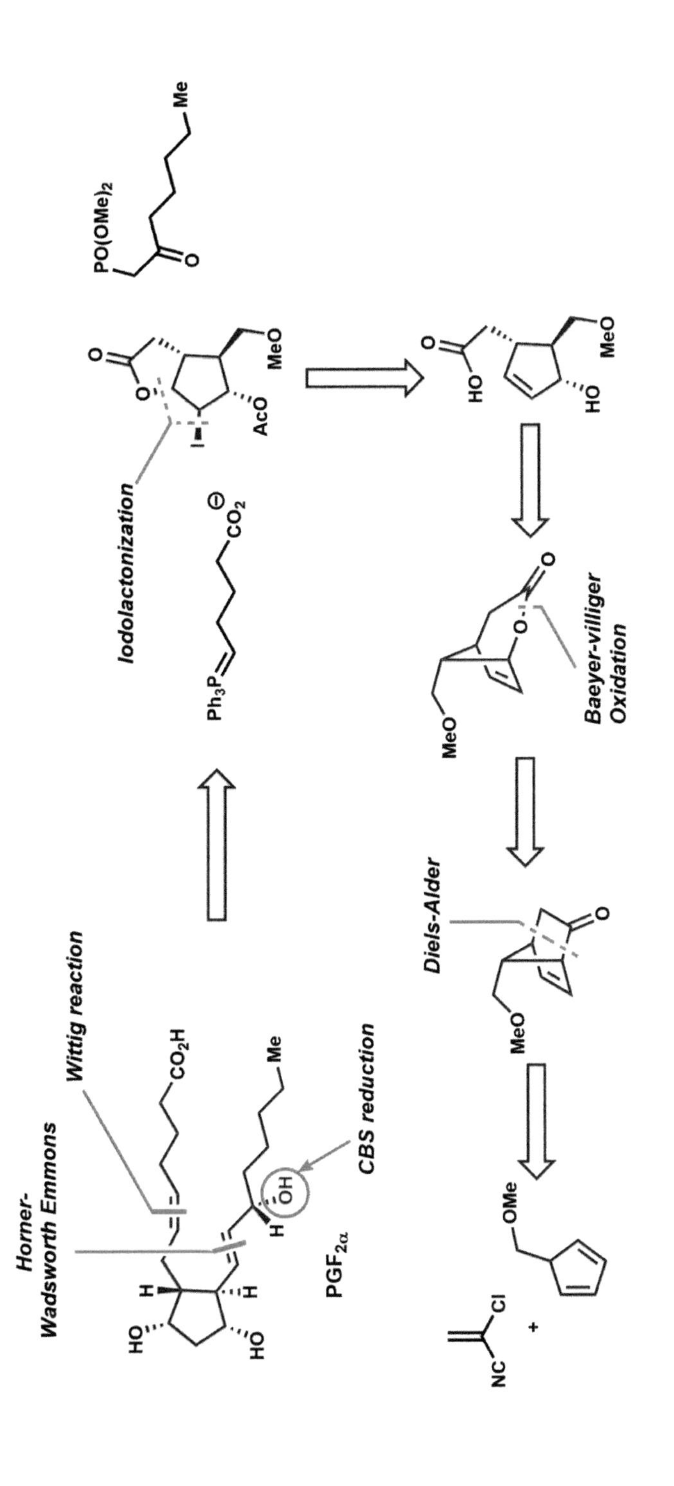

scheme 23

Figure 2.25 Prostaglandin retrosynthesis.[31] Reproduced from https://en.wikipedia.org/wiki/Elias_James_Corey under a CC BY-SA 3.0 license (http://creativecommons.org/licenses/by-sa/3.0)], accessed July 26, 2019.

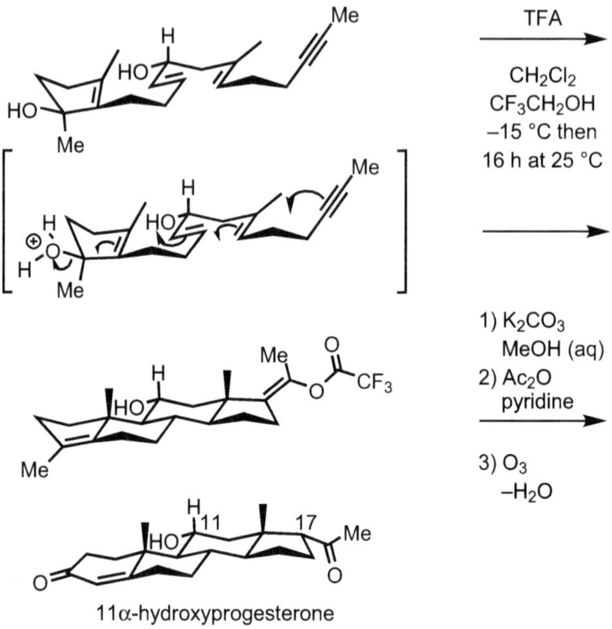

Figure 2.26 Johnson's biomimetic electrophilic cyclization of a polyolefin *en route* to 11α-hydroxyprogesterone.[33]

describe the efficiency of a process that include atom economy, overall yield, and auxiliaries.

The synthesis of single-enantiomer drugs is an important way to improve both the efficiency of production and the effectiveness of the final product. Strategies for more efficient production of drugs taking stereochemistry into consideration focus on: (1) design of achiral drugs, (2) application of asymmetric catalysts, (3) use of asymmetric bio-feedstocks, (4) enzymatic resolution, (5) resolution of diastereomeric salts, and (6) resolution of diastereomeric covalent derivatives.

Designing more efficient syntheses benefits from using the tools: (1) retrosynthetic analysis, (2) convergent synthesis, and (3) biomimetic synthesis.

2.9 Problems: Synthetic Efficiency

For all atom economy, overall yield, and E-factor calculations, give the answer as 2 significant figures. This means you can use just 2 or 3 significant figures for the atomic masses in atom economy calculations (*e.g.* H = 1.0, C = 12, O = 16, Cl = 35.5).

Problem 2.1

a) Balance the S_N2 reaction below by giving all products.

MF:
MW:

b) Calculate the atom economy of the reaction. Show the equation.

Problem 2.2

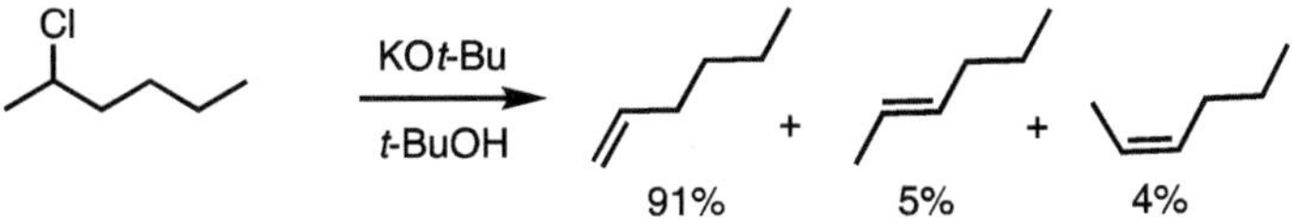

MF

MW

See also ref. 35.

a) Calculate the % atom economy of the elimination reaction to produce 1-hexene above. (Note: *t*-BuOH is the solvent.)
b) Calculate the % atom economy if the base counter ion was Li^+ instead of K^+. Show the equation.
c) Calculate the % atom economy if iodine was substituted for chlorine in the reactant. Show the equation.
d) The use of the bulky base decreases the atom economy. What advantage does the bulky base have over a small base like NaOH?

Problem 2.3

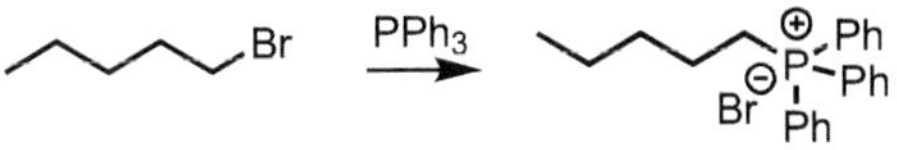

MF
MW

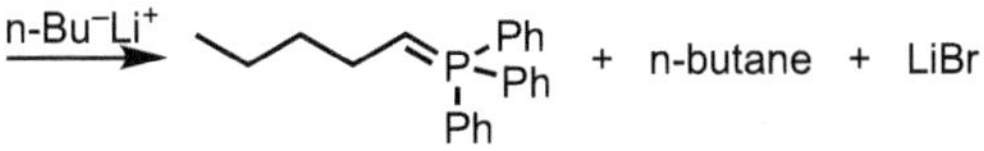

MF
MW

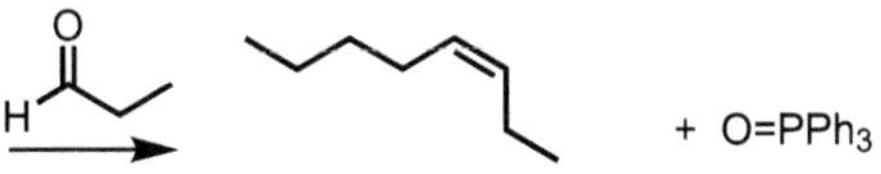

MF
MW

a) Calculate the overall atom economy of the Wittig condensation shown above. Write the MF and MW under each reactant. Show the equation.
b) Calculate the atom economy for a similar reaction that uses NaH as the base instead of *n*-BuLi. What are the by-products in this case?

Problem 2.4

a) Circle the sulfonate that is most atom economical reagent. Why?

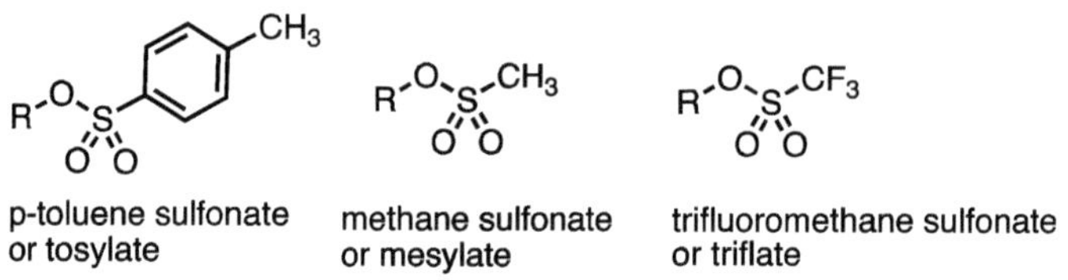

b) Would it be more atom economical to have iodine as a leaving group or mesylate?

Problem 2.5

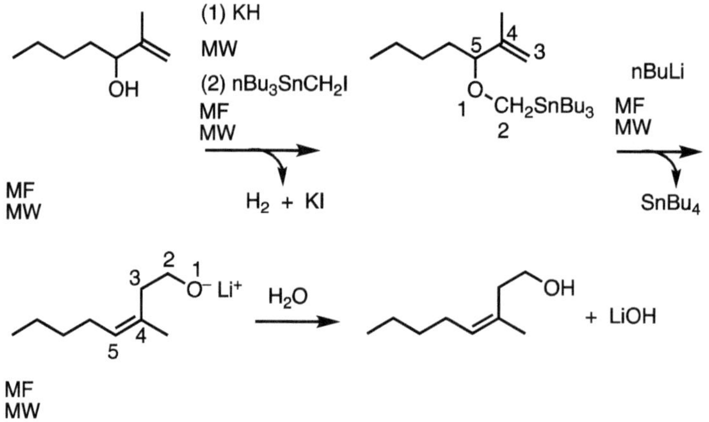

a) Calculate the atom economy of the Still–Wittig rearrangement shown above.
b) Draw the high-energy carbanion intermediate of the second reaction, and add curved arrows to show the mechanism of the rearrangement.
c) What are the four by-products of the reaction?

Problem 2.6

a) Look up the Materials Data Safety Sheet of HF. What hazards are involved?
b) Calculate the atom economy of the BHC synthesis of ibuprofen. Do not include catalysts in your calculation. Show your work in a MF, MW table and an equation.
c) Calculate the atom economy considering that the acetic acid is recycled.

Problem 2.7

Use the experimental description of the asymmetric C–C bond forming reaction of the Winkler synthesis of Ritalin™ (Figure 2.12).

a) Find or calculate the mass of reactants and solvents used in g.
b) Estimate the volume of silica gel used in purification. The density of silica gel is 2.2 $g\,cm^{-3}$.
c) For this size column, one would typically use about 300 mL of the solvent mixture. Find the density of both solvents and calculate the grams used.
d) Calculate the PMI with and without the catalyst included as waste.

Problem 2.8

a) Calculate the overall yield of the Nie synthesis of Tamiflu (Figure 2.20).
b) Draw the curved-arrow (electron-pushing) mechanism of the reaction from compound 7 to **8** in the Nie synthesis of Tamiflu. Designate the mechanism type (S_N1, S_N2, E1, E2 *etc.*).

Problem 2.9

a) The all L-amino acid lipase A gives (3*S*)-stiripentol directly from the transesterification in Figure 2.21. Can you imagine a lipase that would directly produce the desired (3*R*)-stiripentol?
b) Calculate the specific rotations of (3*S*)- and (3*R*)-stiripentol based on the optical rotations and % *ee* (enantiomeric excess) results shown in Figure 2.21.
c) In a synthesis of a chiral drug, how green is this method compared with using an enzyme to catalyze a stereoselective reaction?

Problem 2.10

The Winkler group at the University of Pennsylvania devised a synthesis of the active enantiomer of Ritalin™ using a chiral rhodium catalyst. The ligand is derived from the amino acid, D-proline. Note that the tosyl (Ts) and Boc groups are temporary derivitizations that go against Principle 8 of Green Chemistry. The catalyst can be recycled. The final product is a single enantiomer with two stereocenters.[13]

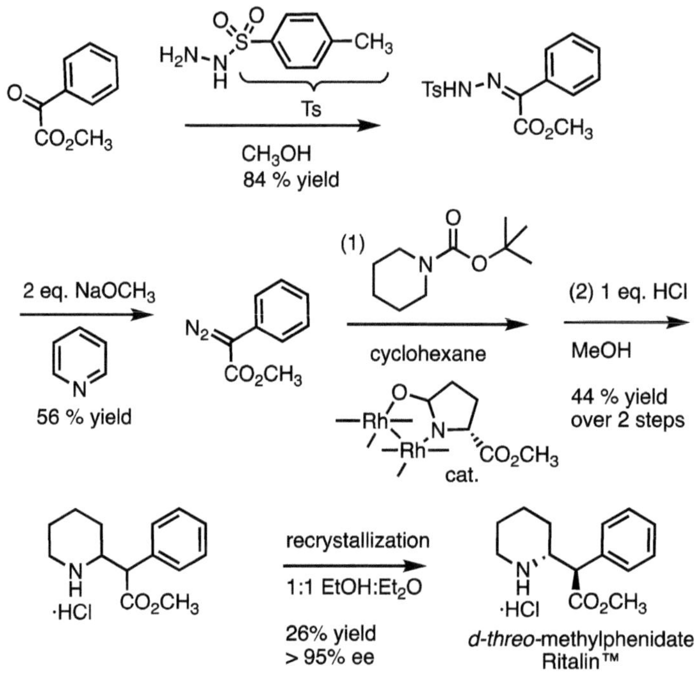

a) Calculate the overall yield.
b) Calculate the atom economy for the Winkler synthesis of Ritalin shown above.
c) For a real challenge, calculate the PMI for the whole synthesis from methylbenzoylformate. You will need to find the experimental details for the first step to make the tosyl (Ts) hydrazine in a reference within the Winkler paper. Find the amounts of each chemical used, find the densities of each solution used, and calculate the masses from the volumes given. Then scale the first two steps to correspond with the amount of methyl phenyldiazoacetate used in the last synthetic step.

References

1. P. T. Anastas and J. C. Warner, *Green Chemistry: Theory and Practice*, Oxford University Press, New York, 1998.
2. Encyclopedia Britannica, *Chemical intermediate*.
3. B. Trost, *Science*, 1991, **254**, 1471–1477.
4. B. M. Trost, *Angew. Chem., Int. Ed. Engl.*, 1995, **34**, 259–281.
5. L. Edward, R. Kosinski and R. R. Soelch, *Low temperature nylon polymerization process*, E. I. Du Pont De Nemours And Company, *U. S. Pat.* US 5,665,854, 1997.
6. W. C. Still and A. Mitra, *J. Am. Chem. Soc.*, 1978, **100**, 1927–1928.
7. Y. R. An, S. J. Kim, H.-W. Park, S. Y. Yu, J. Han, J.-H. Oh, S.-J. Yoon and S. Y. Hwang, *Mol. Cell. Toxicol.*, 2011, **7**, 95–101.
8. J. Mulzer and B. List, *Tetrahedron Lett.*, 1996, **37**, 2403–2404.
9. BHC Company (now BASF Corporation), *BHC Company Ibuprofen Process*, Washington, DC, 1997, http://www2.epa.gov/greenchemistry/1997-greener-synthetic-pathways-award, accessed July 26, 2019.
10. R. A. Sheldon, *Chem. Ind.*, 1992, 903–906.
11. R. A. Sheldon, *Green Chem.*, 2007, **9**, 1273–1283.
12. C. Jimenez-Gonzalez, C. S. Ponder, Q. B. Broxterman and J. B. Manley, *Org. Process Res. Dev.*, 2011, **15**, 912–917.
13. J. M. Axten, R. Ivy, L. Krim and J. D. Winkler, *J. Am. Chem. Soc.*, 1999, **121**, 6511–6512.
14. I. Moritani, T. Hosokawa and N. Obata, *J. Org. Chem.*, 1969, **34**, 670–675.
15. W. C. Still, M. Kahn and A. Mitra, *J. Org. Chem.*, 1978, **43**, 2923–2925.
16. L. Pauling, *Chem. Eng. News Archives*, 1946, **24**, 1375–1377.
17. L. A. Nguyen, H. He and C. Pham-Huy, *Int. J. Biomed. Sci.*, 2006, **2**, 85–100.
18. *Development of New Stereoisomeric Drugs*, 1992, http://www.fda.gov/Drugs/GuidanceComplianceRegulatoryInformation/Guidances/ucm122883.htm, accessed February 17, 2015.
19. S. C. Stinson, *Chem. Eng. News*, 2000, **78**, 55–78.
20. P. Mansfield, D. Henry and A. Tonkin, *Clin. Pharmacokinet.*, 2004, **43**, 287–290.
21. *Frances Oldham Kelsey: Medical reviewer famous for averting a public health tragedy, 2013*, http://www.fda.gov/aboutfda/whatwedo/history/ucm345094.htm, accessed February 17, 2015.
22. A. M. Rouhi, *Chem. Eng. News*, 2003, **81**, 45–55.
23. L. D. L. Lu, R. A. Johnson, M. G. Finn and K. B. Sharpless, *J. Org. Chem.*, 1984, **49**, 728–731.
24. L. D. Nie, X. X. Shi, K. H. Ko and W. D. Lu, *J. Org. Chem.*, 2009, **74**, 3970–3973.
25. M. Federspiel, R. Fischer, M. Hennig, H.-J. Mair, T. Oberhauser, G. Rimmler, T. Albiez, J. Bruhin, H. Estermann, C. Gandert, V. Göckel, S. Götzö, U. Hoffmann, G. Huber, G. Janatsch, S. Lauper, O. Röckel-Stäbler, R. Trussardi and A. G. Zwahlen, *Org. Process Res. Dev.*, 1999, **3**, 266–274.
26. J. Andraos, *Org. Process Res. Dev.*, 2009, **13**, 161–185.
27. R. H. G. P. Arends, K. Zhang, R. H. Levy, T. A. Baillie and D. D. Shen, *Epilepsy Res.*, 1994, **18**, 91–96.
28. E. E. Jacobsen, T. Anthonsen, M. Farrag El-Behairy, E. Sundby, M. Nabil Aboul-Enein, M. Ibrahim Attia, A. A. El-Sattar El-Azzouny, K. M. Amin and M. Abdel-Rehim, *Int. J. Chem.*, 2012, **4**, 7–13.
29. P. Duhamel, L. Duhamel, D. Danvy, J. C. Plaquevent, B. Giros, C. Gros, J. C. Schwartz and J. M. Lecomte, Preparation of enantiomeric derivatives of amino acids as analgesics and cardiovascular agents, EP 318377 A2 19890531, 1989.
30. E. J. Corey and X.-M. Cheng, *The Logic of Chemical Synthesis*, Wiley-Interscience, New York, NY, 1995.

31. E. J. Corey, Wikipedia, https://en.wikipedia.org/wiki/Elias_James_Corey, accessed July 26, 2019.
32. R. A. Yoder and J. N. Johnston, *Chem. Rev.*, 2005, **105**, 4730–4756.
33. W. S. Johnson, R. S. Brinkmeyer, V. M. Kapoor and T. M. Yarnell, *J. Am. Chem. Soc.*, 1977, **99**, 8341–8343.
34. M. C. Cann, and and M. E. Connelly, The BHC Company Synthesis of Ibuprofen, in *Real World Cases in Green Chemistry*, American Chemical Society, 2000.
35. R. A. Bartsch and J. F. Bunnett, *J. Am. Chem. Soc.*, 1969, **91**, 1376–1382.

3 Benign Synthesis

"Principle 3: Wherever practicable, synthetic methods should be designed to use and generate substances that possess little or no toxicity to human health and the environment."[1]

3.1 Eliminating Toxins in Synthesis

Like Principle 2: Synthetic Efficiency, Principle 3 focuses on the *process* of chemical synthesis. In Principle 2, the focus was on eliminating waste in the process, while Principle 3 urges us to eliminate toxins in the process. Why should we focus on the chemical process if only experts will be involved in production of chemicals? Unfortunately, accidents do happen, and toxic substances are released,

Green Chemistry: Principles and Case Studies
By Felicia A. Etzkorn

Published by the Royal Society of Chemistry, www.rsc.org

whether due to failure of design, mechanical or equipment failure, power outages, computer monitoring error, or simply human error. Workers may be poorly trained, distracted, or just tired. We need only look to the accidents in Bhopal, India, Love Canal, NY, Times Beach, MO, Fukushima, Japan, or in your own region. Some accidents are huge, and get all the media attention, while most are small, resulting in harm to a few people. Cumulatively, the latter cause more harm and death than the former, as documented on the Right-to-Know Network site.[2]

Toxicology is the best method to interpret existing data on toxic chemicals, and to know how to avoid them. Yet knowing what to avoid is not enough; what chemicals can we choose in our processes that are safer? A case study that modifies the synthesis of Quantum Dot Light-Emitting Diodes (QLEDs) to use less toxic solvents will show how QD Vision Inc. put these principles into practice.

3.1.1 Ethical Responsibilities

Toxicity is one specific form of hazard of particular concern to chemists. We will address the hazards of fire and explosion separately in Chapter 12, Prevent Accidents. Chemical spills and leaks can occur in transport, in production, in use, and after disposal. When we eliminate toxins from the process, it prevents harm in all these areas. Because a synthesis may depend upon toxic chemicals, we now focus on ways to avoid toxic chemicals in synthesis and process.

Recall that risk is a function of hazard and exposure; it is much easier to eliminate risk by eliminating the hazard than to eliminate exposure of all people, at all times, in all places. None of us would like to think that a chemical process we designed caused harm or death to the truckers who bring in the chemicals, to the workers in the plant, or to the neighbors in the surrounding community. We also have a long-term responsibility to protect people and the environment from hazardous waste in storage. The Super Fund hazardous waste site in Love Canal, NY USA remains a vivid example of how badly the storage of hazardous waste can go wrong in the long run. The ethical solution of Principle 3 is to avoid the use of toxic chemicals in synthesis or process entirely.

3.1.2 Economic Advantages of Eliminating Hazards

From experience, we know that it is also cheaper to prevent toxic releases than to clean up afterwards and deal with consequent litigation. Many companies fail after a major release of toxic chemicals precisely due to these costs. As a consequence, they usually default on their responsibilities to the victims of the accident. Not only does avoiding toxins prevent harm to humans and the environment, it makes the products cheaper, and the profits potentially higher. Recall that Roger Sheldon got the idea for the E-factor because the expense of hazardous waste disposal at a chemical production plant became almost as much as the price of the product.

3.2 Regulatory Frameworks

Exposure of workers to toxic chemicals during production is a major issue that is regulated by the US Occupational Safety and Health Administration (OSHA), the European Agency for Safety and Health at Work (osha.europa.eu/en), the Indian Ministry of Labour and Employment: Industrial Safety and Health, and the Safe Production Law of the People's Republic of China, to name a few.

3.2.1 European REACH Program

The European Union has a growing list of restricted chemicals that is more extensive than current USA restrictions within its Registration, Evaluation, Authorization and restriction of CHemical (REACH) substances program.[3] REACH is a community-wide program involving the European Economic Area (EEA) administered by the European Chemicals Agency (ECHA). Efforts to assess risk are shared by member countries and companies doing business in the EEA. All substances in commerce are subject to the REACH process, including natural substances. ECHA maintains a Community Rolling Action Plan (CoRAP) that tracks chemicals through all stages of reach on a three-year rolling basis. All companies that import, process, resell, or manufacture chemicals must comply with the regulations, "REACH places the burden of proof on companies."[3] All company data on chemical risk is collected, collated, and shared with all companies, countries, and consumers.

3.2.2 USA Toxic Substances Control Act (TSCA)

A series of regulations for toxic chemicals were put in place in the USA in the 1970s, including the Toxic Substances Control Act (TSCA) of 1976.[4] The original TSCA required the USA Environmental Protection

Agency (EPA) to develop a list of toxic chemicals and regulate their import, transport, use and disposal. Unfortunately, chemicals already in use prior to 1979 were exempt, or "grandfathered", from EPA review. Existing chemicals were considered safe because they have already been in use. These exempt chemicals constitute about 99% of all chemicals in commercial use by volume. Ironically, this results in continued use of known toxins, effectively preventing innovation in the use of less toxic substitutes. Clearly, not all chemicals already in use are safe, and several existing toxic chemicals were covered by amendments to TSCA: polychlorinated biphenyls, lead, radon, and asbestos.[5] In almost 30 years of experience with TSCA regulations, environmental organizations have recommended that chemicals with unknown toxicity instead be considered "guilty until proven innocent."

A second difficulty for the EPA is that the evaluation process begins during the discovery phase for new chemicals, even though about 90% of these never go to market. The EPA efforts are thus spread thin, and insufficient effort is given to new chemicals that end up in widespread use.[6]

In 2016, the USA passed into law a major revision of TSCA with bipartisan governmental support; the law was also supported by both industry and environmental organizations. This legislation requires the EPA to base chemical regulations solely on issues of public health instead of cost-benefit considerations.[4]

The Safe Drinking Water Act of 1974 requires the EPA to determine the levels of contaminants at which no adverse health effects are likely to occur over a lifetime of exposure, with an adequate margin of safety. The EPA does not have the power to enforce these health goals, called Maximum Contaminant Level Goals (MCLGs). The EPA then sets the Maximum Contaminant Level (MCL), usually close to the MCLG, as the enforceable regulation.

Most current regulatory processes are reactive, rather than proactive. Matus *et al.* have called upon the EPA to include the principles of green chemistry within the regulatory framework to design products to be benign at the outset, rather than looking for alternatives after a product has been found to be unsafe.[7]

3.2.3 Toxicity Databases

A good source of toxicity data is the European Chemicals Agency (ECHA).

http://echa.europa.eu/web/guest/information-on-chemicals (Accessed July 16, 2019).

The EPA set up two useful databases in 2011: The Toxicity Forecaster database (ToxCastDB) and a database of chemical exposure studies (ExpoCastDB), brought together in the Aggregated Computational Toxicology Resource (ACToR). https://actor.epa.gov/actor/home.xhtml (Accessed October 4, 2019).

The USA Toxics Release Inventory (TRI) is useful to search for all emissions from a specific industry, a locale, or a particular chemical.

http://www.epa.gov/toxics-release-inventory-tri-program (Accessed July 16, 2019).

Finally, the PubChem Database provides a fairly complete chemical portrait, including hazards and toxicity for over 84 million compounds.

http://pubchem.ncbi.nlm.nih.gov (Accessed July 16, 2019).

3.3 Toxicology

Toxicology is an important tool in green chemistry. The study of chemical toxicity involves detecting and reporting the biological effects, chemical structures, mechanism, and treatments for poisons. We will use the words poison or toxin as synonymous with chemical toxin. (Other types of toxins, biological, even physical, are possible.) We restrict our discussion to chemical toxins. To design and implement benign chemical syntheses, we need to know the possible types of toxins.

3.3.1 Acute and Chronic Toxins

Acute toxins have an immediate negative health effect on an exposed person or animal. The acute effect may be as mild and reversible as tearing of the eyes (lachrymator), or as severe as death. Acute refers to the immediate timescale of the toxic effects.

Chronic toxins have slow onset, and usually build up in the system as a result of long-term exposure. Because of this slow onset, establishing the causative agent with an adverse effect is difficult. Chronic toxins may exert difficult to diagnose health effects, such as reproductive, behavioral, or neurological symptoms. Cancer may be the result of chronic exposure to low levels of potent carcinogens that eventually overwhelm the cellular DNA repair mechanisms. For this reason, toxicology studies are essential to demonstrate the safety of new chemical entities.

3.3.2 Severity of Toxins

The severity of a toxin is a qualitative description of what happens to a living being upon exposure to the toxin, the end effect caused by a substance. Different types of chemical toxins have varying degrees of severity from mild to tragic. Irritants span the range from lachrymators, which cause tearing of the mucous membranes, like onions, to strong acids, which cause severe chemical burns. Neurotoxins cause neurological damage in children, as with tetraethyl lead (Et_4Pb), or result in rapid, horrific death, as with chemical warfare agents.[8]

Severity may also be a result of the amount of a chemical in the exposure, so severity and potency may be intimately linked. Organomercury compounds are particularly insidious. Karen Wetterhahn was a professor of chemistry at Dartmouth studying mercury in the environment. She used dimethylmercury (Me_2Hg) as the standard in her NMR samples. She noticed a drop or two had spilled on her glove, and changed gloves quickly. She did not begin to suffer ill effects until three months after exposure, and she died 10 months after the accident.[9] Dimethylmercury is an example of a slow-acting, high potency toxin, with an extremely severe outcome. The non-volatile perchlorate salt, $Hg(ClO_4)_2$, has been suggested as an alternative NMR standard, but it is also extremely toxic. A non-volatile mercury salt, such as $HgCl_2$ dissolved in water in a sealed capillary tube that can be dropped inside an NMR tube, would make a less hazardous Hg NMR standard.

3.3.3 Potency of Toxins

"The dose makes the poison." Paracelsus (1493–1541). Paracelsus wisely understood that toxins are dose dependent. For some, it takes very little, while for others it takes a lot to poison. The potency of a toxin is a quantitative measurement of how much it takes to cause a particular effect. Potency is usually expressed in exposure concentration units of $mg\,kg^{-1}$ body weight. The values are called LD_{50}, the "lethal dose 50," the concentration of toxin at which 50% of the test animals die from exposure (Figure 3.1). Each LD_{50} will also list the species tested, and the route of administration, for example ingestion or intraperitoneal (i.p.). The LD_{50} is often used to extrapolate from mouse or rat to human because it is unethical to perform dose–response experiments on humans. Regulatory agencies are also trying to limit the number of test animals used in toxicity testing by analogy, computer modeling, and statistical methods.

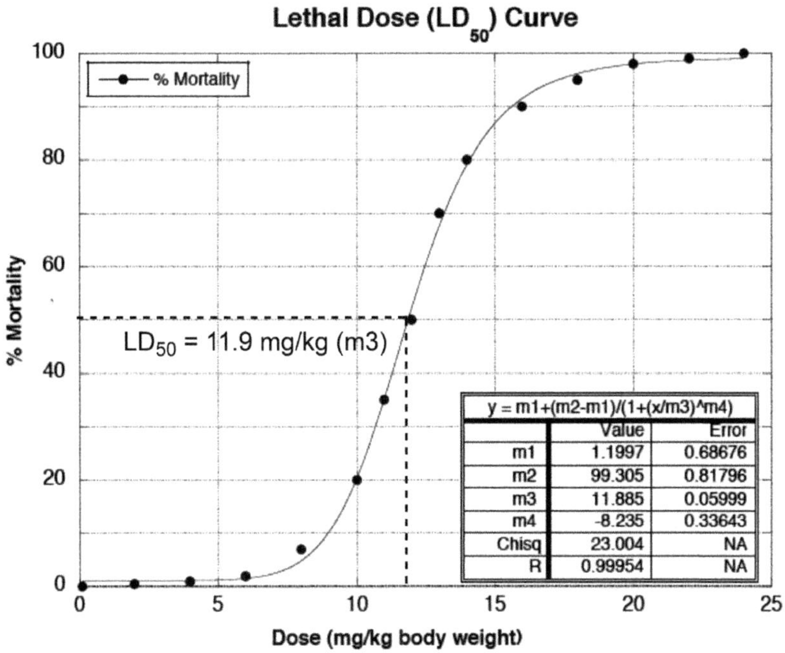

Figure 3.1 Idealized lethal dose 50% (LD_{50}) plot for toxins, showing a sigmoidal equation curve fit.

Other measures of toxicity are possible beyond the LD_{50}. For example, many poisons are volatile, and inhalation is the most direct route of entry. In this case, an LC_{50} (lethal concentration in air) is measured over a specific amount of time. Death need not be the endpoint. If an adverse reaction can be quantified, a dose–response curve can be measured. When it is inappropriate to measure LD_{50} values, such as in cognitive deficits resulting from exposure of children to lead, or cancer resulting from exposure of adult humans to a carcinogen, the value effective dose 50 (ED_{50}) is used instead. As in pharmacology, quantitative toxicology delivers dose–response curves, which are usually plotted as adverse biological effect (y) *vs.* concentration (x).

Ordinarily, these curves are S-shaped, called sigmoidal curves, with a critical threshold for activity, a steep rise in activity around 50%, followed by a leveling off of activity due to saturation of the biological target—100% mortality in the case of the lethal dose curve (Figure 3.1). The LD_{50} is calculated from an equation fitted to the data used to make the curve; in the idealized curve of Figure 3.1 the LD_{50} is m3.

3.3.4 Non-monotonic Dose–Response Curves

This model is not always appropriate; some toxins require different models depending upon the mode of action, and the biological response. Endocrine disrupters are particularly prone to non-monotonic response curves.[10] This is due to the heightened sensitivity of steroid receptors to low doses (parts per trillion) of natural steroid hormones, for example the estrogen receptor is highly sensitive to 17β-estradiol (Figure 3.2). Estrogens bind to the estrogen receptor, causing dimerization and binding to DNA (Figure 4.3). The dose–response curve of estrogen biological activity is non-monotonic. At very low doses (fg mL^{-1}), there is no response, as expected (Figure 3.2). At mid-level doses (pg mL^{-1} to ng mL^{-1}), cellular processes amplify the signal, causing a stronger response. At high doses (>μg mL^{-1}), the response is not receptor-mediated, causing insensitivity to the steroid hormone. Some endocrine disrupters mimic natural steroid hormones giving similar non-monotonic responses. Biological systems often react to high levels of toxins by shutting off the response pathway *via* a negative feedback mechanism. The result is an inverted U-shaped curve that causes extrapolations from high-dose studies to low doses to be very misleading, or even flat-out wrong (Figure 3.2).

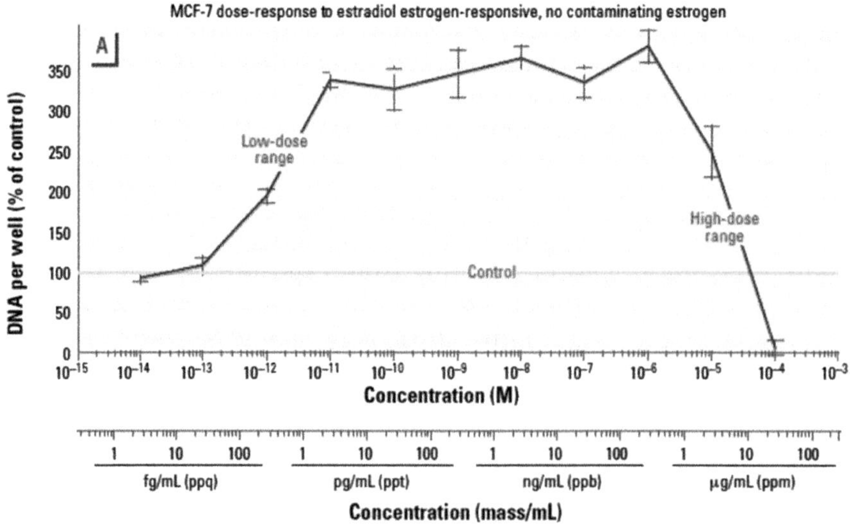

Figure 3.2 Inverted U-shaped curve of breast-cancer cell (MCF-7) biological response to estrogen concentration.[10] Reprinted from W. V. Welshons, K. A. Thayer, B. M. Judy, J. A. Taylor, E. M. Curran and F. S. vom Saal, *Environ. Health Perspect.*, 2003, **111**, 994–1006.

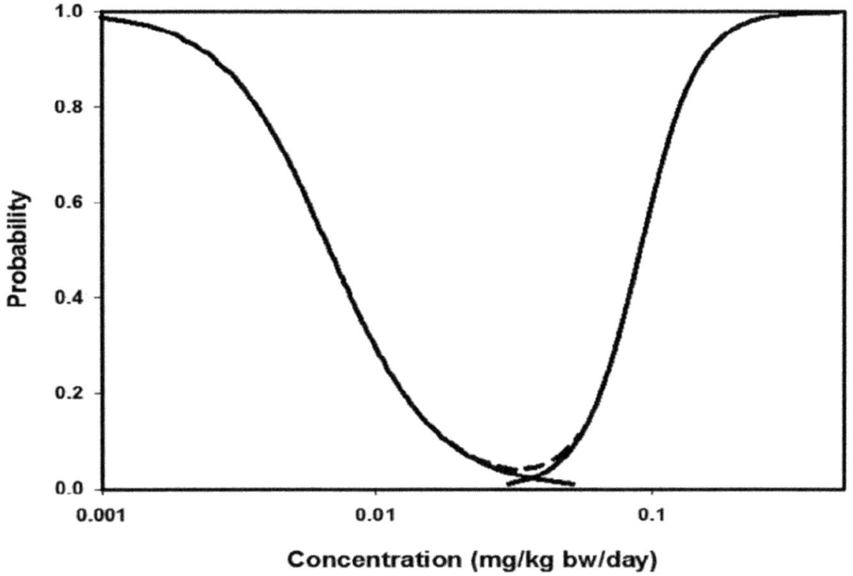

Figure 3.3 U-shaped probability curve of adverse effects for copper deficiency and copper excess.[11] Reprinted from A. Chambers, D. Krewski, N. Birkett, L. Plunkett, R. Hertzberg, R. Danzeisen, P. Aggett, T. Starr, S. Baker, M. Dourson, P. Jones, C. Keen, B. Meek, R. Schoeny and W. Slob, An exposure-response curve for copper excess and deficiency, *J. Toxicol. Environ. Health. B Crit. Rev.* 2010, **13** (7–8), 546–78, with permission from Taylor and Francis.

A third possibility exists when a substance is essential to health at low concentrations, but toxic at high concentrations. Copper is an essential nutrient that follows this type of dose–response curve (Figure 3.3).[11] If the concentration is too low, the body is deficient and certain enzymes that require Cu do not function properly. If the concentration is too high, the Cu creates reactive oxygen species, or Cu can swamp out other metal cofactors for enzymes, such as Zn, or bind to other proteins modifying their biological functions.[12] The optimum intake concentration of Cu in the body is in the range from 0.01–0.1 $mg\,kg^{-1}$ body weight per day.

3.3.5 Reversibility

The reversibility of a toxin is a major issue in toxicology. One problem with lead and mercury toxicity is how slowly these metals are excreted from the body. Karen Wetterhahn, the chemist who had been exposed to dimethylmercury, was treated with chelation therapy to remove the Hg from her system. But it was too late, the poisoning was irreversible; she died 10 months after her initial exposure. Mercury and lead poisoning are nearly irreversible.

One of the major challenges of DDT is that the major metabolite, DDE (*p,p'*-dichlorophenyl-1,1-dichloroethene), is nearly as non-polar as DDT, and it remains in the fat of animals. DDT has been found in the mammary glands of whales and the fat tissues of polar bears. This is one way that the absorption of toxins may be nearly irreversible—they simply stay in the non-polar environment of fat tissues. This can be a particular problem with the neurological system, since the myelin sheaths, necessary for proper function of neurons, are made of fatty tissue. DDT is considered an endocrine disrupter because it affects the proper functioning of steroid hormones (Chapter 4).

3.3.6 Reactivity

The reactivity of a toxin is a second major issue. Cytochrome P450 (CYP) liver enzymes are necessary for excretion of highly non-polar chemicals, such as benzene, hexane, and toluene. The CYPs oxidize non-polar molecules to make them more polar, and hence more water soluble, for excretion through the kidneys. One problem is that sometimes CYP oxidation produces metabolites that are highly reactive in the body.

Benzo[*a*]pyrene undergoes epoxidation by CYPs, resulting in carcinogenic metabolites (Figure 3.4). The metabolite, benzo[*a*]-pyrene-7,8-diol-9,10-epoxide (BP-diol-epoxide), is one of the most potent carcinogens known, because the planar aromatic ring system

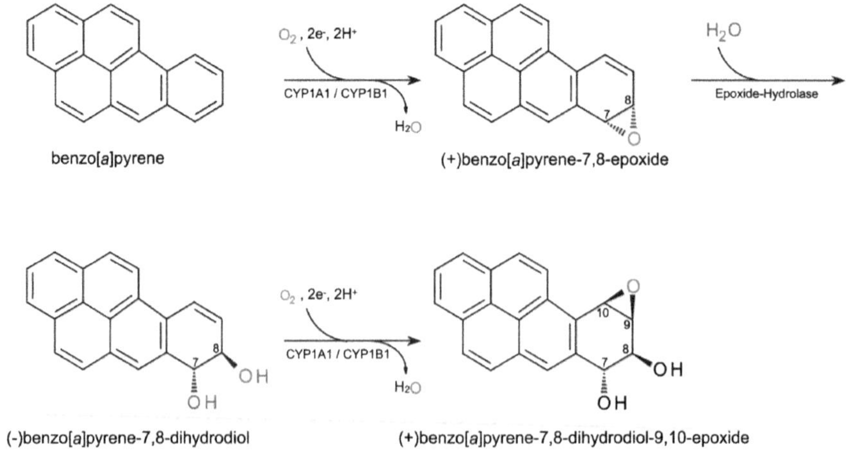

Figure 3.4 Benzo[*a*]pyrene metabolism.[13]

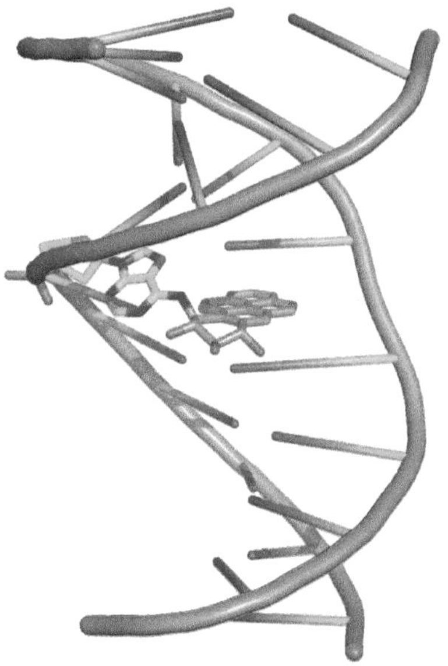

Figure 3.5 X-ray crystal structure of benzo[*a*]pyrene-7,8-diol-9,10-epoxide-DNA complex from the Protein Data Bank entry 1JDG.[14] An adenine base (green) amine nucleophile opened the epoxide ring of BP-diol-epoxide (purple). Figure created with MacPyMol 2006.

intercalates in between π-stacked DNA bases, positioning the epoxide for nucleophilic addition by the DNA base amine groups ($-NH_2$) of adenine (A), guanine (G), or cytosine (C) (Figure 3.5). The BP-base adduct is damaged DNA, which may be repaired (see 2015 Nobel Prize in Chemistry), or BP-diol-epoxide may overwhelm the repair mechanisms, proceeding on from mutagenesis to carcinogenesis.

Avoiding benzo[*a*]pyrene is quite difficult. It is formed during the burning of any organic substance: tobacco, meat, wood, *etc.* Benzo[*a*]pyrene is one of the suspected carcinogens in meat formed by grilling, frying, and other high temperature cooking methods. The World Health Organization declared that processed meats—bacon, sausage, ham, pepperoni, hot dogs, *etc.*—increase the risk of colon and stomach cancer.[15] Unprocessed red meat is probably carcinogenic also.[15]

3.3.7 Additivity and Synergy

The toxicity of chemicals in the real world must be considered in context. Rarely are we exposed to a single, pure toxin, as are laboratory rats when an LD_{50} is determined. Instead, toxicologists have begun to analyze mixtures, because toxins do not always behave the same when tested in more complex chemical environments as when tested in relative isolation. There are at least four models that have been used for mixtures of toxins: (1) the concentration-addition model; (2) the independent-action model; (3) the synergistic model; and (4) the antagonistic model.

The concentration-addition model is used when the mechanism of action is similar for the species under consideration, for example in the toxicity of heavy metals (Figure 3.6).[16] The independent-action model is used to describe the behavior of different toxins with different mechanisms of action, for example liver toxicity and eye irritation. Synergistic describes toxins that act in concert to produce a deleterious effect that is greater than either toxin alone. Antagonistic toxins show less-than-additive toxicity because one of the toxins interferes with the toxicity of the other. Snails exposed to heavy metals mainly exhibited less-than-additive toxicity, possibly by dilution of metals by each other in the binary mixture (Figure 3.6).[16] In one case, Ni and Cd showed more-than-additive, or synergistic, toxicity (Figure 3.6).

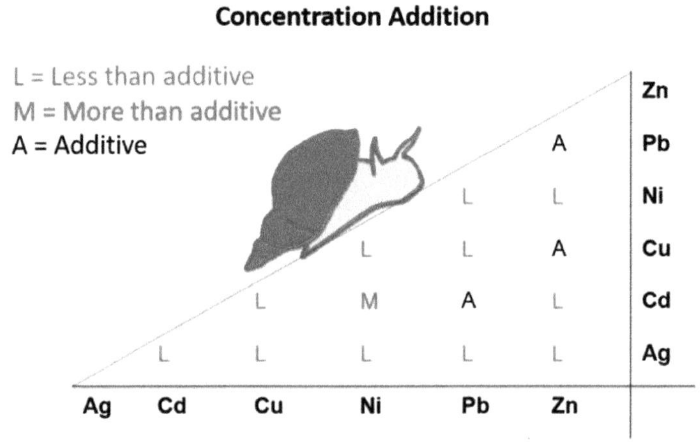

Figure 3.6 Binary metal toxicity in snails.[16] Reprinted from Crémazy, A.; Brix, K. V.; Wood, C. M., Chronic Toxicity of Binary Mixtures of Six Metals (Ag, Cd, Cu, Ni, Pb, and Zn) to the Great Pond Snail Lymnaea stagnalis. Reprinted with permission from *Env. Sci. Tech.*, **2018**, 52 (10), 5979–5988, https://pubs.acs.org/doi/10.1021/acs.est.7b06554, Copyright 2018 American Chemical Society. Further permissions related to this figure should be directed to the ACS.

3.4 Classifications of Chemical Toxins

3.4.1 Globally Harmonized System (GHS)[17]

Toxins can be categorized in different ways depending upon the purpose served. A new system for categorizing toxic chemicals uniformly around the world, the Globally Harmonized System (GHS) of classification and labeling of chemicals was developed at the United Nations Conference on Environment and Development in Rio de Janero, Brazil in 1992, and adopted by the UN in 2003. Although it is not mandatory, it provides a system for the classification of hazards and toxins that is quite useful in international trade. The EU based its REACH system for regulating hazardous substances and mixtures on the GHS. The category of "health hazards," or toxicity, most closely aligns with the goals of this chapter to reduce the use of toxic chemicals in chemical synthesis.

3.4.2 Health Hazards[17]

- **Acute toxicity** – immediate toxicity
- **Skin corrosion** – irreversible damage to the skin
- **Skin irritation** – reversible damage to the skin
- **Serious eye damage** – tissue damage in the eye, physical decay of vision
- **Eye irritation** – changes in the eye, fully reversible
- **Skin sensitizer** – induce an allergic response
- **Germ cell mutagenicity** – mutations in sperm and egg cells
- **Carcinogenicity** – induce cancer
- **Reproductive toxicity** – adverse effects on sexual function and fertility in adults, developmental toxicity (teratogens) in offspring
- **Specific target organ toxicity (STOT)** – impair organ function, reversible and irreversible
- **Aspiration hazard** – severe acute effects such as chemical pneumonia, pulmonary injury or death

3.4.3 Target Organ Toxicity

Table 3.1 gives a more specific listing of the "Specific Target Organ Toxicity (STOT)" categories above. It is interesting to note that skin and eyes are considered organs in physiology, and that aspiration hazards could be considered lung toxins, so that the two listings given are quite similar. Toxicologists often categorize toxins by the target organ, or by non-target organ effects (Table 3.1). Many toxins fall into

Table 3.1 Toxicity by target organ and non-target organ effects.[18]

Target organ	Toxicity	Examples
Blood	Hematotoxin	Acetanilide
Immune system	Immunotoxin	Benzene
Gastrointestinal system	Enterotoxin	Cholera toxin
Liver	Hepatotoxin	Acetaminophen, ethanol
Kidney	Nephrotoxin	Ibuprofen
Respiratory system	Pulmonary toxin	Acrolein (propenal), isocyanates
Nervous system	Neurotoxin	Hg, Pb
Ocular & visual system		Pb, methanol
Heart & vascular system	Cardiotoxin	Rofecoxib (Vioxx)
Skin	Irritant, corrosive	Urushiols (poison ivy), HF
Skin sensitizer	Sensitizer	Dicyclohexylcarbodiimide (DCC)
Reproductive system	Reproductive toxin	Acetonitrile, acrylamide
Endocrine system	Endocrine disrupter	DDT
Sexual organs	Infertility	Anti-depressants[54]
Non-organ directed		
Cell division	Carcinogen	Benzene, dichloromethane
Genetics	Mutagen	Formaldehyde, methyl iodide
Development	Teratogen	1,3-Butadiene

several target organ toxicity categories, so the example toxins given are not exclusive to the category listed. This categorization is strictly by severity (qualitative outcomes); it does not segregate toxins by potency towards those organs.

Categorizing toxins in these ways is helpful, but not if we want to *predict* toxicity of classes of chemicals. Methods for predicting toxicity will be discussed in Chapter 4, Benign Products. For processes, we are choosing to use less toxic reagents.

Many carcinogens, like formaldehyde, react with DNA and thus are mutagens as well. Not all carcinogens are mutagens, and not all mutagens are necessarily carcinogens. Mutagen describes the effect of a toxin on DNA, while carcinogen describes the disease outcome of cancer.

Commercially available chemicals have safety data sheets (SDS, formerly called Materials Safety Data Sheets, MSDS) that describe the hazards involved for particular chemicals. Chemists should consult SDSs for all feasible alternative reagents and solvents in designing the least toxic process. The green chemist must be particularly agile and creative in designing both less wasteful, and less toxic syntheses. With experience, the chemist becomes familiar with the types and degrees of toxicity associated with different classes of chemicals. With this knowledge, we can choose reagents that are less toxic. QD Vision reduced toxicity in the production of QLEDs in the next case study.

3.5 Case Study: Greener Quantum Dot Synthesis (QD Vision)

QD Vision Inc. won a PGCC Award in 2014 for developing a less toxic process to make QLEDs, which are the foundation of modern liquid crystal display (LCD) televisions and computer screens, and light-emitting diode (LED) lighting.[19–21] The market for energy-efficient LCD screens is huge.

The original LEDs were red (GaAs) and green (GaP) emitting semiconductors. To make LEDs emit closer to broad-spectrum white light, blue was added from indium gallium nitride (InGaN), resulting in a cold feel of the light. To offset the blueness, the insides of LED glass capsules were coated with yellow ($Y_3Al_5O_{12}$:Ce^{3+}, YAG:Ce^{3+}), or red ($CaAlSiN_3$) phosphors to give a warmer light. These phosphors absorb the blue light and emit at lower frequencies; they down-convert the light. The problem with phosphors is that the emission bandwidths, or full width at half of maximum (FWHM), are broader than light directly from QLEDs, and there is energy loss in the down conversion.

Nanoparticles are of a size between small molecules and bulk materials; the nanoparticle size range is generally agreed to be about 10 to 100 nm (100 to 1000 Å). QDs are nanoparticles with properties distinct from either single molecules, or bulk materials. The term "QD" refers to the electronic properties of semiconductor nanoparticles. The quantum confinement of the electron wave function increases as the size of the particle decreases, resulting in larger energy gaps with smaller particles. Bulk semiconductors, on the other hand, have electronic wave functions in bands with closely spaced energy levels, resulting in a broader wavelength range of emitted light.

This relationship is analogous to conjugated organic molecules (Figure 3.7). With few conjugated π-bonds, the electrons are confined to quantized molecular orbitals with a large energy gap between the HOMO and LUMO. With more conjugated π-bonds, the bonding MOs are higher in energy and the anti-bonding MOs are lower in energy, resulting in a smaller energy gap. The electrons are less confined.

QLEDs are nanoscale semiconductors that convert incoming energy to emitted color. The emitted color is precisely tunable by controlling the size of the QD (Figure 3.8). In a backlit LCD, the voltage to each pixel of the liquid crystal is controlled. Each pixel has a red, green or blue QLED in front of it. Light is emitted from a pixel only if the

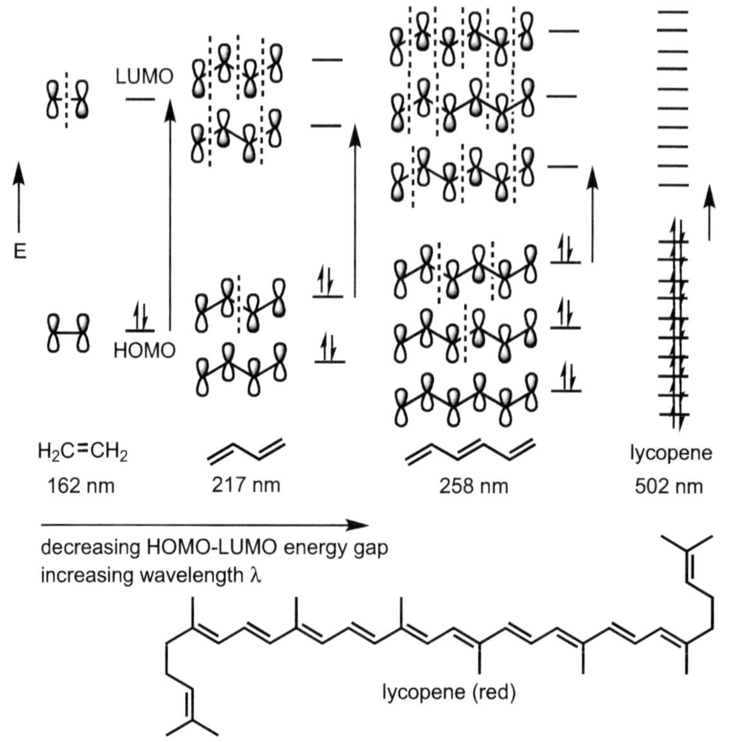

Figure 3.7 Quantum compression in conjugated organic molecules. As the number of conjugated double bonds increases, the energy gap between the HOMO and LUMO gets smaller, resulting in a shift to longer absorption wavelengths ($E = hc/\lambda$). The π-electrons of ethene are more confined than butadiene, and so on. Lycopene is an anti-oxidant that gives tomatoes their red color.

voltage permits the backlight to get through the liquid crystal layer. Sony estimated the use of QLEDs in television screens decreases energy use by 36%. If coal is the electricity source, QLED screens also result in lower CO_2, Hg, and Cd emissions, and lower electricity costs for the consumer.[19]

QD Vision uses a CdSe core coated with a CdZnS or ZnS shell to protect the core from environmental degradation.[20] These quantum dots have ~0.90 quantum yield. Quantum yield is a measure of efficiency, the ratio of the number of photons emitted to the number of photons absorbed. This high quantum yield gives the high-energy efficiencies reported. Cd is currently still the metal of choice for producing narrow bandwidth light with high quantum yields.[19–21] CdS exposure causes dermal irritation, respiratory harm, chronic kidney impairment, and genetic defects; all Cd compounds are

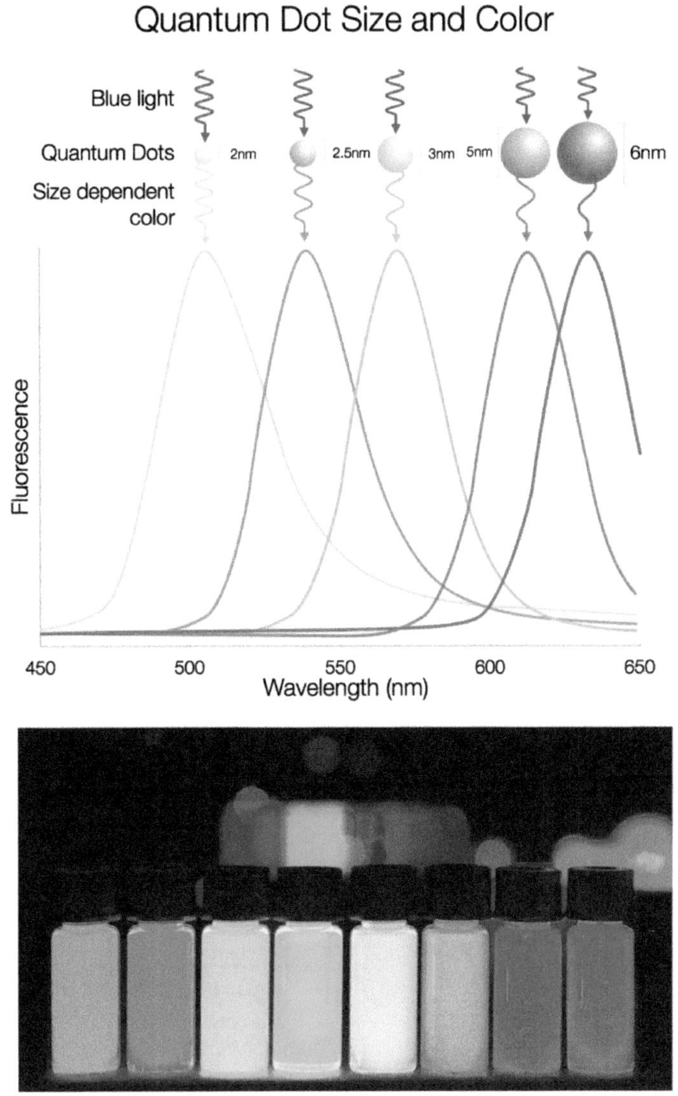

Figure 3.8 Top: Fluorescence emission shift based on quantum dot (QD) size. Reprinted with permission from nanosysinc.com. Copyright 2015 Nanosys Inc.[22] Bottom: Quantum Dots with emission maxima in a 10 nm step are being produced at PlasmaChem in a kg scale" by Antipoff - Photo taken on November 28, 2012, edited in Photoshop on June 29, 2013. Previously published: http://www.plasmachem.com/shop/en/226-zncdses-alloyed-quantum-dots. Licensed under CC BY-SA 3.0 *via* Commons - https://commons.wikimedia.org/wiki/File:Quantum_Dots_with_emission_maxima_in_a_10-nm_step_are_being_produced_at_PlasmaChem_in_a_kg_scale.jpg.

probable human carcinogens. The LD_{50} value for CdS is 7.1 g kg^{-1}, so it is not very toxic by ingestion.

The old QD Vision process used dimethyl cadmium ($CdMe_2$) and diethyl zinc ($ZnEt_2$) as the precursors for the QLEDs (Figure 3.9). Cd compounds are considered to be human carcinogens. The LD_{50} value of $CdMe_2$ is 50 μg kg^{-1}. The route of entry is by inhalation, and $CdMe_2$ is acutely toxic to the lungs, kidneys, liver, brain and nervous system. $CdMe_2$ is on Derek Lowe's list of "Things I Won't Work With."[23] $CdMe_2$ is comparable in potency and severity to $HgMe_2$. $CdMe_2$ and $ZnEt_2$ are both extremely flammable ($CdMe_2$ f.p. −18 °C, $ZnEt_2$ f.p. −40 °C). $ZnEt_2$ is pyrophoric, and it reacts explosively with water, although it leaves a non-toxic residue of ZnO, benign enough to use as sunscreen. It makes good green sense to use different starting materials to synthesize QLEDs.

QD Vision switched to Cd and Zn carboxylates, such as acetates, because they are much less volatile and flammable than the alkyl compounds. Exposure to cadmium carboxylates would be primarily through ingestion, but it is not well absorbed from the gut. $Cd(OAc)_2$ has endocrine disruption and teratogenic activity in rats, but the LD_{50} is 360 μg kg^{-1}, about 5 times less potent than $CdMe_2$.[24] $Cd(OAc)_2$ is also much less volatile than $CdMe_2$, and $Cd(OAc)_2$ is not combustible.

The QD Vision synthesis involves hot, rapid injection of precursors at >200 °C in high boiling solvents (Figure 3.9).[19–21] The two steps are injection of: (1) Cd and Se precursors with formation of

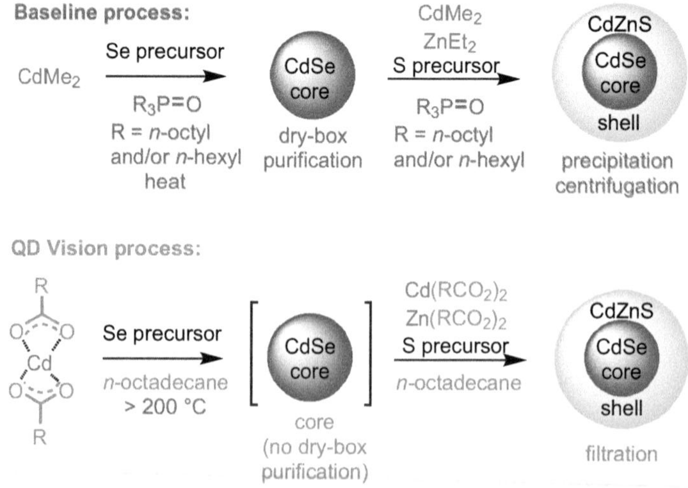

Figure 3.9 QD Vision synthesis of quantum dots for light-emitting diodes before and after improvement for the 2014 PGCC Award.[19]

nanocrystalline CdSe particles for the core, then (2) Cd, Zn, and S precursors for the shell. QD Vision improved the synthesis of their QLEDs by eliminating purification of the CdSe core before applying the shell. This eliminated the step of precipitating the reactive core inside of an inert atmosphere glove-box, with the side benefits of eliminating the energy for vacuum pumping and toxic waste pump oil. Instead of precipitating and centrifuging the nanoparticles after synthesis, they now perform a simple filtration. Streamlining the synthetic process resulted in a two-fold greater incorporation of Cd into the QLEDs, resulting in elimination of 100 kg Cd per year from the hazardous waste stream.

QD Vision also substituted the trialkyl phosphine oxide (TPO) solvents (Figure 3.10),[20] used to dissolve the QLED precursors in both steps, with a long-chain hydrocarbon (C18) solvent thought to be a safer solvent. The boiling point of *n*-octadecane ($C_{18}H_{38}$) is 316 °C. The flash point of TPO is 182 °C, while the auto-ignition temperature of *n*-octadecane is 235 °C. *n*-Octadecane is a flammable solid, with a molecular weight nearly in the paraffin wax range (petroleum derived C20–C40 alkanes). The new process also uses much less solvent, for example 44 grams of C18 solvent per gram of red QDs, instead of 1701 grams TPO solvent per gram.[19]

"Health and Safety: The oral and dermal toxicity of CYANEX 923 extractant is low. CYANEX 923 extractant produces mild eye irritation and severe skin irritation upon contact...CYANEX 923 extractant is highly toxic to fish and invertebrates and great care should be exercised to avoid environmental exposure." (SDS at CYANEX® 923 by Cytec).

Nanoco is a British manufacturer of QLEDs. For the European market, they are required by REACH to eliminate Cd entirely. The maximum permitted concentration of Cd is 10 times lower than Hg or Pb (Table 3.2). The non-cadmium containing QLEDs are made of indium phosphide (InP), coated with ZnS or ZnS/ZnSe. Indium phosphide is a probable human carcinogen, an irritant, and it was found to be phototoxic by generating reactive oxygen species.[25] It gives

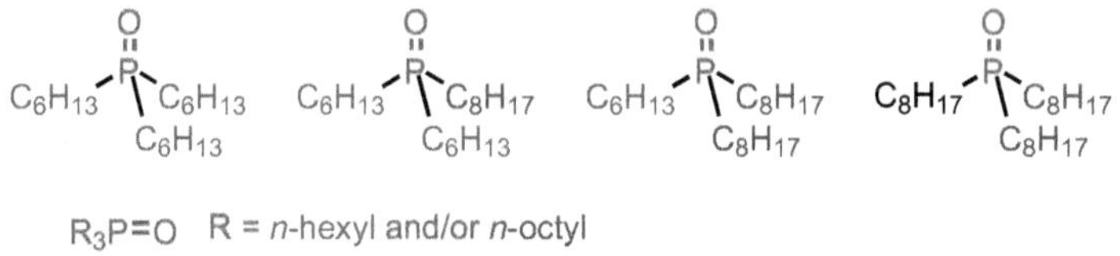

Figure 3.10 Trialkylphosphine oxide composition (CYANEX® 923 by Cytec).

Table 3.2 "Restricted substances referred to in Article 4(1) and maximum concentration values tolerated by weight in homogeneous materials".[26]

Substance	Max wt%
Lead	0.1%
Mercury	0.1%
Cadmium	0.01%
Hexavalent chromium	0.1%
Polybrominated biphenyls (PBB)	0.1%
Polybrominated diphenyl ethers (PBDE)	0.1%

off PH_3, causing central nervous system depression.[25] The LD_{50} of InP is 3.7 $g\,kg^{-1}$, while the LC_{50} (inhalation, rats) is 11 ppm/4 hours.

QD Vision has argued vociferously for an exemption to the European ban against Cd because the life cycle analysis of their process shows that much less Cd is used in the long run when coal is the source of electricity. The goal of both companies is to eliminate toxic substances entirely from QLEDs, making them both non-toxic and highly energy efficient in consumer electronics products.

Nanosys sued QD Vision for patent infringement.[27] Since then, Samsung Display acquired QD Vision's technology for 70M USD.[28]

3.6 Redox Reactions

Oxidation and reduction (redox) reactions remain among the most hazardous and wasteful reactions used in the chemical industries. Oxidation and reduction can be defined in several different ways, the third being most definitive. In organic chemistry, the first two definitions are most helpful in identifying a redox reaction.

1. Oxidation is the gain of oxygen (nitrogen, or halogen), reduction is the loss of oxygen.
2. Oxidation is the loss of hydrogen; reduction is the gain of hydrogen.
3. Loss of Electrons is Oxidation; Gain of Electrons is Reduction. (A mnemonic: LEO the lion says GER).

Dioxins and the related chlorinated dibenzofurans are created as minor, yet highly toxic, by-products of bleaching paper with chlorine (Cl_2), a form of oxidation (Figure 3.11). Bleaching with chlorine dioxide (ClO_2) creates about 10 times less dioxin. The dioxins are produced unintentionally as a by-product of many industrial

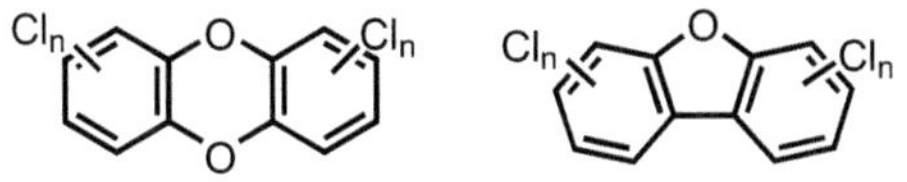

Figure 3.11 Chlorinated aromatic hydrocarbons, dibenzodioxins and dibenzofurans, produced in paper bleaching.

processes, as well as accidentally by combustion. The symmetric 2,4,7,8-TCDD is the most toxic isomer, shown to be teratogenic, mutagenic, carcinogenic, immunotoxic, and hepatotoxic. Measurements of the LD_{50} of dioxin range from 1 to 340 $\mu g\,kg^{-1}$, parts per billion levels (Table 4.1). Any oxidation reaction, like burning, in the presence chlorine and phenoxy-containing compounds, such as lignin, the structural component of wood, is likely to create trace amounts of dioxins and chlorinated dibenzofurans.

3.6.1 Oxidizing Agents

Traditional oxidizing agents typically have been quite hazardous stoichiometric reagents, with chromium(VI) oxidants leading the pack. Chromium(VI) compounds are known carcinogens with a host of other toxic effects. The movie *Erin Brockovich* was about a community leader who fought to have a Cr(VI) waste site cleaned up. On the other hand, Cr(III) is fairly benign because it cannot enter cells readily (0.5–2%), an example of poor bioavailability.[29] Cr(VI) is readily absorbed into cells (2–10%) through sulfate and phosphate transporter proteins.[29] Ironically, once inside cells, Cr(VI) is reduced to Cr(III) by ascorbic acid (vitamin C), glutathione, and/or cysteine, producing free radicals and causing DNA damage.[29] Faraday Technology, Inc. won a PGCC Award in 2013 for a chrome plating process developed to use Cr(III), resulting in a much less hazardous waste stream. This process will be covered in depth as a case study in Chapter 12.

A large number of oxidizing agents work by transferring an oxygen or chlorine atom stoichiometrically to the reactant (Table 3.3). Many of these are inherently toxic because biological systems are very susceptible to oxidation. Even though controlled oxidation of sugars and fats with O_2 is the source of energy for all living things, O_2 is still quite toxic because it is a diradical ($^{\bullet}O{-}O^{\bullet}$). Before 1954, over 10 000 premature babies went blind because they were exposed to high levels of oxygen to keep them alive.[30] Anti-oxidants (oxygen acceptors), like vitamin C (ascorbic acid), vitamin E, butylated hydroxyl toluene (BHT), and sulfites (SO_2, Na_2SO_3, $Na_2S_2O_5$), are big business as preservatives in the food and dietary supplements industries. Less than

Table 3.3 Oxidizing agents.

Element	Reagents
Cr(VI)	CrO_3, pyridinium chlorochromate (PCC), H_2CrO_4, $Na_2Cr_2O_7$
Pb	$Pb(OAc)_4$
Os	OsO_4
Mn	MnO_2, $KMnO_4$
Ag	Ag^+
Cu	Cu^{+2}
Cl	Cl_2, NaClO (chlorine bleach), ClO_2, ClO_2^-, ClO_3^-
Br	Br_2, NaBrO, BrO_2^-, BrO_3^-
I	I_2, IO_3^-, IO_4^-, Dess–Martin periodinane
N	N_2O, HNO_3, NO_3^-
S	H_2SO_4, $H_2S_2O_8$, H_2SO_5, Oxone (2 $KHSO_5 \cdot KHSO_4 \cdot K_2SO_4$), 10% $Na_2S_2O_3$(aq) for phosphite oxidation
B	$NaBO_3$
O	O_2, O_3, H_2O_2, RCO_3H (R = CH_3, CF_3, Ph, *m*-ClPh)

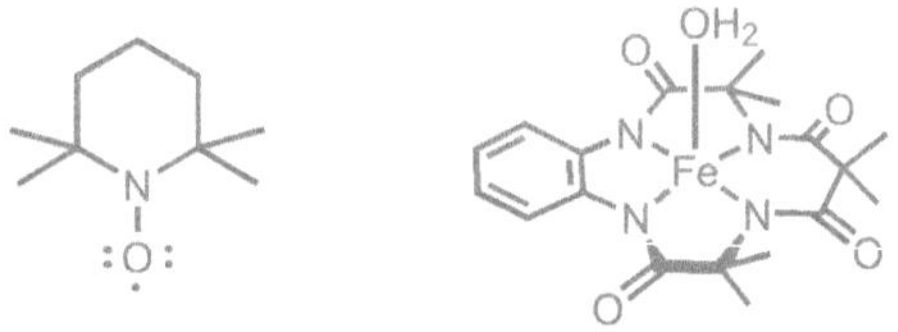

Figure 3.12 L: TEMPO, R: Fe-TAML™ hydrogen peroxide activator.[32,33]

0.05% of the US population is allergic or sensitive to sulfite preservatives, for them, ascorbic acid is a greener substitute.[31]

The heavy metal oxidants Cr, Pb, and Os are inherently toxic because of the toxicity of the metal itself. Some of the oxidants in Table 3.3 are less toxic, for example Ag^+ and Cu^{+2} are very mild oxidants. Bleach (NaOCl) is a relatively safe common household cleaner, but it reacts dangerously with other common household chemicals, such as ammonia, hydrochloric (muriatic) acid, and hydrogen peroxide. Hydrogen peroxide is quite safe at the 3–6% concentrations found in household disinfectant, but it can explode at concentrations as low as 30% from friction, heat or contaminants.

3.6.2 Greener Oxidations

The major efforts in greener oxidation chemistry have been towards using benign, renewable oxidants such as O_2, H_2O_2, and electrochemical oxidation. Since these are fairly unreactive oxidants, which makes them safer, catalysts and oxygen transfer reagents have been developed. Two of the best catalysts are the radical 2,2,6,6-tetramethylpiperidineoxyl (TEMPO),[32] and tetramidomacrocyclic ligand (TAML™) iron complexes (Figure 3.12).[33]

3.7 Case Study: Fe-tetra-amido Macrocyclic Ligand (TAML™) Oxidation (Collins)

Terrence Collins invented catalytic hydrogen peroxide activators, called TAML™-iron complexes. Fe(III)-TAML™ complexes have been used to activate hydrogen peroxide (H_2O_2) in paper bleaching, fabric dye, removal from wastewater, and eliminating dibenzothiophenes from diesel fuel.[34] Collins won a PGCC Award for TAML oxidant activators in 1999. Hydrogen peroxide as the oxygen donor is far less toxic than most other oxidants, and the Fe-TAML™ activator makes the process catalytic. The by-products of paper bleaching with H_2O_2 are water and oxygen, with trace amounts of various oxidized organic compounds, which are likely to be water soluble and biodegradable.

In 2012, wood pulp consumption in the US was 48 billion kg, about 153 kg per person.[35] The brown color of wood is due to lignin, a complex cross-linked structure of aromatic phenoxy compounds that gives wood its strong rigid structure. For many years, chlorine (Cl_2) bleaching was used to produce white paper products. Chlorine reacts with the phenoxides of lignin to give toxic dioxins and chlorinated dibenzofurans. Dioxin production was decreased by 10-fold in paper bleaching by use of ClO_2, which is still a significant amount of dioxin considering the amount of paper bleached. Paper bleaching with H_2O_2 and Fe-TAML™ has the major advantage of not producing any chlorinated dioxins or furans at all.

Since hydrogen peroxide, H_2O_2 can be hazardous in bulk processes, Collins recently published a method to activate O_2 with a TAML™ in aqueous reverse micelles (Figure 3.13).[36] Micelles are

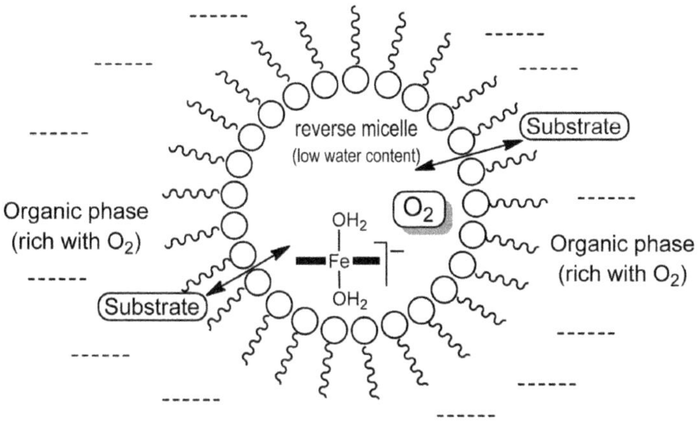

Figure 3.13 Reverse micelle for activation of O_2 by Fe-TAML™ in organic solvent. Reprinted with permission from L. L. Tang, W. A. Gunderson, A. C. Weitz, M. P. Hendrich, A. D. Ryabov and T. J. Collins, *J. Am. Chem. Soc.*, **2015**, 137, 9704–9715, Copyright 2015 American Chemical Society.[36]

spheres formed from chain molecules with a polar head-group and a hydrophobic (or greasy) tail. In a normal polarity micelle, such as laundry detergent, the hydrophobic tails group around a dirt particle with the polar sulfonate groups pointed outwards, dissolving the particle in water. In Collin's reverse micelle, the solvent is octane, and the polar head groups point in, dissolving the Fe-TAML™ in water (Figure 3.13). The O_2 oxidant is more soluble in the non-polar octane, but diffuses into the micelle to be activated by the Fe-TAML™.[36] The substrate also diffuses into the micelle to be oxidized.[36]

3.8 Case Study: Alcohol Oxidation by O_2 with Cu/TEMPO (Stahl)

Oxygen itself is arguably the greenest oxidant in Table 3.3. O_2 is readily available from air, it is atom economical, and it is relatively non-toxic. The lack of toxicity is related to the relatively poor oxidizing power of O_2, so it needs to be activated.

Oxidation of primary alcohols has been done traditionally with toxic Cr(VI) reagents (Table 3.3). In an effort to eliminate Cr(VI) oxidants, Stahl and many others have developed copper-based catalysts that use oxygen from air as the stoichiometric oxygen source.[37,38] One system uses TEMPO as a stable oxidized intermediate that accepts hydrogen atoms from the alcohol substrate (Figure 3.14).[38] The oxidized catalyst regenerates the TEMPO radical. The reaction is catalytic and can be done at atmospheric temperature and pressure.

3.8.1 Reducing Agents

When using petroleum-based feedstocks, reduction is less of a problem than oxidation. Petroleum feedstocks are already in a nearly fully reduced state. On the contrary, to turn petroleum into bioactive or polar products, oxidation is usually necessary. For this reason, one of the earliest reactions presented in organic textbooks is typically free radical halogenation of alkanes, an oxidation reaction.

Instead, if we begin with a biofeedstock, the opposite is true. For example, the common biofeedstock, glucose, has one oxygen at every carbon, a highly oxidized state. Amino acids and nucleic acids are similarly highly functionalized with oxygen and nitrogen. Reduction can thus be more of an issue with biofeedstocks.

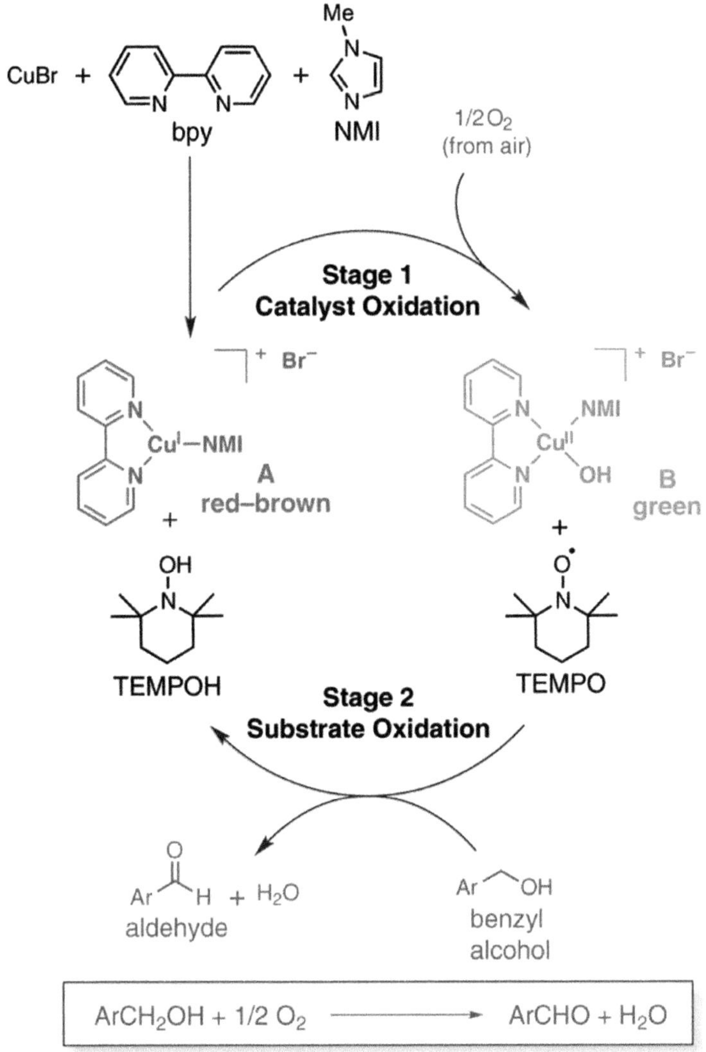

Figure 3.14 Alcohol oxidation by TEMPO with a Cu catalyst. Reprinted with permission from N. J. Hill, J. M. Hoover and S. S. Stahl, *J. Chem. Educ.*, **2013**, 90 (1), 102–105, https://pubs.acs.org/doi/abs/10.1021/ed300368q, Copyright 2013 American Chemical Society. Further permissions related to this figure should be directed to the ACS.[38]

Commonly used reducing agents are given in Table 3.4. The main hazard of reducing agents is typically flammability because our atmosphere is oxygenated. Exposure to oxygen, or even water in the case of metals like sodium, is likely to give spontaneous ignition of the reducing agent.

Table 3.4 Reducing agents.

Category	Examples
Metals	Li, Na, K, Ca, Ba, Mg
Hydrides	KH, NaH, CaH_2, BH_3·THF, $NaBH_4$, $LiAlH_4$, *i*-Bu_2AlH (DIBAL-H), AlH_3
Hydrazine	H_2N-NH_2
Hydrogen	H_2 with catalysts Ni, Pd, Pt
Formic/oxalic salts and acids	HCO_2H, NH_4HCO_2, HO_2C-CO_2H with Pd catalysts
Phosphites	H_3PO_3, H_2PO_2, $Na_xH_{3-x}PO_{2/3}$
Silanes	Ph_2SiH_2, $PhSiH_3$, polysiloxanes
Sulfites	SO_2, Na_2SO_3, $Na_2S_2O_5$
Misc. greener reductants	Electrochemical, photochemical ($h\nu$), biochemical (ascorbic acid, gluconic acid, NADH), graphene oxide (GO)

Although metal hydrides are commonly used in organic chemistry, other reducing agents can be much greener. The four considerations to balance for choosing a reducing agent are: (1) difficulty of the reagent preparation, (2) the hazards of use, (3) the atom economy of the reduction, and (4) the hazards of by-products. The more resistant the substrate is toward reduction, such as amide bonds, the more reactive the reducing agent must be. The more reactive reducing agents are also harder to synthesize and more hazardous to use. Much of the difficulty can be alleviated through the use of catalysts. The typical catalyst, Pd, is not earth abundant and relatively expensive. The use of Pd in medicinal chemistry is undesirable, because traces can remain in the pharmaceutical product. Ni is a trace nutrient, and less toxic than Pd. RANEY® nickel, a Ni–Al alloy, is used commonly to reduce benzene to cyclohexane. The high surface area and high amount of hydrogen gas absorbed into RANEY® nickel also make it quite hazardous to use.

Hydrogen is an excellent reducing agent from the standpoint of atom economy; there are no by-products because both H-atoms are incorporated into the product. Yet the production, transportation, and use of hydrogen gas under pressure are quite hazardous; explosions and fire are real and present dangers. Hydrogen reduction requires a catalyst, usually palladium on carbon (Pd/C), a heterogeneous catalyst easily removed by filtration. But Pd is a precious metal that is not abundant in the earth's crust. (See Earth Abundance Periodic Table, Appendix D).

Hydrazine (H_2N-NH_2) is an interesting reducing agent that is also used as rocket fuel. Production of hydrazine is a fairly safe process, using chloramine ($Cl-NH_2$) and ammonia. When supplied as the

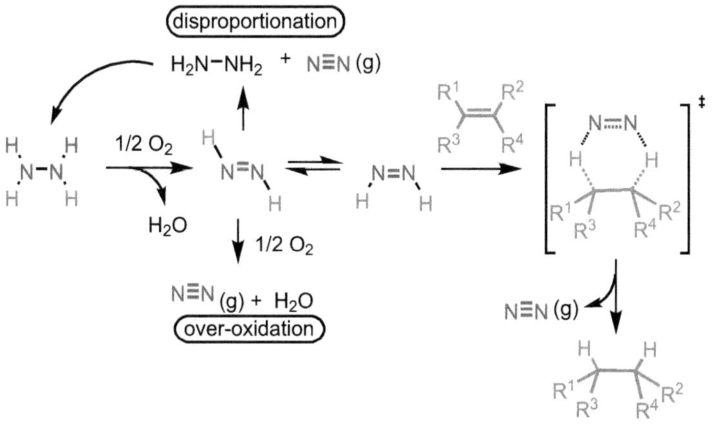

Figure 3.15 Reduction of alkenes with diimide generated *in situ* from hydrazine gives *cis*-hydrogenation. Adapted with permission from B. Pieber, S. T. Martinez, D. Cantillo and C. O. Kappe, *Angew. Chem., Int. Ed.*, 2013, **52**, 10241–10244, © 2013 WILEY-VCH Verlag GmbH & Co. KGaA, Weinheim.[40]

monohydrate, $H_2N–NH_2·H_2O$, hydrazine is relatively stable and safe to use. In the presence of oxygen and a catalyst, such as Cu, Pd, or Fe, hydrazine is first oxidized in air to the unstable diimide (HN=NH), which can then act as a reducing agent for alkynes, unhindered alkenes, and nitro groups (Figure 3.15).[39] (Overly careful organic chemists who rigorously excluded air from their hydrazine reductions have been frustrated by the lack of reaction!) Reduction with diimide gives stereoselective *cis*-hydrogenation (Figure 3.15). The final by-products of hydrazine reduction are simply nitrogen gas and water, both of which are non-toxic. Pieber *et al.* have modernized this process by eliminating the need for a catalyst, improving the reactivity of the *in situ* generated hydrazine, and making the process safer with a flow reactor.[40] Hydrazine is highly toxic and unstable in anhydrous form, but rigorous exclusion of water is usually unnecessary. Hydrazine monohydrate ($H_2N–NH_2·H_2O$) is a safer alternative.

Greener reducing agents often employed as anti-oxidants in food are sulfites and ascorbic acid (vitamin C). Sodium metabisulfite has been used to reduce the disulfide bonds of the protein in the synthesis of an acrylonitrile grafted soy flour (AN-g-SOY) polymer.[41] Ascorbic acid has been used as a reducing agent in organic synthesis,[42,43] in the synthesis of Ag and Cu nanoparticles,[44–47] and in trace elemental analysis of P, Fe and Au.[48–50]

3.9 Case Study: Amide Reductions with Silanes and Fe Catalysts (Nagashima)

Nagashima and coworkers demonstrated 3° amide reduction with 1,1,3,3-tetramethyldisiloxane (TMDS) or poly(methylhydrosiloxane) (PMHS) as the reducing agent catalyzed by iron carbonyl **3.1** or **3.2** (Figure 3.16).[51] The reaction was complete at ambient temperature in 9 hours with photo-assistance from a mercury lamp.[51] Alkenes were not reduced, however, the nitro group was more easily reduced than the amide. The catalyst was absorbed into the residual PMHS and readily removed by filtration.[51] Darcel and coworkers showed that a piano stool iron catalyst **3.3** had tremendous promise in amide reduction (Figure 3.16).[52] The optimal conditions used neat substrate, $PhSiH_3$ as the reducing agent, with light photo-assistance at 100°C for 24 hours.[52] Driess and coworkers developed an Fe N-heterocyclic carbene (Fe-NHC) catalyst **3.4** and showed that it was capable of reducing 3° amides with Ph_2SiH_2.[53] In each of these cases, 3° amides were the substrates tested, though 1° amides could be dehydrated to nitriles by the Fe-NHC catalyst.[52] To date, iron catalysts are the best available choice for amide reduction due to low toxicity of the by-products and the high earth abundance of Fe, although $Fe(CO)_5$ is extremely toxic.

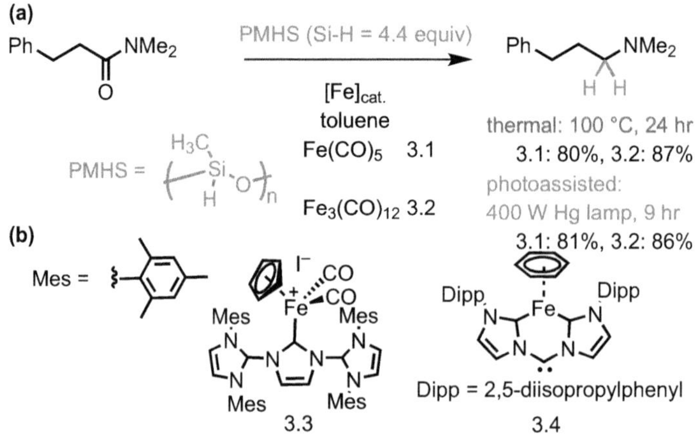

Figure 3.16 a. Hydrogenation of a 3° amide with poly(methylhydrosiloxane) (PMHS) and Fe-carbonyl catalysts, **3.1** and **3.2**. Adapted with permission from Y. Sunada, H. Kawakami, T. Imaoka, Y. Motoyama and H. Nagashima, *Angew. Chem., Int. Ed.* 2009, **48**, 9511–9514, S9511/1-S9511/19, © 2009 WILEY-VCH Verlag GmbH & Co. KGaA, Weinheim.[51] **b**. Fe reduction catalysts **3.3** and **3.4**.

3.10 Summary

Developing a less toxic synthesis requires knowledge of toxicology, and an understanding of why some chemicals are toxic and some are benign. Toxicology concerns the severity and potency of chemicals, and an understanding of the types of dose–response curves. Many databases are available for chemical toxicity to help choose less toxic chemicals. Safety Data Sheets are the most useful for determining the toxicity of specific chemicals. The QD Vision case study described the less toxic synthesis of Cd-based QLEDs for use in energy-efficient televisions. Oxidation and reduction are two types of reactions that involve serious toxicity, but chemists have invented much greener processes for both of these reaction types. Improving redox reactions is currently a very active area of green chemistry.

3.11 Problems: Benign Synthesis

Problem 3.1

The cost of disposal for 1 gallon of waste mixed halogenated solvents is about double the cost for 1 gallon of mixed non-halogenated solvents.

(a) Why is one more expensive than the other?
(b) Give two reasons to choose the solvent with more expensive disposal for a reaction or process.

Problem 3.2

Search the US Toxic Release Inventory (TRI) for all hazardous emissions in the county where your college is located.

(a) Which chemical is released in the largest amount by volume or mass?
(b) 2,4,7,8-Tetrachlorodibenzodioxin (TCDD) is one of the most toxic substances released. What is the LD_{50} (lethal dose for 50% of test animals)?
(c) Which source, if any, in your county releases dioxin? How much is released?

Problem 3.3

The EPA standards for copper in drinking water can be found by searching for "EPA MCLG copper."

(a) What is the MCL established for copper in drinking water?

(b) How does the EPA determine the safe level of copper in drinking water?
(c) What other possible routes of exposure to copper occur in humans?
(d) What is the problem with the uniform MCL for copper?

Problem 3.4

(a) What is the toxicity of cadmium sulfide, both severity and potency?
(b) What is the toxicity of indium phosphide, both severity and potency?
(c) What source of Cd, other than the synthesis, is decreased by QD Vision's QLEDs?
(d) Do you agree with QD Vision's argument that the energy savings and the decreased amount of CdS used in their QLEDs justifies USA EPA or European exemption for Cd in electronic products? Why or why not?

Problem 3.5

(a) What are the boiling points and flash points of $CdMe_2$, and $ZnEt_2$?
(b) What are the LD_{50} values for $CdMe_2$, and $Cd(OAc)_2$?
(c) What are the reported hazards of Se salts?
(d) Se is also a beneficial trace element in the diet. Predict and sketch the shape of the dose–response (adverse effects) curve for Se. Be sure to label the axes.

Problem 3.6

(a) Which of the three PGCC Award Categories: Greener Reaction Conditions, Greener Synthetic Pathways, or the Design of Greener Chemicals, does the QD Vision process for preparing QLEDs fit the best? Explain briefly.
(b) Give the numbers and short titles of two other Principles of Green Chemistry that the QD Vision QLED process embodies, in addition to Principle 3: Less Toxic Synthesis. Explain briefly.

Problem 3.7

Write the reagent typically used for each reaction over the arrow. Balance each of the reactions, including by-products. State whether each reaction is an oxidation, reduction, or neither.

(a) dehydrohalogenation

(b) alcohol to aldehyde

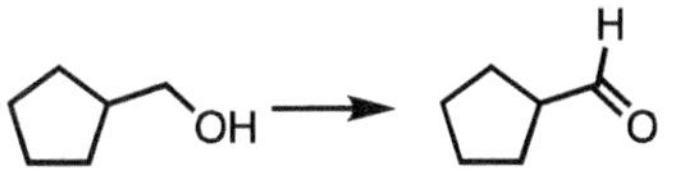

(c) alkyne to alkene

(d) iodoalkane to thioether

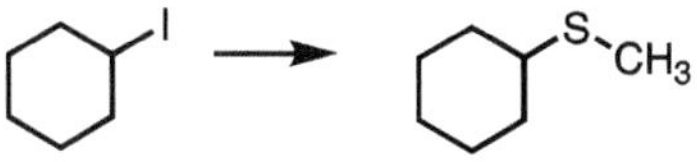

Problem 3.8

(a) Write the balanced reaction for the oxidation of 2-propanol to acetone using H_2CrO_4 as the oxidant. What are the by-products?
(b) Calculate the atom economy of this reaction. Give MWs under the reactants and products above. Show your equation.
(c) Write the balanced reaction for the same oxidation using NaClO (bleach) as the oxidant. What are the by-products?
(d) Calculate the atom economy of this reaction.

Problem 3.9

(a) Draw the mechanism for the formation of the copper-alkoxide intermediate from the "B green" oxidized copper catalyst in Figure 3.12.
(b) Draw the radical mechanism for oxidation of the alkoxide intermediate (α-hydrogen atom abstraction) by TEMPO, and elimination of aldehyde product.
(c) Draw the mechanism for regeneration of the copper catalyst and TEMPO.

Problem 3.10

(a) Balance the reaction for the reduction of 3° amides shown in Figure 3.14.
(b) How many equivalents of hydride are needed for the reaction? ____
(c) What are the by-products?
(d) What are the hazards of the PMHS and the $Fe(CO)_5$?
(e) Name two Principles that apply to the greener reduction of amides. Explain briefly.

Bibliography

Casarett and Doull's Toxicology: The Basic Science of Poisons, ed. C. D. Klaassen, 8th edn, New York, 2013.

List of teratogens: https://www.purdue.edu/ehps/rem/ih/terat.htm accessed October 29, 2015.

References

1. P. T. Anastas and J. C. Warner, *Green Chemistry: Theory and Practice*, Oxford University Press, New York, 1998.
2. Right-to-Know Network, Houston Chronicle, 2019, http://www.rtknet.org/, accessed January 30, 2019.
3. European Chemicals Agency (ECHA), *Understanding REACH*, 2019, https://echa.europa.eu/regulations/reach/understanding-reach, accessed February 6, 2019.
4. US Environmental Protection Agency, *Toxic Substances Control Act*, 2018, https://www.epa.gov/laws-regulations, accessed February 6, 2019.
5. Summary of the Toxic Substances Control Act, U.S. Environmental Protection Agency, 1976, http://www.epa.gov/laws-regulations/summary-toxic-substances-control-act, accessed February 1, 2016.
6. L. Goldman, *Environmental Law Reporter*, 2002, **32**, 11018–11041.
7. K. J. M. Matus, J. B. Zimmerman and E. Beach, *Environ. Sci. Technol.*, 2010, **44**, 6022–6023.
8. *Types of Toxic Effects*, http://www.chemicalspill.org/ChemicalsWorkPlace/types1.html accessed February 1, 2016.
9. D. W. Nierenberg, R. E. Nordgren, M. B. Chang, R. W. Siegler, M. B. Blayney, F. Hochberg, T. Y. Toribara, E. Cernichiari and T. Clarkson, *New Engl. J. Med.*, 1998, **338**, 1672–1676.
10. W. V. Welshons, K. A. Thayer, B. M. Judy, J. A. Taylor, E. M. Curran and F. S. vom Saal, *Environ. Health Perspect.*, 2003, **111**, 994–1006.
11. A. Chambers, D. Krewski, N. Birkett, L. Plunkett, R. Hertzberg, R. Danzeisen, P. Aggett, T. Starr, S. Baker, M. Dourson, P. Jones, C. Keen, B. Meek, R. Schoeny and W. Slob, *J. Toxicol. Environ. Health B: Crit. Rev.*, 2010, **13**, 546–578.
12. M. E. Letelier, A. M. Lepe, M. Faúndez, J. Salazar, R. Marín, P. Aracena and H. Speisky, *Chem. – Biol. Interact.*, 2005, **151**, 71–82.

13. Eleska, *Benzo(a)pyrene metabolism*, 2010, https://commons.wikimedia.org/wiki/File:Benzo(a)pyrene_metabolism.svg#/media/File:Benzo(a)pyrene_metabolism.svg, accessed July 16, 2019.
14. P. Pradhan, S. Tirumala, X. Liu, J. M. Sayer, D. M. Jerina and H. J. C. Yeh, *Biochemistry*, 2001, **40**, 5870–5881.
15. V. Bouvard, D. Loomis, K. Z. Guyton, Y. Grosse, F. E. Ghissassi, L. Benbrahim-Tallaa, N. Guha, H. Mattock and K. Straif, *Lancet Oncol.*, 2015, **16**, 1599–1600.
16. A. Crémazy, K. V. Brix and C. M. Wood, *Environ. Sci. Technol.*, 2018, **52**, 5979–5988.
17. United Nations Economic Commission for Europe, *Globally Harmonized System of Classification and Labelling of Chemicals (GHS), Part 3 Health Hazards*, Geneva, Switzerland, 2005, 216, http://www.unece.org/*trans*/danger/publi/ghs/ghs_rev01/01files_e.html, accessed July 16, 2019.
18. B. Pease, *Scorecard: The Pollution Information Site*, GoodGuide, http://scorecard.goodguide.com/index.tcl accessed Feb 1, 2016, 2011.
19. J. Steckel and S. Coe-Sullivan, *Greener Quantum Dot Synthesis for Energy Efficient, Commercial Display and Lighting Products*, US Environmental Protection Agency, Washington, DC, 2014, https://www.epa.gov/greenchemistry/presidential-green-chemistry-challenge-2014-greener-reaction-conditions-award, accessed July 23, 2019.
20. C. Peyratout and L. Daehne, *Phys. Chem. Chem. Phys.*, 2002, **4**, 3032–3039.
21. S. Coe-Sullivan, M. J. Anc, L. A. Kim, J. E. Ritter, M. Cox, C. Breen, V. Bulovic, I. Kymissis, R. F. Praino and P. T. Kazlas, *Composition including material, methods of depositing material, articles including same and systems for depositing material*, QD Vision Inc., *U. S. Pat.*, US 8470617, 2013.
22. J. Chen, V. Hardev and J. Yurek, *Nanotechnol. Law Business*, 2014, **11**, 4–13.
23. D. Lowe, *Things I Won't Work With: Dimethylcadmium*, 2013, http://blogs.sciencemag.org/pipeline/archives/2013/05/08/things_i_wont_work_with_dimethylcadmium, accessed May 8, 2013.
24. *PubChem Compound Database; CID = 10986*, National Center for Biotechnology Information, https://pubchem.ncbi.nlm.nih.gov/compound/10986 accessed February 1, 2016.
25. H. Chibli, L. Carlini, S. Park, N. M. Dimitrijevic and J. L. Nadeau, *Nanoscale*, 2011, **3**, 2552–2559.
26. Off. J. Eur. Communities, Directive 2011/65/EU of the European Parliament and of the Council of 8 June 2011 on the restriction of the use of certain hazardous substances in electrical and electronic equipment Text with EEA relevance, L: Legis. (Eng. Ed.), E. U. Law, 2011, ELI: http://data.europa.eu/eli/dir/2011/65/oj, accessed July 19, 2019.
27. J. Engel, *Nanosys, QD Vision Patent Fight Could Shake Up Nanotech Sector*, Xconomy, April 19, 2016, https://xconomy.com/boston/2016/04/19/nanosys-qd-vision-patent-fight-could-shake-up-nanotech-sector/, accessed July 23, 2019.
28. R. Young, *Samsung Buys QD Vision's IP for $70M and the Future of Quantum Dots*, Display Supply Chain, Dec 3, 2016, https://www.displaysupplychain.com/blog/-samsung-buys-qd-visions-ip-for-70m-and-the-future-of-quantum-dots, accessed July 23, 2019.
29. E. J. Tokar, W. A. Boyd, J. H. Freedman and M. P. Waalkes, in *Casarett and Doull's Toxicology: The Basic Science of Poisons*, ed. C. D. Klaassen, McGraw Hill Education, New York, 8th edn, 2013, pp. 981–1030.
30. S. Miller, *Doctor Pinpointed Oxygen as Cause of Preemies' Blindness*, The Wall Street Journal, March 16, 2010 https://www.wsj.com/articles/SB10001424052748703909804575124013498575580, accessed July 23, 2019.
31. M. R. Lester, *J. Am. Coll. Nutr.*, 1995, **14**, 229–232.

32. B. L. Ryland and S. S. Stahl, *Angew. Chem., Int. Ed.*, 2014, **53**, 8824–8838.
33. N. Chahbane, D.-L. Popescu, D. A. Mitchell, A. Chanda, D. Lenoir, A. D. Ryabov, K.-W. Schramm and T. J. Collins, *Green Chem.*, 2007, **9**, 49–57.
34. T. J. Collins, *Acc. Chem. Res.*, 2002, **35**, 782–790.
35. *Paper & Paperboard Production & Consumption for USA*, 2012, http://www.paperonweb.com/USA.htm, accessed July 19, 2019.
36. L. L. Tang, W. A. Gunderson, A. C. Weitz, M. P. Hendrich, A. D. Ryabov and T. J. Collins, *J. Am. Chem. Soc.*, 2015, **137**, 9704–9715.
37. S. E. Allen, R. R. Walvoord, R. Padilla-Salinas and M. C. Kozlowski, *Chem. Rev.*, 2013, **113**, 6234–6458.
38. N. J. Hill, J. M. Hoover and S. S. Stahl, *J. Chem. Ed.*, 2013, **90**, 102–105.
39. W. M. N. Ratnayake, J. S. Grossert and R. G. Ackman, *J. Am. Oil Chemists' Soc.*, 1990, **67**, 940–946.
40. B. Pieber, S. T. Martinez, D. Cantillo and C. O. Kappe, *Angew. Chem., Int. Ed.*, 2013, **52**, 10241–10244.
41. V. K. Thakur and M. R. Kessler, *ACS Sust. Chem. Eng.*, 2014, **2**, 2454–2460.
42. J. F. Lambert and T. Norris, *Preparation of sodium-hydrogen exchanger type-1 inhibitors*, Pfizer Products Inc., USA, US6753334, 2004.
43. M. J. Spitulnik, *Synthesis*, 1985, 299–300.
44. S. Raj, P. Rai, S. M. Majhi and Y.-T. Yu, *J. Mater. Sci.: Mater. Electron.*, 2014, **25**, 1156–1161.
45. S. U. Sandhya and A. N. Shetty, *NanoTrends*, 2013, **14**, 1–8.
46. D. Singha, N. Barman and K. Sahu, *J. Colloid Interface Sci.*, 2014, **413**, 37–42.
47. D. Steinigeweg and S. Schluecker, *Chem. Commun.*, 2012, **48**, 8682–8684.
48. R. Moro, M. L. Alv.-Bartolome, M. T. Fernandez and A. Vargas, *Alimentaria*, 1994, **252**, 53–55.
49. S. Shen, M. Li and L. Hao, *Capacity Analysis Method for Titration of Iron with Ascorbic Acid Reduction*, Changzhi University, Peop. Rep. China. CN103063668A, 2013.
50. J. M. Vermeulen, *J. Anal. At. Spectrom.*, 1989, **4**, 77–82.
51. Y. Sunada, H. Kawakami, T. Imaoka, Y. Motoyama and H. Nagashima, *Angew. Chem., Int. Ed.*, 2009, **48**, 9511–9514.
52. D. Bezier, G. T. Venkanna, J.-B. Sortais and C. Darcel, *ChemCatChem*, 2011, **3**, 1747–1750.
53. B. Blom, G. Tan, S. Enthaler, S. Inoue, J. D. Epping and M. Driess, *J. Am. Chem. Soc.*, 2013, **135**, 18108–18120.
54. P. R. Brezina, F. N. Yunus and Y. Zhao, *J. Reprod. Infertil.*, 2012, **13**, 3–11.

4 Benign Products

"Principle 4: Chemical products should be designed to achieve their desired function while minimizing their toxicity."[1]

4.1 Benign by Design

Paul Anastas coined the term "Benign by Design Chemistry" as the title of the first book on the subject,[2] by which he meant that all chemical processes and products should be safe and sustainable. Later he changed the term to Green Chemistry, which caught on and became mainstream. The term benign by design is particularly appropriate for chemical products, because we now design chemical products first, then we figure out how to make them. This chapter is concerned with designing less toxic chemical products.

Green Chemistry: Principles and Case Studies
By Felicia A. Etzkorn

Published by the Royal Society of Chemistry, www.rsc.org

4.2 Cradle-to-Cradle

After the earlier "cradle-to-grave" idea of regulating, tracking, and isolating hazardous materials from the moment they are created until they are no longer useful, another important concept arose. McDonough, an architect at the University of Virginia, and Braungart, a German chemist and environmentalist, wrote a book called *cradle-to-cradle: remaking the way we make things*. Their core idea is that there ought to be two separate product streams, one with entirely biologically sourced, biodegradable materials, such as cotton, and another with only technical or industrial materials, such as nylon. These two streams should not be mixed in the same product, such as shoes, so that the synthetic materials can be recycled, and the natural materials can degrade or be composted. And it is certainly a travesty in their eyes to mix polyester and cotton, plus toxic dyes and polymerization catalysts, into the same shirt. Green chemists aim to make products with single stream, or easily separable materials. The toxicity of products is of even greater concern than the toxicity of processes because the general public and the environment will be directly exposed to these products, not only the workers who make them.

4.3 Toxicology for Products

In the 20th century, a major driving force behind the design of a chemical product was function—whether it did the job it was designed to do. Little thought was initially given to the toxicity of the product, its effect on the environment or people. Many chemical products are brought to market, only to be found later to cause harm in some unforeseen, but perhaps not unforeseeable, way. There is a balance between the useful products that drive our economy, and their health and environmental effects. Products made to satisfy a particular function are gradually evolving into less toxic, and better functioning substitutes.

When medicinal chemistry was born out of the German color dye industry, human toxicity had to be taken into account.[3] This was the origin of animal testing of chemicals. Antibiotics were sought to kill bacterial infections, but they could also kill the human host. When sulfanilamide was discovered, it was considered a miracle drug because it killed the infection without sickening the host.[3] In this way, toxicology as a discipline was born.

Based on extensive knowledge of the structures of compounds that are toxins, and their mechanism of action in many cases, it has become more feasible to predict which physical properties and chemical functional groups to avoid. For example, low molecular weight ethers, such as diethyl ether or tetrahydrofuran, are known to react with atmospheric oxygen over time to produce contact-explosive peroxides. We will see in Chapter 5, Avoid Auxiliaries that these are solvents to avoid.

Methods for designing benign chemicals are in their infancy, yet we are on the right path. We have much to learn from pharmacologists and toxicologists who have long experience in determining toxicity in humans. In recent years, there has been significant growth in our understanding of toxicological mechanisms, metabolic pathways, the application of big data, and the increasing capacity of predictive tools for toxicology.

4.3.1 Formaldehyde

A common environmental pollutant that affects both workers and the public is formaldehyde because of its use in things like plywood. The Occupational Safety and Health Administration (OSHA) set the standard for exposure to formaldehyde in the workplace at 0.75 ppm in air measured as an 8-hour time-weighted average (TWA), or a short-term exposure limit (STEL) of 2 ppm allowed during a 15-minute period.[4] The EPA noted, "One survey reported formaldehyde levels ranging from 0.10 to 3.68 parts per million (ppm) in homes. Higher levels have been found in new manufactured or mobile homes than in older conventional homes."[5] Mobile homes delivered to victims of hurricane Katrina (New Orleans, US 2005) were contaminated with formaldehyde.[6] The Sierra Club, and later the US Centers for Disease Control (CDC), found that the residents of the trailers were sickened by formaldehyde.[6] Clearly, it is important to eliminate formaldehyde from consumer products, including plywood.

4.3.2 DDT

DDT is the most apparent example of a toxic product. DDT killed mosquitos as desired, yet no one thought about effects on wildlife, or long-term effects on humans. DDT was synthesized in 1874, but its pesticide activity was not discovered until 1939, a case of a product waiting for a function to be discovered. The careful observations of biologist Rachel Carson, and her persistence in bringing these off-target effects to the attention of the public, were necessary before DDT was phased out. It was not until *Silent Spring* and the environmental movement of the 1970s that people became aware of the potential hazards of synthetic products other than drugs.

4.3.3 Tetraethyl Lead

Gasoline in the internal combustion engine can ignite too soon, causing small explosions called "knocking" in the engine. Tetraethyl lead ($PbEt_4$) was used as an anti-knock ingredient in gasoline from the 1920s until phase-out for automobile fuel began in the 1970s. As of 2018, $PbEt_4$ was still used in small-piston airplane fuel, "avgas," but phase-out is expected soon. "There is still up to 0.56 $g\,L^{-1}$ of lead in Avgas 100LL."[7] $PbEt_4$ is a neurotoxin associated with significant drops in population intelligence quotient (IQ) values, and violent behavior is strongly correlated with geographical location of lead contamination of the environment.[8] One major anti-knock replacement for $PbEt_4$ was methyl *tert*-butyl ether (MTBE, CH_3-O-*t*Bu), but it has since been phased out and replaced by renewably sourced, biodegradable ethanol.

4.3.4 Potency of Poisons

In toxicology, the distinction made between natural and synthetic toxins is relatively meaningless. Poisonous substances have low LD_{50} values (lethal dose for 50% of test organisms). The lower the LD_{50}, the more potent the poison—it takes very little to have a lethal effect. Of the examples in Table 4.1, both the least toxic (ethanol) and the most toxic (Botulin) substances, over nine orders of magnitude apart, are natural

Table 4.1 Approximate Acute LD_{50} of Some Representative Chemical Agents. Reprinted with permission of McGraw Hill from ref. 11 © 2013.

Agent	LD_{50} ($mg\,kg^{-1}$ body mass)
Ethanol	10 000
Sodium chloride	4000
Ferrous sulfate	1500
Morphine sulfate	900
Phenobarbital sodium	150
Picrotoxin	5
Strychnine sulfate	2
Nicotine	1
D-Tubocurarine	0.5
Hemicholinium-3	0.2
Tetrodotoxin	0.10
Dioxin	0.001
Botulinum toxin	0.00001

products. Synthetic chemicals may also be extremely toxic. The LD_{50} of dioxin, a by-product of chlorine bleaching of paper and other industrial processes, is only 1 $\mu g\,kg^{-1}$. Note that ethanol is still a fairly potent toxin; the LD_{50} value of 10 $g\,kg^{-1}$ body weight is about 2 L of hard liquor (40% ethanol) for the average US woman. Although everyone is susceptible to alcohol poisoning—men, women, children, occasional drinkers—genetically susceptible people are particularly sensitive because they also have low levels of alcohol dehydrogenase, the enzyme required to metabolize ethanol. People do come to the emergency room and sometimes die of acute ethanol poisoning.

4.4 Causation

To determine whether a particular substance is toxic, it is first necessary for toxicologists to show that the substance is the actual cause of an adverse effect. This can be tricky, as we will see in the shift from benzene to hexane/acetone mixtures used by automotive mechanics to clean parts and tools. First benzene was found to be a carcinogen, then part of the substitute solvent mixture, *n*-hexane, was found to cause Parkinson's disease-like neurological effects (see Chapter 5). It was not clear that the neurological effects were due to the occupation of the mechanics or some other cause at first.

4.4.1 Hill's Lines of Evidence

In 1937, the British medical statistician, Sir Austin Bradford Hill, neatly laid out nine criteria necessary to establish causation in epidemiology, the study of disease in human populations.[9,10] Hill clearly stated, "What I do not believe ... is that we can lay down some hard-and-fast rules of evidence that must be observed before we accept cause and effect."[9] While these nine lines of evidence all contribute to establishing a causative relationship, rather than a simple association, it is not necessary for all nine to be true. Sometimes just a few lines of evidence make causation incontrovertible, while for other associations, all lines of evidence may be so weak as to make any assertion of causation trivial and unworthy of attention. Hill developed this approach to provide evidence that a toxin *causes* an adverse effect, and is not simply *associated* or *correlated* with an adverse effect, that is, correlation without causation.[9,11]

1. Strength of association (relationship between independent and dependent variables)

2. Consistency of findings (replication of results by different studies [reproducibility])
3. Specificity of association (cause is tightly linked to an outcome)
4. Temporal sequence (cause before effect)
5. Biological gradient (strength of the dose–response relationship)
6. Biological or theoretical plausibility (mechanism of action)
7. Coherence with established knowledge (no competing hypotheses)
8. Experiment (change in conditions changes outcome)
9. Structural analogy (similar chemicals have similar effects)

Most of these criteria involve chemistry in at least this form—we must know the chemical identity (structure and properties) of the suspect toxin. This is no small feat. I once worked on a project involving neural tube defects in mice. The only thing we knew was that something in the drinking water was the causative agent. Trying to figure out the identity of the chemical agent was a major task, akin to looking for a piece of dust in a haystack! The mystery remains unsolved.

Three of Hill's criteria are more directly within the scope of chemistry. (5) The dose–response measurements use quantitative chemical methods to detect a biological change. The chemical must be weighed and delivered as a solution, solid mixture, or gaseous mixture to the biological target, whether that is a mouse or an enzyme, in a precise and quantitative method. (6) The mechanism of action requires a molecular level experimental investigation of the biochemical mechanism. (9) Analogy, or chemical similarity is based on the chemical principles of structure, polarity, stereoelectronics, acidity, and other chemical properties.

Hill also describes an early version of the Precautionary Principle "All scientific work is liable to be upset or modified by advancing knowledge. That does not confer upon us a freedom to ignore the knowledge we already have, or to postpone the action that it appears to demand at a given time."[9]

4.5 Regulations

Regulatory frameworks of the EU and USA for chemicals are described in more detail in Chapter 3 and the cited web sites. These regulations cover the entire life cycle of chemicals in commerce, including products. The EU REACH program requires companies that import, use, resell, and manufacture products to register, evaluate, assess, and minimize the risk of all chemicals used in commerce.[12] The US Toxic Substances Control Act (TSCA) Modernization Act of 2016 requires all chemicals used in commerce to be subjected to toxicological review and regulation by the EPA.[13] All of the major world economies are developing similar regulatory frameworks.

4.6 Methods for Avoiding Toxic Substances in Products

Benign by design often means to design biodegradable products, which will be taken up in Chapter 10, Degradation or Recovery. Sometimes durability is a more important criterion. It may be feasible someday to design a plastic engine that is lighter and more fuel efficient, but organic polymers do not typically have the durability of metals. Perhaps they will one day, or a self-healing, light-weight, carbon fiber might be invented. For all chemical products, excluding persistent toxic substances and ensuring that degradation will not produce new toxic by-products are both important. Anastas and Warner described four methods that chemists should use to design benign products: (1) mechanism of action, (2) limiting bioavailability, (3) structure–activity relationships, and (4) avoiding toxic functional groups.[1] These four methods are interrelated; avoiding a toxic functional group is also one way of using known structure–activity relationships. Because this is such a complex area, pharmaceutical, biotechnology, and many large chemical companies have toxicologists on staff.

4.6.1 Mechanism of Action

In toxicology, understanding the mechanism by which the body is harmed by toxic substances gives a strong foundation for avoiding substances that cause harm. Each of the steps in toxic mechanisms is an opportunity to design against toxicity. As outlined in Casarett & Doull's *Toxicology*, these steps are: (1) transport from the site of exposure to the target, (2) reaction with the target, (3) cellular dysfunction, and (4) inappropriate repair.[14] Compounds may not be toxic if they do not get into the target site, if they do not bind with the target site, if the binding does not cause cellular dysfunction, or if the body is able to appropriately repair any cellular dysfunction. That is, failure to accomplish any of these steps will result in a lack of toxicity. Thus, chemicals can be designed not to be bioavailable, not to react with any biological target, or not to cause cellular dysfunction, but any of these is very difficult.

Designing products with known non-toxic, natural functional groups, such as the esters in poly(lactic acid), is the simplest solution.

4.6.2 Limiting Bioavailability

Bioavailability falls under the category of transport from the site of exposure in the organism to the target organ or system, for example stomach to liver, or lungs to brain. If it cannot get there, it will not have a toxic effect. Bioavailability is often correlated with the lipophilicity or hydrophobicity of a chemical. For example, suppose a substance causes brain damage, but only if it crosses the blood–brain barrier. Although the blood–brain barrier is not thoroughly understood, a few principles have been developed to guide design. Much is known from pharmacology about the types of molecules that pass through, or are excluded by, the capillaries of the blood–brain barrier. Hydrophilic (water-soluble) compounds and macromolecules, such as proteins, usually do not cross the barrier.

On the other hand, hydrophilic surfactants may be more acutely toxic to aquatic organisms than a highly hydrophobic substance, because surfactants can easily disperse in water and disrupt cell membranes. Cationic surfactants are generally more toxic than anionic surfactants, due to the negative surface charge of the cell membrane. Ideally, it is better to avoid toxins completely, rather than depend on preventing access to target organs. This is analogous to risk avoidance by use of protective gear, rather than by eliminating the hazard completely.

As early as 1863, a French PhD student, Cros, described the correlation between the hydrophobicity of alcohols and their toxicity.[15] As we will see for endocrine disrupters, the hydrophobicity is one important predictor of potential bioactivity. Medicinal chemists often strive for a somewhat hydrophobic compound that can cross the cell membrane and bind to the target, but not so hydrophobic that a compound is retained in liver and fat cells, contributing to toxicity. The standard measure of hydrophobicity or lipophilicity is called "Log*P*," defined as the log of the partition coefficient between two immiscible solvents, typically *n*-octanol and water. The partition coefficient *P* is the ratio of the concentration, *c*, of the chemical in each solvent. In addition to measuring this partition experimentally, predictions can be made by calculating Log*P*, also called "*c*Log*P*" (eqn (4.1)).[16]

$$\mathrm{Log}P \text{ or } \log K_{\mathrm{ow}} = \log \frac{c_{\mathrm{octanol}}}{c_{\mathrm{water}}} \tag{4.1}$$

The focus of environmental or ecotoxicity has been on aquatic organism uptake, the transport aspect of mechanism. In this case, the general property of hydrophobicity is predictive.[17] The more hydrophobic a substance is, the more likely it is to concentrate in fat tissues, and it is more likely to exhibit chronic toxicity.

4.6.3 Structure–activity Relationships (SAR)

Structure–activity relationship (SAR) studies are the way medicinal chemists correlate the structure of a specific chemical with its biological activity, good or bad. SAR is a more direct way to avoid toxicity than avoiding a mechanistic pathway. If the biological target of the toxin is known and well characterized, we can design something that will not bind to the target to avoid that specific toxicity. Either the 3-dimensional structure on a computer *in silico*, or an experimental bioassay for the target must be known to test for toxicity *in vitro* (biochemical), or *in vivo* (in living organisms).

The difficulty lies in avoiding all other biological targets as well. SARs are so much a part of medicinal chemistry that computer programs for 3D quantitative analysis (QSAR) have been widely adopted. In medicinal chemistry, more attention had been traditionally paid to the SAR of the binding of the drug to the target than to off-target effects, which are often the cause of drug toxicity. Off-target effects are much more difficult to analyze, yet very important, because there are many thousands of potential targets in the body. Toxicological SARs are even more challenging because even the potential target is usually unknown. This is the reason animal studies, unfortunately, continue to be crucial to toxicology, so that the target organ or biomolecule(s) can be identified.

Toxicologists have adopted QSAR methods as well.[17,18] These methods are best at comparing groups of similar compounds, for example alcohols. This analog-based approach suffers when the mechanism of action is different. Now QSARs have been developed for compounds that have the same mechanism of action.[17] Differences between model animals and human biology can lead to false negative or positive results, in medicinal chemistry or toxicology.

4.6.4 Avoiding Toxic Functional Groups

Chemists must rely on toxicology to demonstrate that particular substances are toxic. Indeed, the last of Hill's lines of evidence, analogy, is of great use to chemists trying to design benign products.

If one mercury salt is toxic, it is highly likely that all mercury salts are toxic. If one organic nitrile is toxic, all nitriles should at least be suspect and tested in toxic screenings before bringing a product to market. If one bisphenol mimics a steroid hormone, we should not substitute a similar bisphenol without evidence of safety.[19]

Acute toxicity is correlated with chemical structure in both general and specific ways. For example, most commodity small organic chemicals, including solvents, have narcotic action, such as alcohols, ketones, and ethers. Narcosis is a general mechanism that correlates well with cLog*P*.[17] Narcosis is an overall shutting down of respiratory and cardiovascular function, or anesthesia. The anesthetic compounds are thought to bind at the membrane bilayer interface, affecting membrane ion channel functions.

Determining an electro(nucleo)philic mechanism of toxicity in QSARs relies on knowledge of specific reactivity. For example, methyl iodide and methyl bromide are potent DNA-methylating agents, mutagens, and probable carcinogens. These volatile alkyl halides act by means of S_N2 reactions with DNA bases.[17] Another class is Michael acceptors—α,β-unsaturated carbonyls or enones—electrophiles that accept endogenous nucleophiles, such as thiols of the amino acid cysteine (Cys) in glutathione, or Cys in enzyme active sites. Nitriles are to be avoided in designing benign products. Nitriles (R–CN) can release cyanide (CN^-), which binds to the heme center of cytochrome c oxidase, inhibiting oxidative metabolism, and may cause cardiovascular arrest.[20,21] Thus, toxins that release cyanide upon metabolism are also of great concern.

Receptor-mediated toxicity, such as endocrine disrupters (see below), must be treated separately from acute toxicity. Receptors are proteins that bind signaling molecules, such as steroid hormones. As we saw with the dose–response curve for estrogen (Figure 3.2), these effects can be highly non-monotonic, resulting in poor predictions from dose–response curves (Chapter 3).

4.7 Persistent Organic Pollutants (POPs)

One of the most challenging classes of products to make benign by design are pesticides. By definition, pesticides must be toxic to the target organism, so they are toxic by design. The key word though is "target," and the way out of this dilemma is to design highly specific pesticides that are toxic only to the target organism. We will consider the lack of specificity for pesticides historically and currently, then look at a case study of Spinosad, a greener pesticide for mosquitos.

In the mid-20th century, Julius Hyman at Velsicol Corp. in Memphis, TN invented the pesticides aldrin, dieldrin, and chlordane. All were synthesized by Diels–Alder reactions of chloroalkenes (Figure 4.1). These pesticides were designated persistent organic pollutants (POPs) by an international treaty in Stockholm, Sweden in 2001. In addition, endrin, heptachlor, hexachlorobenzene, toxaphene, mirex, polychlorinated biphenyls (PCBs), *p,p′*-dichlorodibenzo-1,1,1-trichloroethane (DDT), polychlorinated dibenzodioxins, and polychlorinated dibenzofurans, a total of 12 substances, were banned by the treaty (Table 4.2). More have been added since then. A total of 152 nations were signatories for the treaty; most of the major populated countries have ratified and put the treaty into effect by now. The USA has not yet ratified the treaty, yet most of the provisions have been implemented in other ways.

Dioxins are produced unintentionally as a by-product of many industrial processes (Table 4.2). 2,3,7,8-Tetrachlorodibenzodioxin (TCDD) is the most toxic isomer—it was shown to be teratogenic, mutagenic, carcinogenic, immunotoxic, and hepatotoxic.

To be classified as a POP under the Stockholm Treaty, the following four criteria must all be met.

1. Persistence
    - Half-life >2 months in water
    - or half-life >6 months in soil
    - or evidence chemical is otherwise persistent
2. Bioaccumulation
    - Bioconcentration factor >5000
    - or log$P$ (logK_{ow}) >5

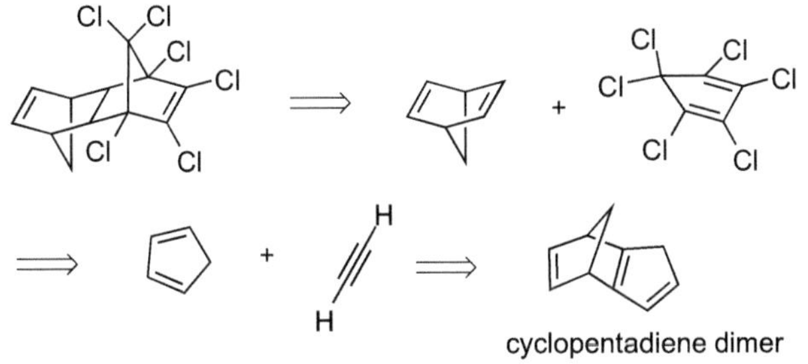

Figure 4.1 Retrosynthetic analysis of the persistent organic pollutant, Aldrin. The name derives from being synthesized by a Diels–Alder [2+4] cycloaddition.

Table 4.2 Persistent Organic Pollutants (POPs) banned by the 2001 Stockholm Convention.[34]

Compound	Structure	Use	Exemptions
Aldrin	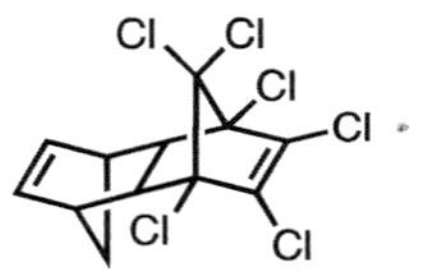	Insecticide	Local ecto-parasiticide, insecticide
Chlordane	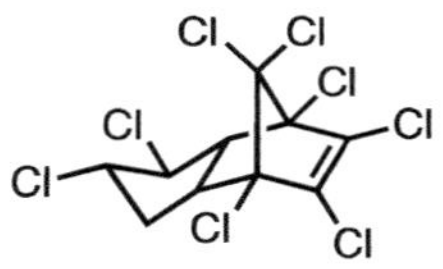	Insecticide	Local ectopocide, insecticide, termiticide, in plywood adhesives
Chlordecone	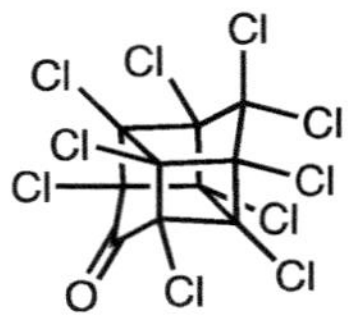	Pesticide	None
Decabromodiphenylether	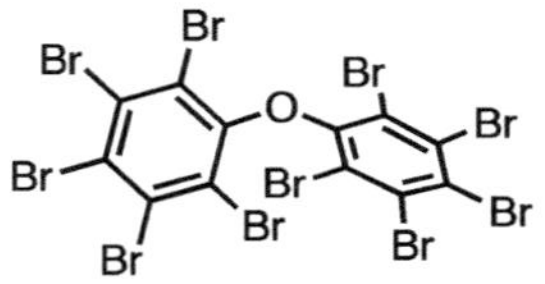	Flame retardant	Removal during recycling
p,p′-Dichlorodiphenyl-1,1,1-trichloroethane (DDT)	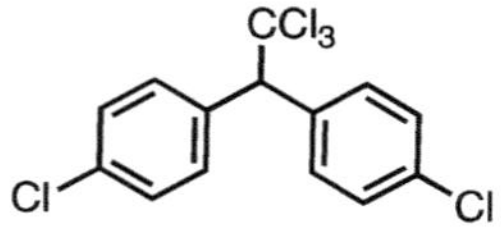	Insecticide, malaria control	Non-US malaria control
Dieldrin	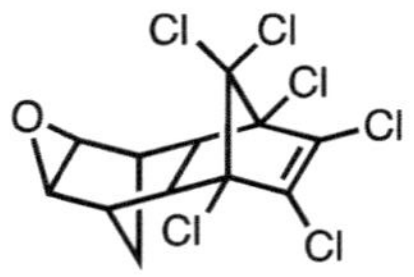	Insecticide	Agricultural operations

Endosulfan	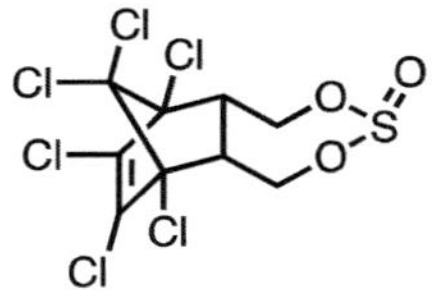	Insecticide, wood preservative	Listed crops
Endrin	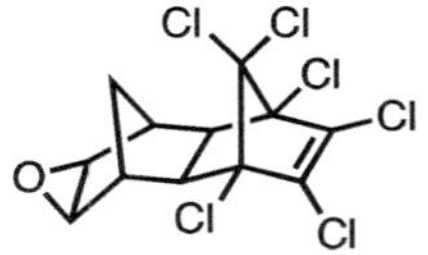	Rodenticide and insecticide	No exemptions
Heptachlor	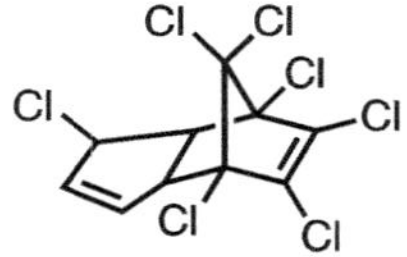	Insecticide	Termiticide
Hexabromocyclododecane	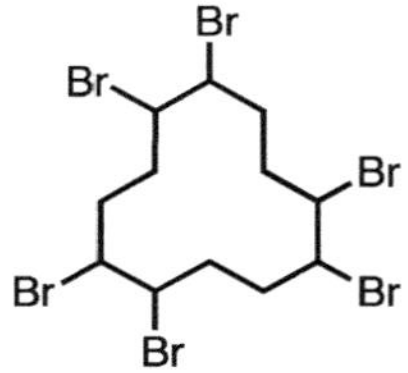	Flame retardant	Expanded polystyrene
Hexabromodiphenyl, heptabromodiphenyl	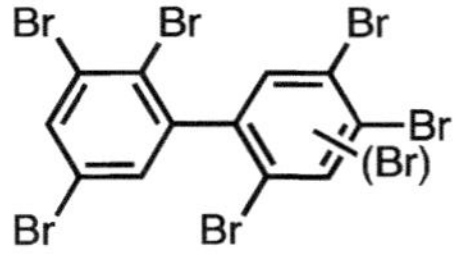	Flame retardant	Aircraft, textiles excluding clothing and toys, plastic housings and parts for heating home appliances, polyurethane foam for building insulation

Table 4.2 *(Continued)*

Compound	Structure	Use	Exemptions
Hexachlorobenzene	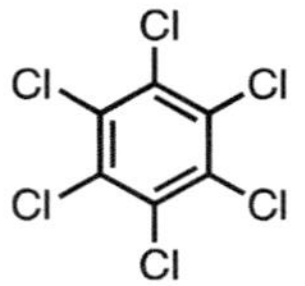	Fungicide	Listed parties for intermediates, solvent in pesticides, closed-system site
Hexachlorobutadiene	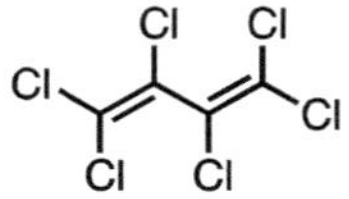	By-product in manufacture of chloroform and tetrachloroethene	None
α-Hexachlorocyclohexane	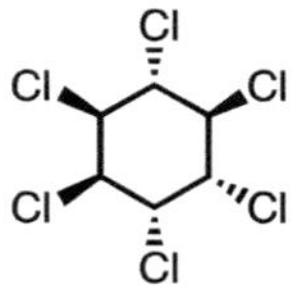	By-product of lindane manufacture	None
Mirex	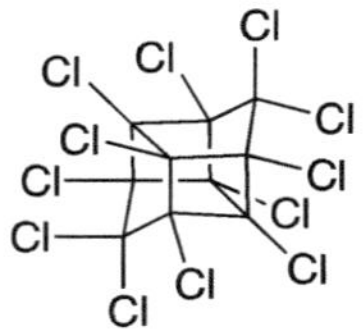	Insecticide for ants and termites	Production – listed parties Use – termiticide

Perfluorooctane sulfonic acid (PFOS), salts, and fluoride	CF_3-$(CF_2)_7$-SO_3H (Na, K, F)	Electronics, fire-fighting foam, hydraulic fluids, textiles	Listed uses
Polychlorinated biphenyls (PCBs):	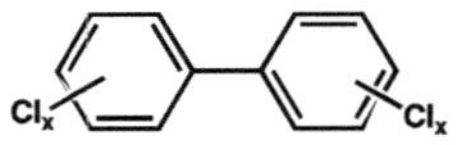	Capacitors and transformers, adhesives and plasticizers	Production – none Use – in accord with Part II of annex
Polychlorinated dibenzodioxins (dioxins)	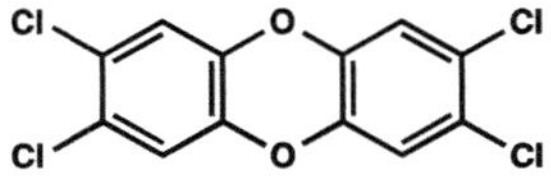 2,3,7,8-tetrachloro-dibenzodioxin (TCDD)	By-product of combustion, paper bleaching, waste incinerators, Cu, Fe, steel, Al, Zn production, motor vehicles, crematoria, leather & textile dyeing, waste oil refineries, 2,4-D pesticide manufacture	None
Polychlorinated dibenzo-furans	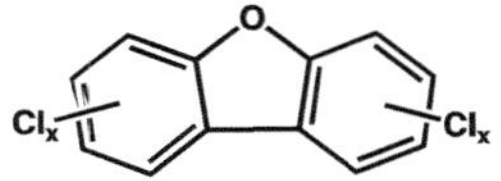	By-product of PCB manufacture, often found with dioxin	None
Polychlorinated naphthalenes	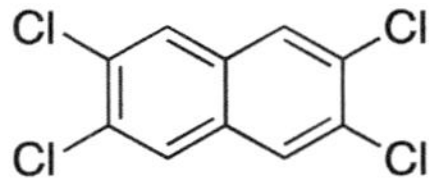	Mothproofing	Use in production of polyfluorinated naphthalenes
Toxaphene	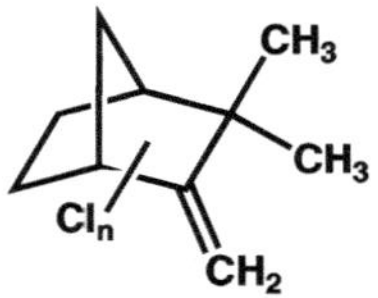	Insecticide for ticks and mites	None

3. Transported long-range
   - Measurable levels distant from site of release
   - Transport *via* air, water or migratory species
4. Toxicity
   - Adverse effects to human health or environment

Many of the listed POPs are heavily chlorinated, leading Greenpeace to propose a ban on all chlorinated organic substances. This carries the "analogy" line of evidence perhaps a bit too far. Aromatic halides are typically more stable than alkyl halides. There are many chlorinated natural products, and not all of these are toxic. Chlorination occurs primarily by oxidative mechanisms in nature.[22] Natural mechanisms for degradation include dehalogenation, photolysis, hydrolysis, and biodegradation.[23]

In the US TSCA Modernization Act of 2016, these chemicals are referred to as "Persistent, Bioaccumulative and Toxic (PBT)," instead of POPs. This is the essence of the restrictions, notably missing the transport criterion.[13]

4.7.1 Incineration

If green chemistry methods cannot be used to avoid toxic by-products, like dioxins, they should be destroyed. Incineration should be the method of choice to completely destroy hazardous waste that does not contain toxic heavy metals, such as Hg, Pb, or Cd, or halogenated hydrocarbons, such as polychlorinated biphenyls (PCBs), DDT, brominated diphenyl ethers (BDPEs), chlorophenols, or chlorinated bleaching by-products. If the waste contains any chlorine, more dioxin is created during incineration unless extremely high temperatures, high residence time, and high turbulence are used.[24] Most waste streams, such as municipal solid waste, are complex mixtures that may contain heavy metals and halogenated organics in batteries, electronics, preserved wood, paint, pesticides, and furniture (flame retardants). Any method that avoids the formation of dioxins during incineration can be considered green chemistry or engineering.

To lower the temperature required for complete incineration of halogenated organics, a catalyst could be used. One of the challenges is that chloride ions can poison catalysts by binding to the metal surface. Van der Avert and Weckhuysen have developed a lanthanum oxide (La_2O_3) catalyst that is capable of completely destroying carbon tetrachloride, a surrogate for chlorinated hydrocarbons like dioxins, at reasonably low temperatures (Figure 4.2).[24] The La_2O_3 catalyst is not poisoned in their study.

$$CCl_4 + 4\,H_2O \xrightarrow[\substack{10\ \text{wt\%} \\ La_2O_3/Al_2O_3 \\ 350\ ^\circ C}]{} 4\,HCl + 2\,H_2O + CO_2$$

Figure 4.2 Destruction of chlorinated hydrocarbons at lower temperatures with a benign and earth-abundant catalyst, La_2O_3.[24]

4.8 Endocrine Disrupters

One of the most challenging mechanisms of action in toxicology is endocrine disruption. Endocrine disrupters (EDs) alter the chemical messenger systems of the body, blatantly as in sex reversal in amphibians, or subtly as in the thinning of bird eggshells. EDs are frequently small, hydrophobic molecules that mimic binding of steroid hormones to nuclear hormone receptors. A model of how the estrogen receptor works in the nucleus is given in Figure 4.3. The estrogen receptor binds 17β-estradiol (E2) and dimerizes in the cytoplasm, called a receptor-mediated response. After the complex is imported into the nucleus, it binds to DNA and recruits a co-activator and RNA polymerase for transcription.

EDs can act in both receptor-mediated responses and non-receptor-mediated responses. Enzyme inhibition is another possible mechanism of action. Some EDs block the enzyme-catalyzed conversion of estrogen to androgens. Atrazine was found to inhibit cyclic-AMP-dependent phosphodiesterases that are important in pituitary gland signaling, a part of the endocrine system.[25]

4.8.1 Bisphenols

Another suspect endocrine disrupter is bisphenol A (BPA), a plasticizer used primarily in polycarbonates and epoxy resins to make them less brittle (Figure 4.4). The US National Toxicology Program evaluated the risk of public exposure to BPA in 2010. The report listed "some concern" for developmental toxicity in fetuses, infants, and children due to the potential for adverse effects on brain, behavior, and the prostate gland, and "minimal concern" for effects on the mammary glands, early puberty in females, and reproductive toxicity in workers exposed to high levels.[26] However, Frederick vom Saal at the University of Missouri reviewed the existing studies and concluded that a new risk assessment was necessary because studies at very low doses (ppt) showed greater adverse effects than the high dose studies.[27] This is probably due to negative feedback inhibition of hormone signaling at high doses. In 2011, BPA was shown to have a non-monotonic

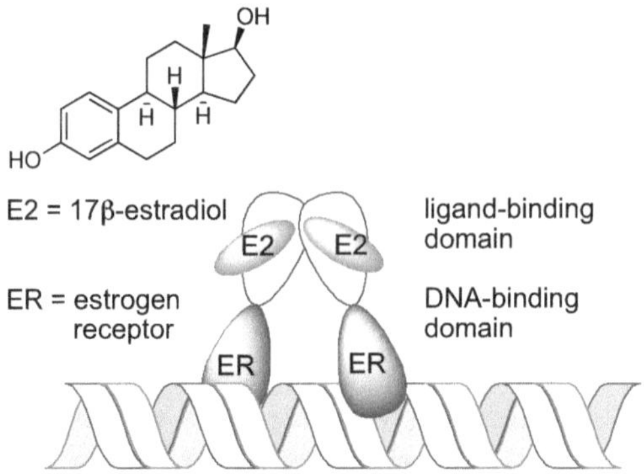

Figure 4.3 Above: Structure of estrogen, 17β-estradiol, E2. Below: Cartoon of 17β-estradiol (E2) binding to the nuclear estrogen receptor (ER) ligand binding domain, causing dimerization, binding to DNA, and transcriptional activation.

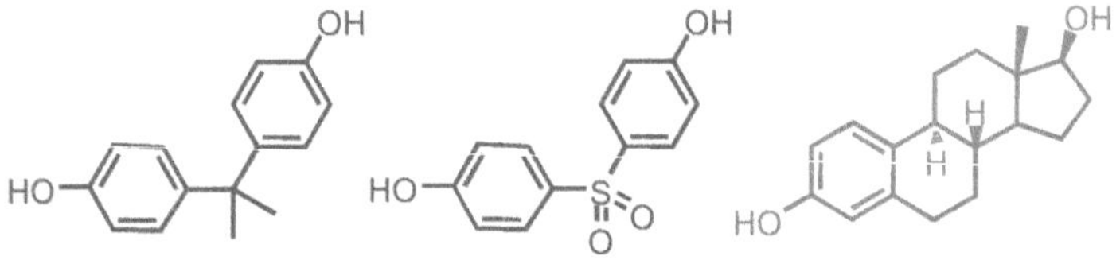

Figure 4.4 The structures of bisphenol A, bisphenol S, and 17β-estradiol.

dose–response curve in cancer cells at levels relevant to human exposure. The highest incidence of lung cancer metastases in mice was at the lowest level of exposure to BPA.[28]

Companies have marketed "BPA-free" water bottles. But Zimmerman and Anastas pointed out in *Science* that a drop-in substitute (meaning the process did not need much alteration), bisphenol S (BPS), has very similar shape, hydrophobicity, and polar group placement as BPA, so there are hazard concerns for BPS (Figure 4.4).[19] BPA and BPS have very similar structures, so Hill's 9th criterion for causative agents applies, and this makes BPS a candidate for analysis as an endocrine disrupter. As of July 2019, Nalgene is making a BPA- *and* BPS-free water bottle. Zimmerman and Anastas thus advise us to work "towards substitution with no regrets."[19] This means using our knowledge of toxicology to avoid making the same mistake twice.

4.8.2 Fluorinated Hydrocarbons

Per- and polyfluorinated alkyl substances are generally referred to as PFAS. Perfluorooctanoic acid (PFOA) is used in the synthesis of

Teflon™ and Gore-Tex™. Small amounts are found in the serum of most people in the developed countries (US median 7.0 ng mL^{-1}).[29] A review of epidemiological studies found inconclusive evidence of association with any suspect diseases.[29] PFOA was found to cause an increase in several types of cancer in rodents, and it was shown to cause an increase in estradiol in men.[29] Perfluorooctanesulfonic acid (PFOS) is similarly suspect and under investigation (Table 4.2).[29] In the early 2000s, 3M discontinued the manufacture of PFOS that was used in Scotchguard, a waterproofing product. PFOS is also used in fire-fighting foam. A green alternative will be described in the Solberg case study in Chapter 10. 3M is now making perfluorinated sulfonamides as replacement perfluorinated compounds without evidence of safety, not complying with the new TSCA law. In June 2019, 3M admitted to illegally dumping a substitute for PFAS compounds, perfluoro-1-butane-sulfonamide (FBSA) into the Tennessee River.[30] We must remain vigilant about drop-in replacements that may be just as toxic as their predecessors.

4.8.3 Flame Retardants

From 1975 to 2013, California had a regulation called Technical Bulletin 117 (TB 117) that required the use of flame retardants in children's pajamas and furniture to meet the open-flame test of flame suppression for 12 seconds. The current version, TB 117-2013, requires a smolder test that can be met without flame retardants. The main flame retardant once used in pajamas, and now furniture, tris(1,3-dichloro-2-propyl)phosphate (TDCPP or chlorinated tris), is a mutagen and suspect carcinogen.[31,32] On July 1, 2007 the US Consumer Products Safety Commission (CPSC) required all manufacturers to include flame retardants in mattresses. Antimony (III) oxide (Sb_2O_3) is used as a flame retardant in polyurethane foam along with halogenated organic compounds, such as polybromodiphenylethers (PBDEs, Table 4.2). According to California Proposition 65, Sb_2O_3 is a teratogen and carcinogen. PBDEs are restricted by the European Union to $<0.1\%$ in consumer products because of their toxicity (Table 3.2).

A common flame retardant is a mixture of antimony oxide and PBDEs. PBDEs are EDs, and they bioaccumulate in humans and persist in the environment.[33] PBDEs have now been classified as POPs by the Stockholm Convention.[34] Even as PBDEs are being phased out, new flame retardants that are halogen-free are being marketed as replacements, in most cases without significant regulatory oversight. The revised TSCA may require all new consumer product chemicals to be reviewed by the EPA, although the 2015 version of the bill restricts testing of imported products.[35]

Flame retardant-treated furniture burns just as fast as furniture without flame retardants in CPSC testing in 2009.[36] We do not need flame retardants in furniture if we use natural materials such as cotton, linen, leather and wool. These natural materials are already partly oxidized, so they do not burn as easily. Petrochemicals, such as polyester and polyurethane, are very highly reduced, and they burn much more rapidly. Burning is a process of oxidation, so partly oxidized natural materials burn less quickly than petrochemicals. The main reason for flame retardants in furniture is the use of petrochemical fabric and foams, and green chemistry guides us to use biofeedstocks instead of petrochemical feedstocks.

4.9 Pesticides

4.9.1 DDT Reprise

DDT refers to a commercial mixture of compounds (Figure 4.5). DDT has been found to be a potent carcinogen,[37] and a significant developmental toxin, as Rachel Carson observed in the thinning of bird eggshells. DDT is considered an endocrine disrupter.[38,39] DDT and DDE (*p,p*′-dichlorophenyl-1,1-dichloroethene), the major metabolite of

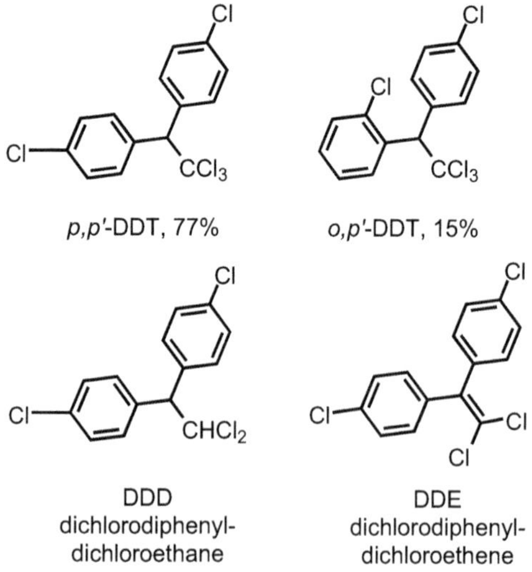

Figure 4.5 Commercial DDT mixture components, *p,p*′-DDT and *o,p*′-DDT. DDD and DDE are the major metabolites and degradation products of DDT.

DDT, mimic the structure of the major steroid hormone, 17β-estradiol, but it binds to the estrogen receptor with 10^4-fold lower affinity.[40] DDT shows endocrine disrupter activity in both *in vitro* DNA assays, and breast cancer cell assays. The *o,p′*-DDT isomer has more potent estrogenic activity, and it is present at about 15% in commercial DDT.[41]

DDT is among the most persistent of pollutants; the half-life in soil can range from 22 days to 30 years. However, even the breakdown products, DDD and DDE, are persistent and harmful. The bioconcentration factor (BCF) of DDT was determined experimentally to be 61 600, which is well above the 5000 minimum for POPs.[42]

Because DDT is very stable towards environmental or biological degradation, and very non-polar, it is transported long-range to northern latitudes and higher altitudes by a process of "global distillation." It has been found in the northern Rocky Mountains, in polar bear and whale milk, and indigenous people of the Arctic. DDT fits all four criteria for POPs.

There is considerable controversy over whether DDT should be used to combat malaria in endemic areas. Carrier mosquitos have become DDT resistant in India, Africa, and other parts of the world. As a consequence, strong international efforts are being made to design insecticides that are highly specific for the malaria mosquito. Others are working towards drugs to target the malaria parasite directly *in vivo*.

4.9.2 Pyrethroids

Pyrethroid insecticides have largely replaced DDT as a result of resistance, but these too have resistance problems.[43] Permethrin is a natural product of chrysanthymums (Figure 4.6). A wide variety of pyrethroid derivatives have been synthesized. Resmethrin, with a variant ester group is 20–50 times more effective than permethrin (Figure 4.6), but less potent toward mammals (LD_{50} (rat, oral) = 2000 mg kg^{-1}). Although some contain chlorine, the pyrethrins, both natural and synthetic, are

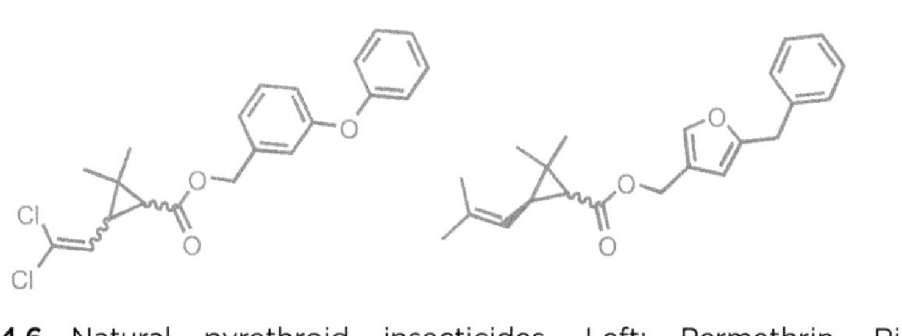

Figure 4.6 Natural pyrethroid insecticides, Left: Permethrin, Right: Resmethrin.

readily biodegradable by sunlight in water and soil with a half-life of 3 to 430 days.[44]

4.9.3 Acetyl Cholinesterase Neurotransmission

Mosquito-borne diseases, such as malaria, dengue fever, and now the Zika virus, are significant health problems in Africa, Latin and South America. POPs, such as DDT, are a problem as insecticides because they are not biodegradable, and they affect fertility and development in wildlife and humans *via* endocrine disruption. Another major class of insecticides in common use is acetylcholinesterase (AChE) inhibitors. Acetylcholine (ACh) is an important neurotransmitter in all animals, so these inhibitors are not selective, with significant toxicity to humans and wildlife. Acetylcholine signals across the gap between nerve cells to muscle cells (Figure 4.7). After the signal is received, the ACh must be deacetylated by AChE to turn off the signal, and choline is recycled.

The chemical mechanism of AChE is substitution at the acetyl carbonyl by addition-elimination, with the formation of a tetrahedral intermediate (Figure 4.8). The catalytic triad of Ser, His, and Glu amino acid side chains in the AChE active site act as a proton relay system. The Ser –OH is temporarily acetylated until water diffuses into the active site and hydrolyzes the enzyme Ser-acetyl intermediate.

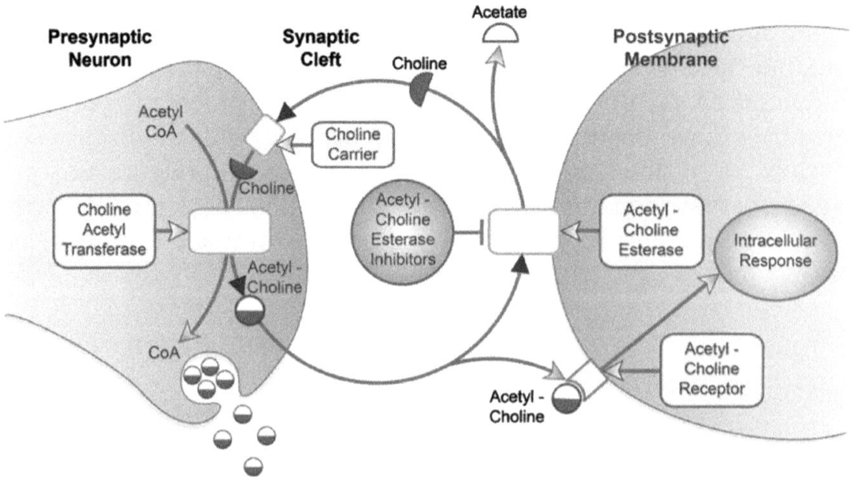

Figure 4.7 Diagram of neurotransmission with acetylcholine, showing the significance of deacetylation by AChE to turn off the signal. Reprinted with of Oxford University Press from ref. 45, Copyright 2013.

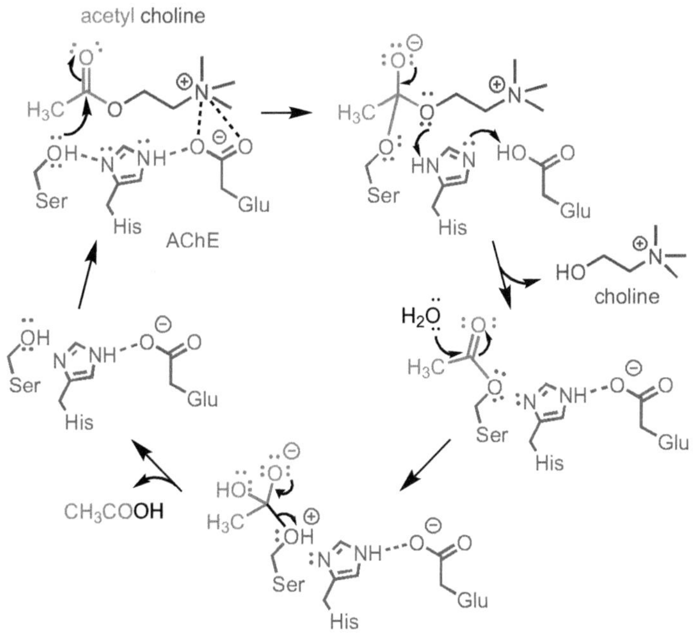

Figure 4.8 The mechanism of acetylcholine hydrolysis by AChE.

4.9.4 Organophosphate and Carbamate Insecticides

Organophosphates mimic the transition state of AChE because they are stable tetrahedral phosphonates, with long bonds that mimic the new bonds forming or bonds breaking in the tetrahedral intermediates (Figure 4.9). We will see in Chapter 9, Catalysis that enzyme inhibitors that mimic the transition state bind very tightly to the enzyme. Thus, organophosphates are very potent and irreversible inhibitors of AChE.

Inhibition of AChE causes continuous neurotransmission, resulting in seizures and death. Chemical warfare agents, such as soman and saran, and organophosphate and carbamate insecticides are potent, irreversible AChE inhibitors (Figure 4.9). When chemical warfare agents bind irreversibly to AChE, nerve signaling is turned on permanently, resulting in a horrible death by muscle paralysis, convulsions, and asphyxiation. Organophosphates are less potent, yet they cause neurological problems in people exposed to them directly.

The Carlier group at Virginia Tech has made significant strides to design AChE inhibitors that are more toxic to mosquitos than to humans. The team targeted the AChE of *Anopheles gambiae* (*A. gambiae*), a malaria carrying mosquito, and succeeded in making a carbamate

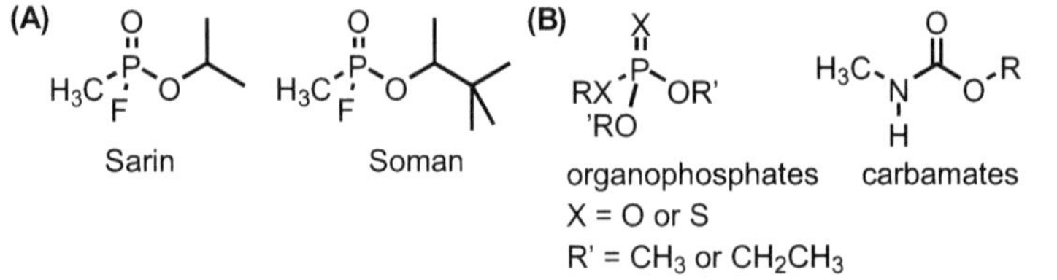

Figure 4.9 (A) Chemical warfare agents, and (B) insecticides that covalently modify AChE, organophosphates and carbamates.

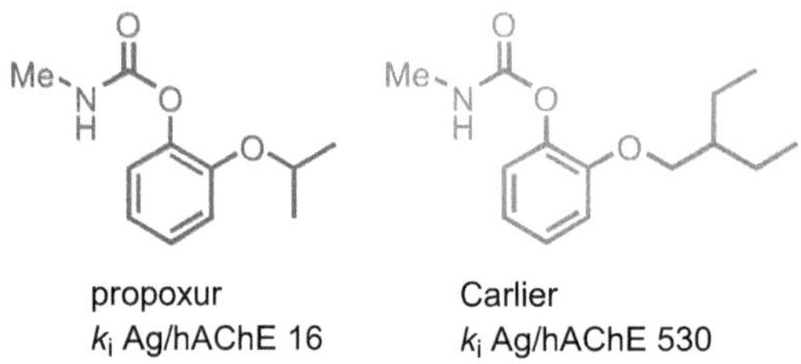

Figure 4.10 *Anopheles gambiae/human* ratio of AChE inhibition rate constants (k_i) for two selective inhibitors.[46]

insecticide based on propoxur that is 530 times more toxic to *A. gambiae* than humans, while propoxur is only 16 times more toxic (Figure 4.10).[46] A major remaining challenge is the bioavailability of these inhibitors through the mosquito foot exoskeleton.

It is good that AChE inhibitor insecticides, organophosphates and carbamates, are readily degradable in the environment. Even so, farm workers and by-standers should not be exposed to them because of acute toxicity. In addition, the manufacture of carbamate insecticides, like carbaryl (Sevin), has used reactive isocyanate intermediates, the cause of the Bhopal disaster that was mentioned in Chapter 1, and will be described in more detail in Chapter 12. By the risk principle, rather than try to prevent exposure of workers and the public to AChE inhibitors, it would be better not to make them at all. Avoiding organophosphates is an example of avoiding toxic functional groups as a means of designing benign chemicals.

4.10 Case Study: Spinosad and Natular (Clarke)

Clarke, a subsidiary of Dow Agrosciences LLC, won a PGCC Award in 1999 for the development of Spinosad, an environmentally friendly insecticide.[47] Spinosad is not chemically synthesized, it is synthesized by a soil microorganism in the Caribbean. Spinosad is a mixture of spinosyn A and D, that is active against mosquito larvae (Figure 4.11).

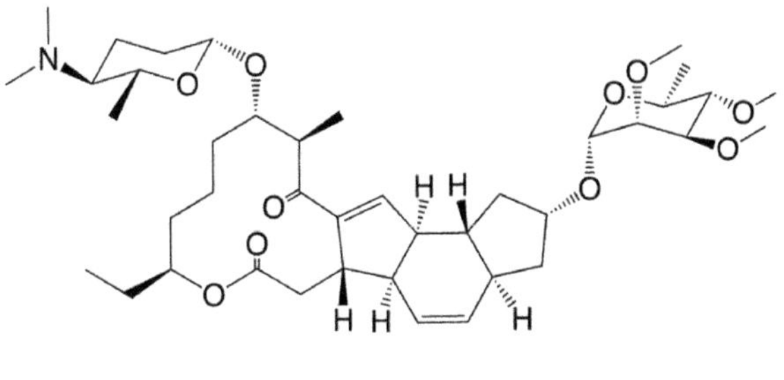

Spinosyn A

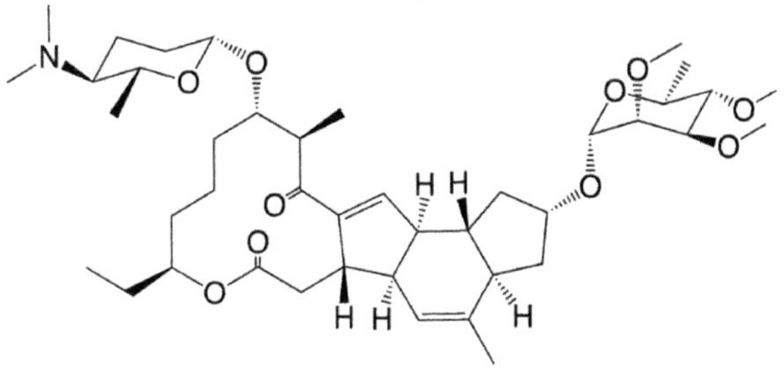

Spinosyn D

Figure 4.11 Spinosad is a mixture of Spinosyn A and D derived from a soil microorganism. Reprinted from http://chembl.blogspot.com/2011/01/new-drug-approvals-2011-pt-i-spinosad.html, accessed May 28, 2019, licensed as CC0 by EMBL-EBI.

Insects exposed to Spinosad show symptoms of neurotoxicity. Spinosad does not act through AChE inhibition, but rather as a new class of insecticide acting on nicotinic and γ-butyric acid receptors.[47] By acting on a different target than known insecticides, resistance is less likely. Rotation through different insecticide chemical classes could potentially prevent resistance entirely.

Spinosad is not persistent, bioaccumulative, or transported long distances in the environment, so it is not a POP.[47] The spinosyns are hydrolyzed in water and degraded by sunlight. Spinosad has low toxicity towards mammals and birds. It has moderate toxicity towards fish, but the risk to fish is lower than many synthetic insecticides. Spinosad has high selectivity towards pests, leaving 70 to 90% of beneficial insects—pollinators, such as bees and lacewings, and wasps that prey on other insect pests—unharmed.[47]

The key limitation of Spinosad as an insecticide is one of the very things that makes it less toxic, its hydrolytic instability. Clarke won a PGCC Award in 2010 for developing a sequential matrix mechanism to deliver Spinosad slowly in aqueous applications.[48] The "Sequential

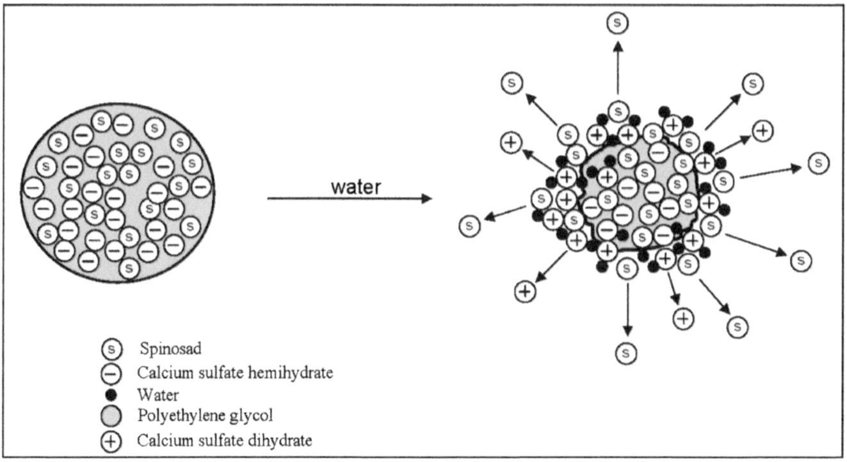

Figure 4.12 Sequential Matrix Mechanism of Natular. Reprinted with permission of Clarke from ref. 48, Copyright 2009.

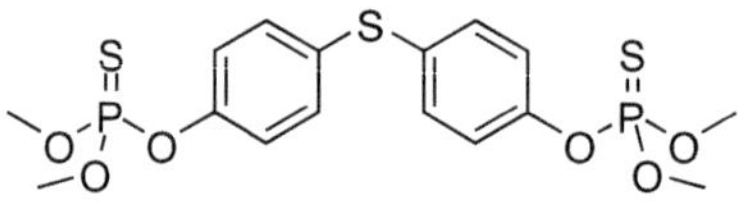

Figure 4.13 The structure of temephos, an organophosphate insecticide replaced by Spinosad-Natular.

Plaster Matrix" releases the spinosyns in water for up to 180 days, almost 6 months (Figure 4.12).[48] It can control mosquito larvae in a variety of aqueous environments, from catch basins to salt marshes. The tablet product, called Natular, is more effective at 2 to 10 times lower use rates than traditional synthetic insecticides, and 15 times less toxic to mammals than the organophosphate alternative temephos, which is the active ingredient in the pesticide Abate (Figure 4.13). Because so much less Spinosad can be used in application, considerable savings in packaging and transportation are attained. A 1.34 g tablet of Natular treats the same volume of water as 20 g of Abate (Figure 4.13), and it lasts twice as long.

Clarke wanted to use only US EPA-approved inert ingredients, which also conform to the US Department of Agriculture (USDA) National Organic Standards, for compound Spinosad. They settled on a mixture of plaster and polyethylene glycol to compress with Spinosad to make the dustless tablets of Natular. "Upon application to water, the polyethylene glycol in the first matrix progressively dissolves, permitting water to react with the calcium sulfate hemihydrate (plaster). This forms a second matrix of solid calcium sulfate dihydrate (gypsum), which then slowly dissolves to release the Spinosad molecule from the matrix into the aqueous environment"[48] (Figure 4.11). The mixture was adjusted to provide the desired dispersion and slow release rate in aqueous environments. No toxic reagents and no solvents are used in the Natular process. All waste from tablet production is directly reused on site.[48]

The use of Natular is projected to decrease the use of organophosphates for the dengue mosquito, *Aedes Aegypti*, market alone by 200 000 pounds internationally over the decade from 2010 to 2020.[48] Thus, Natular extends the impact of Spinosad with the plaster matrix slow release formula. The slow release tablets are prepared with benign materials, plaster and polyethylene glycol. The Spinosad pesticide itself is derived from natural biofeedstock from a soil microbe, and it is biodegradable by hydrolysis.[48]

4.11 Toxic Heavy Metals

The metals, As, Be, Cd, Cr, Hg, Ni and Pb, are considered major toxic metals because of a combination of high potency and high exposure.[49] Cd and As are readily absorbed by plants and found in food crops at elevated levels. In Chapter 3, we saw that Cd is a critical component of QLEDs. Hg and Pb are potent neurotoxins. In the 18th and 19th centuries, $Hg(NO_3)_2$ was used in hat felting, resulting in occupational exposure and dementia. The Mad Hatter of *Alice's Adventures in Wonderland* by Lewis Carroll is thought to be an allusion to Hg poisoning—hatters used mercury to make the pelts pliable for shaping. Now the primary source of Hg in the environment is due to burning coal. Pb was formerly widespread in commerce, such as in paints and plumbing. Pb lingers in pre-1980s pipes and solder in plumbing systems, and becomes particularly problematic if the water source is corrosive.[50]

Toxic metals Sb, Ba, Cs, Fe, Ge, In, Pd, Ag, Te, Tl, Sn, Ti, U, and V are considered minor, either because of low abundance, low usage, or low toxicity.[49] Antimony (III) oxide, Sb_2O_3, was discussed earlier as a component of flame retardants. This list does not exclude any other metals from suspicion, but these are the most widely used. Even essential nutrients, such as Fe and Cu, are toxic at high enough levels.

4.12 Case Study: Yttrium as a Lead Substitute in Electrodeposition (PPG Industries)

Now we turn to something entirely different, production of a benign inorganic product, car paint primer. Most automobile bodies are still made from iron-based sheet metal.[51] Corrosion (oxidation) of iron results in rust, iron(III) oxides ($Fe_2O_3{\cdot}nH_2O$ and $Fe(OH)_3$). The main way to prevent rust is the use of an anti-corrosive primer, which used to be lead-based. US federal law permitted the use of lead in automobile primer applications many years after lead was banned for other uses because there were no cost-effective substitutes.

PPG Industries won a PGCC Award in 2001 for developing a primer with yttrium (Y) substituted for lead (Pb). Electrocoating with Y gives $Y(OH)_3$, the metal part is baked, and a hard, resistant coating of Y_2O_3 is formed. Yttrium has several advantages over lead: (1) Y_2O_3 is much less toxic than Pb;[52] (2) Y is more earth abundant than Pb (Appendix D); (3) the reduction potential of Y is more favorable than Pb (Table 4.3); and (4) Y weighs less than Pb. The oral ingestion toxicity of Y is low (LD_{50} (rat) $>10\ g\ kg^{-1}$), in part because Y_2O_3 is a ceramic and very insoluble. Yttrium (33 $mg\ kg^{-1}$) is about 2-fold more earth abundant than Pb (14 $mg\ kg^{-1}$), and it is widely distributed across the earth.[53] The price and availability of Y are resistant to geopolitical factors as a result.

Oxidation of Y is more energy efficient than oxidation of Pb, or even Cr, because of its low reduction potential (Table 4.3).[54] More positive reduction potentials show more favorable reduction reactions because $\Delta G = -nFE^\circ$. Note that iron has a two-step *oxidation* process to give Fe^{+3}, the first is easier ($E^\circ = -0.44$) than the second ($E^\circ = +0.77$). Since Y is more easily oxidized than either oxidation state of Fe, the Y coating develops a thin film of insoluble Y_2O_3 upon

Table 4.3 Selected standard reduction potentials in acidic aqueous solution at 25 °C.[54] The most easily reduced, and difficult to oxidize, species have positive E° values.

Half reaction	E° (volts)
$Au^+(aq) + e^- \rightarrow Au(s)$	+1.68
$Pt^{2+}(aq) + 2e^- \rightarrow Pt(s)$	+1.2
$Fe^{+3}(aq) + e^- \rightarrow Fe^{+2}(aq)$	+0.771
$Pb^{2+}(aq) + 2e^- \rightarrow Pb(s)$	−0.126
$Fe^{2+}(aq) + 2e^- \rightarrow Fe(s)$	−0.44
$Cr^{+3}(aq) + 3e^- \rightarrow Cr(s)$	−0.74
$Y^{+3}(aq) + 3e^- \rightarrow Y(s)$	−2.37

baking that protects the metal from further oxidation. Using Y also avoids the necessity of pretreatment with Cr and Ni, used to make the lead primer stick. Electrodeposition of Y as a primer results in a 2-fold lower weight of the primer than Pb. Over millions of cars and the lifetime of those cars, this gives significant weight and fuel savings.

4.13 Summary

Designing more benign chemical products is a significant challenge for chemists. Toxicology provides useful tools to assist in the design processes that rely on chemical principles and chemical intuition. The 2003 Stockholm treaty banning persistent organic pollutants (POPs) has identified a growing number of toxins, mainly pesticides, based on four criteria: (1) persistence, (2) bioconcentration, (3) toxicity, and (4) long-range transport. Four methods for avoiding toxic products were described: mechanism of action, bioavailability, structure–activity relationships, and avoiding toxic functional groups. Endocrine disruption is one mechanism of action resulting in toxicity. Halogenated aromatic flame retardants are mutagens, carcinogens, and EDs. Pesticides are a major class of toxins that presents the most challenge to the benign by design concept. Selective toxicity towards only the target organism is the key. Spinosad pesticide and Natular compounding are examples of targeting a specific pest with effective green principles. Toxic heavy metals should be avoided in products. The use of yttrium as a primer for car coatings has resulted in significantly lower use of toxic Pb, Cr, and Ni.

4.14 Problems: Benign Products

Problem 4.1

(a) Choose three of Hill's criteria for evidence of causation that are best addressed by chemistry. Justify your choices briefly.
(b) How can one of these criteria be used to design less toxic products?

Problem 4.2

(a) What are the uses and exemptions for Dieldrin and Endrin?
(b) Why do you think Dieldrin has exemptions, but Endrin does not? (Hint: Look up oral and dermal LD_{50} values.)

(c) What is the stereochemical relationship of Dieldrin and Endrin?
(d) Propose a retrosynthetic analysis of Dieldrin.

Problem 4.3

(a) Draw the structures and note similarities of bisphenol A and S (Figure 4.4).
(b) What are the log K_{OW} values of bisphenol A and bisphenol S?
(c) Would this difference make bisphenol S more or less of an endocrine disrupter (fat soluble) than A?

Problem 4.4

(a) In the mechanism of AChE hydrolysis (Figure 4.8), what is a second role of the Glu residue besides proton transfer?
(b) Which transition state could organophosphate AChE adducts mimic other than the *formation* of the tetrahedral intermediate?
(c) Draw the covalent adduct of AChE with Sarin. What is the leaving group?
(d) Why are organophosphates such potent AChE enzyme inhibitors? (Hint: Linus Pauling said that enzymes evolve to bind the transition state more tightly than the ground state.)

Problem 4.5

Examine the structure of the spinosyns

(a) Which bond of a spinosyn is *most* likely to react with water?_______________
(b) What *size ring* contains the hydrolytically unstable bond of spinosyns? _____________
(c) Certain size ring systems are less stable than others; for example, 3 and 4-membered rings are more reactive than 5- and 6-membered rings. Why?
(d) Why is the hydrolyzed ring of spinosyn less stable than 5 or 6-member rings?

Problem 4.6

(a) Draw the mechanism of hydrolysis of the ring of spinosyn.
(b) What other reactive functional groups in the spinosyns might lead to environmental degradation?

Problem 4.7

(a) Give the number and name of two Principles that Spinosad best fulfills. Explain briefly.
(b) Look up the three PGCC Award areas. Which of the three areas does the Spinosad case fit best? Explain briefly.

Problem 4.8

(a) Draw the structure of methoprene, an alternative to organophosphate insecticides.
(b) What is its mode of action?
(c) What molecule does it mimic? Draw the structure.
(d) Does methoprene affect any non-target economically important species?

Problem 4.9

(a) What is the LD_{50} in rat for lead oxide (PbO)?
(b) What is the severity of lead ingestion?
(c) Why isn't the LD_{50} a good measure of toxicity for PbO?

Problem 4.10

Refer to Table 4.3 or Appendix E Standard Reduction Potentials.

(a) Which is harder to oxidize, Au or Fe? ____
(b) Which is harder to oxidize, Pb or Y? ____
(c) Write the half reactions for the oxidation of Y and the reduction(s) of Fe. Sum the reactions and the reduction potentials.
(d) Which is the favorable direction?

Bibliography

Casarett and Doull's Toxicology: The Basic Science of Poisons, ed. C. D. Klaassen, New York, 8th edn, 2013.

References

1. P. T. Anastas and J. C. Warner, *Green Chemistry: Theory and Practice*, Oxford University Press, New York, 1998.

2. P. T. Anastas and C. A. Farris, *Benign by Design*, American Chemical Society, Washington, DC, 1994, vol. 577.
3. T. Hager, *The Demon Under the Microscope*, Three Rivers Press, Crown Publishing Group, Random House, Inc., New York, 2006.
4. *OSHA Fact Sheet: Formaldehyde*, 2011, https://www.osha.gov/OshDoc/data_General_Facts/formaldehyde-factsheet.pdf, accessed July 23, 2019.
5. U.S. Environmental Protection Agency, *Health and Environmental Effects Profile for Formaldehyde*, Cincinnati, OH, 1988, https://ntrl.ntis.gov/NTRL/dashboard/searchResults.xhtml?searchQuery=PB88174958&starDB=GRAHIST, accessed July 23, 2019.
6. B. Keim, *FEMA Katrina trailer formaldehyde study: too little, too late*, Wired, 12.13.07, https://www.wired.com/2007/12/fema-katrina-tr/, accessed 02/04/2019.
7. Shell Oil Co., *Avgas*, https://www.shell.com/business-customers/aviation/aviation-fuel/avgas.html, accessed February 4, 2019.
8. L. K. Wolf, *Chem. Eng. News*, 2014, **92**, 27–29.
9. A. B. Hill, *Proc. R. Soc. Med.*, 1965, **58**, 295–300.
10. V. Farewell and T. Johnson, *Stat. Med.*, 2010, **29**, 1459–1476.
11. D. L. Eaton and S. G. Gilbert, in *Casarett and Doull's Toxicology: The Basic Science of Poisons*, ed. C. D. Klaassen, McGraw Hill Education, New York, 8th edn, 2013, pp. 13–48.
12. European Chemicals Agency (ECHA), *Understanding REACH*, 2019, https://echa.europa.eu/regulations/reach/understanding-reach, accessed February 6, 2019.
13. US Environmental Protection Agency, *Toxic Substances Control Act*, 2018, https://www.epa.gov/laws-regulations, accessed February 6, 2019.
14. Z. Gregus, in *Casarett and Doull's Toxicology: The Basic Science of Poisons*, ed. C. D. Klaassen, McGraw Hill Education, New York, 8th edn, 2013, pp. 49–122.
15. A. F. A. Cros, *Action de l'alcool amylique sur l'organisme*, University of Strasbourg, 1863.
16. R. Mannhold, G. I. Poda, C. Ostermann and I. V. Tetko, *J. Pharm. Sci.*, 2009, **98**, 861–893.
17. T. W. Schultz, M. T. D. Cronin, J. D. Walker and A. O. Aptula, *J. Mol. Struct.: THEOCHEM*, 2003, **622**, 1–22.
18. J. D. McKinney, A. Richard, C. Waller, M. C. Newman and F. Gerberick, *Toxicol. Sci*, 2000, **56**, 8–17.
19. J. B. Zimmerman and P. T. Anastas, *Science*, 2015, **347**, 1198–1199.
20. H. B. Leavesley, L. Li, S. Mukhopadhyay, J. L. Borowitz and G. E. Isom, *Toxicol. Sci*, 2010, **115**, 569–576.
21. M. P. Grillo, *Expert Opin. Drug Metab. Toxicol.*, 2015, **11**, 1281–1302.
22. F. H. Vaillancourt, E. Yeh, D. A. Vosburg, S. Garneau-Tsodikova and C. T. Walsh, *Chem. Rev.*, 2006, **106**, 3364–3378.
23. G. W. Gribble, *Natural Organohalogens*, Eurochlor, 2004, https://www.eurochlor.org/wp-content/uploads/2019/04/sd6-organohalogens-final.pdf, accessed July 23, 2019.
24. P. Van der Avert and B. M. Weckhuysen, *Angew. Chem., Int. Ed.*, 2002, **41**, 4730–4732.
25. M. Kucka, K. Pogrmic-Majkic, S. Fa, S. S. Stojilkovic and R. Kovacevic, *Toxicol. Appl. Pharmacol.*, 2012, **265**, 19–26.
26. US National Toxicology Program, *Bisphenol A (BPA)*, National Institute of Environmental Health Sciences, 2010, https://www.niehs.nih.gov/health/topics/agents/sya-bpa/, accessed July 23, 2019.
27. F. S. vom Saal and C. Hughes, *Environ. Health Perspect.*, 2005, **113**, 926–933.
28. S. Jenkins, J. Wang, I. Eltoum, R. Desmond and C. A. Lamartiniere, *Environ. Health Perspect.*, 2011, **119**, 1604–1609.
29. K. Steenland, T. Fletcher and D. A. Savitz, *Environ. Health Perspect.*, 2010, **118**, 1100–1108.

30. WHNT News 19, *3M admits to illegal chemical release in Tennessee River*, 2019, https://whnt.com/2019/06/14/3m-admits-to-illegal-chemical-release-in-tennessee-river/, accessed July 22, 2019.
31. A. Blum and B. N. Ames, *Science*, 1977, **195**, 17–23.
32. M. Mergel, *Chlorinated Tris (TDCPP)*, 2012, http://www.toxipedia.org/pages/viewpage.action?pageId=9175065, accessed February 18, 2016.
33. H. R. Pohl, S. Bosch, R. J. Amata and C. J. Eisenmann, *Toxicological profile for polybrominated biphenyls and polybrominated diphenyl ethers*, Agency for Toxic Substances and Disease Registry, Atlanta, GA, 2004, 594, https://play.google.com/store/books/details?id=3LKUCS5Hd5YC&rdid=book-3LKUCS5Hd5YC&rdot=1, accessed July 23, 2019.
34. *All POPs listed in the Stockholm Convention*, Secretariat of the Stockholm Convention, 2016, http://chm.pops.int/TheConvention/ThePOPs/AllPOPs/tabid/2509/Default.aspx, accessed February 4, 2019.
35. *Frank R. Lautenberg Chemical Safety for the 21st Century Act*, U. Congress, Washington, DC, 2015, https://www.congress.gov/bill/114th-congress/house-bill/2576/text, accessed July 23, 2019.
36. M. Hawthorne, *CPSC considers ban on toxic flame retardants in household products*, Chicago Tribune, Sept 28, 2015, https://www.chicagotribune.com/investigations/ct-flame-retardants-toxic-chemicals-met-20150925-story.html, accessed July /23, 2019.
37. U. G. Ahlborg, L. Lipworth, L. Titus-Ernstoff, C.-C. Hsieh, A. Hanberg, J. Baron, D. Trichopoulos and H.-O. Adami, *Crit. Rev. Toxicol.*, 1995, **25**, 463–531.
38. T. Colborn, D. Dumanoski and J. P. Myers, *Our Stolen Future: Are We Threatening Our Fertility, Intelligence, and Survival? – A Scientific Detective Story*, Dutton, New York, 1996.
39. W. J. Rogan and A. Chen, *Lancet*, 2005, **366**, 763–773.
40. G. G. J. M. Kuiper, J. G. Lemmen, B. Carlsson, J. C. Corton, S. H. Safe, P. T. van der Saag, B. van der Burg and J.-Å. Gustafsson, *Endocrinology*, 1998, **139**, 4252–4263.
41. M. R. Bratton, D. E. Frigo, H. C. Segar, K. P. Nephew, J. A. McLachlan, T. E. Wiese and M. E. Burow, *Environ. Health Perspect.*, 2012, **120**, 1291–1296.
42. G. G. Briggs, *J. Agric. Food Chem.*, 1981, **29**, 1050–1059.
43. C. Strode, S. Donegan, P. Garner, A. A. Enayati and J. Hemingway, *PLoS Med.*, 2014, **11**, e1001619.
44. D. A. Laskowski, *Rev. Environ. Contam. Toxicol.*, 2002, **174**, 49–170.
45. R. V. Jeger, *Eur. Heart J.*, 2013, **34**, 2580–2581.
46. J. A. Hartsel, D. M. Wong, J. M. Mutunga, M. Ma, T. D. Anderson, A. Wysinski, R. Islam, E. A. Wong, S. L. Paulson, J. Li, P. C. H. Lam, M. Totrov, J. R. Bloomquist and P. R. Carlier, *Bioorg. Med. Chem. Lett.*, 2012, **22**, 4593–4598.
47. Dow AgroSciences LLC, *Spinosad: A New Natural Product for Insect Control*, U.S. Environmental Protection Agency, 1999, http://www.epa.gov/greenchemistry/1999-designing-greener-chemicals-award, accessed July 26, 2019.
48. Clarke, *Natular: Developing a Sequential Matrix Mechanism Using Green Chemistry for Aqueous Application of Spinosad*, Washington, DC, 2009, https://www.epa.gov/greenchemistry/presidential-green-chemistry-challenge-2010-designing-greener-chemicals-award, accessed July 26, 2019.
49. E. J. Tokar, W. A. Boyd, J. H. Freedman and M. P. Waalkes, in *Casarett and Doull's Toxicology: The Basic Science of Poisons*, ed. C. D. Klaassen, McGraw Hill Education, New York, 8th edn, 2013, pp. 981–1030.
50. S. Roy and M. A. Edwards, *Curr. Opin. Env. Sci. Health*, 2019, **7**, 34–44.
51. A. Lovins, *A 40-Year Energy Plan*, 2012, https://www.ted.com/talks/amory_lovins_a_50_year_plan_for_energy?language=en, accessed July 23, 2019.

52. Agency for Toxic Substances & Disease Registry, *Toxic Substances Portal - Lead*, US Centers for Disease Control and Prevention, 2007, http://www.atsdr.cdc.gov/toxfaqs/tf.asp?id=93&tid=22, accessed July 26, 2019.
53. T. Helmenstine, *Periodic Table of the Elements Abundance of Elements in Earth's Crust*, 2012, http://chemistry.about.com/od/periodictables/ig/Printable-Periodic-Tables/Periodic-Table—Abundance.htm, accessed July 25, 2019.
54. A. J. Bard, R. Parsons and J. Jordan, *Standard Potentials in Aqueous Solutions*, IUPAC, Marcel Dekker, New York, USA, 1985; S. G. Bratsch, *J. Phys. Chem. Ref. Data*, 1989, **18**, 1.

5 Avoid Auxiliaries

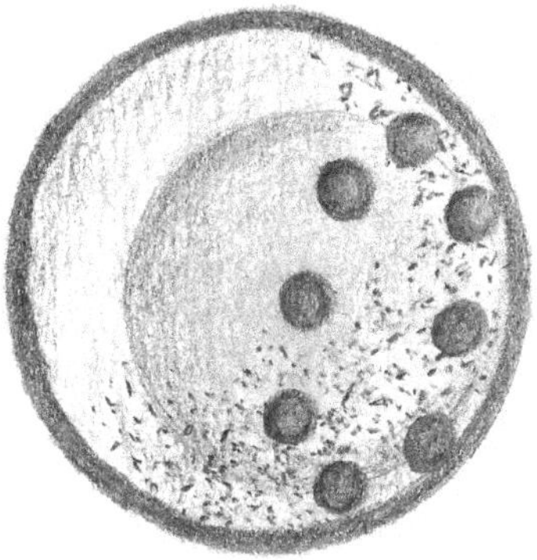

"Principle 5: The use of auxiliary substances (*e.g.*, solvents, separation agents, *etc.*) should be made unnecessary wherever possible and innocuous when used."[1]

5.1 Auxiliaries Defined

Substances other than reactants are necessary to make almost any chemical reaction go, or to make the product pure. Polylactic acid (PLA, Chapter 1) polymerization is a rare example of a reaction done without a solvent. Chemicals involved in purification include: acids, bases, salts, buffers, adsorbents, and solvents. Under the broadest definition of the term, even the plastic, glass or stainless steel of a reactor, or pump oil may be considered auxiliaries. QD Vision noted that the elimination of the purification step in the synthesis of QLEDs eliminated many gallons of pump oil in the process (Chapter 3).

Green Chemistry: Principles and Case Studies
By Felicia A. Etzkorn

Published by the Royal Society of Chemistry, www.rsc.org

Disposable plastics, such as weighing boats, micropipet tips, and Eppendorf tubes are a major source of waste that might be considered auxiliaries, especially in molecular biology processes. However, we focus only on the disposables used in reactions in this chapter. Solvents for all purposes will be considered together.

5.2 Solvent Auxiliaries

Solvents comprise the largest category of auxiliaries that are the target of efforts to make processes greener. The role of solvents in synthesis is to liberate molecules from a viscous liquid, solid matrix, or crystal, and to allow them to come into contact with other reactants, reagents, and catalysts.

Any liquid may serve as a suitable solvent for a reaction or purification depending upon the reaction, the polarity of the solutes, and the temperature of the process. Even a substance that is a solid at room temperature may be used as a solvent for a reaction that is run above the solvent melting point. QD Vision used *n*-octadecane as a solvent at elevated temperature for the production of QLEDs (Chapter 3). The solvent must be unreactive under the reaction conditions. The principle of "like dissolves like"—polar solvents for polar solutes, non-polar solvents for non-polar solutes—is a good guide to choosing a solvent for a reaction. The octanol–water partition coefficient, log K_{OW}, and water miscibility are reasonable ways to assess the polarity of a solvent (Table 5.1). It gets a bit tricky when one reactant is very polar and another is non-polar. Tetrahydrofuran (THF) has often served well as a solvent with different polarity reactants, but THF is not a recommended green solvent. Otherwise, a well-stirred reaction mixture with two solvents can be used. For liquid reactants, solvent-free reactions can be done, which is the most efficient of all. Reactions can also be run at elevated temperatures so that the reactants become liquid, called reaction or polymerization "in the melt." Of course, reactions at high temperatures are more energy intensive, in violation of Principle 6, Energy Efficiency.

5.2.1 Solvent Hazards

Flammability is the most common hazard associated with organic solvents. The key parameter associated with flammability is the flash point, the temperature at which the substance can vaporize and form a flammable mixture with air. Solvents with very low flash points are

Table 5.1 Commonly used solvent data arranged in order of increasing log K_{OW}.[a]

Solvent (abbreviation or common name)	b.p. (°C)[b]	f.p. (°C)[c]	H_2O sol. ($g\ L^{-1}$)[d]	Log K_{OW}[e]	LD_{50} ($g\ kg^{-1}$)[f]	LC_{50} sp. Inhalation[f]
Glycerol	290	176	M	−1.76	12.6	NF[g]
Water	100		M	−1.38	>90	
1,2-Ethanediol (ethylene glycol)	197	111	M	−1.36	4.7	NF[g]
Dimethyl sulfoxide (DMSO)	189	87	M	−1.35	14.5	NF[g]
Dimethylformamide (DMF)	153	58	M	−1.01	2.8	7.3 ppt/2 h[m]
Methanol	65	9	M	−0.77	5.6	64 ppt/4 h[r]
N-Methylpyrrolidinone (NMP)	202	86	M	−0.38	3.9	NF[g]
Acetonitrile	82	2	M	−0.34	2.5	2.7 ppt/1 h[m]
Ethanol	78	12	M	−0.31	7.1	30 ppt/4 h[m]
Dioxane	101	12	M	−0.27	4.2	29 ppt/2 h[m]
Acetone	56	-18	M	−0.24	5.8	62 ppt/4 h[r]
2-Propanol	82	12	M	0.05	5.0	43 ppt/2 h[m]
t-Butanol (*t*-BuOH)	83	11	M	0.35	2.7	NF[g]
Tetrahydrofuran (THF)	66	−14	M	0.46	1.6	62 ppt/2 h[r]
Ethyl acetate	77	−4	64	0.73	5.6	1.5 ppt/4 h[m]
1-Butanol	117	29	63	0.88	0.79	8 ppm/4 h[r]
Diethyl ether	35	−45	60	0.89	1.2	152 ppt/90 m[m]
Methyl *t*-butyl ether (MTBE)	55	−28	51	0.94	4.0	23.6 ppt/4 h[r]
Dichloromethane (DCM)	40		13	1.25	1.6	1 600 ppt/15 m[r]
Chloroform	61		8	1.97	0.70	3.6 ppt/4 h[r]
Benzene	80	−11	1.8	2.13	0.93	10 ppt/7 h[r]
Trichloroethylene (TCE)	87	>200	1.3	2.61	4.9	1.0 ppt/4 h[r]
Toluene	111	4	0.5	2.73	0.64	0.4 ppt/24 h[m]
Carbon tetrachloride	77		0.79	2.83	2.35	8 ppt/4 h[r]
2-Methyltetrahydrofuran (MeTHF)	78	−11	156	3.12	0.3–2	NF[g]
p-Xylene	138	27	I	3.15	5	4.6 ppt/4 h[r]
n-Pentane	36	−49	0.04	3.39	0.4	282 ppt/4 h[r]
Cyclohexane	81	−18	0.05	3.44	12.7	NF[g]
n-Hexane	69	−22	0.009	3.90	25	48 ppt/<4 h[r]
n-Heptane	98	−7	0.003	4.66	5	0.08 ppt/4 h[r]

[a]S. Kim, J. Chen, T. Cheng, A. Gindulyte, J. He, S. He, Q. Li, B. A. Shoemaker, P. A. Thiessen, B. Yu, L. Zaslavsky, J. Zhang and E. E. Bolton, PubChem 2019 update: improved access to chemical data, *Nucleic Acids Res.*, 2019, **47**(D1), D1102–D1109.
[b]b.p. = boiling point.
[c]f.p. = flash point.
[d]Solubility in water, M = miscible, I = immiscible.
[e]K_{OW} = octanol/water partition coefficient. C. Hansch, A. Leo, D. Hoekman. Exploring QSAR – Hydrophobic, Electronic, and Steric Constants. Washington, DC: American Chemical Society, 1995.
[f]LD_{50} oral in rat, heptane in mouse; LC_{50} inhalation in parts per thousand (ppt), sp. m for mouse, r for rat.
[g]NF = not found.

the most flammable, while solvents with high flash points are the least flammable (Table 5.1). Low-boiling ethers, such as diethyl ether and THF, are extremely flammable. Ethers also form explosive peroxides in storage, so they must not be kept longer than 6 months. THF is hygroscopic (water absorbing) so it must be dried for many reactions by distillation from sodium metal or $LiAlH_4$, both pyrophoric solids, adding to the high risk of using THF as a solvent. Yet THF is a polar aprotic solvent that will dissolve almost anything, so it is still in frequent use. Renewably sourced 2-methyltetrahydrofuran (MeTHF), with a higher flash point (f.p.), has been proposed as a greener alternative that can be synthesized from sugars. MeTHF it is not yet widely available and it is expensive. Table 5.1 gives the flash point f.p. and other relevant data for the most commonly used solvents.

Toxicity is the second hazard to consider in choosing a solvent. Halogenated solvents have typically very low flammability. In hazardous waste disposal, they must be kept separate from flammable solvents, which may be incinerated instead of disposing in landfill. Halogenated solvents are much harder to burn, and dioxins may be created in burning, so disposing of halogenated solvents is more expensive. Most halogenated solvents are carcinogenic, but this is not reflected in their LD_{50} values (Table 5.1). Dichloromethane (CH_2Cl_2 or DCM), chloroform ($CHCl_3$), carbon tetrachloride (CCl_4), trichloroethylene (TCE), and trifluoroethanol (CF_3CH_2OH) are the most commonly used in this class. DCM is toxic to the respiratory system, the nervous system, and is a probable carcinogen.[2] $CHCl_3$ has a similar toxicity profile, and it is also a reproductive and developmental toxin, or teratogen.[2] The toxicities listed in Table 5.1 are misleading because only the *oral* toxicity in rat is given. The normal route of entry for humans using volatile organic solvents is by inhalation. We will see that *n*-hexane is a potent neurotoxin by inhalation when we consider a case of auto mechanics in California.

The reactivity of a solvent is a third consideration. Ethanol, a fairly benign solvent, is often not suitable as a reaction solvent because its acidic –OH proton (pK_a 16) may react with other components of the reaction. Similarly, ketones and esters have reactivity considerations; they cannot be used when the functional group might react with any other components of the reaction mixture. For example, a ketone like acetone would be a poor choice for a Grignard reaction because all of the Grignard reagent would react with the solvent. Yet acetone is the solvent of choice for alcohol oxidation because it does not further oxidize under the reaction conditions.

Renewability of solvents is a fourth consideration. Solvents derived from renewable sources rather than petroleum are beginning to make their way into chemistry. EtOH and MeOH are the most common renewable solvents. EtOAc could be renewably sourced if both the HOAc and EtOH were both derived from fermentation. Supercritical CO_2 is widely used in chromatography, and it is becoming more common as a reaction solvent. Dimethyl carbonate is already used as a solvent in lithium ion batteries, yet flammability is an issue which will be discussed later in this chapter. MeTHF is a renewable solvent that can be made from sugars, however it is still quite expensive at this time.

The fifth consideration is the environmental degradability of the solvent. Solvents that are biodegradable, or break down with UV sunlight into benign products, have a distinct advantage. Even if they leak from a waste site they will eventually break down into benign products. Ketones, esters and alcohols, the more reactive solvents, are most likely to be readily biodegradable. Aromatic and halogenated solvents, the more stable solvents, are least likely to be biodegradable, as well as the most toxic.

A sixth consideration is the energy required to evaporate solvents to remove them from the desired product. Water is particularly difficult to remove; typically, water is removed by lyophilizing (freeze-drying in high vacuum), an energy intensive process. Dimethyl sulfoxide (DMSO) is even more difficult to remove by evaporation, but it too can be lyophilized. It is easy to freeze with a melting point of 18 °C, very close to room temperature. DMF has a high f.p., but it also has a high b.p., so it is difficult to remove after a reaction. Toluene is often used as a less-toxic substitute for benzene, but toluene has a higher b.p. This is a difficult trade-off; high boiling solvents are less flammable, yet inherently more energy intensive to evaporate or recycle. If the desired product precipitates under the reaction conditions, it can be collected by filtration to avoid evaporating the solvent. It may also be possible to extract the product into a solvent of different polarity.

5.2.2 Solvent Uses and Recovery

Recycling solvents on-site by distillation is an excellent method for recovery. The HF used as the solvent and catalyst for the BHC ibuprofen synthesis is highly toxic, but non-flammable. When it was used, HF was recycled in a closed system on-site at BASF. Flammable solvent waste may be burned as fuel to generate heat or energy for a chemical plant. Halogenated solvents are not suitable for this purpose because they must be incinerated at very high temperatures, with long burn times, and under specific conditions of turbulence, usually combined with flammables to prevent dioxin formation.

5.2.3 Hexane Toxicity

A mixture of *n*-hexane and acetone was used to replace benzene in many automotive repair cleaning products after benzene was found to be a carcinogen.[3] In the 1960s, workers in several industries reported peripheral neuropathy, which is weakness, and tingling or pain, usually in the hands and feet. Several studies in the 1980s linked the peripheral neuropathy to *n*-hexane, and elucidated the mechanism (Figure 5.1).[3]

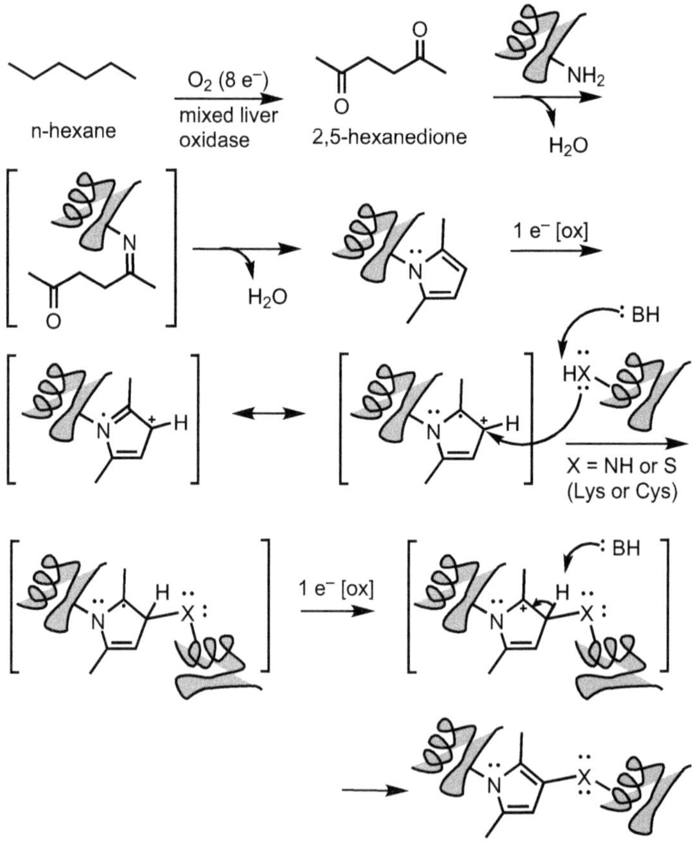

Figure 5.1 Mechanism of *n*-hexane toxicity causing peripheral neuropathy. Proteins are represented by the squiggly line with the amine over the second arrow.[3]

Metabolic oxidation, with liver cytochrome P450 (CYP) enzymes, is a common mechanism of toxicity (see benzo[a]pyrene, Chapter 3). Production of 2,5-hexanedione, the reactive intermediate in hexane toxicity, requires an 8 e^- oxidation at the second carbon from each end (Figure 5.1).[3] The CYP enzymes are effectively specific for oxidizing the second carbon from the end of a chain (ω 2) oxidation. The 2,5-hexanedione, a γ-diketone, then reacts with an amine of Lys side chains in proteins, losing two water molecules, to give a 2,5-dimethylpyrrole (Figure 5.1). Other γ-diketones such as 3,4-dimethyl-2,5-hexanedione give pyrroles, but α- or β-diketones cannot form the 5-membered ring. (The Greek letter designation refers to the number of carbons between the two ketones: α is for adjacent ketones, β is for ketones two carbons away, and γ is for three carbons away.) Further oxidation activates the pyrrole towards nucleophilic addition by Lys and Cys side chains on other proteins to cross-link two proteins together. The structural and transport protein, tubulin, is thought to be involved in the neuropathology of *n*-hexane.[3] Swelling in the axons nearer the body from myelin sheath gaps, called nodes of Ranvier, cause the protective fatty myelin sheath of the neurons to retract, resulting in painful peripheral neuropathy.[3]

Hexane is used as a degreaser in automotive shop products, in glues for shoes, leather and roofing, to extract non-polar cooking oils such as canola and soy oils from plants, and many other purposes. Hexanes comprise a significant fraction of gasoline. Chemists routinely use hexanes in column chromatography to purify reaction products. Many companies are phasing out the use of hexanes in their products because of the hazards involved, and substituting *n*-heptane, which does not form pyrroles *in vivo*, but it is higher boiling and slightly more difficult to remove by evaporation. This is an example of understanding the mechanism of toxicity to choose a less-toxic alternative.

5.3 Extraction Auxiliaries

A second class of chemicals is used in purification to extract reactant by-products, leftover starting materials, and undesired products from the desired products. This is commonly referred to as the reaction "work-up." Colored impurities may be adsorbed onto activated charcoal. Then aqueous extraction is used to remove charged molecules by dissolving them in the aqueous layer—acids in aqueous base, bases in aqueous acid. Separatory funnels are typically used to mix the organic reaction solution and an aqueous solution, allow separate layers to form, and then separate the layers by draining the higher density layer from the bottom. In chemical industries, such separations are often done in large stainless steel tanks. The reaction tank can be equipped with a valve at the bottom to drain the higher density layer.

Sometimes the opposite is done; the desired product is charged to dissolve it in the aqueous layer, and non-polar impurities are extracted with an organic solvent that is water immiscible (Table 5.1). After extracting the impurities, the aqueous layer is either acidified or made basic to neutralize the charge of desired molecule, which can then be extracted into clean organic solvent or recrystallized. With the desired product remaining in the organic solution, saturated sodium chloride (brine) may be used to pre-dry the solution to remove large amounts of water. Then the organic solution of product is dried over a drying agent—an inorganic salt with an affinity for water, typically sodium sulfate (Na_2SO_4) or magnesium sulfate ($MgSO_4$)—which is then removed by filtration. Some common extraction and drying auxiliaries are listed (Table 5.2).

If the work-up is planned carefully, the chromatography step may be eliminated. This would be desirable on an industrial scale. The aqueous waste from a work-up may contain mixtures of organic salts and water miscible solvents. Chromatography consumes large quantities of solvent and silica gel, both of which are inhalation hazards. In a research lab, chromatography is usually done after each step to ensure high-purity intermediates that can be clearly identified as new chemical entities by elemental analysis and spectroscopy (NMR and mass spectroscopy (MS)). Once characterization of a new compound is complete, it is only necessary to obtain product that is pure enough to carry on to the next step successfully.

Table 5.2 Parameters and uses of common extraction auxiliary chemicals.

Extraction auxiliary	d[a] ($g\,mL^{-1}$)	pH	Removes
1 M HCl (aq)	1.04	0.5	Bases, amines
sat'd NH_4Cl (aq) (7.2 M)	1.5	4.2	Bases, amines
5% $NaHCO_3$ (aq) (0.6 M)	1.04	5.2	Acids, stronger bases
1 M NaOH (aq)	1.04	14	Acids
sat'd NaCl (aq) (6.2 M)	1.20	7.0	water
Na_2SO_4 (anhydrous)		—	Water
$MgSO_4$ (anhydrous)		—	Water

[a]Densities from https://www.engineeringtoolbox.com/density-aqueous-solution-inorganic-sodium-salt-concentration-d_1957.html, accessed February 22, 2018.

5.4 Chromatography Auxiliaries

Chromatographic separation depends upon the differential solubility of the components of a mixture in two different phases, a stationary phase and a mobile phase. In gas chromatography (GC), the stationary phase is a viscous liquid and the mobile phase is a gas. GC is used most often for analysis on capillary columns, rather than for preparative separation using large packed columns. The material of greatest concern in analytical GC is the He gas used as the mobile phase, because He is becoming scarce—floating into outer space.[4] In liquid chromatography (LC), the stationary phase is solid, and the mobile phase is liquid. LC is the powerhouse separation method for modern pharmaceutical chemistry. The use of supercritical CO_2 as a greener LC solvent will be described later in this chapter.

For reasons similar to their use in reactions, solvents are also a major consideration in preparative separations: LC, crystallization, or recrystallization. In LC, the basis of separation is that molecules interact differentially with the solid adsorbent and the solvent. In order to do so, individual molecules must be isolated from each other by solvent. In crystallization, the impurities must remain dissolved in the solvent, while the desired material slowly aggregates with itself in crystallizing out of the solution. Solvent mixtures are typically used in both chromatography and crystallization to differentiate between the desired and undesired substances.

The adsorbents used in LC are a major consideration in disposal. Adsorbents are most often silica gel (SiO_2), and many organic derivatives of silica gel, such as C18, an n-$C_{18}H_{37}$ alkyl chain linked to a silica support. Other less frequently used adsorbents are alumina (Al_2O_3), florasil (magnesium silicate), and cellulose; agarose and polyacrylamide are used in electrophoresis. Some chromatography adsorbents are reused again and again, for example in HPLC (high performance liquid chromatography) columns. Considerable amounts of solvents are typically used to clean and regenerate a column after a purification step. The amount of solvent required for cleaning must be balanced against disposing of the solid adsorbent as hazardous waste. In HPLC, the most insoluble impurities are removed using a guard column containing a minimal amount of adsorbent, which can then be discarded as needed. This permits the column adsorbent to be reused for hundreds of separations.

The SiO_2 adsorbent in flash chromatography, a medium-pressure method typically performed after a reaction, is usually discarded after use. Modern labs often use medium-pressure chromatography systems in which the adsorbent cartridges may or may not be reused.

5.5 Minimize Auxiliary Substances

The goal of Principle 5 is to minimize the amounts, the toxicity, and the hazards of auxiliaries used in synthesis: solvents, extraction chemicals, and adsorbents. We have seen that auxiliaries are the greatest contributor to the E-factor or PMI (Chapter 2). Elimination of auxiliaries is the ultimate goal because this eliminates waste, hazards, and thus risk.

5.5.1 Minimize Number of Reaction Steps

The efficiency of a reaction or a multi-step synthesis was analyzed by three methods in Chapter 2: atom economy, overall yield, and Sheldon's E-factor. Only Sheldon's E-factor considers the waste due to the use of auxiliaries. It is obvious, but bears repeating, that the amounts of auxiliaries are decreased dramatically by decreasing the number of synthetic and purification steps in a synthesis. As in the BHC synthesis of ibuprofen, a 3-step process is much more efficient than a 6-step process (Chapter 2).

5.5.2 Minimize Reaction Volume

Minimizing reaction volumes by eliminating solvent allows the reaction to be run in much smaller reactors. Eliminating solvent can also lead to shorter reaction times, particularly if the rate-determining step of the reaction is bimolecular. Industry refers to these two factors as improving the "space-time yield" of a process. Heat transfer is also more efficient in smaller reactors, saving energy and potentially increasing safety.

The pharmaceutical industry has begun to use smaller disposable plastic reactors for protein drugs produced by bacteria, plant, or animal cells, which improves space-time yield.[5] However, anti-oxidants used to stabilize the polyethylene can leach into the product creating purification challenges. The anti-oxidant additive tris(2,4-di-*tert*-butylphenyl)-phosphite (TBPP) can degrade under the gamma-ray sterilization to an extractable—bis(2,4-di-*tert*-butylphenyl)-phosphate (bDtBPP)—that kills or slows the growth of cell cultures (Figure 5.2).[5]

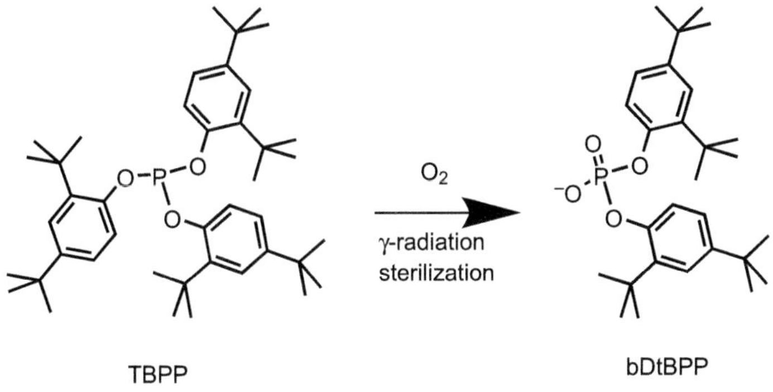

Figure 5.2 Anti-oxidant TBPP in plastic reactors forms water extractable bDtBPP upon sterilization.[5]

5.5.3 Minimize Purification

Solvent-free synthesis can often give a high-purity product at a faster rate, making extractions and chromatography unnecessary. If purification is necessary, we must consider ways to minimize the number of purification steps and decrease the amounts of auxiliaries used. Perhaps the product can be precipitated or crystallized directly from the reaction mixture by adding a solvent in which it is insoluble, while leaving all of the by-products in solution. Conversely, perhaps all by-products can be precipitated and removed by filtration, leaving only a solution of pure product. Even if extraction steps are necessary, clever choice of the right pH may allow a single aqueous extraction of by-products. Recrystallization is less solvent intensive than chromatography, and it eliminates the adsorbent. While no process is perfect, even small steps add up to big strides towards green chemistry. Industrial process chemistry strives to create the most efficient and cost-effective chemistry possible. Often this means reducing or eliminating extraction and chromatography steps.

5.6 Solvent-free Synthesis

Of course, the safest solvent is no solvent at all. Not only is solvent-free synthesis greener, it is faster, and cheaper. In project management, there is a common saying, "Good, fast, cheap, pick any two." If you want it fast and cheap, it will be poor quality. If you want it good and fast, it will be expensive. If you want it good and cheap, it will be very slow. But solvent-free synthesis can achieve the impossible: good, fast *and* cheap (Figure 5.3)!

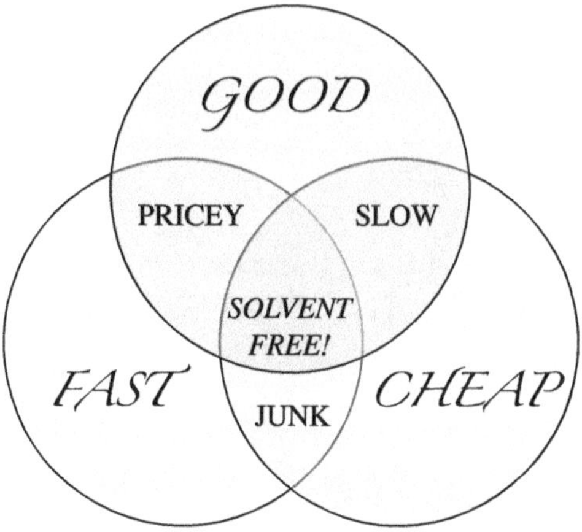

Figure 5.3 Good, fast, cheap, solvent-free synthesis.

When a reactant is a liquid, the reaction may be carried out under solvent-free conditions. For bimolecular reactions, the kinetics are much faster in a solvent-free system because of the high concentration of both reactants. On the other hand, diffusion of substrates to a catalyst may be the rate-determining step if the catalyst increases the rate of the chemical steps dramatically. In the case of diffusion-controlled reactions, the viscosity of the solution is an important parameter; more viscous solutions result in slower reactions. Solvent-free reactions are likely to be more viscous.[6]

Polymer syntheses are often run under solvent-free conditions, which can be either in the melt phase or solid state.[7] The monomer may be a liquid, or the polymer product may be in a liquid state by running the reaction at elevated temperatures. The polymer product of melt-phase or solid-state polymerization is typically of high purity, requiring no purification.[7] Solvent-free polymerization decreases the likelihood of side reactions and colored impurities. A high production volume polymer, poly(ethyleneterephthalate) (PETE), is produced in the solid state.[7]

5.6.1 Mechanical Mixing

Reactions can be run under solvent-free conditions by high-speed mixing in a ball mill, a traditional method in inorganic chemistry.[8] James Mack and coworkers adapted this method for organic chemistry in the reduction of aldehydes and esters with $NaBH_4$ (Figure 5.4a), and a Pd-catalyzed homocoupling.[9,10]

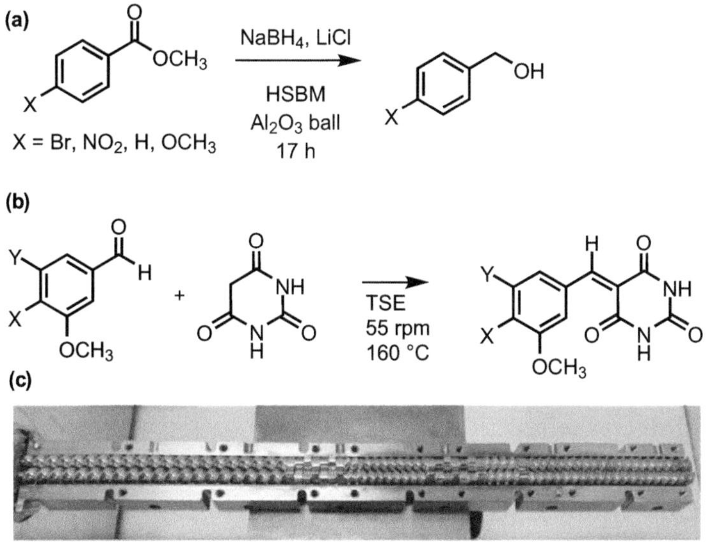

Figure 5.4 Mechanochemistry: (a) High-speed-ball-milling (HSBM) solvent-free reduction of esters.[9] (b) Twin-screw extrusion (TSE), solvent-free Knoevenagel reaction.[11] (c) Twin-screw extrusion device.[11] Reproduced from ref. 11 with permission from the Royal Society of Chemistry.

Another mechanical method is twin-screw extrusion (TSE) (Figure 5.4c). In a Knoevenagel reaction, TSE delivered high-purity product in nearly 100% yield within two minutes, an extremely efficient reaction in addition to being solvent-free (Figure 5.4b).[11]

5.7 Case Study: A Solvent-free Biocatalytic Process for Cosmetic and Personal Care Ingredients (Eastman Chemical Co.)

Eastman Chemical Co. won a PGCC Award in 2009 for developing a versatile esterification process for cosmetic ingredients.[12] The solvent-free process uses natural acids and alcohols, and an immobilized enzyme catalyst. Because the reactions are run at lower temperatures, there are fewer side products. The catalyst is removed by filtration, and no further purification is necessary.

Consumers view cosmetics as discretionary purchases, so there is increasing consumer pressure on cosmetic companies to use all natural, non-toxic ingredients in their products. Esters are used as emulsifiers, emollients, surfactants, and skin illuminators. Emulsifiers create emulsions of oil and water mixtures. Emollients are used to make the skin or hair soft and supple, as in lotions and conditioners. Surfactants are amphiphilic (polar and non-polar) molecules that decrease the surface tension between polar and non-polar substances; detergents and soaps are charged surfactants that surround grease or dirt and make them soluble in water. Skin illuminators create an even skin tone by reducing undesirable skin pigmentation, such as age spots. Examples of naturally derived esters used in cosmetics are sorbitan oleate, glyceryl stearate, isopropyl myristate, and retinyl palmitate.[12]

The problem has been that chemical esterification processes use strong acid catalysts, typically sulfuric acid, which can oxidize delicate natural ingredients, producing by-products that affect the color and odor, and decrease reaction yields.

5.7.1 Chemical Esterification

Emil Fischer developed the oldest and simplest method for esterification in 1895, still called the Fischer esterification (Figure 5.5a).[13] A strong acid, such as HCl or H_2SO_4, is used in catalytic amounts, and the reaction is run at higher temperatures (Δ), usually refluxing at the boiling point of the alcohol component, R^2–OH. The R^2–OH alcohol is frequently the solvent, such as methanol, which is used in large excess to drive the reversible reaction to completion. Alternatively, water may be removed from the reaction to drive it to completion. Both methods work because of Le Chatelier's Principle. Transesterification (swapping alkoxy groups) may be used if the desired acid is available in the form of an ester (Figure 5.5b). If R^3–OH is volatile, such as the

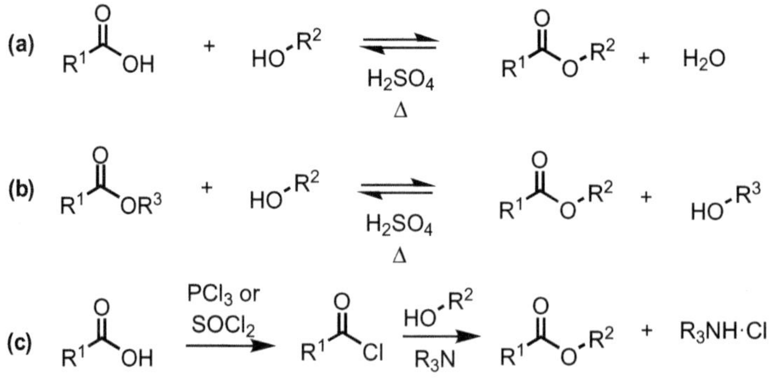

Figure 5.5 General chemical methods for synthesis of esters from carboxylic acids and alcohols. (a) Fischer esterification with a strong acid catalyst. (b) Transesterification with $R^3 = CH_3$, C_2H_5, C_3H_7, C_4H_9 volatile alcohol products. (c) Preparation of esters *via* intermediate acyl chlorides.

C_1–C_4 alcohols, methanol to butanol, and the R^2–OH reactant is a less volatile long-chain alcohol, the R^3–OH alcohol can be removed by distillation from the reaction to drive it to completion. Esters are often prepared *via* intermediate acyl chlorides, which are reactive enough for weak alcohol nucleophiles (Figure 5.5c). The acyl chloride esterification is irreversible, so it is not necessary to drive the reaction by either of the methods described above. Acyl chlorides are used for larger, or more expensive alcohols. Making the acyl chloride is an extra step. A full equivalent of an amine base and either toxic PCl_3 or $SOCl_2$ are required, and a solvent is used, so the synthetic efficiency of the third method is much lower. The use of solvent decreases the "space-time yield" of esterification.

5.7.2 Lipase Enzyme Catalyzed Transesterification

Lipase enzymes catalyze the hydrolysis of lipids (fats, or fatty esters). The names of enzymes indicate their primary substrate followed by the suffix "ase." Lipases may also be used to synthesize esters, the reverse reaction. Providing an excess of carboxylic acid and/or alcohol reactants can drive the reverse reaction. Catalysts do not affect the thermodynamic position of equilibrium; the ratio of reactants to products does. In a transesterification, 1 : 1 ratio of substrate ester to product ester is likely under equilibrium conditions ($K_{eq} \approx 1$), because the reactants and products are chemically similar.

Eastman uses a lipase-catalyzed transesterification to synthesize cosmetic esters (Figure 5.5).[12] Removal of the low-boiling C_1–C_4 alcohol product of transesterification by sparging (bubbling) with nitrogen gas is used to drive the reaction.[12] Clearly, the reactant alcohol, in the example retinol, must be significantly less volatile than the product alcohol, MeOH, so that it does not evaporate (Figure 5.6).[12]

The key to this process is the use of an immobilized enzyme catalyst (Chapter 9). Eastman uses an immobilized enzyme for esterifications, which has the advantage of removal by simple filtration.[12] There are several ways to immobilize enzymes: (1) non-covalent adsorption or deposition onto a carrier, (2) covalent immobilization, either to a carrier or by cross-linking crystals or aggregates of the enzyme, and (3) encapsulation within a porous matrix (Figure 5.7).[14] Immobilization of enzymes by inclusion with a surfactant during polymerization of polyurethane was patented by Eastman Chemical Co.[15] This is a covalent modification because surface enzyme lysine $-NH_2$ groups become part of the polyurethane during polymerization.

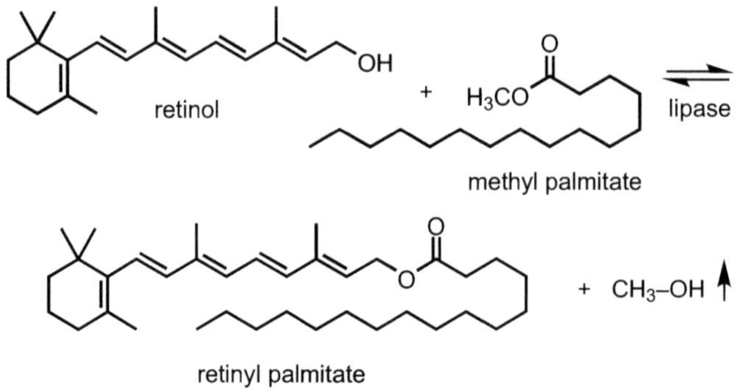

Figure 5.6 Example of an Eastman lipase-catalyzed transesterification to make retinyl palmitate driven by methanol removal.[12]

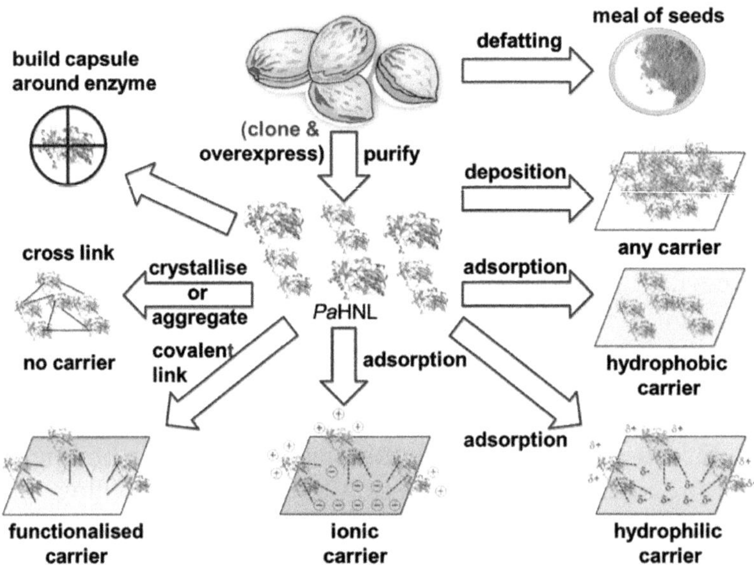

Figure 5.7 Enzymes can be immobilized in many different ways, by deposition, adsorption, covalent binding (with and without a carrier) and by encapsulation. To illustrate all these methods [*Prunus amygdalus* hydroxynitrile lyases] PaHNL is depicted here. Reproduced from ref. 14 with permission from the Royal Society of Chemistry.

5.7.3 Solvent-free Esterification

The Eastman method is a broadly applicable, solvent-free esterification. A wide variety of product esters can be produced, specific for each client's application. The biofeedstocks used are natural esters, acids, and alcohols. The ester reactants have low molecular weight (C_1–C_4) alkoxide groups (Figure 5.6).[12] Many of the reactants are liquids at relatively low temperatures, enabling solvent-free esterification. The enzymatic reaction can be run at lower temperatures, generally <60 °C, and neutral pH, so delicate plant esters, such as unsaturated fats and alcohols do not decompose. High-purity products can be obtained directly from the reaction without purification.

5.7.4 Plant Fatty Acids

Eastman uses plant fatty acids in their process.[12] Plant fatty acids have multiple unconjugated double bonds, with allylic positions in between that are highly susceptible to oxidation (Figure 5.8). Once oxygen reacts with a fatty acid alkene, a cascade of reactions can occur, creating a mixture of damaged molecules, causing rancid odors and negative health effects.

Hydrogen atom abstraction by O_2 occurs at the allylic positions due to the stabilization of the allylic radical intermediate. The extra stability of allylic radicals can be envisioned with either a resonance model or a molecular orbital (MO) model (Figure 5.8). The MO model is built from atomic orbitals for the three carbons in the allylic system. Nodes are mathematical points in space at which the MOs

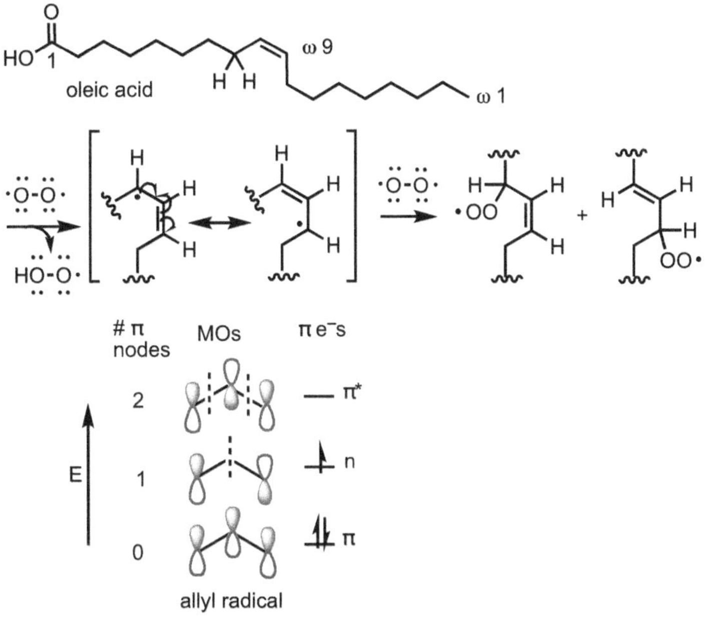

Figure 5.8 Oxidation of the natural unsaturated fatty acid, oleic acid, at allylic positions.

change symmetry. There is zero probability of an electron existing at a node in the MO.

The lowest energy MO has no vertical nodes (Figure 5.8). This π MO is completely bonding across all three carbons, and it contains one pair of electrons. The second MO from the bottom contains one vertical node at the central carbon, where the MO changes symmetry. The MO is a non-bonding *n* MO because there are no bonding and no anti-bonding interactions across the node; the anti-symmetric lobes of the MO are not on adjacent carbons. Since there are three π electrons in an allyl radical, this middle energy MO contains one unpaired electron, a free radical. The single unpaired electron density is at the two terminal carbons simultaneously (a quantum quirk), just as we would predict from the Lewis resonance structures shown above the MO model (Figure 5.8). The third highest energy MO has two vertical nodes, one between each pair of carbons giving two anti-bonding relationships, so this is a net anti-bonding π* MO, and it contains no electrons. Notice the alternating shading across each node. Both the resonance and MO models explain the particular reactivity at the allylic positions of unsaturated plant fatty acids. The free radical intermediates of alkene allylic oxidation are resonance stabilized.

Natural unsaturated fatty acids exist exclusively in the *cis* configuration (Figure 5.8), and they are named by the position of the alkene counting from the end, or ω-carbon. Omega-3 fats, which have unconjugated alkenes at ω-3, ω-6, and ω-9 positions, are found in plants and fish oil, and they are widely recognized for positive health effects. However, omega-3 fats are even more susceptible to oxidation because of their doubly allylic positions in between two alkenes.

A second problem with strong acid and heating in traditional chemical esterification reactions is the isomerization of the *cis* alkenes to the more stable, unnatural *trans* alkenes, and to conjugated polyenes. Trans alkenes also result from hydrogenation of natural unsaturated fats. Trans fats have a highly ordered solid-state structure within cell membranes, resulting in adverse health effects, "...sabotaging the flexible and porous functionality required by the body."[16,17]

5.7.5 Skin-illuminating Ingredients

An interesting example described in the Eastman PGCC Award proposal is the synthesis of 4-hydroxybenzyl alcohol (4-HBA) acetates (Table 5.3).[12] Chemical acetylation exclusively on the phenol is called "chemical 4-HBA acetate," while acetylation by Eastman's

Table 5.3 Tyrosinase inhibition values of acetate esters of 4-hydroxybenzyl alcohol. Reprinted with permission of Eastman Chemical Co. from ref. 12, © 2008.

Compound	EC_{50} (mM)	Structure
4-HBA	0.19	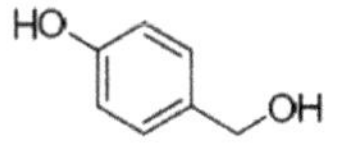
Enzymatic 4-HBA acetate	0.038	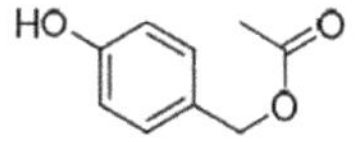
Chemical 4-HBA acetate	0.92	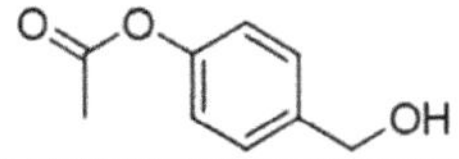

immobilized lipase enzyme only at the alkyl alcohol is called "enzymatic 4-HBA acetate."

The enzyme tyrosinase (TYR), also called tyrosine hydroxylase, catalyzes the first two steps in the formation of the skin pigment eumelanin from the amino acid tyrosine (Figure 5.9). TYR adds a hydroxyl to the ring of the tyrosine substrate, and then oxidizes both hydroxyls of the dopamine (DOPA) intermediate to the *ortho*-quinone, dopaquinone (Figure 5.9). Tyrosinase inhibitors are now used as "skin illuminators" in lotions. (A review by Chang is much more blunt, calling tyrosinase inhibitors "skin-whitening" agents.[18]) Enzymatic 4-HBA acetate is an inhibitor of TYR, and therefore inhibits the production of the eumelanin pigment (Table 5.3).

The potency of enzyme inhibitors can be measured by *in vitro* assays. The EC_{50} is the concentration of inhibitor that is effective at inhibiting 50% of normal TYR activity. The lower the EC_{50}, the more potent the compound. The enzymatic 4-HBA acetate is 24-times more potent than the chemical 4-HBA acetate as a tyrosinase inhibitor (Table 5.3).[12] A simple structure–activity relationship (SAR) can be postulated from the EC_{50} values.

The phenol group of the enzymatic 4-HBA acetate probably serves as a better mimic of the tyrosine phenol side chain, since the chemical 4-HBA acetate with the acetyl-blocked phenol has a higher EC_{50} value. The enzymatic 4-HBA acetate is 5-fold more potent than the unacetylated 4-HBA, so the acetyl group contributes in some way to the inhibition, possibly by decreasing the polarity. The less polar enzymatic 4-HBA acetate could have increased partitioning between the non-polar enzyme environment and aqueous biological solutions.

The enzymatic 4-HBA acetate would be much more difficult to synthesize by a chemical route. The lipase naturally acetylates

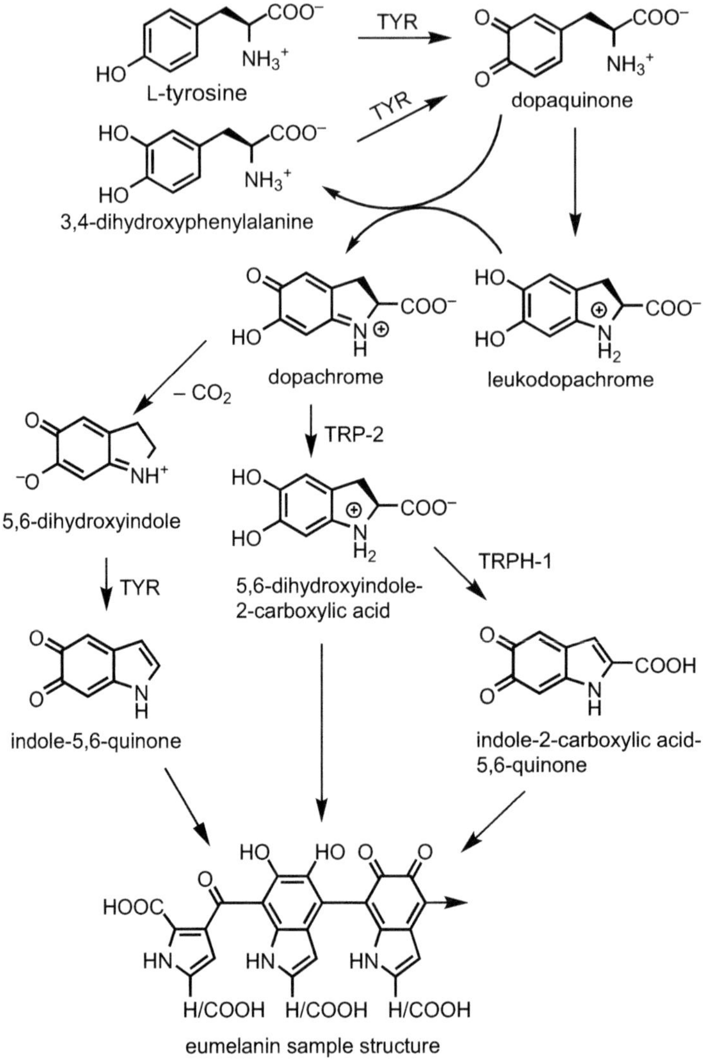

Figure 5.9 Biosynthesis of eumelanin. TYR: tyrosinase; TRP: tyrosinase related protein.[18] Adapted from https://commons.wikimedia.org/wiki/File:Eumelanine.svg#/media/File:Eumelanine.svg, licensed as CC0 by Roland Mattern.

1° alcohols, and not phenols. Protection of the phenol hydroxyl group, followed by acetylation and deprotection would be required for chemical acylation, adding two steps to the process. This would count against Principle 8, Avoid Protecting Groups. Thus, the immobilized lipase provides much greener access to this skin-illuminating ingredient.

The health effects of tyrosinase inhibitors are unclear. On one hand, melanin protects skin from UV radiation, which can cause melanoma cancers, so inhibition of TYR leaves skin cells unprotected.[18] On the other hand, tyrosinase inhibitors have potential therapeutic value against existing melanoma cancer cells.[19] Staying out of the sun and tanning salons while using skin-illuminating lotions is probably advisable.

5.8 Selecting Conventional Solvents

The Innovative Medicines Institute (IMI)-CHEM21 is a group of six major European pharmaceutical companies and ten universities that have joined together to educate chemistry students in sustainable practices.[20] Four or these pharmaceutical companies, Astra-Zeneca, Glaxo-Smith-Kline (GSK), Pfizer, and Sanofi have evaluated the most commonly used "classical" solvents for three green chemistry issues: health, safety, and environment, on various semi-quantitative and qualitative scales. CHEM21 compared these four tables with a solvent selection guide published by the ACS Green Chemistry Institute Pharmaceutical Roundtable (GCI-PR) and combined them into a single guide (Appendix F).[20] The solvents are divided into categories to aid in choosing the appropriate characteristics for the reaction desired: water, alcohol, ketones, esters, ethers, hydrocarbons, halogenated, aprotic polar, miscellaneous, acids, and amines.[20] Solvents were scored from 1 as the best to 10 as the worst for safety, health, and environment, and these scores were divided into three regions: 1–3 green, 4–6 yellow, and 7–10 red.[20] Hazardous solvents have one red score ≥ 8 or two red scores, problematic solvents have one score $= 7$ or two yellow scores, and recommended solvents are neither of the above, usually having at least two green scores and one yellow score.[20] Hazard ranking statements in Appendix F refer to the Globally Harmonized System (GHS) for materials in commerce (Chapter 3).

By examining the solvent selection guide, a less hazardous classical solvent can be chosen for a particular process (Appendix F).[20] For example, ethanol, ethyl acetate, and acetonitrile were each "recommended" with two out of three green scores and one yellow score. Acetonitrile was ranked "problematic" after discussion, while anisole was the only ether ranked "recommended." Some of the recommended solvents are high boiling, which contributes to their safety ranking, but removing the solvent at higher temperature and lower pressure is energetically costly. The CHEM21 analysis led to a summary table of solvents in order from recommended to highly hazardous (Table 5.4).[20]

Table 5.4 Overall ranking of solvents. Adapted from ref. 20 with permission from the Royal Society of Chemistry.

Recommended	Water, EtOH, i-PrOH, *n*-BuOH, EtOAc, i-PrOAc, *n*-BuOAc, anisole, sulfolane.
Recommended or problematic?	MeOH, *t*-BUOH, benzyl alcohol, ethylene, glycol, acetone, MEK, MIBK, cyclohexanone, MeOAc, AcOH, Ac_2O.
Problematic	Me-THF, heptane, Me-cyclohexane, toluene, xylenes, chlorobenzene, acetonitrile, DMPU, DMSO.
Problematic or hazardous	MTBE, THF, cyclohexane, DCM, formic acid, pyridine.
Hazardous	Diisopropyl ether, 1,4-dioxane, DME, pentane, hexane, DMF, DMac, NMP, methoxy-ethanol, TEA.
Highly Hazardous	Diethyl ether, benzene, chloroform, CCl_4, DCE, nitromethane.

Some rankings could not be completely reconciled, so they are listed in between the more certain rankings with a question mark (Table 5.4). None of the very non-polar hydrocarbons are listed as recommended due to their high environment scores.[20] Yet heptane is safer than hexane because of its higher f.p., and its lower health score due to the neurotoxicity of hexane described earlier in this chapter.

The CHEM21 solvent selection guide does not include rankings for energy considerations nor greenhouse gas emissions for manufacture or removal by evaporation.[20] These are complicated metrics still under investigation. If the product can be crystallized and isolated by filtration, the energy cost need not be considered. If evaporation *is* a consideration, solvents such as acetone, ethanol, ethyl acetate, and methyl acetate are reasonable choices, even though they have lower flash points and are less safe. In the final analysis for a particular process, all factors of reactant solubility, solvent reactivity, the green chemistry rankings, renewable sourcing, recyclability, as well as the method of solvent removal, should be weighed to choose a solvent.

THF is a powerful polar aprotic solvent that is standard for many types of reactions (Figure 5.10). But purification of THF involves hazardous chemicals, and it forms explosive peroxides over time. 2-Methyltetrahydrofuran (MeTHF) has been suggested as a substitute for THF. It can be synthesized from biofeedstocks *via* furfural

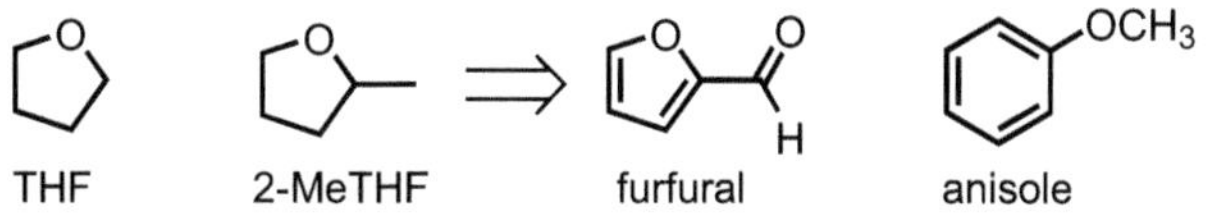

Figure 5.10 Structures of the ether solvents, THF, 2-MeTHF derived from furfural, and anisole.

(Figure 5.10). MeTHF has a higher flash point than THF, and it has been approved as a gasoline additive. MeTHF is still prone to peroxide formation, and also requires drying with pyrophoric solids for use in water-sensitive reactions, so it is classified as problematic. MeTHF separates cleanly from water, so no additional non-polar solvent is required during aqueous extractions. More energy is required to evaporate MeTHF than THF because it is higher boiling. MeTHF is quite expensive currently. Anisole (methoxybenzene) may be a substitute for THF, but it is also high boiling and expensive, and currently produced from petroleum (Figure 5.10).

CHEM21 also analyzed the characteristics of less common solvents to assess the safety, health and environment scores, including some renewable solvents.[20] Some of these results are surprising—glycerol, D-limonene, and cyrene have low health effects scores, but they have high environmental scores (7), and they were classified as hazardous.[20] Glycerol has similar properties to ethylene glycol (Table 5.1). Glycerol is a high boiling triol that is obtained as a by-product from biodiesel production. Because it is high boiling and oxidized, glycerol is much less flammable than traditional solvents. It scored very well on safety and health (1), but it scored high on environment (7). In general, solvents remain undesirable, if not on safety, health, or environment, because they are wasteful.

5.9 Greener Substitutes for Solvents

5.9.1 Renewable Solvents

Most commodity chemicals, including solvents, are produced and separated very efficiently for the market in petroleum refineries; commodity chemicals have the lowest E-factor as a class (Table 2.2). Biorefineries have adopted the efficiencies of petroleum refineries based on biofeedstocks (Chapter 7). Instead of petroleum, feedstocks from plants, agricultural waste such as corn stover, or municipal waste are used to prepare a wide variety of products derived from cellulose, sugars, lignin, and oils (Figure 5.11). Europe, China, India, and the US are all establishing biorefineries.[21–24]

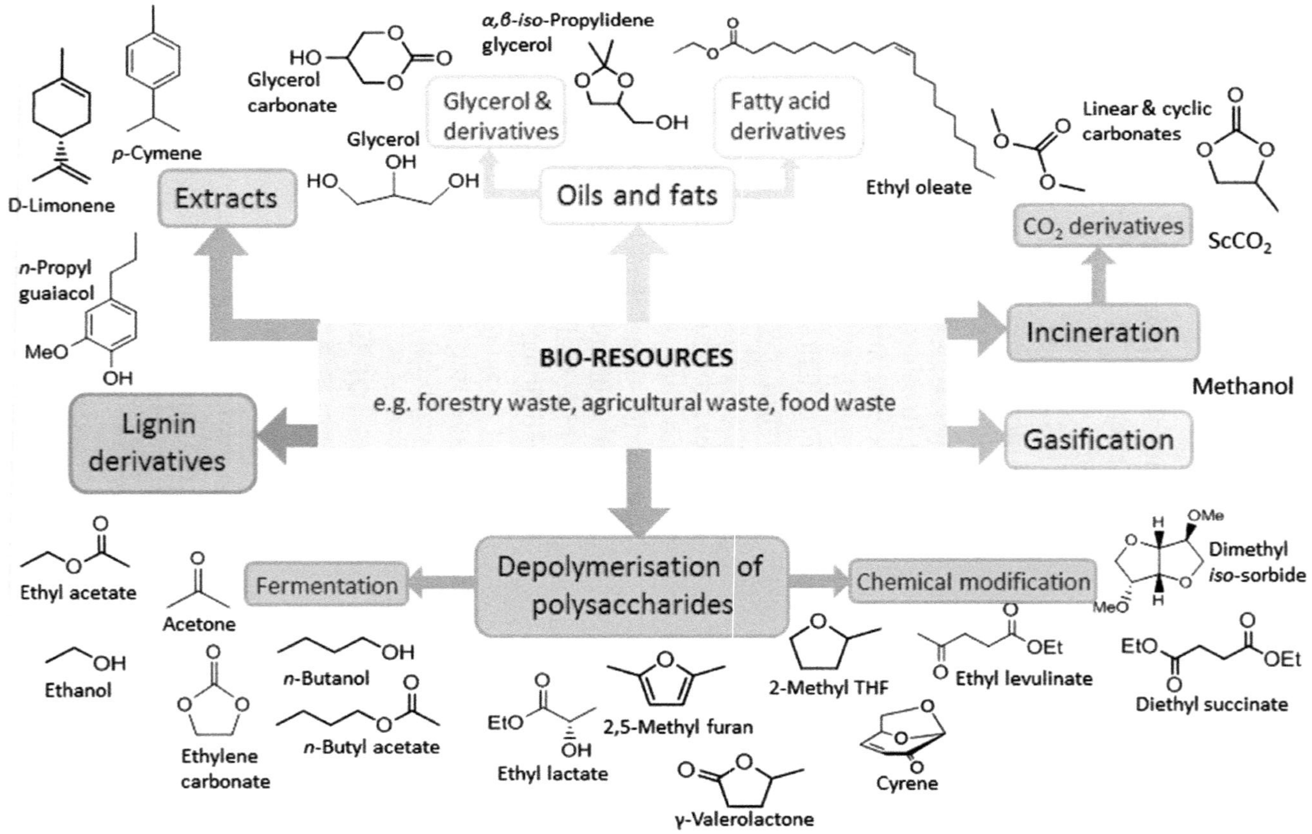

Figure 5.11 Renewable solvents derived from a variety of bio-sources. Reprinted with permission from CHEM21 http://learning.chem21.eu/solvents/sol/alternative-solvents/bio-derived-solvents/.

Bio-based solvents are included among the products from plant sources by fermentation at biorefineries, including methanol, ethanol, and acetic acid (Figure 5.11). EtOAc can be made from bioethanol and bio-acetic acid. MeTHF can be prepared from furfural obtained from glucose or cellulosic biomass. More unconventional solvents from biorefineries, like ethyl oleate and ethyl lactate, are gaining popularity in green industrial processes. These two are safer than more conventional solvents because they are high boiling and less flammable. Bio-based waxes, like the *n*-octadecane solvent used in the QD Vision process (Chapter 3), can be derived from fats and oils by reduction.

5.9.2 Water as a Solvent

Water is the safest solvent of all: non-flammable, non-toxic, and highly renewable, on our planet anyway. The properties that make it suitable for life are high polarity, hydrogen bonding, and the fact that solid ice is less dense than liquid water. Unfortunately, water is also highly reactive as an acid or base—it ionizes easily into H^+ and OH^-, and it can be oxidized to H_2O_2. Much of organic and inorganic synthesis was developed to exclude water during chemical reactions because of its reactivity. Yet, the biochemical processes of life evolved to run the same types of chemical reactions by excluding water in elegantly green ways. Enzyme active sites often exclude water by binding substrates with high selectivity in hydrophobic binding pockets. Enzymes may make use of specific bound water molecules in catalyzing the reaction. Besides its reactivity, water can also be a poor choice as a solvent for green chemistry because of its high boiling point. The energy to evaporate water is considerable; in the laboratory, water is usually lyophilized, as described near the beginning of this chapter.

5.9.3 Reactions in Water

Water can serve as a solvent for reactions in which none of the reactants nor the catalyst is water sensitive. Development of basic organic (non-metal) catalysts has resulted in a variety of reactions that can be done with water as the solvent.[25] Rare earth, and earth-abundant, base-metal catalysts that are stable in water have also been developed.[25]

Font *et al.* developed a heterogeneous proline catalyst prepared by attaching 4-azidoproline to a Wang polystyrene resin *via* a Huisgen click

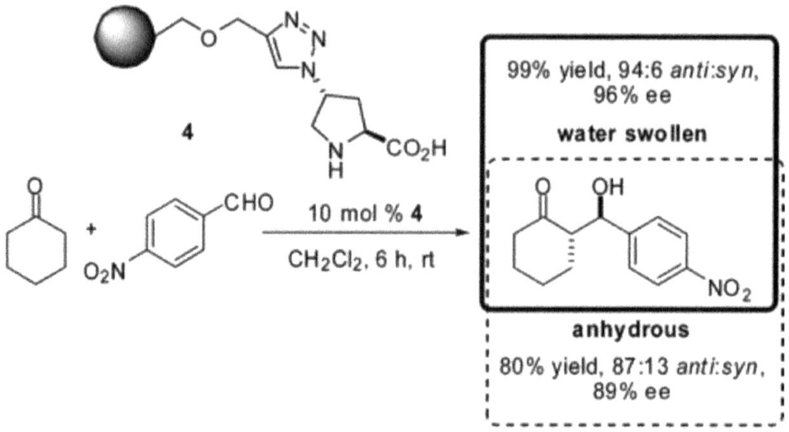

Figure 5.12 Heterogeneous proline-polystyrene catalyst for high-yielding, high-stereoselectivity aldol reactions with water. Adapted with permission of American Chemical Society from ref. 26, Copyright 2008.

reaction (Figure 5.12).[26] Better yields and stereoselectivity were obtained with a multi-phasic reaction by including water to swell the resin, and allowing access of the polar substrates to the catalyst attached to the resin.[26] Notice, though, that DCM is used as the main solvent.

5.10 Case Study: Water-based Acrylic-alkyd Technology (Sherwin-Williams Co.)

Sherwin-Williams Co. won a 2011 PGCC Award for a new water-based paint.[27] Alkyd (oil-based) paint is a concern for both the painters and the building occupants because the solvents are normally volatile, organic, and toxic, called volatile organic compounds (VOCs). Alkyds are polyesters composed of phthalic acids and polyols. In a typical alkyd paint, mineral spirits composed of lighter petroleum-distillate fractions, hydrocarbons of about C_4 to C_8, are used as the solvent. VOCs are also greenhouse gases, albeit short-lived. The C-eq of butane is 7-times greater than CO_2, but much lower than methane (21 C-eq) because methane is more stable. Even attempts to move toward greener coatings, such as water-based polyurethane, has resulted in the use of *N*-methylpyrrolidinone (NMP) (Table 5.1) as a co-solvent with water. NMP is a known teratogen (fetal toxin), and ranked "hazardous" in Appendix F. At the time of the award, most water-based paints were acrylics with properties that were not as high quality as the oil-based alkyds.

Sherwin-Williams created an acrylic-alkyd composite paint that is a water dispersion instead of petroleum solvent-based (Figure 5.13).[28] Soybean oil fatty acids, such as linolenic, linoleic, and oleic acid, were added to improve the biofeedstock content, as well as the flexibility, gloss, and cure time (Figure 5.13).[27] The polymer also includes polyethylene terephthalate (PET) from recycled plastic bottles. The PET segments confer rigidity, hardness, and resistance to hydrolysis by water.[27] A third component is an acrylic polymer,[27] typically a hydroxyethylmethacrylate with hydrogen-bonding capability (Figure 5.13). The polymer is synthesized by reacting the fatty acids with a polyol, such as glycerol, and then with a diacid, such as phthalic acid derived from PET, at temperatures up to 193 °C.[28] An acrylate with a free monoalcohol is added, which can react with any of the free acids to create crosslinks between the alkyd and the acrylate polymer chains (Figure 5.13). Any VOCs used in the synthesis of the polymer are 99.9% removed under high vacuum.[28] The polymer is then dispersed in water with the colorants and other components that make it paint.

The major green advantage of the Sherwin-Williams paint is that it is entirely water-based when delivered to the consumer. It is safer in transport, for the end-use painter, and for the consumer. Recycled PET and soybean oil, a biofeedstock, are included.

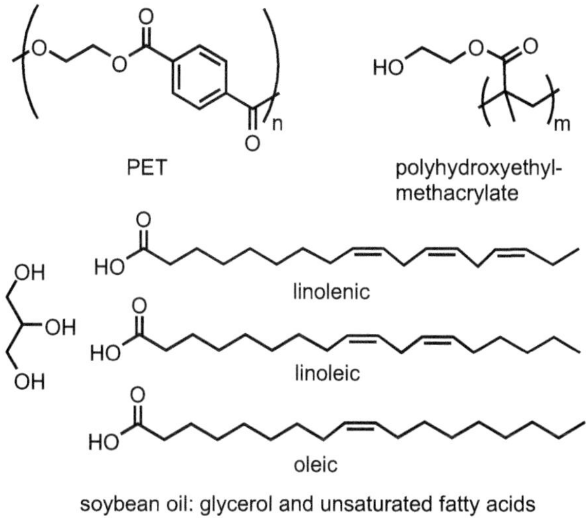

Figure 5.13 Sherwin-Williams acrylic-alkyd water-based paint polymer components.[28]

5.10.1 Supercritical Fluids

Supercritical fluids are very promising greener solvent substitutes. At temperatures and pressures above a level particular to each compound, the compound is no longer a solid, nor yet a liquid, existing in a state called a supercritical fluid. While ordinary liquids have narrow ranges of diffusivity, viscosity, and solvation ability, supercritical fluid properties can be varied greatly by mechanical changes in pressure and temperature. Supercritical fluids have greater diffusivity, lower viscosity, and negligible surface tension compared with normal liquids. Supercritical fluids have already found a multitude of applications, for example as reaction solvents, in dry cleaning fluids, and in chromatography. The supercritical fluid most studied for greener applications is carbon dioxide. Supercritical carbon dioxide ($scCO_2$) is the most useful because the critical temperature and pressure are easily attainable ($T_c = 31$ °C, $P_c = 73$ atm).[29] Supercritical water exists under conditions that are less practical ($T_c = 374$ °C, $P_c = 218$ atm).

5.10.2 Supercritical Carbon Dioxide ($scCO_2$)

Both vilified and essential, carbon dioxide occupies a unique niche in the realm of solvents. CO_2 in the atmosphere is the major driver of global climate change, yet it is the ultimate source of carbon for all life on earth. Harnessing CO_2 as an inert, non-toxic, readily evaporated solvent for chemical reactions is a green chemistry solution to the problem of solvent auxiliaries. Normally, carbon dioxide is a gas, or condensed as a solid (dry ice) below its sublimation temperature −78.5 °C. At atmospheric pressure, CO_2 sublimes directly from a solid to a gas, never becoming a liquid. At higher pressure and temperature, CO_2 exists in a supercritical fluid state. The phase diagram for CO_2 shows the lowest temperature ($T_c = 31$ °C) and pressure ($P_c = 73$ atm) at which CO_2 exists as a supercritical fluid, called the critical point (Figure 5.14).[29] An early success was the use of $scCO_2$ to extract caffeine from coffee, which has made the use of organic solvents to produce decaffeinated coffee obsolete.

5.10.3 Reactions in $scCO_2$

Supercritical CO_2 has been used as the solvent in several types chemical reactions, including radical, decarbonylation, and hydrogenation.[31–34] Hydrogenation is particularly advantageous because $scCO_2$ and H_2 are totally miscible.[35]

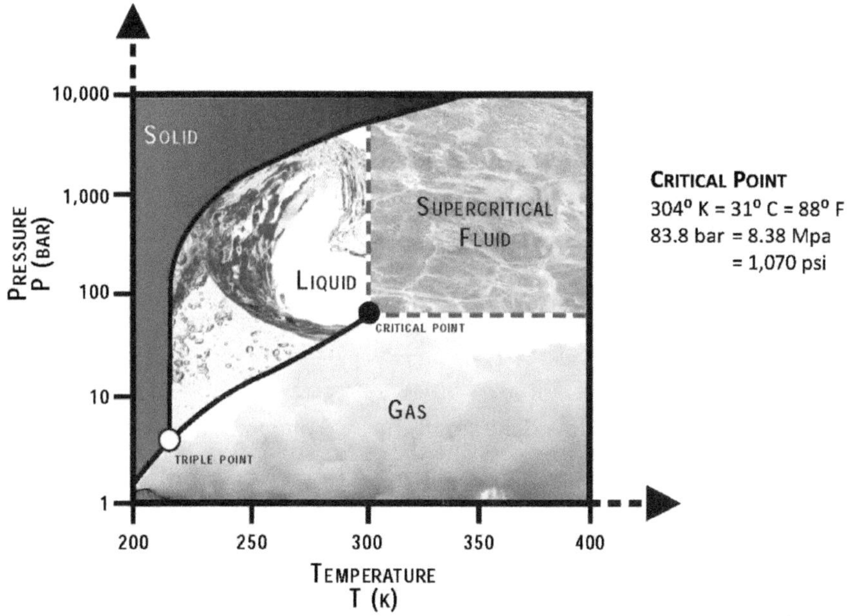

Figure 5.14 Carbon dioxide phase diagram, showing the region above the critical point where CO_2 exists as a supercritical fluid. Reprinted with permission of US Department of Energy from ref. 30, Copyright 2019.

5.11 Case Study: Hydrogenation of Isophorone in $scCO_2$ (Nottingham-Swan)

An academic-industrial partnership was formed by the University of Nottingham and Thomas Swan & Co. in 1995 to develop a multi-reaction, supercritical, continuous-flow reactor at production scale.[36] The team won first prize as the Royal Society of Chemistry (RSC) 2002 Industrial Innovation Team for "developing production scale chemistry in supercritical carbon dioxide."[36]

Continuous-flow reactors are amenable to the use of heterogeneous catalysts in a fixed bed. They work well with $scCO_2$ because simply releasing the pressure allows the solvent to evaporate (Figure 5.15). The $scCO_2$ solvent can be recovered and recompressed for reuse. The system is quite versatile for a variety of hydrogenations at multi-tonne-per-year scale production. Thomas Swan & Co. designed and built a production scale continuous-flow plant that is readily switched between reaction types (Figure 5.15).

The Nottingham-Swan team chose to develop the hydrogenation of isophorone to trimethylcyclohexanone (TMCH) in $scCO_2$

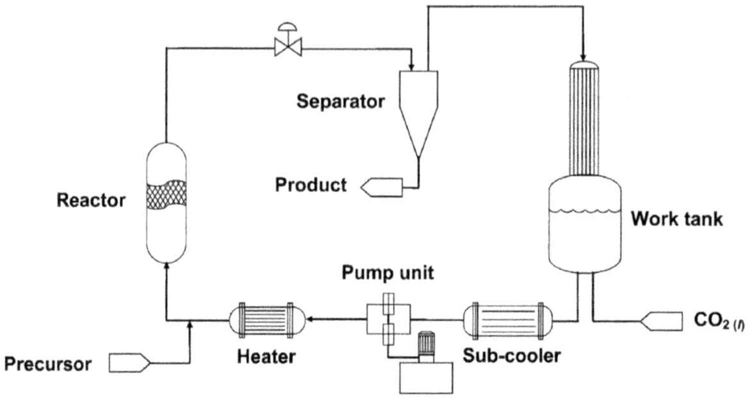

Figure 5.15 Schematic flow diagram of the SCF plant at Thomas Swan & Co. Ltd., constructed by Chematur Engineering. Reproduced from ref. 36 with permission from the Royal Society of Chemistry.

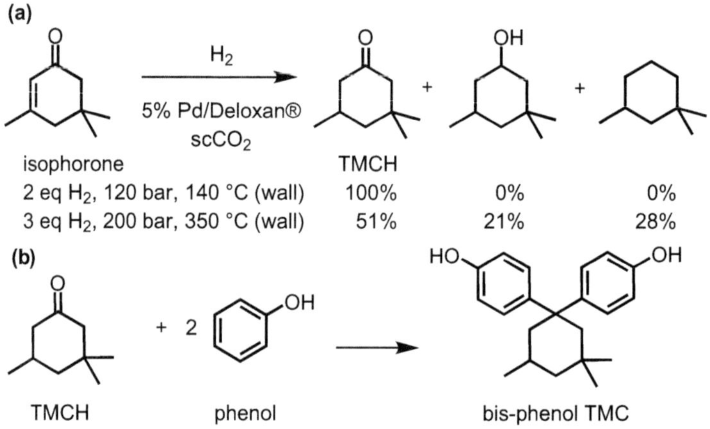

Figure 5.16 (a) Nottingham-Swan production of trimethylcyclohexanone (TMCH) by reduction of isophorone in $scCO_2$ in a continuous-flow reactor.[37] (b) Bis-phenol TMC synthesis.[39]

(Figure 5.16a).[37] Isophorone is a trimer of acetone, produced by an aldol condensation with KOH.[38] TMCH is a fine chemical of commercial value to prepare an analog of bis-phenol A in the production of polycarbonates (Figure 5.16b).[39] TMCH is required to be of very high purity by the commercial end-users.[37] Two side products of hydrogenation, trimethylcyclohexanol, and trimethylcyclohexane result from over-reduction (Figure 5.16a).[37] The catalyst was optimized for the ratio of the mass of Pd to mass of product; the polyaminosiloxane support, Dexolan®, could be replaced by a number of other supports.[36] Optimal production of TMCH was obtained at a catalyst

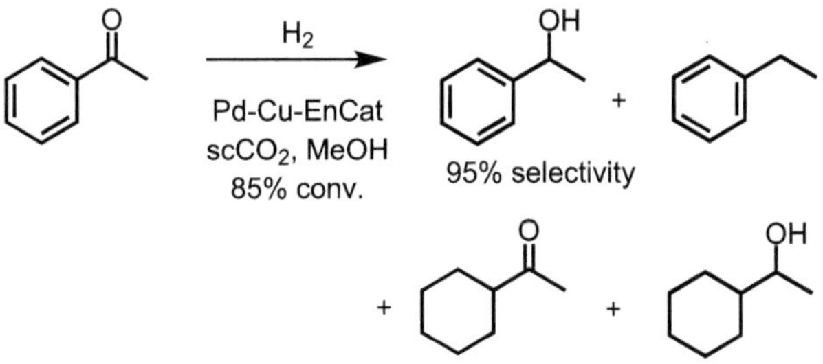

Figure 5.17 Reduction of acetophenone to give 1-phenethanol in $scCO_2$.[40]

loading of 2% Pd/support.[36] The isophorone : hydrogen : $scCO_2$ system did not need to be monophasic; optimal production was at about 50% isophorone in a biphasic flow mixture.[36] The team found that over-reduction could be eliminated by careful control of the substrate-to-hydrogen ratio, pressure, and temperature of the reactor (Figure 5.16a).[37] With 100% yield of the desired TMCH product, purification steps were completely eliminated.[36]

Recently, these principles have been applied to the hydrogenation of acetophenone in $scCO_2$ to make high-purity, pharmaceutical-grade 1-phenethanol, a strawberry flavoring agent (Figure 5.17).[40] To prevent contamination of the product with heavy metal catalysts leached from the resin, More *et al.* developed a polyurea-encapsulated bi-metallic catalyst, Pd-Cu-EnCat.[40] Hydrogenation (20 bar H_2) of acetophenone with Pd-Cu-EnCat (0.015 $g\,cm^{-3}$) gave 85% conversion with 95% selectivity for 1-phenethanol in $scCO_2$ (80 bar total) with MeOH as the co-solvent (Figure 5.17).[40]

5.11.1 Surfactants for $scCO_2$ in Dry Cleaning

Since the 1930s, the primary organic solvent used for dry cleaning clothing has been tetrachloroethylene, called "perc" for perchloroethylene. Prior to that, flammable solvents such as gasoline were used, and perc was considered a great advance in safety. Dry cleaning still uses perc today, although it is being phased out. EPA has classified tetrachloroethylene as "likely to be carcinogenic to humans."[41] Therefore, alternatives to perc for dry cleaning are highly sought after.

The EPA hazard summary says,

> Effects resulting from acute (short term) high-level inhalation exposure of humans to tetrachloroethylene include irritation of the upper respiratory tract and eyes, kidney dysfunction, and neurological effects such as reversible mood and behavioral changes, impairment of coordination,

> dizziness, headache, sleepiness, and unconsciousness. The primary effects from chronic (long term) inhalation exposure are neurological, including impaired cognitive and motor neurobehavioral performance. Tetrachloroethylene exposure may also cause adverse effects in the kidney, liver, immune system and hematologic system, and on development and reproduction. Studies of people exposed in the workplace have found associations with several types of cancer including bladder cancer, non-Hodgkin lymphoma, [and] multiple myeloma.[41]

The potential for practical applications of $scCO_2$ was realized in research from the DeSimone and Lacroix-Desmazes laboratories involving the use of perfluorinated surfactants. In 2001, they patented a method for dry cleaning with $scCO_2$ and a perfluorinated polymer surfactant.[42,43] Perfluorinated hydrocarbons have a strong affinity for $scCO_2$. The surfactants were synthesized by a nitroxide-mediated radical polymerization (NMRP) (Figure 5.18).[43] The initiator for the polymerization was azobis(isobutyronitrile) (AIBN), and the radical mediator was (2,2,6,6-tetramethylpiperidin-1-yl)oxyl (TEMPO). The block copolymer has regions that are CO_2-phobic (polystyrene PS), labeled "p", and regions that are CO_2-philic (perfluoroalkylstyrene PFS), labeled "q" in Figure 5.18.

Surfactants are molecules like soap and detergents with charged head groups ($-COO^-$ and $-SO_3^-$), and long non-polar alkyl tails. In water, surfactants form micelles with the non-polar tails aggregating inside a roughly spherical shape with the charged or polar head groups on the surface to solubilize the micelle. Surfactants decrease

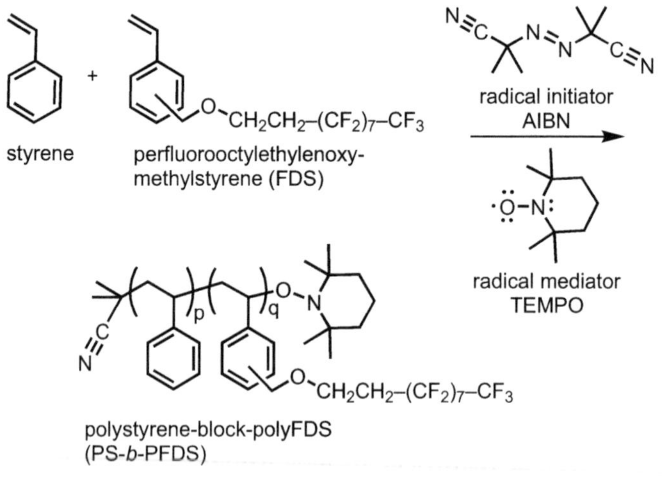

Figure 5.18 Synthesis of a perfluorinated polymer surfactant for $scCO_2$ dry cleaning by nitroxide-mediated radical polymerization.[43]

the surface tension of water to allow dissolution of grease and dirt particles.

In $scCO_2$, surfactants are completely different. Hydrocarbon chains are completely insoluble, while perfluorinated alkyl chains are soluble in $scCO_2$. Micelles form with polystyrene p-blocks on the interior, and PFS q-blocks on the exterior solubilizing the micelle in $scCO_2$ (Figure 5.18). In the Lacroix-Desmazes work, the block copolymer with a q/p block ratio of 16.6 was found to be soluble at 4 wt wt^{-1}% in $scCO_2$ at 380 bar, and 25–65 °C (Figure 5.18).[43] Some of the fluorinated surfactants were shown to be useful in $scCO_2$ dry cleaning and have been commercialized. The expensive infrastructure to set up an $scCO_2$ dry cleaner requires a large enough population (~100 000 people) to support it (J.M. DeSimone, personal communication).

5.11.2 Supercritical Fluid Chromatography (SFC)

Since chromatography is one of the largest uses of organic solvents (typically EtOAc, hexanes, acetonitrile, and methanol), $scCO_2$ has been proposed as a greener solvent for separations in supercritical fluid chromatography (SFC). Petrochemical solvents are flammable, non-renewable, and toxic, while carbon dioxide is inert, renewable, and non-toxic. In addition, CO_2 is easily recycled by evaporation just by decreasing the pressure.

The solubility properties of $scCO_2$ are very different from water or organic solvents. Less polar molecules are more soluble in $scCO_2$, as in organic solvents, but perfluorinated alkanes are the most soluble in $scCO_2$. Because $scCO_2$ is more compressible than ordinary solvents, solubility can be finely tuned by changes in pressure or temperature.

LC with solid adsorbents is already typically done under high pressures (HPLC), so the transition to using $scCO_2$ for chromatography is fairly straightforward. Separation of medicinal compounds by SFC using modified commercially available instrumentation, was demonstrated as early as 1985.[44,45] Most medicines are fairly polar compounds, and have low solubility in $scCO_2$.[46] Thus, eluent systems for SFC are typically modified with methanol, a very polar solvent that is renewable, but flammable (Table 5.1).

SFC initially found greatest use in chiral separations. Today it is being used in all phases of pharmaceutical development: analytical separations, pharmacology, toxicology, and preparative separations. Using SFC for preparative separations gives the most dramatic savings in solvent and energy use.

Astra-Zeneca scaled up the separation of a synthetic intermediate to kilogram quantities in a research-scale laboratory built to accommodate only multi-gram separations (Figure 5.19).[47] The intermediate has only the weak amide and ester UV chromophores, so it was advantageous to monitor the purification by mass spectrometry (MS) (Figure 5.19a). Normal phase (NP)-HPLC gave reasonably good separation, but the retention time was 7.0 minutes (Figure 5.19b). Reverse-phase (RP)-HPLC did not retain the compound enough to separate it sufficiently; it came out too fast at 0.2 min (Figure 5.19c). The same column used for normal phase LC was used for SFC. SFC gave excellent separation, with sharper peaks than NP-HPLC and a much shorter retention time of 2.0 min (Figure 5.19d). The purification could be done on an 800 g scale instead of a 250 g scale with the same small column, with minimal use of flammable solvent (10% 2-propanol/scCO_2), and much faster than the NP-HPLC.[47] The SFC resulted in a much greener process overall.

5.11.3 Ionic Liquids

Ionic liquids are popular solvent substitutes in green chemistry. They are much safer than typical organic solvents because of their high flash points. The most common, commercially available, ionic liquids are *N*-methyl-*N′*-alkyl imidazolium salts of various anions (Figure 5.20). The upper limit of melting temperature for a salt to be considered an ionic liquid is set by convention to 100 °C. Imidazolium ionic liquids are reported to have the best melting point and stability characteristics.[48] Room temperature ionic liquids have been useful in a wide variety of processes: organic synthesis,[49] cellulose processing,[50] nuclear fuel reprocessing,[51] and as electrolytes in less flammable lithium ion batteries.[48] Many examples have shown that ionic liquids are readily separated from the desired product of a process. However, most ionic liquids are hygroscopic (water absorbing), a major disadvantage for recycling.

5.11.4 Toxicology of Ionic Liquids

The environmental toxicology of ionic liquids has been studied in a variety of assays.[52] One of the most commonly used ionic liquids is [bmim][Cl] (Table 5.5). The flash points of imidazolium ILs are typically greater than 100 °C (Table 5.5). The LD_{50} of few ionic liquids have been reported. Toxicity depends as much on the negative counter-ion as on the imidazolium. For example,

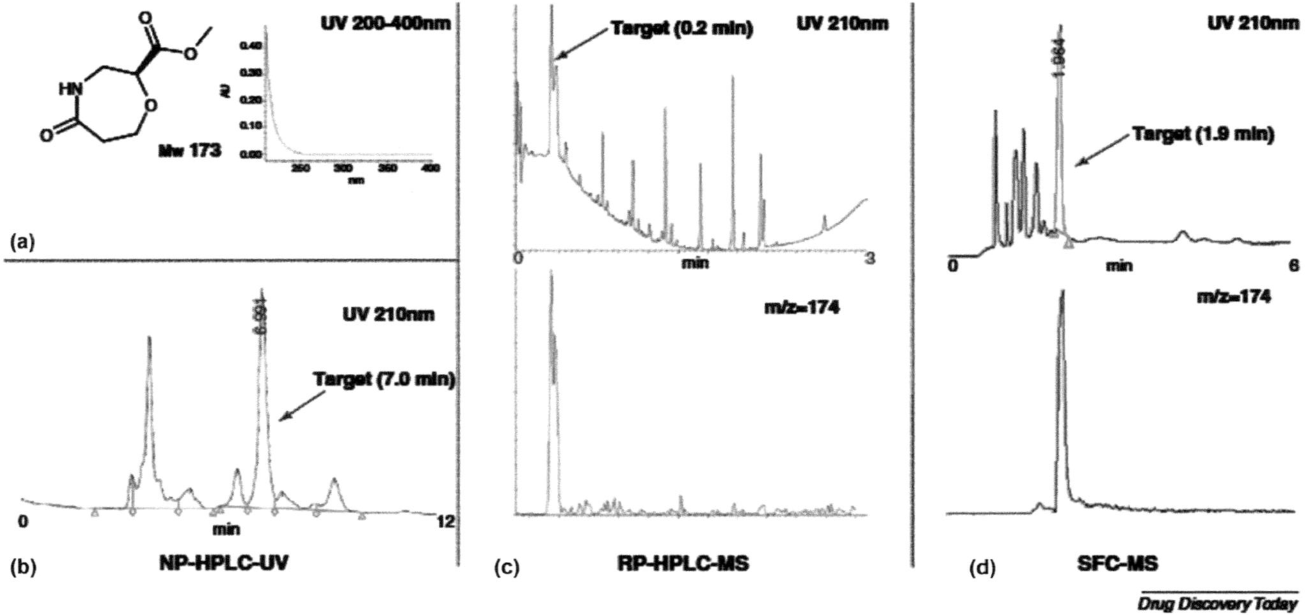

Figure 5.19 (a) Structure of target in the crude product and the UV spectrum (200–400 nm). (b) Normal-phase liquid chromatography (NP-HPLC)-UV, (c) Reverse-phase chromatography (RP-HPLC)-UV and -MS, (d) Supercritical fluid chromatography (SFC)-UV and -MS. Abbreviation: Mw, molecular weight (measured in Da). Reprinted with permission of Elsevier from ref. 47, Copyright 2014.

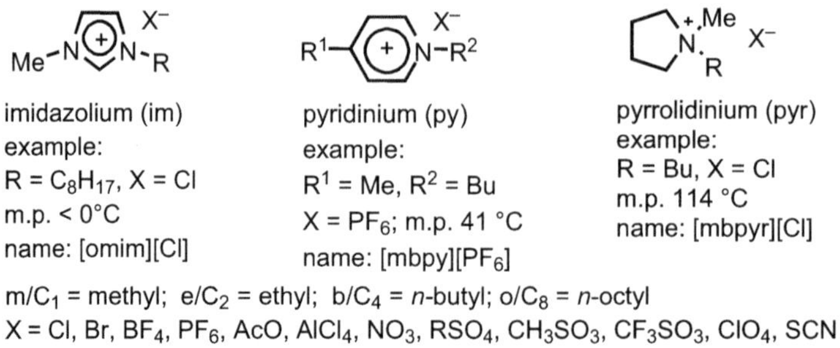

Figure 5.20 Commonly used ionic liquid structures and nomenclature.

Table 5.5 Some imidazolium ionic liquids, melting points, and LD_{50} values. Data are from Sigma Aldrich safety data sheets (SDS), and ref. 53.

Ionic Liquid	m.p. (C)	f.p. (C)	LD_{50} (mg kg^{-1} body wt)
[bmim][Cl]	41°	192°	>50 >300 (female mice) >2000 (oral, rat)
[emim][BF_4]	NA[a]	113°	>300–2000 (oral, rat)
[bmim][$MeSO_3$]	75–80°	119°	NA[a]
[cholinium][AcO]	80°	NA[a]	MIC[b] >1.5 M *P. brevicompactum*

[a]NA = not available.
[b]MIC = minimum inhibitory concentration.

hexafluorophosphate (PF_6^-) and tetrafluoroborate (BF_4^-) are unstable towards hydrolysis, creating HF as a by-product. In general, ionic liquids have low mammalian toxicities by ingestion. Developmental abnormalities in mice have been reported, and there is cause for concern about toxicity to aquatic species.[52] A major advantage of ILs is negligible vapor pressure, resulting in virtually non-existent inhalation toxicity, which is a major problem for volatile organic solvents.

Choline is a natural quaternary ammonium ethanol that is found in neurotransmission. A study of cholinium alkanoates demonstrated very low toxicities, ranging from a minimum inhibitory concentration (MIC) of 1 mM to 1.5 M in fungal microorganisms (Figure 5.21). These fungal toxicities were directly related to mammalian toxicity (Figure 5.21).[53] Toxicities correlate directly with the length of the alkanoate chain (Figure 5.21).[53] ILs represent a case where the scientific community is trying to approach the use of a potentially greener method with eyes wide open. The entire life cycle analysis of ILs should be considered in comparison with volatile organic solvents in choosing for a particular process. Again, if possible, the best solvent for a reaction is no solvent.

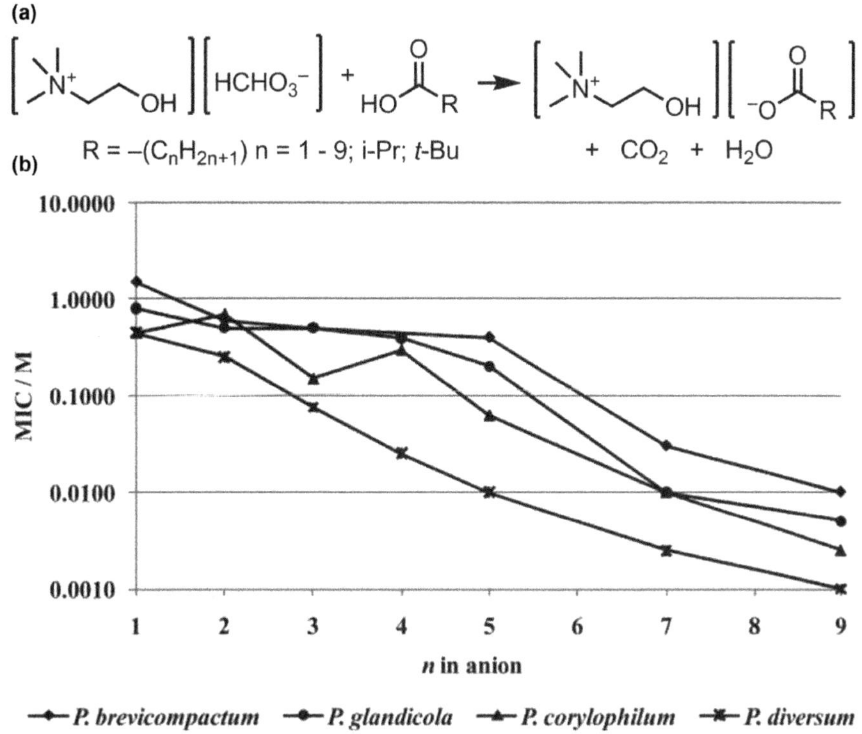

Figure 5.21 (a) Cholinium alkanoate ionic liquids synthesis. (b) MIC *vs* anion length for cholinium alkanoates in four fungal species. (a) Adapted from, and (b) reproduced from ref. 53 with permission from the Royal Society of Chemistry.

5.11.5 Deep Eutectic Solvents

Deep eutectic solvents (DESs) are mixtures of two compounds—one is usually ionic and the other a hydrogen bond donor—that melts at a much lower temperature than either compound alone. Both the ionic choline chloride and the hydrogen-bonding glycerol are renewable and biocompatible. When combined into DES choline chloride–glycerol (DES-Ch12), the DES is a room temperature liquid with excellent solvation properties, and it remains as non-toxic as the two components. DES-Ch12 was found to efficiently extract lignins to pretreat cellulosic biomass.[54] Lignins are cross-linked polyphenolic compounds that are very resistant to degradation (Chapter 7). Downstream, a mixture of DES-Ch12 (4.5 g) in 1 M citric acid buffer (40 mL) has a biocompatible pH of 5 that allows efficient growth of *Saccharomyces cerevisiae* yeast for fermentation of the remaining polysaccharides to produce ethanol.[54] Use of biocompatible DES-Ch12 permitted the entire multi-step conversion of

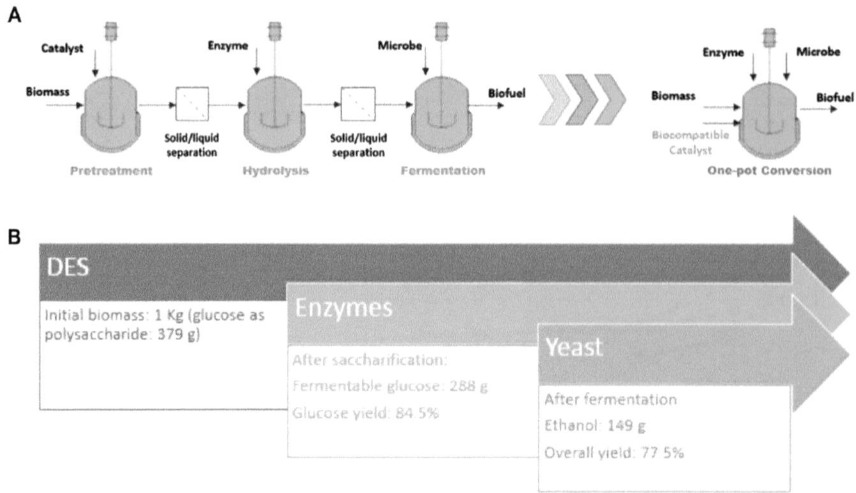

Figure 5.22 (A) One-pot bioethanol production with biocompatible deep eutectic solvents (DES-Ch12) and (B) glucose balance of the one-pot bioethanol conversion. Reprinted with permission of American Chemical Society from ref. 54,Copyright 2018.

corn stover to ethanol to be performed in a single reactor (Figure 5.22).[54] Thus, DES-Ch12 made from biofeedstocks is used as an enzyme-catalyst-compatible solvent to make biofuels in a one-pot process!

5.12 Summary

Auxiliaries are chemicals other than reactants or catalysts required for a reaction or purification. Solvents, extraction chemicals, and chromatography adsorbents are all auxiliaries that should be minimized in greener chemical processes. Solvent characteristics to be considered in choosing a greener solvent are: (1) flammability, (2) toxicity, (3) reactivity, (4) environmental degradability, and (5) energy required for evaporation (boiling point). Auxiliaries should be minimized or eliminated to minimize risk. High-speed-ball-milling and twin-screw extrusion have been used to eliminate solvents in both inorganic and organic reactions. Solvent-free chemistry is the safest, greenest choice. Eastman eliminated solvent from a process to make cosmetic esters with immobilized enzymes. Greener substitutes for solvents are renewable solvents, water, supercritical CO_2, and ionic liquids. Water can be a greener solvent in many cases, though it is energy intensive to evaporate. Supercritical CO_2 has been used as a solvent in diverse chemical reactions—hydrogenation of isophorone was described as an example. Astra-Zeneca scaled up pharmaceutical separations using $scCO_2$. Ionic liquids based on choline are non-toxic, room temperature liquids that have been used as solvents. The deep eutectic solvent, choline chloride–glycerol, was an excellent solvent for processing biomass into ethanol.

5.13 Problems: Avoid Auxiliaries

Problem 5.1

Selecting a greener solvent.

(a) Compare the characteristics of EtOAc, DCM, DMF, THF, and MeTHF in Tables 5.1 and Appendix F. Sum the scores for each of these solvents in Appendix F.
(b) Which of these solvents has the lowest overall score?

____

(c) Choose one of these as a solvent for an amino acid coupling reaction. The reactants are very polar (R-NH_2 and R′-COOH). All four solvents are polar enough to dissolve the reactants. The solvent must be unreactive, be readily removed, and easily disposed of at the end of the reaction.
(d) Which parameter in Appendix F did you weigh most heavily? Justify your choice.

Problem 5.2

Auxiliaries in context.

(a) What part of a synthesis typically uses the most auxiliaries: reaction, work-up, or purification?
(b) Which part of a synthesis uses the most aqueous solutions?
(c) Which type of purification method uses the least amounts of auxiliaries: chromatography, distillation, or recrystallization? Justify your choice.

Problem 5.3

Consider the Fischer esterification (Figure 5.5a).

(a) Draw the curved arrow mechanism for an HCl catalyzed esterification. Include lone pairs on all reacting atoms. Include reversible arrows for mechanistic steps and square brackets for high-energy intermediates as necessary.

(b) How is the reaction driven to completion?

Problem 5.4

Linoleic acid is an ω-6 fatty acid, meaning the unsaturated positions start at the 6th carbon from the end.

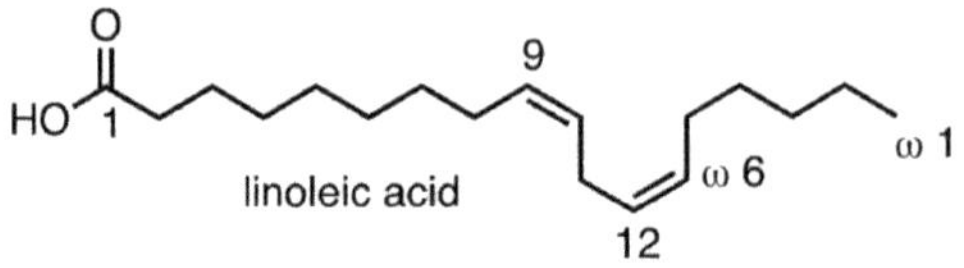

(a) Which positions (numbered from carbon 1, not ω-1) of linoleic acid are most likely to react with atmospheric oxygen?

____

(b) Draw the resonance structures of the radical species resulting from hydrogen atom (H·) abstraction by O_2 at the single most reactive allylic carbon in C9 through C13 of linoleic acid. (Draw a 5-carbon piece. Add H's and R groups.)

(c) Draw the MOs and the energy level diagram of the same radical species as in part b). Include the number of π-nodes and the energy arrow on the left as in Figure 5.8. (Hint: with 5 carbons, 5 atomic p-orbitals give 5 π-MOs.) Add the correct number of π electrons to the five energy levels next to each of the MOs of the radical. Label the MOs bonding (π), non-bonding (*n*), or anti-bonding (π*).

Problem 5.5

Esterification of unsaturated fatty acids with strong acid causes double bond rearrangements in addition to oxidized products.

(a) Draw a curved arrow mechanism for isomerization of *cis*-2-butene *via* strong acid protonation of a double bond, followed by deprotonation to give a new alkene. (Hint: isomerization to the more stable alkene is thermodynamically favored.)

(b) Draw the structure of the most likely rearranged product of linoleic acid (Problem 5.4). Why is this bad for human health?

Problem 5.6

Consider the tyrosinase inhibitors of Table 5.3 and the eumelanin biosynthesis of Figure 5.9.

(a) What class of chemical reaction does tyrosinase catalyze?

____

(b) Draw three resonance structures of enzymatic 4-HBA acetate (Table 5.3).

(c) Based on the resonance structures, which is more electron withdrawing, the $-CH_2OH$ or $-CH_2OAc$ group? ____________________

(d) Compare the EC_{50} values of chemical and enzymatic 4-HBA acetate. Which inhibits the tyrosinase reaction better, the chemical or enzymatic 4-HBA acetate? Which mimics the tyrosine substrate better?

Problem 5.7

Give the number and short name for the two most significant Principles of Green Chemistry embodied by the Eastman esterification case study. Explain briefly.

Problem 5.8

Consider the alkyd-acrylic water-based paint of Sherwin-Williams (Figure 5.13).

(a) Draw a glycerol cross-link between different strands of PET.

(b) Draw a mechanism of transesterification between glycerol and PET.

Problem 5.9

Describe three advantages of continuous-flow hydrogenations performed in $scCO_2$.

Problem 5.10

Consider the ionic liquid, [bmim][Cl] (Figure 5.21).

(a) Draw three resonance structures of the aromatic ring of [bmim][Cl]. Put a star by the greatest resonance contributor(s) and explain briefly.

(b) Draw the π-type MO diagram with energy levels and 6 π-electrons for the aromatic ring of [bmim][Cl]. (Hint: aromatic MOs are different from linear molecule MOs. See Appendix B for the benzene MO diagram.)
(c) Give two structural reasons why [bmim][Cl] is a low m.p. ionic liquid (41 °C).

References

1. P. T. Anastas and J. C. Warner, *Green Chemistry: Theory and Practice*, Oxford University Press, New York, 1998.
2. Agency for Toxic Substances and Disease Registry (ATSDR), *Toxic Substances Portal*, U.S. Department of Health and Human Services, Atlanta, GA, http://www.atsdr.cdc.gov/substances/index.asp, accessed July 23, 2019, 2014.
3. J. V. Bruckner, S. S. Anand and D. A. Warrren, in *Casarett & Doull's Toxicology: The Basic Science of Poisons*, ed. C. D. Klaassen, McGraw Hill Education, NY, 8th edn, 2013, pp. 1031–1112.
4. M. S. Reich, *Chem. Eng. News*, 2017, **95**, 11.
5. C. H. Arnaud, *Chem. Eng. News*, 2015, **93**, 10–13.
6. S. L. James, C. J. Adams, C. Bolm, D. Braga, P. Collier, T. Friscic, F. Grepioni, K. D. M. Harris, G. Hyett, W. Jones, A. Krebs, J. Mack, L. Maini, A. G. Orpen, I. P. Parkin, W. C. Shearouse, J. W. Steed and D. C. Waddell, *Chem. Soc. Rev.*, 2012, **41**, 413–447.
7. T. E. Long and M. O. Hunt, *Solvent-Free Polymerizations and Processes*, American Chemical Society, Washington, DC, 1999.
8. C. Suryanarayana, *Prog. Mater. Sci.*, 2001, **46**, 1–184.
9. J. Mack, D. Fulmer, S. Stofel and N. Santos, *Green Chem.*, 2007, **9**, 1041–1043.
10. L. R. Chen, B. E. Lemma, J. S. Rich and J. Mack, *Green Chem.*, 2014, **16**, 1101–1103.
11. D. E. Crawford, C. K. G. Miskimmin, A. B. Albadarin, G. Walker and S. L. James, *Green Chem.*, 2017, **19**, 1507–1518.
12. Eastman Chemical Co., *A Solvent-free Biocatalytic Process for Cosmetic & Personal Care Ingredients*, Washington, DC, 2009, https://www.epa.gov/greenchemistry/presidential-green-chemistry-challenge-2009-greener-synthetic-pathways-award, accessed July 23, 2019.
13. E. Fischer and A. Speier, *Chem. Ber.*, 1895, **28**, 3252–3258.
14. U. Hanefeld, *Chem. Soc. Rev.*, 2013, **42**, 6308–6321.
15. K. E. LeJeune and A. J. Russell, *Enzyme-containing polyurethanes*, Agentase, LLC, Pittsburgh, PA (US), *U. S. Pat.*, US6759220B1, 2004.
16. S. Bhardwaj, S. J. Passi and A. Misra, *Diabetes Metab. Syndr.: Clin. Res. Rev.*, 2011, **5**, 161–164.
17. D. Mozaffarian, M. B. Katan, A. Ascherio, M. J. Stampfer and W. C. Willett, *N. Engl. J. Med.*, 2006, **354**, 1601–1613.
18. T.-S. Chang, *Int. J. Mol. Sci.*, 2009, **10**, 2440–2475.
19. S. K. Kudugunti, N. M. Vad, A. J. Whiteside, B. U. Naik, M. A. Yusuf, K. S. Srivenugopal and M. Y. Moridani, *Chem.-Biol. Interact.*, 2010, **188**, 1–14.
20. D. Prat, J. Hayler and A. Wells, *Green Chem.*, 2014, **16**, 4546–4551.
21. T. Tan, F. Shang and X. Zhang, *Biotechnol. Adv.*, 2010, **28**, 543–555.

22. P. D. Sankar, M. A. A. M. Saleh, C. I. Selvaraj, V. Palanichamy and R. Mathew, *Res. Biotechnol.*, 2013, **4**, 26–35.
23. B. I. Consortium, *Mapping European Biorefineries*, 2017, https://biconsortium.eu/news/mapping-european-biorefineries, accessed February 8, 2019.
24. US Department of Energy, *Integrated Biorefineries*, 2019, https://www.energy.gov/eere/bioenergy/integrated-biorefineries, accessed February 8, 2019.
25. T. Kitanosono, K. Masuda, P. Xu and S. Kobayashi, *Chem. Rev.*, 2018, **118**, 679–746.
26. D. Font, S. Sayalero, A. Bastero, C. Jimeno and M. A. Pericàs, *Org. Lett.*, 2008, **10**, 337–340.
27. Sherwin-Williams Co., *Water-based Acrylic Alkyd Technology*, Washington, DC, 2011, https://www.epa.gov/greenchemistry/presidential-green-chemistry-challenge-2011-designing-greener-chemicals-award, accessed July 25, 2019.
28. K. A. Koglin, J. K. Marlow, P. J. Ruhoff and R. F. Tomko, *Low VOC Aqueous Polymer Dispersions*, Sherwin-Williams Co., *U. S. Pat.*, US20100160586A1, 2010.
29. National Institute of Standards and Technology, *Carbon Dioxide*, 2011, http://webbook.nist.gov/cgi/cbook.cgi?ID=C124389&Mask=4, accessed July 23, 2019.
30. Supercritical CO_2 Tech Team, *Clean Coal & Natural Gas Power Systems*, US Department of Energy, 2019, https://www.energy.gov/supercritical-co2-tech-team, accessed June 24, 2019.
31. C. M. Wai, F. Hunt, M. Ji and X. Chen, *J. Chem. Educ.*, 1998, **75**, 1641.
32. R. Pacut, M. L. Grimm, G. A. Kraus and J. M. Tanko, *Tetrahedron Lett.*, 2001, 1415–1418.
33. S. Mayadevi, *Indian J. Chem., Sect. A: Inorg., Bio-inorg., Phys., Theor. Anal. Chem.*, 2012, **51A**, 1298–1305.
34. M. Chatterjee, T. Ishizaka and H. Kawanami, *Green Chem.*, 2018, **20**, 2345–2355.
35. C. Y. Tsang and W. B. Streett, *Chem. Eng. Sci.*, 1981, **36**, 993–1000.
36. P. Licence, J. Ke, M. Sokolova, S. K. Ross and M. Poliakoff, *Green Chem.*, 2003, **5**, 99–104.
37. M. G. Hitzler, F. R. Smail, S. K. Ross and M. Poliakoff, *Org. Process Res. Dev.*, 1998, **2**, 137–146.
38. H. Siegel and M. Eggersdorfer, *Ketones*, Weinheim, Wiley-VCH, 2005.
39. V. Serini, *Polycarbonates*, Weinheim, Wiley-VCH, 2000.
40. S. R. More and G. D. Yadav, *ACS Omega*, 2018, **3**, 7124–7132.
41. *Tetrachloroethylene (Perchloroethylene)*, U.S. Environmental Protection Agency, 2012, https://www3.epa.gov/airtoxics/hlthef/tet-ethy.html, accessed March 13, 2016.
42. J. B. McClain, T. J. Romack, J. P. DeYoung, R. B. Lienhart, J. M. DeSimone and K. L. Huggins, *Methods for carbon dioxide dry cleaning with integrated distribution*, MiCell Technologies Inc, *U. S. Pat.*, US6248136B1, 2001.
43. P. Lacroix-Desmazes, P. Andre, J. M. Desimone, A. V. Ruzette and B. Boutevin, *J. Polym. Sci., Part A: Polym. Chem.*, 2004, **42**, 3537–3552.
44. J. B. Crowther and J. D. Henion, *Anal. Chem.*, 1985, **57**, 2711–2716.
45. L. J. Mulcahey and L. T. Taylor, *J. High Resolut. Chromatogr.*, 1990, **13**, 393–396.
46. D. Suleiman, L. A. Estevez, J. C. Pulido, J. E. Garcia and C. Mojica, *J. Chem. Eng. Data*, 2005, **50**, 1234–1241.
47. M. A. Lindskog, H. Nelander, A. C. Jonson and T. Halvarsson, *Drug Discovery Today*, 2014, **19**, 1607–1612.
48. M. Armand, F. Endres, D. R. MacFarlane, H. Ohno and B. Scrosati, *Nat. Mater.*, 2009, **8**, 621–629.

49. T. Welton, *Chem. Rev.*, 1999, **99**, 2071–2084.
50. R. P. Swatloski, S. K. Spear, J. D. Holbrey and R. D. Rogers, *J. Am. Chem. Soc.*, 2002, **124**, 4974–4975.
51. P. Giridhar, K. A. Venkatesan, T. G. Srinivasan and P. R. V. Rao, *Electrochim. Acta*, 2007, **52**, 3006–3012.
52. T. P. Thuy Pham, C.-W. Cho and Y.-S. Yun, *Water Res.*, 2010, **44**, 352–372.
53. M. Petkovic, J. L. Ferguson, H. Q. N. Gunaratne, R. Ferreira, M. C. Leitao, K. R. Seddon, L. P. N. Rebelo and C. Silva-Pereira, *Green Chem.*, 2010, **12**, 643–649.
54. F. Xu, J. Sun, M. Wehrs, K. H. Kim, S. S. Rau, A. M. Chan, B. A. Simmons, A. Mukhopadhyay and S. Singh, *ACS Sustainable Chem. Eng.*, 2018, **6**, 8914–8919.

6 Energy Efficiency

"Principle 6: Energy requirements should be minimized, and processes should be conducted at ambient temperature and pressure whenever possible."[1]

6.1 Energy

Abundant energy drives our modern, comfortable lifestyles. Yet the devastating global effects of burning fossil fuels lead to the 6th Principle: Energy Efficiency. Burning coal produces particulates, which cause asthma and cancer, and releases mercury, which causes neurological and developmental damage, toxic arsenic, and sulfur dioxide, which leads to acid rain. Production of gasoline and diesel

Green Chemistry: Principles and Case Studies
By Felicia A. Etzkorn

Published by the Royal Society of Chemistry, www.rsc.org

from petroleum entails great risk of spills in the environment, such as the Exon spill in Valdez, Alaska in 1989, and the British Petroleum Deepwater Horizon spill in the Gulf of Mexico in 2010. All fossil fuel combustion, including oil, natural gas, and coal, produces carbon dioxide leading to climate change. Climate change describes an overall warming of the planet, but with dramatically different effects geographically dispersed around the globe. For example, the eastern US was cooler than average in 2014, while the western US suffered much warmer temperatures and drought. US President George H. W. Bush said, "America is addicted to oil." Just as consumers are withdrawing from fossil fuels, so must the chemical enterprises. We now address energy efficiency, primarily in chemical *processes*, but also in energy *products*, such as batteries, biofuels, transformer fluids, and solar photovoltaics.

Minimizing energy use in chemical processes includes conservation, cogeneration, renewable energy, and catalysts, an "all of the above" approach. Conservation means using ambient temperatures and pressures for reactions as much as possible, and ensuring efficient transfer of heat or cooling to the reaction. Efficient heat transfer can be accomplished using microwaves and flow reactors. Energy savings can also be achieved by eliminating unnecessary synthetic steps, as we saw in Chapter 2, Synthetic Efficiency, or by reducing purification steps, as we saw in Chapter 5. Cogeneration uses waste chemical or energy by-products as fuel to run the process, or energy to heat, cool, or light the production plant.

Renewable energy sources, such as solar, wind, hydroelectric, and geothermal, are well known. Direct use of solar energy *via* photochemical reactions is another significant greener direction. Other potential uses of renewable energy might include direct mechanical stirring or agitation with water, wind, or wave energy, and direct heating for reactions and distillations by concentrating heat from the sun.

Chemists also have a great deal to contribute to energy-related products: biofuels, photovoltaic materials, batteries, and transformers. Amory Lovins, the Rocky Mountain Institute physicist, said, "Two-thirds of the energy used to move a typical car is caused by its weight."[2] Thus, chemists can help save energy by designing strong, lightweight materials for cars, trucks, trains, and planes. Whether these alternatives are used in chemical manufacturing, transportation, or consumer products, these more efficient energy products contribute to decreasing our global use of fossil fuel energy to fulfill the intent of Principle 6.

6.2 Conservation of Energy

6.2.1 Kinetics and Thermodynamics

The use of energy in chemical processes is often necessary because we want reactions to be fast, and high temperature increases the rate of most reactions. Traditionally, if a reaction was too slow because the barrier ($\Delta G^{\ddagger}$) is high, it was heated to increase the rate of the reaction. A good catalyst lowers the barrier that leads to the desired product without itself being consumed in the chemical reaction. Much of modern chemistry has been devoted to finding more selective catalysts that give single pure products at lower temperatures (Chapter 9).

Another way to increase the rate of a reaction is by creating more reactive substrates. For example, elimination of acetates to give alkenes by an E_i (elimination intramolecular) mechanism is a high temperature process called "pyrolysis," which means to use heat (pyro) to break (lysis) bonds in Latin (Figure 6.1).[3] On the other hand, elimination of sulfonamides by the Burgess reaction can be done near ambient temperature.[4] Both processes give the same desired alkene product, but the Burgess reaction uses less energy for the reaction (Figure 6.1). However, the Burgess reaction is less atom economical, and synthesis of the reactant requires more steps. Thus, the energy required is "pre-paid" in the synthesis of the reactant. In this way, we must examine the entire life cycle of a process to assess the overall energy requirements.

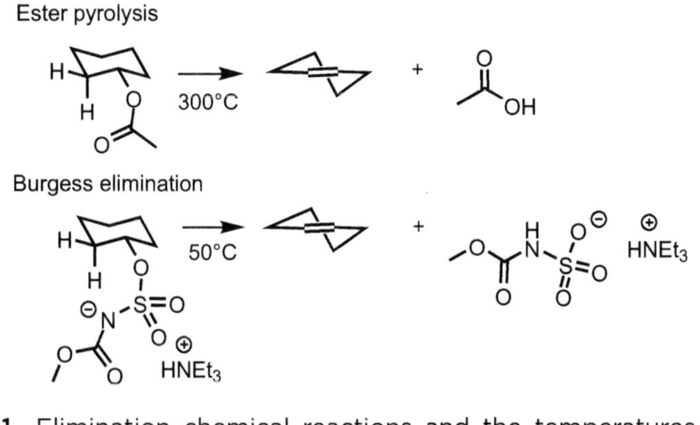

Figure 6.1 Elimination chemical reactions and the temperatures required to give the same product, cyclohexene, using substrates with differing reactivities.

The relationship between temperature and reaction rate (kinetics) is described by the Arrhenius equation:

$$k = Ae^{-E_a/RT}$$

where k is the rate constant, A is a constant for a specific reaction that gives the probability of collision between molecules in the right orientation to react, E_a is the experimental activation energy (rather than the theoretical activation barrier, $\Delta G^{\ddagger}$, which includes the entropy of the reaction), R is the gas constant, and T is the temperature. Rate constants are typically found by measuring the rate of the reaction by the disappearance of a reactant or the appearance of product as a function of temperature. Measuring the disappearance of reactants is preferred to obtain initial rates because their concentration is high at the start of the reaction, and therefore easier to determine than the low concentrations of products. Measuring the rate of appearance of the desired product has the advantage of eliminating error due to side product formation.

The thermodynamics of a reaction characterize the difference in thermodynamic properties such as internal energy, entropy, or Gibbs energy between reagents and products. The ratio of the concentration of products to reactants at equilibrium (K_{eq}) is related to the standard reaction Gibbs energy (ΔG°, also called free energy). Since high yields of a single product are the most economical, chemists often go to great lengths to avoid unfavorable equilibria—even a 90 : 10 ratio of products to reactants is undesirable. Often this is done by raising the energy (activating) the reactant so that the reaction is more "downhill," or thermodynamically favorable. The Burgess elimination is an example of this, in which the energy is pre-paid by activating the alcohol precursor as a sulfonamide (Figure 6.1).

$$\Delta G^{\circ} = -RT \cdot \ln K_{eq}$$

Consider two extremes: kinetic and thermodynamic control of a reaction (Figure 6.2). If product-1 (P_1) is more stable than P_2 as shown in (A), then P_1 would be both the thermodynamically and the kinetically favored product. Under many reaction conditions, the thermodynamic and kinetic product is the same—the lower energy kinetic pathway leads to the more thermodynamically stable product P_1. The reaction can be run at a lower temperature to obtain desired product P_1. The temperature could be ambient, which requires no energy to heat or cool, or a lower temperature, which would require energy to cool the reaction.

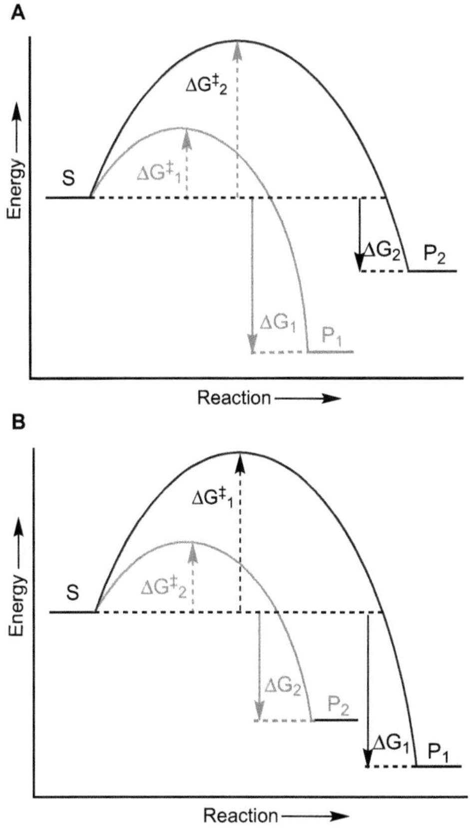

Figure 6.2 Thermodynamic *vs.* kinetic control hypothetical reaction coordinate diagrams: (A) high or low temperature gives P_1, both the kinetic and thermodynamic product; (B) high temperature gives mainly thermodynamic product P_1, low temperature gives mainly the kinetic product P_2.

In Figure 6.2(B), the kinetic (P_1) and thermodynamic (P_2) products are different. If selectivity in a reaction like (B) is necessary to produce the desired product, temperature control over the product ratio may be exerted. Suppose that P_1 is the desired product. Kinetic control can be achieved by running the reaction at a lower temperature to avoid the formation of undesired side product P_2. Low temperature prevents most molecules of substrate (S) from gaining enough energy to surmount the $\Delta G_2^{\ddagger}$ barrier. Thus, P_1 is called the kinetic product. In this case, the temperature should be just high enough to overcome the $\Delta G_1^{\ddagger}$ barrier for the reaction to proceed sufficiently fast, but not so high that the $\Delta G_2^{\ddagger}$ barrier can be crossed (Figure 6.2). Sometimes this entails running reactions at low temperature and for longer periods of time, which is also energy intensive. A standard method of cooling a reaction to −78 °C is easily achieved with a dry ice–acetone bath, but making, transporting, and storing dry ice and acetone also incur energy costs.

Under thermodynamic control, the reaction is heated to a high enough temperature to overcome the barrier to P_2, and allowed to run for a sufficiently long time to fully equilibrate. P_1 will be formed quickly as well because the P_1 barrier is lower. But if the reaction is fully reversible at temperatures high enough to overcome the P_2 barrier, an equilibrium mixture of P_1 and P_2 will form, with P_2 dominating because it is more stable, and lower in energy. Thus, P_2 is called the thermodynamic product. The ratio of products, $K_{eq} = [P_1]/[P_2]$, depends upon the difference in the reaction Gibbs energies of the products $(\Delta G_2^\circ - \Delta G_1^\circ)$. P_1 and P_2 could be *cis* and *trans* alkene isomers resulting from elimination in an open chain alcohol for example. The *trans* isomer is more thermodynamically stable, but the *cis* isomer might be obtained by kinetic control under the right conditions.

A method frequently used to obtain a desired product with an unfavorable equilibrium is to use Le Chatelier's Principle. An excess of a cheap reagent, for example methanol in a Fischer esterification (Chapter 5), will drive the reaction. Alternatively, removal of a product by precipitation or evaporation will also drive the reaction forward to completion.

Chemical methods can also be used to control selectivity for the desired product in the course of a reaction. We can think of this as redirecting the reaction along a different kinetic pathway to specify the product. Elimination of HBr from 2-bromobutane with a small, hard base like potassium ethoxide (KOEt) gives more of the thermodynamically stable 2-butene as the major product.[5] The product ratio can be tilted towards giving a higher yield of less stable 1-butene by using a sterically bulky base, such as lithium tetramethylpiperidide (LiTMP) (Figure 6.3).[6] In this example, steric control is a form of kinetic control. The cost of this selectivity is a higher E-factor, and lower atom economy for the reaction because of the size of the base, the auxiliary 12-crown-4 used to capture the Li^+ counter ion, and THF as the solvent instead of ethanol.[6] Selectivity is critical to obtaining the desired product, and selectivity often comes at a cost of both materials and energy.

6.2.2 Temperature and Pressure

Heating a reaction is a well-established method to increase the rate. As a rough approximation for organic reactions, increasing the

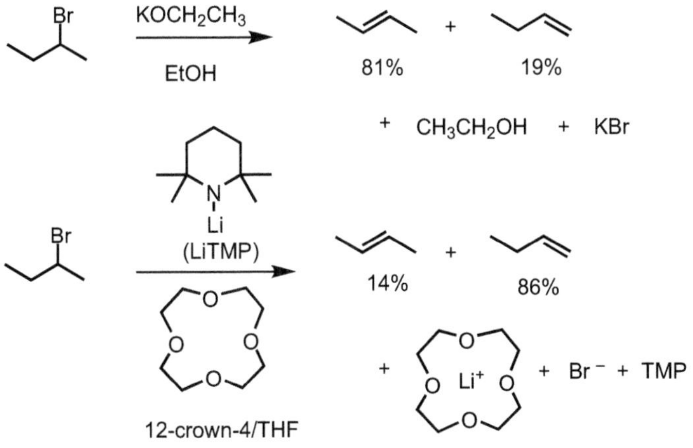

Figure 6.3 Chemical control over the product ratios of elimination reactions.[5,6]

temperature by 10 °C will increase the rate about 2-fold. Unfortunately, heating is energetically costly, but so is stirring a reaction for extended periods of time. The concept of the "space-time yield" of a process was discussed in the Eastman case study (Chapter 5). Optimizing space-time yield means getting the highest yield of the desired product from the smallest possible reaction volume, in the least amount of time possible. Eastman was able to optimize the space-time yield of plant-based esters by using a solid-phase enzyme catalyst in solvent-free esterifications (Chapter 5).

Increasing the pressure above atmospheric pressure can accelerate reactions with more gaseous reactants than gaseous products. A thick-walled reaction vessel can be pressurized with the reactant gas, hydrogen for example, or by sealing the vessel and heating it. These types of vessels, called reaction "bombs," can be pressurized by mechanically compressing the gas phase of the reaction. But the very name "bomb" serves as a caution about the potential for accidents when reactions are run under pressure. Any method for pressurizing a reaction requires energy input. One downside of reactions run in $scCO_2$ as the solvent is that the CO_2 must be pressurized to attain the supercritical fluid state. However, the necessary pressure is easily achieved under ordinary laboratory conditions, as we saw in Chapter 5. The real advantage is that the supercritical temperature is very close to ambient, so there are no sizable heating costs.

Reactions that give the desired structural and stereochemical selectivity at ambient temperature and pressure are most desirable from the perspective of Principle 6. Shorter reaction times are also desirable. Often mechanical stirring is required to maintain a homogeneous reaction mixture, which also requires energy even for room temperature reactions. Other methods for lowering energy inputs while achieving good space-time yields will now be considered.

6.2.3 Eliminating Auxiliaries

Minimizing or eliminating solvent, as recommended by Principle 5, allows smaller reaction volumes, increasing the space-time yield. Smaller reaction vessels also require less energy to heat, cool, or pressurize. Eliminating purification steps also decreases energy input. The QD Vision case study in Chapter 3 was an example of reducing energy use in the production of QLEDs. QD Vision eliminated the purification of the cadmium selenide (CdSe) core, which had to be done in dry boxes using energy for the vacuum pumps. The liquid crystal display products also saved electrical energy because QLEDs produce color displays that are more energy efficient for the consumer.

6.2.4 Catalysts

Catalysts are vitally important in lowering the barrier of a reaction, decreasing the energy needed to obtain the desired product (Chapter 9). In addition, catalysts often lead to greater selectivity and higher yield of the desired product, thus increasing the space-time yield of processes. Enzymes are biological catalysts that can be used in a purified form, as we saw in the case study of Eastman production of plant-based esters. Alternatively, engineered bacteria, such as *E. coli*, can be cultured and used directly as catalytic systems in the production of chemicals. Enzyme catalysts give the kinetically controlled product whether or not it is also the thermodynamic product because they evolve to lower the barrier for a specific product (Chapter 9).

6.3 Case Study: Succinic Acid Through Metabolic Engineering (BioAmber)

BioAmber won a PGCC Award in 2011 for their process to make the commodity chemical and natural product, succinic acid (butanedioic acid).[7] Succinic acid is a biodegradable, 4-carbon dicarboxylic acid used in the production of pharmaceuticals, cosmetics, solvents, and food. Succinate salts can replace corrosive acetates in deicing products. Succinic acid can be used as a replacement for adipic acid (hexanedioic acid) in the production of polyurethanes, or directly as a monomer in polybutylene succinate (PBS), a biodegradable plastic.[7]

BioAmber's process uses biofeedstocks to make succinic acid by a greener synthetic pathway catalyzed by a genetically modified bacteria (Figure 6.4). The process reduces energy consumption by 60% compared to the production of succinic acid from petrochemicals.[7] The cost of BioAmber succinic acid is economically viable even if the cost of oil drops below $40 per barrel; in 2011 this was 40% less than fossil feedstock derived succinic acid.[7]

The former industrial process to synthesize succinic acid began with benzene, a probable carcinogen derived from refined petroleum (Figure 6.5).[8] Petroleum refining is a high-energy process. Hydrogen gas is explosive and difficult to handle safely, under pressure, and on a large scale. Even though it is a relatively short synthesis, these steps are energy intensive and carbon wasteful—two equivalents of CO_2 are produced as waste.

BioAmber uses modified bacteria, whole-cell *E. coli* K12, as the *catalysts* to make succinic acid (Figure 6.4). Pyruvate is the entry point to the Krebs cycle (Figure 6.6).[9] Modification of the bacteria is required to run the Krebs cycle in reverse direction, and to stop halfway through at the succinic acid intermediate.[7] This reverse left half of the Krebs cycle uses a series of reactions: reduction with nicotinamide adenine dinucleotide hydride (NADH), dehydration, and reduction with coenzyme QH_2, that result in succinate under anaerobic

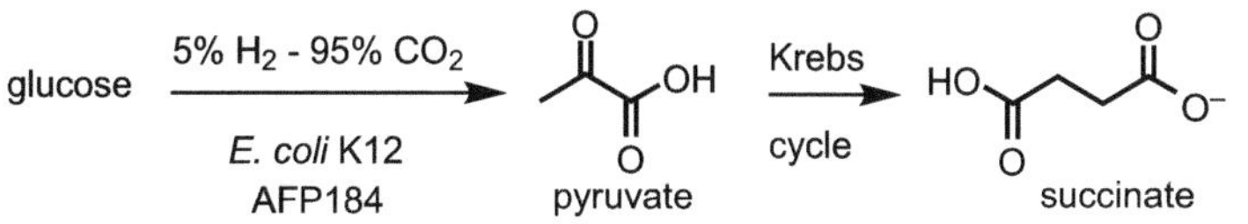

Figure 6.4 BioAmber process to make succinic acid with mutated *E. coli* K12 under anaerobic conditions.[7]

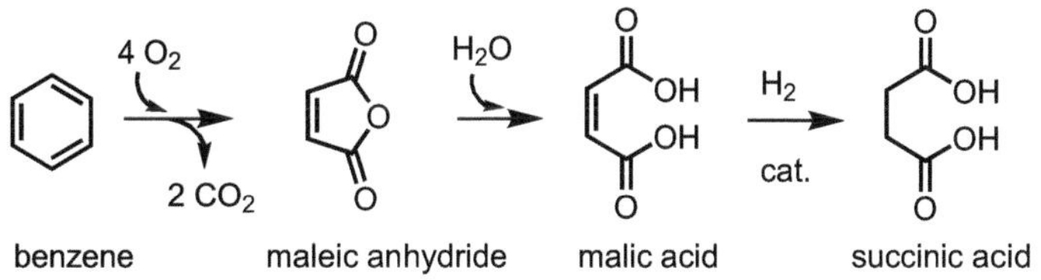

Figure 6.5 Petroleum-based process to make succinic acid.[8]

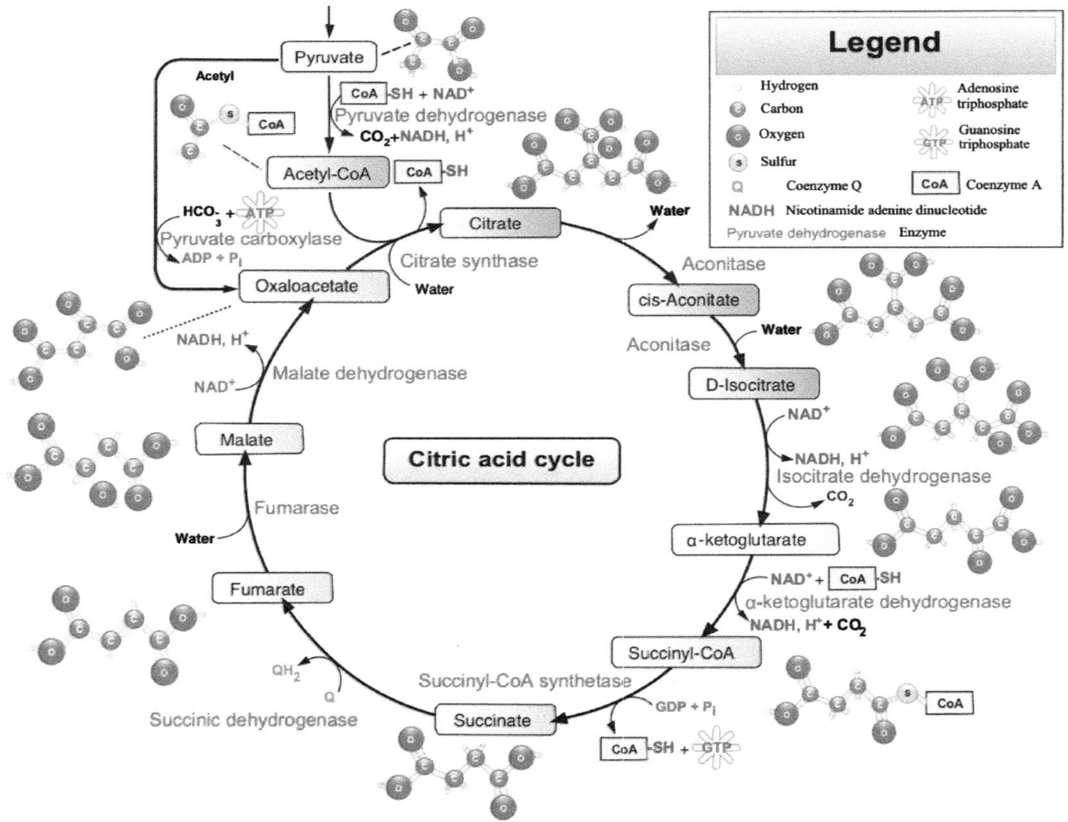

Figure 6.6 In the BioAmber process, the Krebs cycle (also called tricarboxylic acid (TCA) or citric acid cycle) is run in reverse under anaerobic conditions, down the left side, and blocked halfway through to synthesize succinate in bacteria. Reproduced from ref. 9 under the terms of the CC BY-SA 3.0 licence, https://creativecommons.org/licenses/by-sa/3.0.9.

conditions (Figure 6.6).[9] The oxidized form of NADH, NAD^+, has lost a hydride from the nicotinamide ring to form a pyridinium aromatic ring (Figure 6.7). The oxidized form of coenzyme Q-H_2 has lost two hydrogen atoms (H·), net H_2, to make the quinone (Figure 6.7). Each of these reducing agents has very poor atom economy, but the bacteria are able to easily recycle them back to the active reduced forms under the reducing atmosphere conditions.

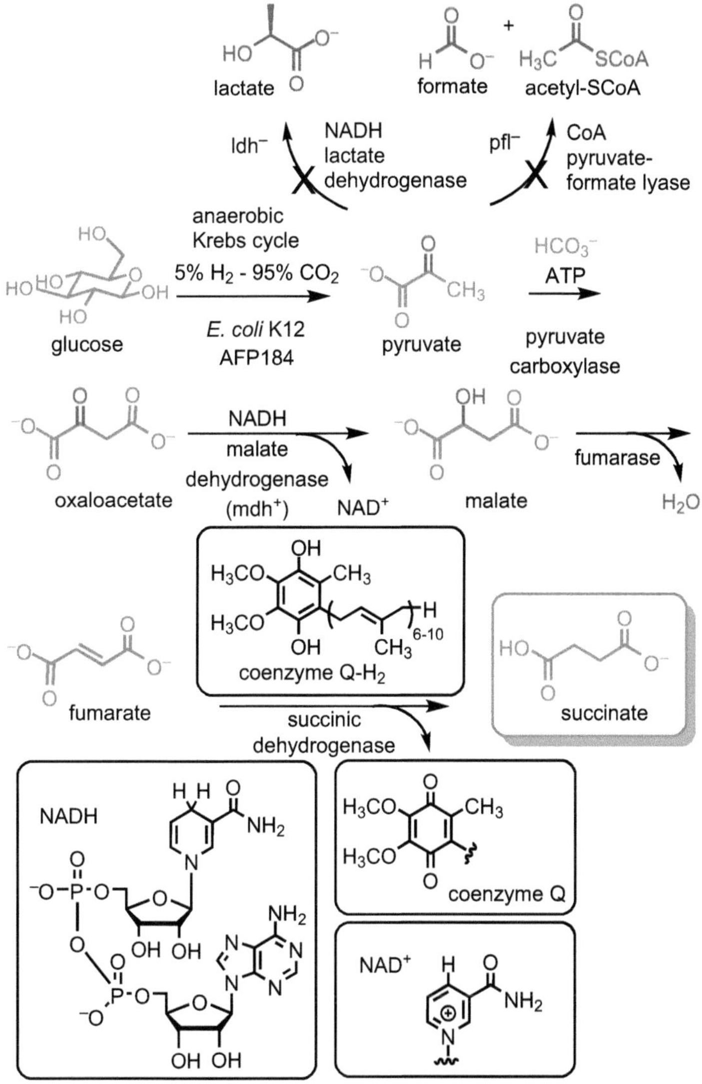

Figure 6.7 BioAmber process to make succinate under anaerobic conditions using whole-cell *E. coli* K12 AFP184 as the catalyst system.[7]

The BioAmber process essentially uses CO_2 and H_2O as feedstock for the bacteria (Figure 6.7).[7] (Note that glucose is also made from CO_2 and H_2O in plants.) This is a carbon-neutral process, using either CO_2 captured from combustion, or directly from the atmosphere. Under anaerobic conditions, *E. coli* normally make formate, lactate, and ethanol from pyruvate (Figure 6.7). Genetic mutations were made to the *E. coli* to deactivate the lactate dehydrogenase (*ldh*) and pyruvate-formate lyase (*pfl*) genes to prevent the formation of lactate, formate, and acetate side-products (Figure 6.7).[7] A gene encoding malate dehydrogenase (*mdh*) was added in to produce malate under anaerobic conditions (Figure 6.7).[7] A spontaneous mutation in the phosphotransferase transporter system (*ptsG*), found by screening, permitted this strain to grow on glucose under anaerobic conditions: 5% H_2/95% CO_2.[7] The final strain was called *E. coli* K12 AFP184, which gave a succinate yield as high as 99%, and a continuous productivity of 0.87 $g\,L^{-1}\,h^{-1}$.[7]

BioAmber scaled up production to a 350 000-liter fermenter. First, the bacteria are grown under aerobic conditions to increase the biomass. Then the culture is transferred to the large-scale anaerobic fermenter under axenic (single strain) conditions to prevent contamination of the culture, as well as to avoid releasing the modified organism into the environment. Purification of the succinate product is an integrated, continuous process, a distinct space-time yield advantage, which also saves energy. Discontinuous purification normally requires stopping the fermentation, chemically precipitating the product, which could kill the organism, and cleaning the fermenter. The product often has to be further purified as well.

BioAmber has used their succinic acid as a monomer in polybutylene succinate (PBS), a biodegradable plastic.[7] BioAmber modified PBS has many improved characteristics: (1) distortion resistance up to 110 °C, (2) the strength of polypropylene or polyvinylchloride (PVC), (3) drop-in extrusion and injection-molding, (4) 90-day compostability, and (5) over 50% renewable content.[7]

Succinic acid can be used to make a wide variety of commodity chemicals. Most of the products shown in Figure 6.8 are 4-carbon products; two have added methyl esters, and one has a methyl amide. These other products are derived from succinate by a variety of oxidation, reduction, amination, methylation, and dehydration reactions. Succinate esters are proposed as replacements for phthalate esters as plasticizers. Phthalate esters are suspect endocrine disrupters. The BioAmber process uses renewable feedstocks, CO_2 and H_2O, an

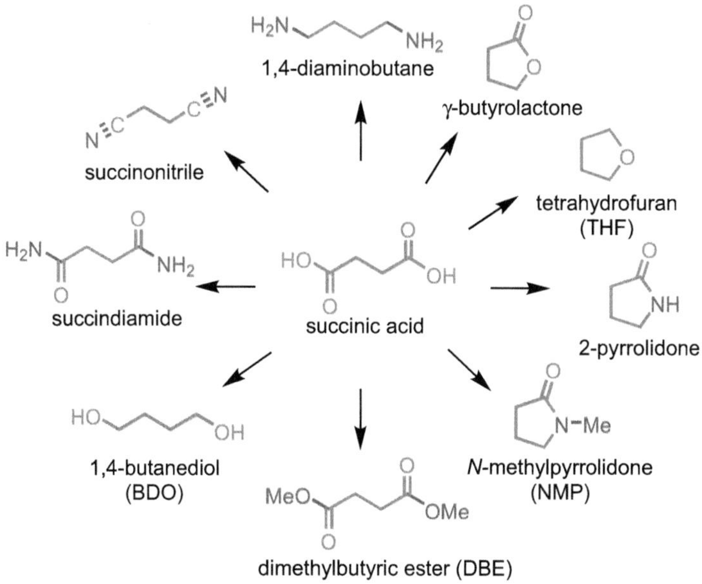

Figure 6.8 Products made from 4-carbon succinic acid.[7]

E. coli biocatalyst system, and water as the solvent, to make commodity chemicals and biodegradable polymers with 60% less energy.[7]

6.4 Microwaves

The use of microwaves to drive reactions can be effective in saving energy. Microwaves heat solvents and reactants by transferring electromagnetic radiation directly into the molecules, while leaving the reactor unheated. This is more energy efficient than traditional heating, since the mass of the glass or steel reactor is never heated more than the contents.

Microwaves use low frequency oscillating electric and magnetic fields. Microwave heating occurs by three mechanisms: (1) polarization of dipoles, (2) ionic conduction, and (3) interfacial polarization, which is less important.[10] Solvent dipoles become polarized in the oscillating electromagnetic field. The dipoles do not quite keep in synch with the frequency of the microwave radiation due to inertia, which results in heat generation.[10] Ionic conduction is the result of the charge carriers (dipolar solvent) moving through the material and inducing electrical currents. Resistance to the electrical currents also causes heating.[10] Microwaves also accelerate chemical reactions by superheating the solvent.[11] Superheating of 4 to 26 °C above the boiling point has been measured for a variety of polar solvents: water, methanol, ethanol, isopropanol, THF, CH_2Cl_2, $CHCl_3$, CH_3CN, and CH_3NO_2.[11] Because of the centimeter-scale penetration depth of microwaves, continuous flow reactors (Chapter 11) that include microwave heating are more effective than microwave batch reactors.[12]

The energy used in conventional *vs.* microwave reactors must be compared on a case-by-case basis, considering all energy used in the processes, including stirring, pumping, *etc.*[12] Microwave reactors are not inherently more energy efficient than conventional heating. The advantage of microwave reactions is more likely the result of the selectivity for desired products as a result of the uniformity of heating, and the higher and more selective temperatures attainable.[12] Typical laboratory-scale microwave reactors are "appallingly energy inefficient," while large-scale optimized magnetrons (industrial scale microwaves) are more promising.[12]

6.5 Case Study: Cellulose Processing by Microwave with an Ionic Liquid (Rogers)

Cellulose is a linear polymer of glucose with tightly packed chains that is resistant to solvation by water and often requires harsh conditions to process. The viscose method, an older way to process cellulose to make rayon, requires toxic and corrosive chemicals: NaOH, CS_2, and H_2SO_4.[13]

A great alternative way to process cellulose is by dissolving it in an ionic liquid using microwave energy. In 2005, the Rogers group at the University of Alabama won the PGCC Award for the use of [bmim]Cl ionic liquid (Chapter 5) and microwave energy to dissolve cellulose up to 25 wt% without covalent modification (Figure 6.9).[14] The chloride ions are thought to break up the hydrogen bonding that gives cellulose its structure. The cellulose can be covalently modified or processed into fibers. The cellulose could be precipitated from the ionic liquid solution by addition of just 1% water, dissolving the chloride ions, and causing the cellulose to form a powdery flocculent.[14] Thin fibers and rods could be formed by extrusion into water.[14] Removing the dissolved water from the ionic liquid is somewhat energy intensive, but in this way the [bmim]Cl can be reused. The toxic chemicals

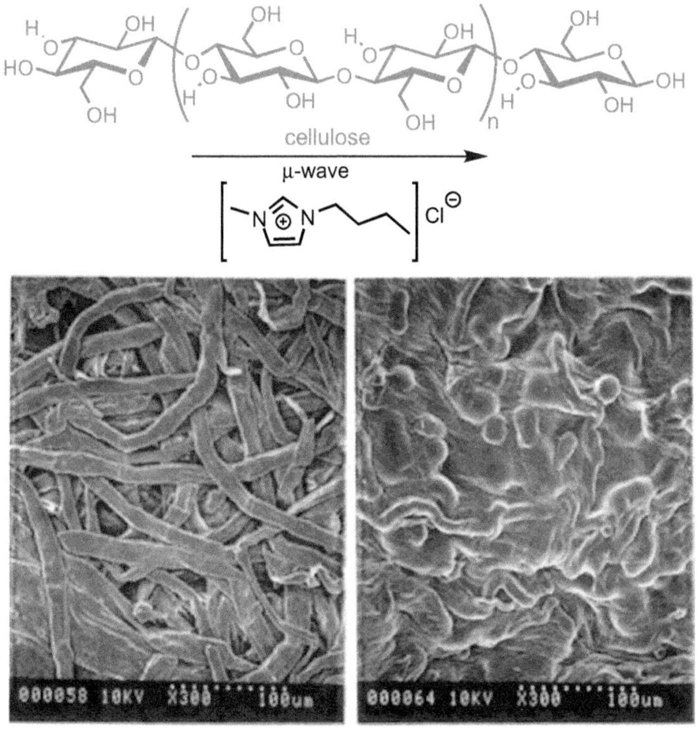

Figure 6.9 Greener process to make rayon from cellulose using the ionic liquid [bmim]Cl and microwave energy. Reprinted with permission of American Chemical Society from ref. 14 Copyright 2002.

of the viscose process can be avoided using an ionic liquid solvent and microwave energy.

6.6 Photochemistry

Energy directly from the sun can be used to drive chemical reactions. For example, photoisomerization of alkenes favors *cis* over *trans* alkenes, while *trans* alkenes are favored in thermodynamic equilibria. Chemical sensing of light by retinal bound to the rhodopsin protein uses this alkene isomerization in vision. Photosynthesis includes several photoreactions. In this sense, biofeedstocks and biofuels from plants use photochemical energy in their synthesis. Most photoreactions require high-energy ultraviolet light, which can be unreliable from the sun, so traditionally, photochemistry has been done with electrically powered UV lights since their invention.

A typical photoreaction is the [2+2] cycloaddition with formation of a 4-membered ring from two π-bonds, usually alkenes (Figure 6.10). Multiple regioisomers and stereoisomers can be obtained from a photocycloaddition. If the alkenes undergo *cis–trans* isomerization under the reaction conditions, even more stereoisomers can result. For example, if *cis–trans* isomerization of the **A** alkene occurs during the reaction and there are three stereocenters in the product, each with two options: (*R*) or (*S*), the number of possible stereoisomeric products is 2^3 or 8. Only the four of these isomers resulting from the *cis*-disubstituted alkene **A** are shown in Figure 6.10. In addition, there are four possible structural isomers in which the R^3 group of reactant **B** is next to R^2 instead of R^1 (Figure 6.10). Again, there would be 8 of these if *cis–trans* isomerization occurred during the reaction. In total, there could be *16 possible products* resulting from this [2+2]

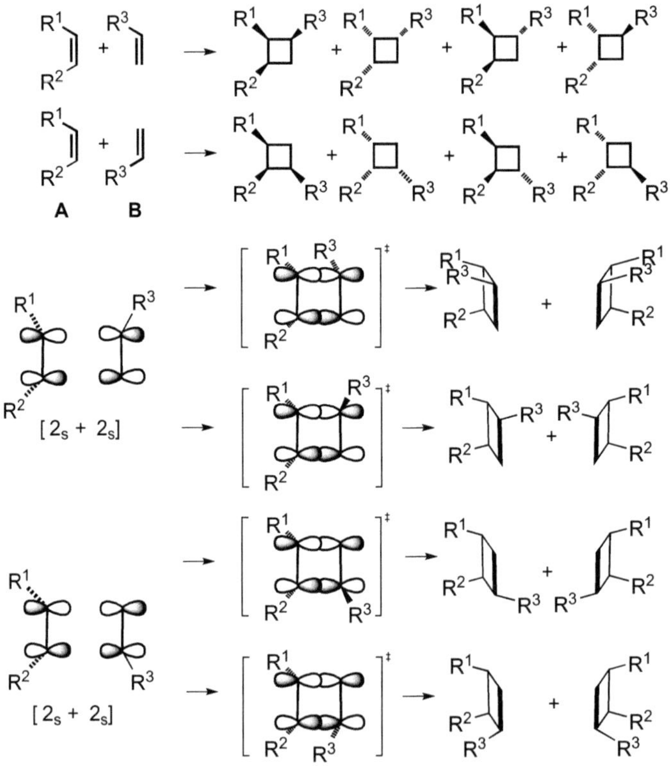

Figure 6.10 Stereochemical and regiochemical possibilities of a [2+2] photocyclization between a *cis*-disubstituted (A), and a mono-substituted (B) alkene. Four pairs of enantiomers may be produced without stereocontrol. Molecular orbital models show the transition states of the [2+2] cycloadditions.

cycloaddition. The homodimers of each alkene would also be possible unless their reactivity is different.

Woodward and Hoffman developed qualitative molecular orbital (MO) theory to explain why various organic reactions occur under certain conditions and not others.[15] For [2 + 2] cyclizations of two alkenes, it was known that these occur in the presence of light, but not thermally, in the absence of light. The Woodward–Hoffman rules explain this in terms of overlap between molecular orbitals with the same symmetry, called "Conservation of Orbital Symmetry."[15] One orbital must be the electron donor (nucleophile), so it must be occupied with electrons, typically the highest occupied MO (HOMO), and the other orbital must be an electron acceptor (electrophile), typically the lowest unoccupied MO (LUMO). In the familiar case, the Diels–Alder [4 + 2] cycloaddition reaction is thermally allowed, but photochemically disallowed.

In an allowed [2 + 2] photocyclization, one π-electron of alkene **A** is excited to the π* orbital (Figure 6.11). The photoexcitation of one electron from the HOMO (green) produces a singly unoccupied orbital (SUMO) with the same symmetry as the LUMO (blue) of alkene **B**. These two orbitals overlap in the photoreaction to produce an excited state of the product with two unpaired electrons (center energy levels). Relaxation of the higher energy unpaired electron produces a cyclobutane with two new σ-bond orbitals (green), lower right (Figure 6.11).

To broaden the scope of photoreactions, Teshik Yoon (University of Wisconsin, Madison) has developed photocatalysts that use energetically less costly visible light with greater control of stereochemistry.[16]

6.7 Case Study: Ruthenium Photocatalyst (Yoon)

Yoon's photocatalysts are based on ruthenium tris-bipyridyl ($Ru(Bpy)_3$) (Figure 6.12).[17] Visible light drives the reaction by exciting the Ru photocatalyst, which transfers energy to the alkene reactants. This transfer of lower energy, visible light from the catalyst avoids direct UV activation of the reactants, which leads to *cis–trans* alkene isomerization, and thus poor stereochemical control.[17] By avoiding *cis–trans* isomerization, the stereochemistry of reactants like **4.1** is retained in the product; *trans*-**4.1** gives 2,3-*trans*-**4.4**, while *cis*-**4.1** would give 2,3-*cis*-**4.5** (Figure 6.12). The C=O of ketone **4.1** is conjugated with the aromatic Ph, so it is lower in energy than the C=O of methyl vinyl ketone; the excitation energy of the aromatic ketone **4.1**

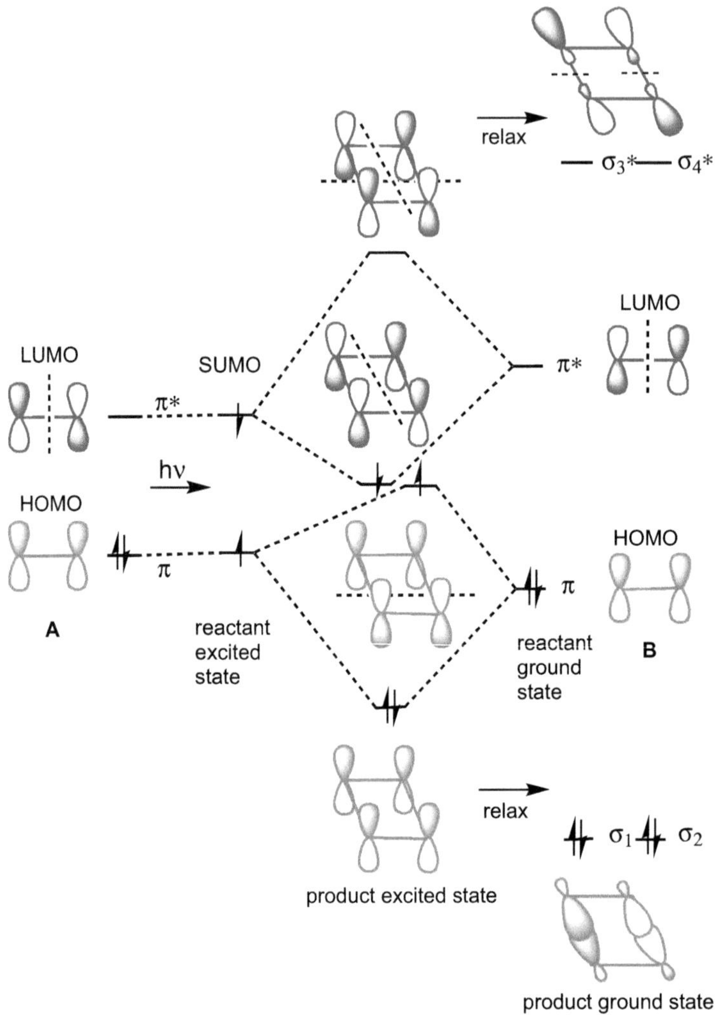

Figure 6.11 Molecular orbital diagram of a Woodward–Hoffman photochemically allowed [2 + 2] cycloaddition reaction. The symmetry of the SUMO π* excited-state orbital of alkene **A** (blue) is the same as the LUMO of alkene **B** (blue), and the symmetry of the π excited-state orbital of alkene **A** (green) is the same as the HOMO of alkene **B** (green).

is thus higher than that of methyl vinyl ketone. The energy transferred from the Ru catalyst should primarily excite methyl vinyl ketone.

Lanthanide Lewis acids, such as Eu, exchange ligands rapidly, which facilitates catalytic turnover.[17] The $Eu(OTf)_3$ Lewis acid binds the asymmetric ligand **4.2**, which then coordinates the carbonyl (C=O) oxygens of both Lewis basic ketones, so they end up next to

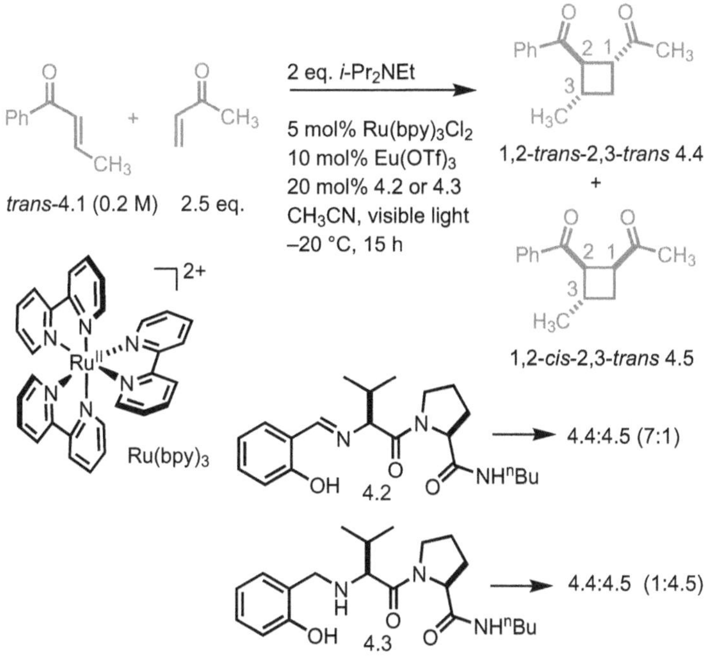

Figure 6.12 Yoon's stereoselective [2+2] photocycloaddition of two enones catalyzed by $Ru(bpy)_3$ and $Eu(OTf)_3$.[17]

each other (1,2-substituted) on the cyclobutane ring. The chiral ligand **4.2** for the Eu catalyst favors the single enantiomer of the 1,2-*trans* product **4.4** over the 1,2-*cis* **4.5** by 7 : 1, with 92% enantiomeric excess (Figure 6.12).[17] The chiral ligand **4.2** for Eu is derived from the L-Val-L-Pro dipeptide with the natural stereochemistry. Reduction of the imine (C=N) of ligand **4.2** to the secondary amine **4.3** results in the opposite diastereoselectivity, favoring 1,2-*cis* **4.5** over 1,2-*trans* **4.4** by 4.5 : 1 (Figure 6.12).[17] Structurally diverse, aromatic ketones like **4.1** react successfully with alkyl ketones in this reaction.[17] Not only does the photocatalyst use lower cost, abundant, visible light energy for catalysis, it gives much tighter stereochemical control of the desired product.

6.8 Battery Technology

Renewable energy is inherently variable. The sun shines only during the day, and its energy can only be harvested when it is not too cloudy. The wind blows strongly only in some places, only some of the time.

The tides come in and out on the moon's schedule. To replace fossil fuels with renewables, we need superior battery technology to store the abundant renewable energy when the sun is shining, the wind is blowing, or the tides changing. Unfortunately, we have come to rely mainly on cheap lead-acid batteries for cars, which use toxic Pb, lead dioxide (PbO_2), and corrosive sulfuric acid as the electrolyte. Like most batteries, completely draining the battery, such as after leaving the lights on all night, changes the chemistry and wrecks the battery. Consumer electronics currently use lithium ion rechargeable batteries. These are lightweight because lithium is a very light element, unlike Pb. Lithium is not a particularly abundant element at 20 mg kg^{-1} in the earth's crust, while Na and K are much more abundant (Appendix D, Abundance of Elements in Earth's Crust). More abundant elements, like Na, are likely to be a better element for stationary major power sources, such as solar and wind farms.

6.9 Case Study: Vanadium Redox Flow Battery (UniEnergy Technologies)

UniEnergy Technologies, LLC (UET) won the 2017 PGCC Award for inventing a grid-scale multi-hour energy storage system based on vanadium oxides.[18] The vanadium-redox flow battery (VFB) was designed to be used for stationary electric storage, and could be used to optimize delivery of power generated by solar or wind energy. The VFB operates in a broad temperature range (−40 to 120 °F) and uses only one-fifth the footprint of previous flow batteries.[18] The VFB is very durable because contamination of the electrolyte with V ions is less problematic than with mixed species batteries. The electrolyte is water-based, and thus non-flammable and more stable than organic electrolytes used in lithium ion batteries, and much less toxic than the sulfuric acid used in lead batteries.[18] The VFB is recyclable, conserving vanadium ores that must be mined.

The VFB is a dual circuit, all-vanadium battery that uses all four stable redox states of vanadium (Table 6.1). The catholyte (negative electrolyte) uses the V^{3+}/V^{2+} couple, and the anolyte (positive electrolyte) uses the $VO_2{}^{+}/VO^{2+}$ couple (Figure 6.13).[20] Power scales with the number of electrochemical cells, and capacity scales with the volume of electrolyte, so these systems can be easily adapted for various applications and scales.[20] The advantages of the VFB are that it has a rapid charge/discharge response rate, a long cycle lifetime, and less toxic, earth-abundant vanadium (Appendix D).[21]

Table 6.1 Reduction potentials of aqueous vanadium ions. Data obtained from M. Skyllas-Kazacos, *Introduction to Redox Flow Batteries*, Laboratory of Physical and Analytical Electrochemistry (LEPA), 2012, vol. 2017.[20,38]

Reduction reaction	Oxidation state	Color	Reduction potential (V)
$V^{2+} + 2e^- \rightarrow V(s)$	2^+	violet	−1.13
$V^{3+} + 1e^- \rightarrow V^{2+}$	3^+	green	−0.26
$VO^{2+} + 2H^+ + 1e^- \rightarrow V^{3+} + H_2O$	4^+	blue	+0.34
$VO_2{}^+ + 2H^+ + 1e^- \rightarrow VO^{2+} + H_2O$	5^+	yellow	+1.00

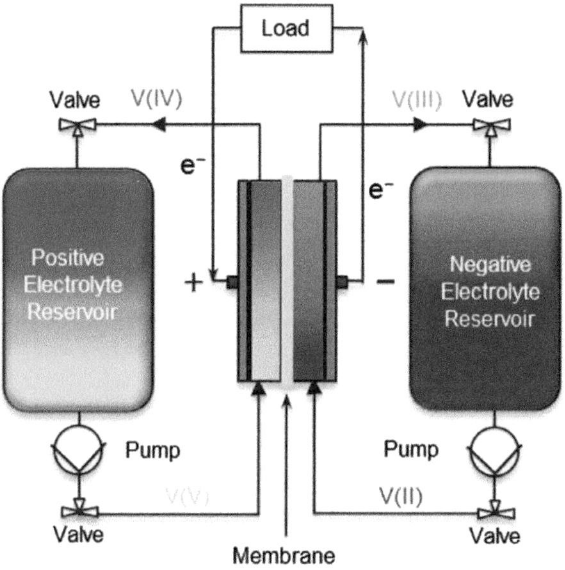

Figure 6.13 Vanadium redox flow battery schematic. Catholyte: V^{2+} (violet)—V^{3+} (green); Anolyte: VO^{2+} (blue)—$VO_2{}^+$ (yellow). Reproduced from ref. 19 with permission from the Royal Society of Chemistry.

6.9.1 Lithium Ion Batteries

The electrolytes of lithium batteries have been a serious fire hazard because they are primarily flammable organic alkyl carbonates. When lithium metal dendrites grow between the two electrodes, the electrical short can cause sparks and ignite the electrolyte (Figure 6.14). We will see in Chapter 12 that the DeSimone and Balsara groups are working toward fluorinated electrolytes that are non-flammable.

Louis Madsen's group at Virginia Tech created an electrolyte gel that physically prevents the Li metal dendrites from growing and shorting out the battery (Figure 6.15). Their electrolyte gel is prepared

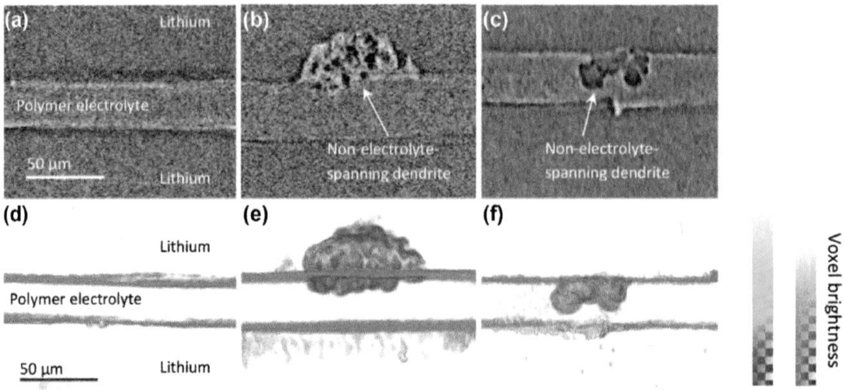

Figure 6.14 Lithium metal battery growth of Li dendrites may cause the battery to short. The short can cause an organic electrolyte to catch on fire. Use of a solid separator prevents formation of Li dendrites. Reproduced from ref. 22, http://dx.doi.org/10.1149/2.0511503jes, under the terms of the CC BY 4.0 license, https://creativecommons.org/licenses/by/4.0/.

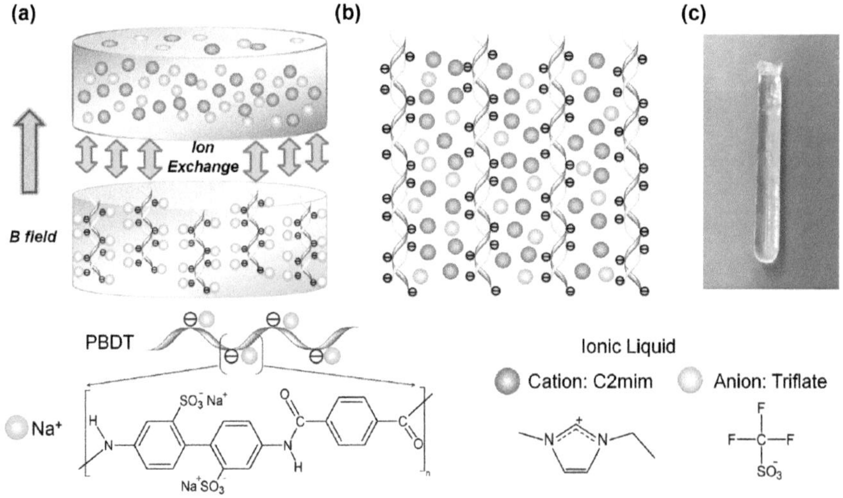

Figure 6.15 Liquid crystalline ion gels formed from a rigid-rod sulfonated aramid polymer and an ionic liquid. (a) Formation scheme of the anisotropic PBDT [poly(2,2′-disulfonyl-4,4′-benzidine terephthalamide)] LC ion gel through ion exchange between the top IL and the bottom PBDT aqueous seed solution ($C \geq 2\%$) in a magnetic field. (b) Schematic of dried PBDT IL gel with IL dispersed in aligned PBDT polymer matrix. (c) PBDT IL gel with water present as formed in the bottom PBDT aqueous seed solution. Reprinted with permission of WILEY-VCH Verlag GmbH & Co. KGaA from ref. 23, Copyright 2016.

by mixing an ion conducting polymer, poly(2,2′-disulfonyl-4,4′-benzidine terephthalamide) (PBDT), and the ionic liquid [C_2mim]·OTf, in a magnetic field to align the polymer (Figure 6.15).[23] Diffusion of the ionic liquid ions occurs preferentially in the direction parallel to the polymer above 10 wt% PBDT, allowing ions to diffuse freely to and from the electrodes during use or recharging.[23] The rigidity of the PBDT gel provides superior blocking of Li dendrites.

6.10 Transformer Technology

Electrical transformers have been notoriously polluting. Since 1929, the electrical grid in the US began using polychlorinated biphenyls (PCBs) produced by Monsanto Co. as a non-flammable replacement for mineral oil (hydrocarbon) transformer fluid. PCBs are good heat conductors, yet good electrical insulators, and have a good temperature range with low viscosity, making them ideal as transformer fluids. From the beginning, PCBs were known to have adverse health effects, yet Monsanto hid the evidence of adverse effects until well into the 1970s.[24] Anniston, Alabama, the site of the Monsanto PCB production plant, has experienced severe health and environmental effects since 1929 due to releases of PCBs to air, water, and soil. PCBs are now listed in the 2001 Stockholm Persistent Organic Pollutants (POPs) Treaty (Table 4.2).

Transformers are electrical devices that transfer electricity between circuits by electromagnetic induction (Figure 6.16). They must be well insulated and durable over many years to function effectively in protecting the electrical power supply grid. In the winter of 1984, Kansas City, MO had an epic ice storm. Power was out across the city for 7 to 10 days. As the electrical system was being brought back up, transformers exploded due to overload on particular circuits all over the city.[25] I lived there, and it was quite the fireworks display in many different colors! This was likely due to chemical differences in the transformer fluids used.

6.11 Case Study: Vegetable Oil Dielectric Insulating Fluid for High Voltage Transformers (Cargill)

Cargill, Inc. won the PGCC Award in 2013 for inventing Envirotemp FR3 transformer fluid.[26] At that time, several hundred thousand transformers filled with Envirotemp FR3 had already been in use

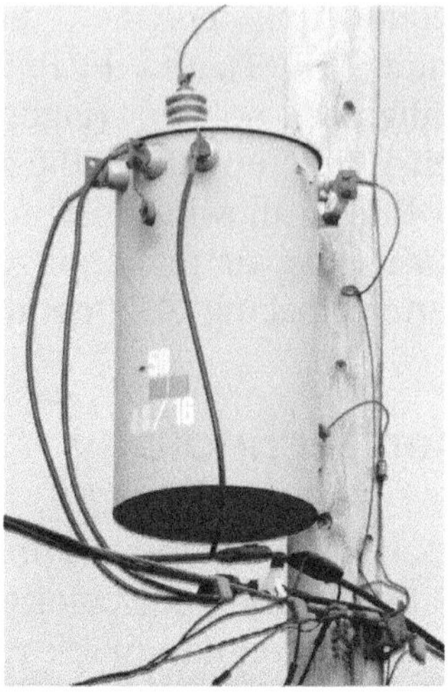

Figure 6.16 Electrical transformer. Reproduced from https://commons.wikimedia.org/w/index.php?curid=17514694, accessed July 25, 2019, under the terms of the CC BY-SA 3.0 license, https://creativecommons.org/licenses/by-sa/3.0/.

since 2003 with no accidents reported. The fluid is mainly soybean oil, a natural product that is non-toxic and biodegradable, fulfilling Principles 4, 7 and 10. The critical aspect of Principle 4, that products achieve their desired function, is exceeded by Envirotemp FR3 because it performs even better than previous fluids. Envirotemp FR3 transformer fluid has twice the flash and fire point temperatures of mineral oil, so it is also much safer.

Cellulose is the primary stable insulator used in transformers. The problem is that cellulose degrades over time due to the heat of the transformer, giving off water (Figure 6.17). Cellulose is a polyol, so there are many ways to eliminate water. If the water remains in close contact with the cellulose, it further degrades cellulose by hydrolysis at the anomeric position, eventually breaking it down completely.

The Cargill Inc. transformer fluid is composed of 98% soybean oil that is a mixture of C16 and C18 fatty acids: 10% C16 palmitic, 5% stearic, 23% oleic, 52% linoleic, and 7% linolenic (Figure 6.18).[26] The remaining components are a branched polymer to prevent crystallization of the oil, a phenolic antioxidant, and trace quantities of a green dye to distinguish Envirotemp from mineral oil transformer fluids.[26] The vegetable oil absorbs water released from cellulose because it is more polar than mineral oil. The hydrophobicity of mineral oil keeps the water in the cellulose insulation, causing degradation. In addition, two sites in the vegetable oil can potentially

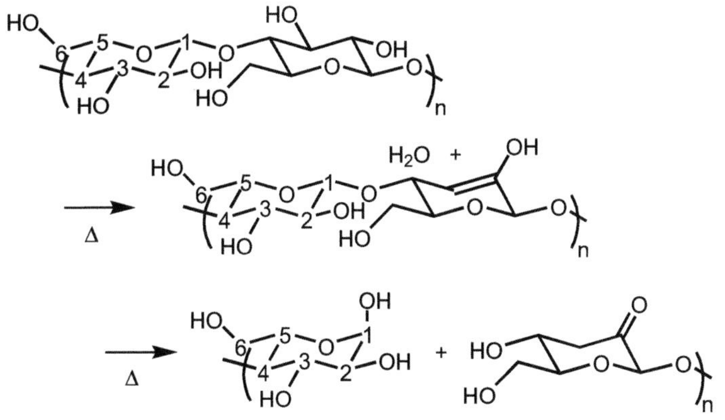

Figure 6.17 Example reactions that could cause cellulose degradation by loss of water.

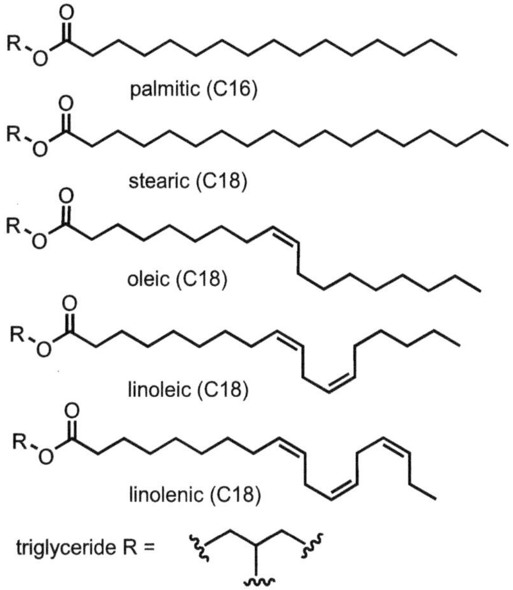

Figure 6.18 Fatty esters that make up soybean oil used in Cargill Inc. Envirotemp FR3 transformer fluid. The R group is the glycerol of the oil triglycerides.

react with water, the ester bonds and the alkenes of the unsaturated chains, preventing breakdown of the cellulose. Natural vegetable oils absorb significantly more water (>200 ppm) than mineral oil (20–30 ppm). Vegetable oils have slightly higher breakdown voltage as a function of humidity than mineral or silicone oils.[27]

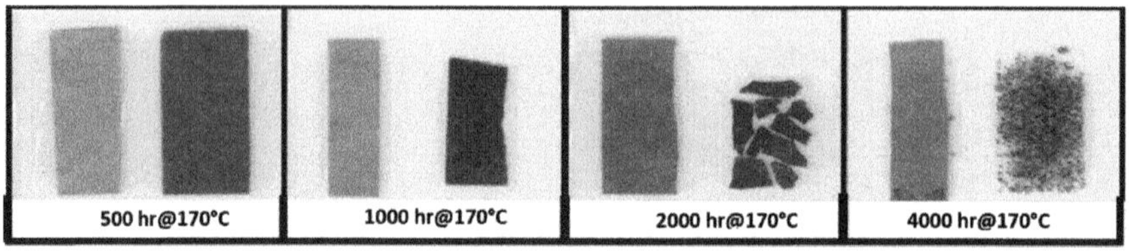

Figure 6.19 Thermally upgraded papers aged in Envirotemp FR3 fluid and mineral oil for varying times at 170 °C. Reprinted with permission of Cargill Inc. from ref. 26, Copyright 2013.

The water absorption and reactivity of Envirotemp FR3 oil led to improved protection of the cellulose insulation from degradation (Figure 6.19). Envirotemp FR3 compares very favorably with mineral oil transformer fluid in protecting cellulose at 170 °C, well above the 110 °C upper limit of performance required for the lifetime of transformer fluid (Figure 6.19).[26] The standard of the Institute of Electrical and Electronics Engineers established one transformer unit of life as 20.6 years at 100% electrical usage with a 110 °C hotspot.[26] The breakdown voltage as a function of relative moisture content of natural esters is comparable to mineral and silicone oils.[27] The consensus International Electrotechnical Commission (IEC) breakdown voltage limit is 30 kV.[27] Transformers with Envirotemp FR3 are designed to withstand up to 130 °C at a smaller size with 15% less fluid and 3% less construction materials.[26] Both the function and the environmental performance of transformers are thus improved with Envirotemp FR3.

A life cycle assessment of the production of greenhouse gases was performed comparing mineral oil and Envirotemp FR3. Because plants synthesize cellulose from atmospheric CO_2, and the materials used in production of the transformer fluid are plant-based, Envirotemp FR3 actually absorbs more greenhouse gases than it produces, net −0.263 tons CO_2 equivalents (CO2e) per 1000 gallons.[26] Envirotemp FR3 is highly biodegradable in an aerobic aquatic environment—it is completely converted to CO_2 within 45 days, while silicone and mineral oil are much more persistent (Figure 6.20).[26]

6.12 Renewable Liquid Fuels

6.12.1 Ethanol

The original liquid fuel for cars was a mixture of natural products, ethanol and turpentine. Henry Ford's first automobile in 1896, and the high-production Model T in 1904, both ran on pure ethanol.[28]

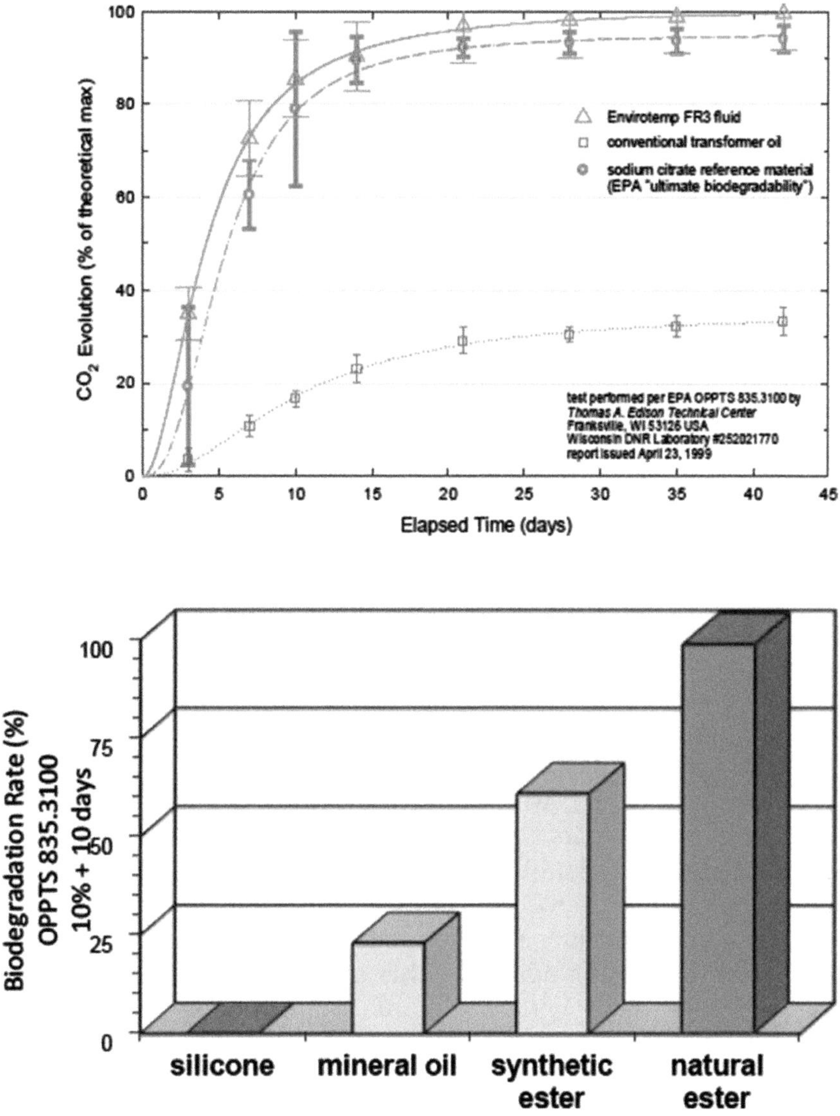

Figure 6.20 Biodegradation rate comparison of Envirotemp FR3. Reprinted with permission of Cargill Inc. from ref. 26, Copyright 2013.

Although petroleum has been used since ancient times, the first modern drilling of Drake's well in Pennsylvania in 1859, and the development of commercial-scale refining techniques brought about a temporary lapse in the use of ethanol as a liquid fuel.[29]

Ethanol can be readily obtained from carbohydrates, primarily glucose polymers, by fermentation of a number of biofeedstocks.

Since 1931, Brazil has mandated the use of sugarcane for the extensive production of ethanol for transportation. By 2015, their fuel was up to E27, or 27% ethanol. In the USA, the second leading producer of ethanol, corn is primarily used for the production of ethanol, which is added to gasoline at 10 to 15% for three reasons: (1) as a substitute for toxic tetraethyl lead ($PbEt_4$), and later methyl *tert*-butyl ether (MTBE), to increase the octane of gasoline and prevent engine knocking; (2) to increase the use of domestic fuel sources and decrease dependence on foreign oil; and (3) to lower net carbon dioxide emissions. Because ethanol is obtained from plants, which essentially "fix" CO_2 from the atmosphere while growing, it is potentially a carbon-neutral fuel.

Much controversy surrounds the issue of whether corn-based ethanol is carbon-neutral, due to the fossil fuel used in farming, transportation, and processing of corn into ethanol. In addition, ethanol competes with corn for food production, driving the prices of corn higher. There is much active research into the production of ethanol from alternative, non-food-based crops, including switchgrass that grows on marginal agricultural land, and corn stover, the leaves, husks, and stalks left after harvesting corn kernels. Production of ethanol is relatively green, using enzymes in bacteria, moderate temperatures, and an aqueous system. In comparison, petroleum refining uses very high temperatures and multiple steps to produce liquid fuel, with sulfuric acid, nitric acid, and various heavy metal by-products that must be trapped and discarded safely. One problem with ethanol is that it has to be separated from water. Distillation is expensive and ineffective, because ethanol and water form an azeotrope, a liquid mixture of fixed composition that boils below the boiling point of either pure component, in this case 95% EtOH:5% water with a b.p. of 78.2 °C, only 0.2 °C below pure EtOH.

6.12.2 Biodiesel

Biodiesel is another liquid fuel in current use. Biodiesel is made by transesterification of fats, which are triesters of glycerol (Figure 6.21). Biodiesel synthesis is used as an example of how to draw a mechanism in Appendix B. The methanol required for biodiesel can be made from wood waste, so it is also a biofeedstock. Biodiesel can also be made from waste vegetable oil from restaurants. A number of US companies collect vegetable oil as a free service to restaurants and institutions, and convert it to biodiesel for profit.

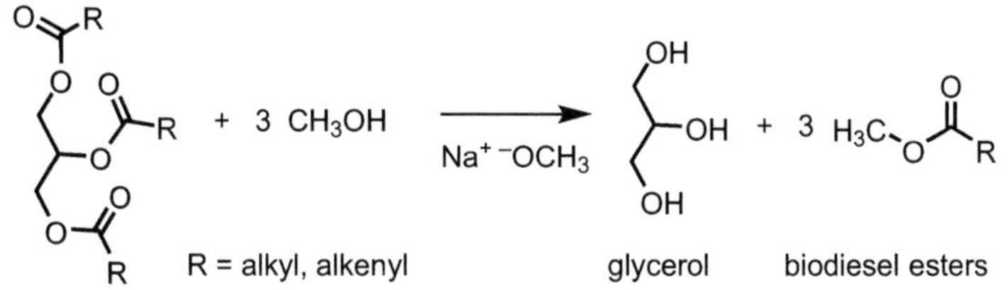

Figure 6.21 Synthesis of biodiesel from vegetable oil. See Figure 6.18 for some possible fatty acid R groups.

6.12.3 Syngas

Syngas is a mixture of carbon monoxide (CO) and hydrogen (H_2). Syngas is an important feedstock for the production of hydrogen, ammonia, methanol, and synthetic liquid fuels *via* the Fischer–Tropsch process. Although syngas has been made primarily from fossil fuels, it can now be made from waste glycerol from biodiesel production or from carbohydrate fermentation (Figure 6.22).[30] The "liquid hydrocarbons" or alkanes are high-energy-content liquid fuels, like gasoline and diesel, that can be made from syngas (Figure 6.22). The energy involved in syngas processing from start to finish, a life cycle analysis, must be considered for any comparison with petroleum-based liquid fuels to see if there is a net energy benefit, but certainly there is an advantage to using biofeedstocks as net neutral in CO_2 production.

6.12.4 Propylene Glycol

Glycerol is the major by-product of biodiesel synthesis. Galen Suppes (University of Missouri) won a PGCC Award in 2006 for a process to make propylene glycol from biodiesel waste glycerol (Figure 6.23).[31] Although it is not a fuel, propylene glycol can be used in transportation engines as an anti-freeze, and in a number of other applications, instead of the far more toxic ethylene glycol (LD_{50} human 1.4 mL kg^{-1}).[31,32] The process is solvent-free since glycerol is a liquid at the reaction temperature. Because it is a catalytic process, the reaction can be run at a relatively low temperature of 200 °C.[31] The copper chromite catalyst ($Cu_2Cr_2O_5$) is less toxic because it is less bioavailable in the Cr(III) oxidation state. But in consideration of the life cycle analysis, the catalyst is derived from copper chromate ($CuCrO_4$), the more toxic Cr(VI) oxidation state. In the absence of hydrogen gas, the acetol intermediate from the dehydration step can be diverted and used as a commodity chemical intermediate for a variety of other purposes.

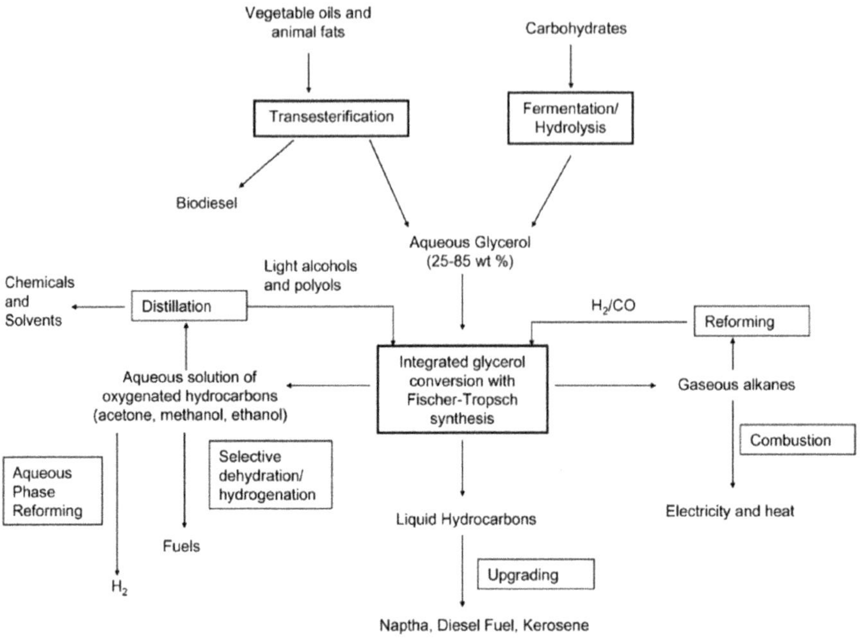

Figure 6.22 Process pathway for production of liquid fuels from biomass by integrated glycerol conversion to synthesis gas and Fischer-Tropsch synthesis. Reproduced from ref. 30 with permission from the Royal Society of Chemistry.

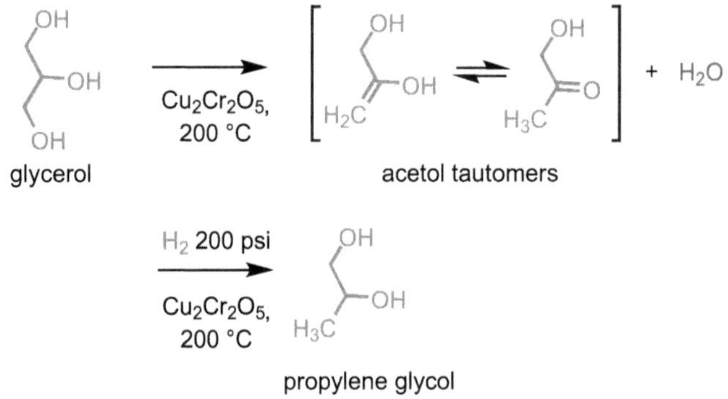

Figure 6.23 Suppes group synthesis of propylene glycol from glycerol.[31]

6.13 Solar Photovoltaics

Solar energy is estimated to be the largest contributor of renewable energy in the world. Photovoltaics convert sunlight energy directly into electricity, so they are one of the least polluting energy sources.

Chemists have a strong presence in the development of photovoltaic materials. Photovoltaics fall into two broad categories, inorganic and organic semi-conductors. Semi-conductors are materials halfway between conducting metals, with low electrical resistance, and insulating ceramics with high electrical resistance.

6.13.1 Inorganic Photovoltaics

Pure elements in group 4A, such as silicon or germanium, have four valence electrons, so they are readily oxidized or reduced to empty or fill their valence, acting as semi-conductors. The high earth abundance of Si at 282 $g\,kg^{-1}$ earth's crust, makes it far greener and cheaper than the much less abundant Ge at 1.5 $mg\,kg^{-1}$ (Appendix D). Great strides have been made in recent years to produce Si solar photovoltaics with a photoconversion efficiency as high as 26%, which is amazing since the theoretical efficiency of Si is 29%.[33] Binary compounds between groups 3A and 5A, such as gallium arsenide (GaAs) are also common semi-conductors. Chapter 2 described quantum dots made from cadmium selenide (CdSe) or indium phosphide (InP) that are light-emitting diodes, undergoing excitation–emission of light energy. Quantum dots can also be used in photovoltaics.

In order to create semi-conductors from Si, small amounts of impurities are introduced, called doping. Doping Si with electron-poor elements, like group 3A gallium, creates electron "holes" or p-type (positive) semi-conductors; doping with electron-rich elements like group 5A arsenic, creates an excess of electrons called n-type (negative) semi-conductors. A solar photovoltaic cell is made with a layer of n-type and a layer of p-type semi-conductor, with electrical contact material on the front and back. The device is protected by glass or a polymer, preferably non-yellowing such as silicone (Figure 6.24). Photons from the sun excite electrons in the conducting band of the n-type semi-conductor, which flow into the p-type semi-conductor, creating an electrical current. The electricity can be used directly. In consumer applications, it is converted to alternating current with a DC-to-AC converter. Then the electricity is either fed into the electrical grid or stored in batteries.

Materials for photovoltaic cells are optimized for the efficiency of conduction *vs.* voltage output (Figure 6.25). The higher the current at the highest voltage possible produces the most power. Unfortunately, the most efficient solar cells are not cost effective because they often use rare and expensive materials. Thus, most commercial cells are

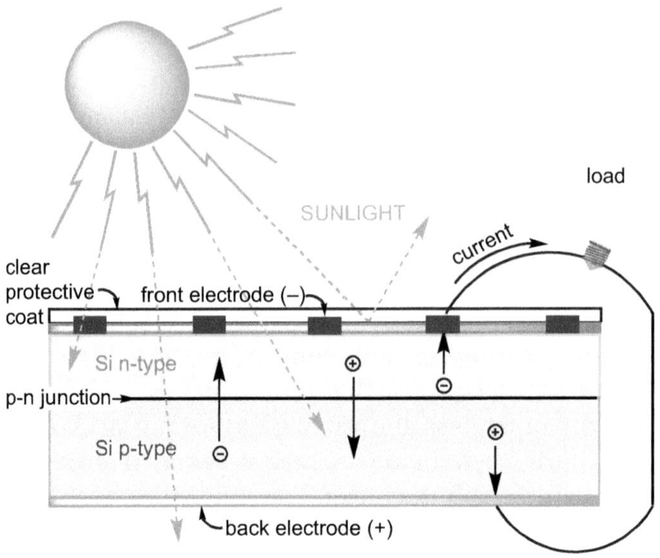

Figure 6.24 Diagram of a silicon-based solar photovoltaic cell.

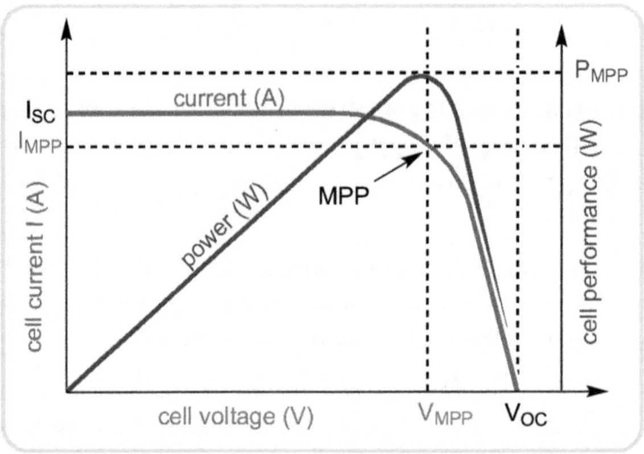

Figure 6.25 Idealized solar photovoltaic cell plot of current (A) *vs.* voltage (V) (blue curve), and power (W) *vs.* voltage (V) (purple curve) showing that the maximum power lies at the upper right corner of the current curve. I_{SC} is the short-circuit current. I_{MPP} is the current at the maximum power point (MPP). V_{MPP} is the voltage at the MPP. V_{OC} is the open-circuit voltage. P_{MPP} is the power production at the MPP. Adapted from ref. 34, with permission from Elsevier, Copyright 1985.

silicon-based. There is a wide range of efficiency among Si cells because of different types and levels of doping to obtain n-type and p-type materials. Processing crystalline silicon into solar cells is

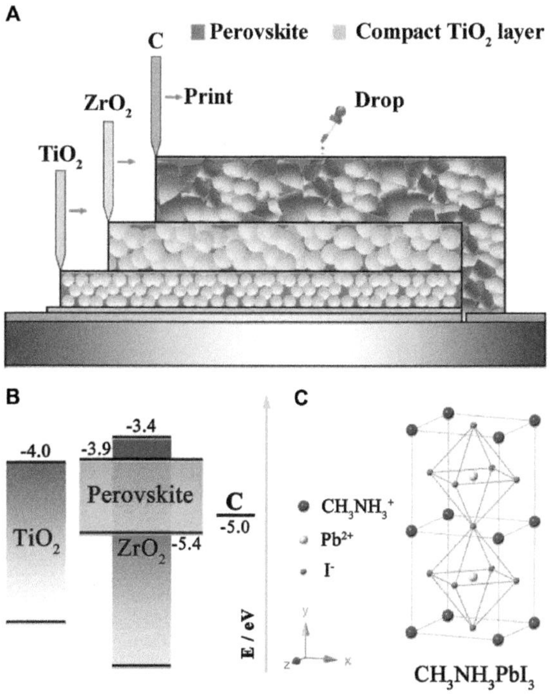

Figure 6.26 Triple-layer printed perovskite solar cell with the crystal structure of a perovskite. Reprinted with permission of American Association for the Advancement of Science from ref. 36, Copyright 2014.

energy intensive because it requires a special "clean room," with high vacuum, and high temperatures.

Perovskites are an ABX_3 crystal structure type of material, where there are two different size cations interlocked with an anion (Figure 6.26). Perovskites such as magnesium silicate and calcium titanate are abundant in the earth's crust, and therefore cheap (Appendix D). Although perovskite solar cells are not yet commercialized, they hold great promise, and their solar conversion efficiencies zoomed from 4% to 19% in the five years from 2009 to 2014.[35] A recent advance was fabrication by printing a triple layer of $TiO_2/ZrO_2/C$ and drop casting $CH_3NH_3PbI_3$ onto it to infiltrate through the layers (Figure 6.26).[36] This resulted in formation of a functioning solar cell without a hole-conducting layer, simplifying production and making it cheaper.[36] The cell was stable in air for 1000 hours with an efficiency of 12.8%.[36] Although Pb is toxic, these perovskite photovoltaics are exciting because they are simple to manufacture from cheap, abundant materials.

6.13.2 Organic Photovoltaics

Similar principles apply to organic photovoltaic (OPV) materials: there must be donor and acceptor materials to produce electricity when a photon is absorbed. Most OPVs are highly conjugated aromatic systems, whether they are polymers like PCDTBT, or medium-sized molecules like fullerenes (C_{60} and derivatives) (Figure 6.27). In these highly conjugated systems, the HOMO–LUMO gap is quite small, so that promotion of an electron to higher energy is facile (Figure 3.7).

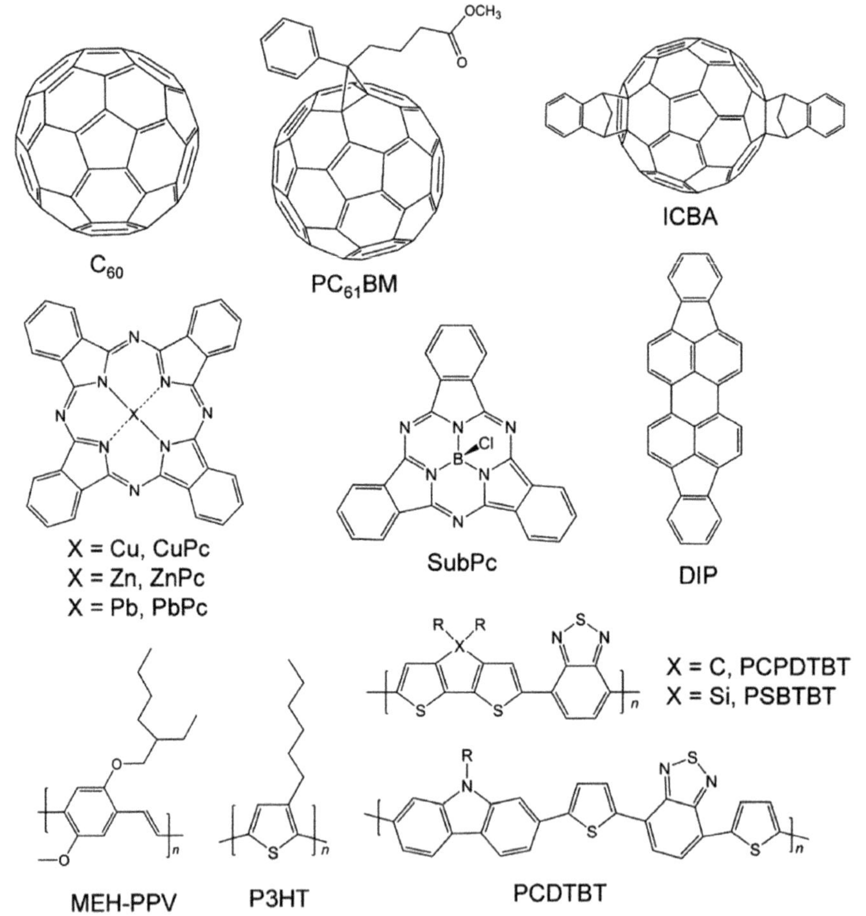

Figure 6.27 Molecular structures of several widely used organic donor and acceptor materials (small molecules and polymers). Reproduced from ref. 37 with permission from the Royal Society of Chemistry.

With organic molecules, the electron donor (HOMO) and acceptor (LUMO, holes) must be separated to achieve reasonable efficiencies. When an electron is excited from the HOMO to the LUMO by a photon, the electron must be transferred to a neighboring molecule to produce a charge-transfer exciton.[37] This separated electron/hole pair is held together by Coulombic attraction. The HOMO (or LUMO) level offset at the interface between donor and acceptor, called the bulk heterojunction, must be greater than the Coulombic attraction in order to separate the exciton pair.[37] The electron and the hole must then move towards their respective electrodes to produce a photocurrent or photovoltage.[37] OPV materials with bulk heterojunctions are more efficient than thin-film heterojunctions. The stability of the organic material is a critical issue for commercialization. Cao and Xue suggest that better encapsulation of OPVs is necessary for device stability.[37] In addition, the bulk heterojunction must remain phase-separated during operation.[37] Another promising area is the development of mixed organic and inorganic photovoltaic materials.

6.14 Summary

Chemists play an enormous role in the creation of new processes that minimize the use of energy. A fundamental understanding of the role of kinetics and thermodynamics in optimizing the space-time yield of a desired product is essential. Processes run at ambient temperatures and pressures minimize the use of energy. BioAmber invented a microbial catalyst to produce commodity succinic acid. Microwave energy is a mixed bag—laboratory-scale microwaves are incredibly inefficient, while large-scale industrial microwave processes are more efficient than conventional heating. Yoon's photocatalyst system uses abundant visible light energy, giving high yields of single-enantiomer cyclobutanes. UniEnergy has commercialized a vanadium water-based redox flow battery for stationary uses, and the Madsen group invented a conducting polymer/ionic liquid electrolyte for lithium batteries that prevents fire. Cargill developed a soybean oil-based transformer fluid that is more efficient and less toxic than PCB fluids. Chemists are also central in converting biomass into liquid transportation fuels and inventing new solar photovoltaic materials for capturing abundant solar energy directly.

6.15 Problems: Energy Efficiency

Problem 6.1

(a) Describe three ways that chemists can use energy **conservation** in chemical processes.
(b) Name three **renewable energy** sources that can be used in chemical processes.
(c) Describe three ways chemists can contribute to energy-saving commercial and consumer **products**.

Problem 6.2

Consider the elimination reactions of Figure 6.1.

(a) Using the Arrhenius equation and the temperatures given in the figure, calculate the **ratio of activation energies** (E_a) between the two processes. Assume A is identical in both reactions, and $k_1 = k_2$. (Both are unimolecular processes. Assume the time to completion, therefore the rate, is the same for both reactions.)
(b) Given that both processes ultimately begin with cyclohexanol to give cyclohexene, how does the Burgess reaction "pay" for the energy difference?
(c) Draw the curved arrow mechanism for the acetate pyrolysis of cyclohexyl acetate.

Problem 6.3

(a) Give the number and short name of two Principles that the BioAmber process best fits, other than Principle 6 Energy Efficiency. Explain briefly.
(b) Which of the three areas of the PGCC Awards does the BioAmber succinic acid process best fit? Explain briefly.

Problem 6.4

Examine the products derived from succinic acid in Figure 6.8.

(a) Draw the products and circle the four carbons in each product that came from succinic acid.
(b) Which products required reduction of one carboxylic acid? Both carboxylic acids?
(c) Choose one product other than BDO and propose reagents and a synthesis of it from succinic acid.

Problem 6.5

Three enzyme genes were mutated in the *E.coli* K12 AFP184 strain to attain the highest yield of succinate.

(a) Give the gene names and full names of the enzymes.
(b) Draw the reactions catalyzed by each of these enzymes.
(c) What gases and percentages gave the highest yield of succinate?

Problem 6.6

BioAmber subsidiary, Sinoven Biopolymers, has the proprietary technology to produce poly-butylene succinate (PBS).

(a) Draw the structure of PBS and circle the two monomers.
(b) One monomer is succinic acid. How can the other monomer can be derived from succinic acid?
(c) Draw the curved arrow mechanism for the neutral polymerization of PBS.
(d) What would be the easiest way to drive the polymerization forward?

Problem 6.7

Consider the Yoon photoreaction in Figure 6.12.

(a) What is the advantage of using the $Ru(bpy)_3$ in [2 + 2] cycloadditions?
(b) What is the function of the $Eu(OTf)_3$?
(c) Why do the PhC=O and Me groups end up *trans* in products **4.4** and **4.5**? (Hint: Look at the stereochemistry of the reactants.)
(d) Draw the enantiomer of product **4.4**. What is the function of ligand **4.2**?

Problem 6.8

Refer to the UniEnergy vanadium redox flow battery case study.

(a) Compare the earth abundance of vanadium to sodium.
(b) What is the toxicity of vanadium (v) oxide by ingestion?
(c) Calculate the potential energy storage in $kJ\,mole^{-1}$ of vanadium in the catholyte (Table 6.1) using the Nernst equation.

Problem 6.9

Consider the Cargill Inc. Envirotemp FR3 transformer fluid case study.

(a) Draw the equilibrium between the cyclic and open forms of glucose.
(b) Draw the structure of cellulose with the chair conformations, including the 1,4-linkage.
(c) Draw two reactions of an oleic fat (ester) that can absorb the water to keep cellulose from degrading.

Problem 6.10

Photovoltaics

(a) What life cycle considerations should go into assessment of the energy-producing potential of photovoltaic materials?
(b) What energy-level characteristics are common to the inorganic and organic photovoltaics?
(c) Draw a diagram of a simple Si-based photovoltaic cell with labels.

References

1. P. T. Anastas and J. C. Warner, *Green Chemistry: Theory and Practice*, Oxford University Press, New York, 1998.
2. A. Lovins, *A 40-Year Energy Plan*, 2012, https://www.ted.com/talks/amory_lovins_a_50_year_plan_for_energy?language=en, accessed July 23, 2019.
3. F. A. Carey and R. J. Sunburg, *Advanced Organic Chemistry: Part B: Reaction and Synthesis*, Springer, New York, 5th edn, 2010.
4. E. M. Burgess, H. R. Penton and E. A. Taylor, *J. Am. Chem. Soc.*, 1970, **92**, 5224–5226.
5. H. C. Brown, I. Moritani and Y. Okamoto, *J. Am. Chem. Soc.*, 1956, **78**, 2193–2197.
6. I. E. Kopka, M. A. Nowak and M. W. Rathke, *Synth. Commun.*, 1986, **16**, 27–34.
7. BioAmber, *Integrated Production and Downstream Applications of Biobased Succinic Acid*, Washington, DC, 2011, https://www.epa.gov/greenchemistry/presidential-green-chemistry-challenge-2011-small-business-award, accessed July 25, 2019.
8. K. Lohbeck, H. Haferkorn, W. Fuhrmann and N. Fedtke, *Maleic and Fumaric Acids*, Wiley-VCH, Weinheim, 2000.
9. Narayanese, WikiUserPedia, YassineMrabet and TotoBaggins, *Citric acid cycle*, https://en.wikipedia.org/wiki/Citric_acid_cycle, accessed July 25, 2019.
10. G. Whittaker, *A Basic Introduction to Microwave Chemistry*, 2007, http://tan-delta.com/basics.html, accessed January 7, 2017.
11. A. G. Whittaker and D. M. P. Mingos, *J. Microwave Power EE*, 1994, **29**, 195–219.
12. J. D. Moseley and C. O. Kappe, *Green Chem.*, 2011, **13**, 794–806.
13. M. C. Cann and T. P. Umile, *Real-world Cases in Green Chemistry*, American Chemical Society, vol. II, 2008.

14. R. P. Swatloski, S. K. Spear, J. D. Holbrey and R. D. Rogers, *J. Am. Chem. Soc.*, 2002, **124**, 4974–4975.
15. R. B. Woodward and R. Hoffman, *The Conservation of Orbital Symmetry*, Academic Press, Germany, 1971.
16. T. P. Yoon, M. A. Ischay and J. Du, *Nat. Chem.*, 2010, **2**, 527–532.
17. J. Du, K. L. Skubi, D. M. Schultz and T. P. Yoon, *Science*, 2014, **344**, 392–396.
18. UniEnergy Technologies LLC, *The UniSystemTM: An Advanced Vanadium Redox Flow Battery for Grid-Scale Energy Storage*, Washington, DC, 2017, https://www.epa.gov/greenchemistry/green-chemistry-challenge-2017-small-business-award, accessed July 25, 2019.
19. P. Peljo, H. Vrubel, V. Amstutz, J. Pandard, J. Morgado, A. Santasalo-Aarnio, D. Lloyd, F. Gumy, C. R. Dennison, K. E. Toghill and H. H. Girault, *Green Chem.*, 2016, **18**, 1785–1797.
20. M. Skyllas-Kazacos, *Introduction to Redox Flow Batteries*, Laboratory of physical and Analytical Electrochemistry (LEPA), 2012, accessed 12/5/17.
21. Agency for Toxic Substances and Disease Registry (ATSDR), *Vanadium*, Public Health Service, U.S. Department of Health and Human Services, Atlanta, GA, 2017.
22. N. Schauser, K. Harry, D. Parkinson, H. Watanabe and N. Balsara, *J. Electrochem. Soc.*, 2014, **162**, A398–A405.
23. Y. Wang, Y. Chen, J. W. Gao, H. G. Yoon, L. Y. Jin, M. Forsyth, T. J. Dingemans and L. A. Madsen, *Adv. Mater.*, 2016, **28**, 2571–2578.
24. M. Grunwald, Monsanto Hid Decades of Pollution, *The Washington Post*, 1 January 2002, https://www.washingtonpost.com/archive/politics/2002/01/01/monsanto-hid-decades-of-pollution/244d1820-d49d-4145-9913-35644a734936/?utm_term=.cc43d9e62e66, accessed July 25, 2019.
25. J. M. Vaughn, *PICTURES: 10 Year Anniversary of Epic Ice Storm*, fox4kc.com, https://fox4kc.com/2012/01/30/pictures-10-year-anniversary-of-epic-ice-storm/, accessed July 25, 2019.
26. Cargill Inc., *Vegetable Oil Dielectric Insulating Fluid for High Voltage Transformers*, Washington, DC, 2013, https://www.epa.gov/greenchemistry/presidential-green-chemistry-challenge-2013-designing-greener-chemicals-award, accessed July 25, 2019.
27. P. Boss, *Insulating fluids for power transformers*, CIGRE International Council on Large Electric Systems, http://a2.cigre.org/Publications/SC-A2-Position-Papers, accessed March 22, 2016.
28. Fuel-Testers, *Ethanol Fuel History*, 2017, http://www.fuel-testers.com/ethanol_fuel_history.html, accessed July 25, 2019.
29. Wikipedia, *Petroleum*, https://en.wikipedia.org/wiki/Petroleum#Early_history, accessed January 25, 2017.
30. D. A. Simonetti, J. Rass-Hansen, E. L. Kunkes, R. R. Soares and J. A. Dumesic, *Green Chem.*, 2007, **9**, 1073–1083.
31. M. A. Dasari, P.-P. Kiatsimkul, W. R. Sutterlin and G. J. Suppes, *Appl. Catal., A*, 2005, **281**, 225–231.
32. J. Brent, *Drugs*, 2001, **61**, 979–988.
33. K. Yoshikawa, H. Kawasaki, W. Yoshida, T. Irie, K. Konishi, K. Nakano, T. Uto, D. Adachi, M. Kanematsu, H. Uzu and K. Yamamoto, *Nat. Energy*, 2017, **2**, 17032.
34. S. Ashok and K. P. Pande, *Sol. Cells*, 1985, **14**, 61–81.
35. J. Wimberley, *Perovskite Solar Cells Beat New Records (In the Lab)*, 2014, https://cleantechnica.com/2014/08/14/perovskite-solar-cells-beat-new-records-in-the-lab/, accessed July 25, 2019.
36. A. Mei, X. Li, L. Liu, Z. Ku, T. Liu, Y. Rong, M. Xu, M. Hu, J. Chen, Y. Yang, M. Grätzel and H. Han, *Science*, 2014, **345**, 295–298.
37. W. Cao and J. Xue, *Energy Environ. Sci.*, 2014, **7**, 2123–2144.
38. P. Vanýsek, *CRC Handbook of Chemistry and Physics*, Taylor and Francis Group, LLC, 96th edn, 2016.

7 Renewable Feedstocks

"Principle 7. A raw material or feedstock should be renewable rather than depleting whenever technically and economically practicable."[1]

7.1 Fossil Feedstocks

In chemical synthesis the raw materials used to make a product are called "starting materials" or "feedstocks." In general, fossil feedstocks are derived from the lighter petroleum fractions, but they can also come from coal or natural gas. These are called "depleting" because the timescale for regeneration of petroleum or coal is marked in geological time—it takes eons to regenerate them—and they become

Green Chemistry: Principles and Case Studies
By Felicia A. Etzkorn

Published by the Royal Society of Chemistry, www.rsc.org

depleted. These will be referred to as fossil feedstocks. While they remain in the ground, oil and coal trap carbon as high molecular weight graphite or hydrocarbons. Once petroleum is pumped, or coal is mined, from the ground and processed, the carbon will eventually be released as carbon dioxide (CO_2) or methane (CH_4). Both of these are greenhouse gases that cause global warming. Atmospheric methane absorbs radiated heat from the earth in a different part of the infrared spectrum than carbon dioxide, so the combined effects are particularly potent (Figure 7.1). In addition, methane is about 24-fold

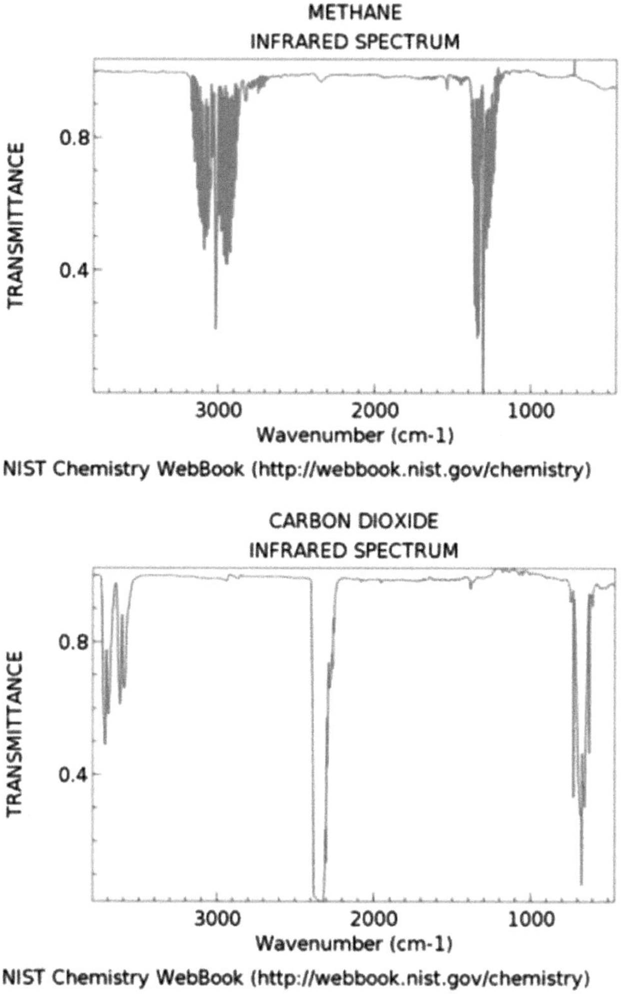

Figure 7.1 Infrared spectra of top: methane (CH_4), and bottom: carbon dioxide (CO_2). Reprinted from National Institute of Standards and Technology, http://webbook.nist.gov/chemistry, NIST Standard Reference Data (SRD); COBLENTZ SOCIETY. Used with permission. © United States of America as represented by the Secretary of Commerce.

more potent than carbon dioxide as a greenhouse gas. This is integrated over time because methane decomposes gradually in the atmosphere, whereas carbon dioxide does not.

Carbon dioxide is unique as a feedstock because it results from both the burning of fossil fuels, and the respiration of animals and plants. Of course, CO_2 is the ultimate feedstock for plants, which synthesize the majority of biofeedstocks (Figure 7.2).

Methane is produced by microbes in the upper gut of cattle and released, primarily by cows burping. One could make the case that using CO_2 and CH_4 from either biological or industrial sources as feedstocks is beneficial to keep them out of the atmosphere and prevent global warming.

The chemical industry uses approximately 8% of the world's fossil fuels: coal, natural gas, and petroleum. Fossil feedstocks take hundreds of thousands to millions of years to produce hydrocarbons from once living matter that has captured CO_2 from the atmosphere. Climate change is thus a global kinetic problem. We burn fossil fuels very quickly, but they take millions of years to form. The system is far out of equilibrium, and it will take eons of time to rebalance. Drilling and fracking for oil and gas, petroleum refining, and transportation by pipeline, rail, or boat are also problematic for the environment.

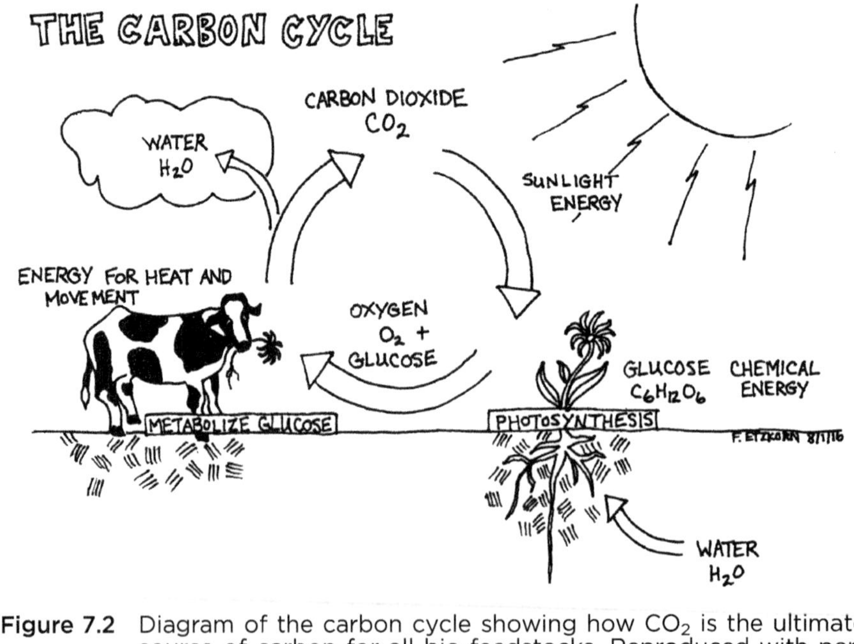

Figure 7.2 Diagram of the carbon cycle showing how CO_2 is the ultimate source of carbon for all bio-feedstocks. Reproduced with permission, Copyright © 2016 Felicia A. Etzkorn.

Therefore, in green chemistry we try to avoid using all fossil feedstocks. It is critical that we stop mining fossil fuels and feedstocks out of the ground and begin using renewable feedstocks wherever possible.

In Chapter 6, we saw that fuel can be derived from biofeedstocks—biodiesel from fats, and ethanol from carbohydrates. This chapter will focus on the use of renewable feedstocks for chemical production.

7.2 Renewable Feedstocks

Most renewable feedstocks are biological in origin, and biofeedstocks are more diverse than fossil feedstocks. Biofeedstocks must be isolated or synthesized from recently living organic matter, usually of plant origin, that has captured CO_2 from the atmosphere by photosynthesis, or from animals that have eaten plants (Figure 7.2). Plant matter is, of course, the most common feedstock, and most of us are familiar with corn-based ethanol in gasoline. Other common biological sources are ethanol from sugar cane in Brazil, cellulose from wood, amino acids from agricultural waste, microbial fermentation, and natural products from plant sources. Polymers from bio-based feedstocks are becoming more economically viable as the cost and scarcity of petroleum rises.[2] The demand for plastics is expected to rise from current levels of 200 million tons per year to ~1000 million tons per year by the end of the 21st century.[2] Using biofeedstocks also frequently results in more environmentally degradable plastics, which will prevent pollution both microscopic and macroscopic, such as the Great Pacific Garbage Patch (Chapter 10).[3]

7.2.1 Earth Abundance

The earth abundances of carbon (200 mg kg^{-1}) and hydrogen (1400) pale in comparison with oxygen (461 000), silicon (282 000), aluminum (82 300), and iron (56 300) (Appendix D). In this sense, silicon, aluminum, and iron could all be considered renewable. Certainly, since these elements are mostly found in nonvolatile compounds, they will not evaporate from the earth, and they can be recovered and reused as feedstocks. Some metals are inherently toxic, or readily converted into the toxic form (Pb, Cd, Hg), and these should be avoided as much as possible (see Chapters 3 and 4). Iron and copper are relatively non-toxic unless ingested at high concentrations. Steel and copper are already recycled, and more of this could be done. Silicon in the form of glass is recycled. Thus, organic chemistry does not have a corner on the renewable feedstock market!

7.2.2 Oxidation State Matching

One major advantage of biofeedstocks is that they already contain elements other than carbon and hydrogen, such as oxygen, nitrogen, and sulfur, which might be desired in the final product. In other words, biofeedstocks are often already near the oxidation state of the desired product.

Feedstocks from different sources have different characteristic oxidation states. Fossil feedstocks are generally hydrocarbons, a highly reduced state of organic matter. Biofeedstocks are typically much more oxidized. For example, glucose has the molecular formula $C_6H_{12}O_6$, with an oxygen attached to every carbon. On the other hand, fats and oils are much more reduced esters of long-chain alkayl or alkenyl fatty acids. Proteins and amino acids lie somewhere in between. Plants generate a large variety of secondary metabolites, classified as natural products, which are a great resource for pharmaceuticals as feedstocks, intermediates, and products.

Matching the starting material oxidation state to desired product minimizes the need for redox reactions, which are troublesome reactions with toxic and hazardous reagents, as seen in Chapter 3. If we choose biofeedstocks that closely match oxidation state to the desired product we can avoid redox reactions as much as possible.

7.2.3 Commercial Availability

A critical feature of the best renewable feedstocks is their commercial availability. Commercial availability and low cost are good indicators that the material might otherwise go to waste, and that it is easy to isolate key components from the renewable source. One of the major challenges to producing ethanol from grass or other fibrous plant is hydrolyzing the cellulose into glucose. Once that hurdle has been cleared, the glucose derived from non-food sources, such as switchgrass, becomes a viable feedstock for liquid fuel. Agricultural waste from local sources is the best renewable feedstock of all. For example, coffee bean hulls have a very high protein content, and could be used as a renewable source of proteins and amino acids in coffee producing regions.

7.2.4 Physical State: Solid, Liquid, or Gas

The physical state of a feedstock can be an asset or detrimental. Natural gas, primarily methane, is a highly flammable gas. This makes it difficult to use as a feedstock due to the potential for leaks and spills during fracking, transportation, and manufacturing. Newlight Technologies won a 2016 PGCC Award for developing a way to use waste methane mixtures with air to make a thermoplastic they call AirCarbon™.[4] Since the methane can be used in mixtures with air, dairy barns and other indoor livestock operations can be used as a feedstock source. Newlight sends their trucks to the sources to capture the methane/air mixture. They developed whole-cell bacteria to catalyze the synthesis of a polyhydroxyalkanoate thermoplastic.[4] The bacteria used as the catalyst has three redundant polyhydroxyalkanoate synthase genes.[4] The major obstacle, product inhibition to the bacterial growth, was by-passed by disabling the negative feedback receptors on the polyhydroxyalkanoate synthase enzyme to increase the yield to an outstanding 9 : 1 product to biocatalyst ratio.[4] This yield makes the polymer cost competitive with petroleum-based polymers. AirCarbon™ is now being used to make everything from disposable forks to furniture.[4]

7.2.5 Agricultural Waste

Food crops, except for agricultural waste, should be avoided as bio-based feedstocks. The corn "stover," the leaves, stalks, and cobs of corn, is a much better source that does not interfere as much with the food supply, although corn stover is an important winter feed for beef cattle.[5] Other promising sources of cellulose and starch are switch-grass, hemp, sawdust, and wood scrap.[6,7] Although most of the work in this area has been done on production of biofuels, there is a similar high potential for biofeedstocks.

One of the major challenges to using agricultural waste as feedstock is the high water content, incurring costly drying and isolation processes. However, that high water content can be turned around and used advantageously in an enzyme or bacteria catalyzed process.

7.3 Cellulose

Cellulose, the main structural polymer of all plants, is the most abundant biomaterial on earth. Cellulose can be processed to make a variety of useful materials: paper, cardboard, textiles, building materials, biofuel, adhesives, cellophane, and celluloid film. Renewed interest in using biofeedstocks has led to greener ways to process cellulose into well-defined materials.

Cellulose is not water-soluble; previously, a strong base had to be used to dissolve it. Rayon, a silk substitute, was first made in France from cellulose nitrate in 1884, then in England by the viscose process in 1894.[8] The chemical steps for viscose involve dissolving cellulose in

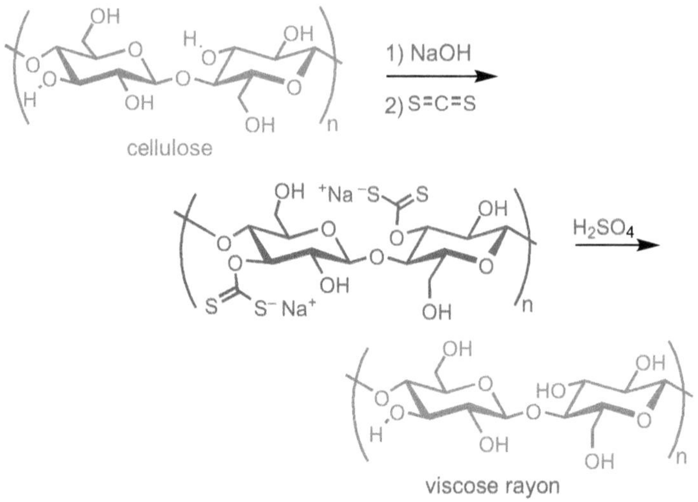

Figure 7.3 Chemical steps in the viscose process of making Rayon from cellulose.[9] Green indicates renewable feedstock, and brown indicates depleting feedstock.

a strong base (17.5% NaOH), followed by a reaction with toxic carbon disulfide (CS_2) to produce cellulose xanthate, which is processed into fibers (Figure 7.3).[9] The xanthate is then removed by reaction with sulfuric acid (Figure 7.3).[9] The final viscose cloth product is free of CS_2, but the strong base, strong acid, and CS_2 used in the process are very polluting and harmful to workers. They are gradually being phased out in favor of more benign processes, such as lyocell.

Lyocell (Tencel®) is a greener cellulosic fabric than rayon. Lyocell is processed by dissolving cellulose in a non-toxic radical amine oxide (R_2N–O·) solvent.[10] The solvent is recycled in a closed-loop process by Lenzing of Austria to make the brand name fiber, Tencel®. The process won the *European Award for the Environment* from the European Union in 2002.[11] Although the fiber production is a green process, the chemical treatment and dyeing to produce fabrics for clothing and upholstery may or may not be green, depending upon the producer.

7.4 Sugars

The monomers of polysaccharides like cellulose and starch are simple sugars, and glucose is the most abundant of these. Yeast and microbes can be used to easily ferment glucose as a feedstock to produce a broad range of products from ethanol to complex antibiotics. Yeast fermentation to produce ethanol is thought to be the oldest chemistry done by

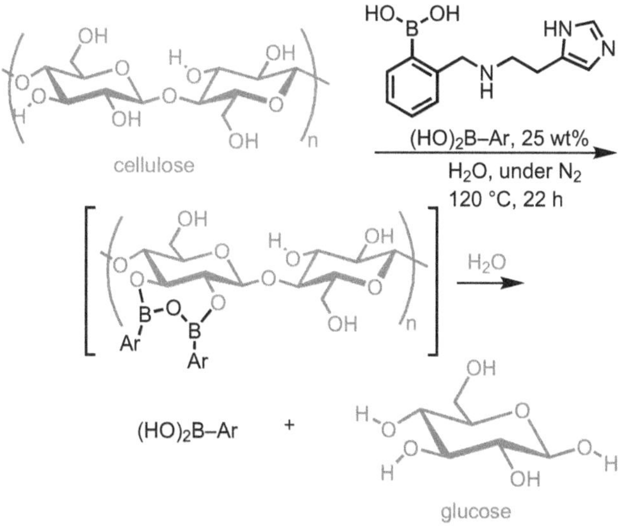

Figure 7.4 Hydrolysis of cellulose catalyzed by boronic acids in water.[12]

humans. Processing sugars into biofuel is an intense area of current research. However, hydrolysis of cellulose into glucose is challenging because of the tight hydrogen-bonding between strands of the linear polymer. Cellulose depolymerization has been achieved under very mild conditions with water-solubilized *o*-aminomethyl-phenylboronic acids (Figure 7.4).[12] The most efficient catalysts were morpholine and imidazole derivatives that both gave 48% cellulose dissolution (Figure 7.4).[12]

7.5 Case Study: Cost-advantaged Production of Intermediate and Basic Chemicals from Renewable Feedstocks (Genomatica)

Genomatica won a PGCC Award in 2011 for the production of commodity chemicals from renewable feedstocks. Their first product was 1,4-butanediol (BDO). Like BioAmber (Chapter 6), their goal was the production of 4-carbon intermediates from the tricarboxylic acid (TCA) biosynthetic pathway, also called the Kreb's cycle. However, instead of stopping at succinic acid, Genomatica produced the reduced diol, BDO, avoiding the highly flammable reducing agents BH_3 or $LiAlH_4$ required to chemically synthesize alcohols from carboxylic acids.

The older process for making BDO was developed by Walter J. Reppe. The Reppe processes use acetylene (ethyne), an explosive, petroleum-based gas, the strong base, sodium amide ($NaNH_2$), and hydrogen gas,

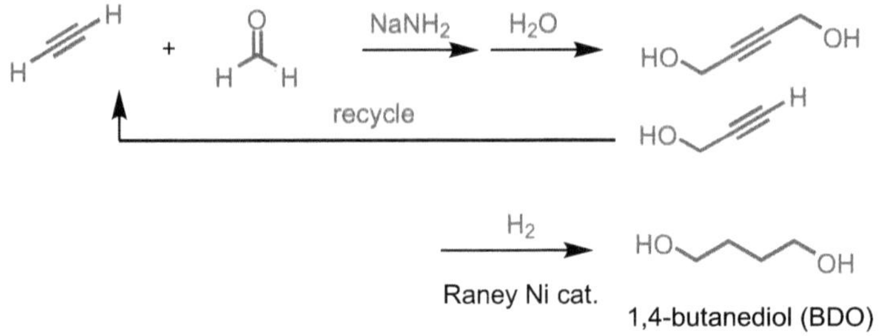

Figure 7.5 The Reppe process for synthesizing BDO.

also explosive (Figure 7.5). The intermediate propargyl alcohol is recycled to achieve yields up to 95%. Acetylene and hydrogen gases require working under pressure. Reppe was known for his design of spherical stainless-steel reactors to minimize the risk of working with acetylene.

Genomatica uses sugars, in particular a patented mixture of sucrose and glucose, as the renewable feedstock for their process; biomass and syngas are also possibilities (Figure 7.6).[13] Syngas, a mixture of CO, CO_2, and H_2, which can also be produced from renewable sources, is an alternative feedstock (Chapter 6). Life cycle analysis shows that the process uses 60% less energy and produces 70% less CO_2 emissions.[14] Genomatica projected that start-up capital costs will be 50% lower, and operating costs 15 to 30% lower, than petroleum-based processes.[14]

Sugar is fed to *E. coli* engineered with enzymes from other organisms to sidetrack the Krebs cycle (Figure 6.6) to make BDO, an unnatural reduced product, from the ordinary succinyl-CoA metabolic precursor (Figure 7.6).[14] The process requires four equivalents of the biological reducing agent, nicotinamide adenine dinucleotide hydride (NADH). Hundreds of enzymes were screened to find those that produced pathways to BDO. After discovering functional routes to BDO, the organisms were engineered to optimize growth rates, the number of NADH equivalents, and the throughput of carbon in the system.[14] The *E. coli* was engineered to be dependent on the BDO pathway for survival, eliminating undesired by-products. Since the engineered organism cannot survive without the BDO pathway, safety is ensured if it were to be accidentally released in the environment. The fermentation broth is processed by cell separation, salt separation, water removal, and BDO purification.

The Genomatica BDO process recycles the coenzyme-A (CoA) carboxyl activating reagent and the NADH reducing agent, so it is relatively atom economical even though these are high MW reactants (Figure 7.6). It is a very efficient synthesis since all the chemical steps are done in one reactor, without protecting groups, and with enzyme catalysts. The synthesis is also benign through the use of biological

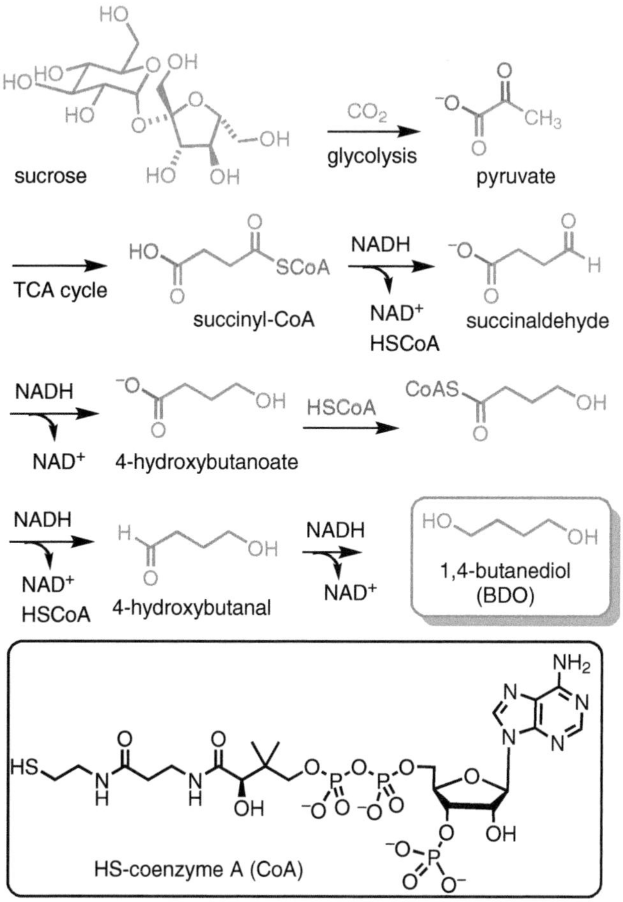

Figure 7.6 Genomatica biosynthesis of 1,4-butanediol (BDO) from sucrose using the TCA cycle (Figure 6.6) in genetically engineered *E. coli*. Structure of the HSCoA cofactor for the reduction of carboxylic acids.[13,14] (See Figure 6.7 for the structures of NADH and NAD^+).

catalysts and reducing agents, water as the solvent, low temperature, and ambient pressure.

At the time of the application for the 2011 PGCC Award, Genomatica had successfully increased BDO production by 20 000-fold and validated the process at a pilot plant scale of 3000 L fermentation.[14] Products made globally from BDO are polybutylene succinate (PBS), thermoplastic polyurethane, spandex, and solvents like THF.[14]

7.6 Lignins

Lignins are the highly cross-linked polyphenolic compounds that give wood its hardness and its brown color (Figure 7.7). Lignins are

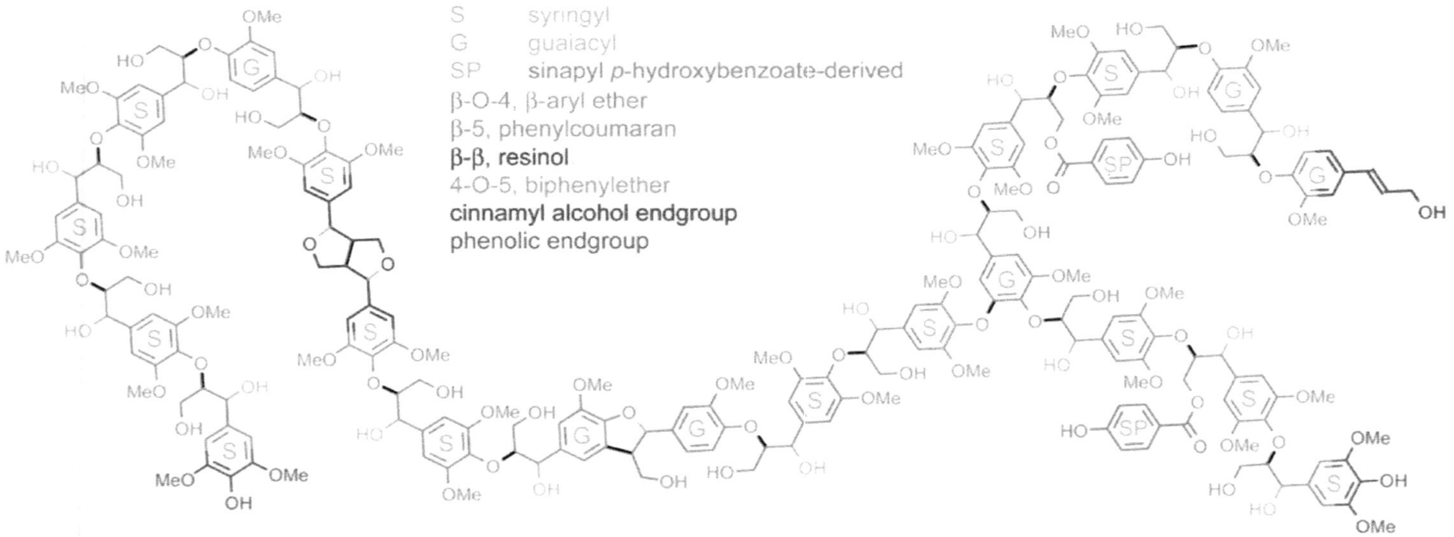

Figure 7.7 Structure of lignin polyphenolic polymer. Republished with permission of American Society of Plant Physiologists from ref. 38; permission conveyed through Copyright Clearance Center.

chemically resistant, so that paper production requires bleaching to remove them. In the past when chlorine was used for bleaching, dioxin was created in the process. In processes that use wood as a source of cellulose or sugars, lignins are typically removed from the cellulose prior to hydrolysis and fermentation of sugars. Lignins are partially reduced to aromatic hydrocarbons relative to sugars, so they have great potential as sources of feedstocks for polymers, or other phenolic commodity feedstocks.

7.7 Nitrogen: Proteins, Amino Acids, and Nucleic Acids

Nitrogen is a crucial element in medicinal, agricultural, and polymer chemistries. Of course, lipids and saccharides have no nitrogen content, which makes them better feedstocks for liquid fuels. Nitrogen-containing fuels create a lot of toxic NO_x gases upon burning. Medicines and pesticides require biological activity, so their composition necessarily reflects the composition of the cell. The most important component of fertilizer is nitrogen due to its necessity in creating the nucleic acids and proteins of living cells. Plants require microbes in the soil to "fix" nitrogen gas from the atmosphere—essentially to reduce it to a reactive amine form. The Haber–Bosch process, which uses 1–2% of the world's energy supply, is used to make nitrogenous fertilizer—ammonia, ammonium nitrate, and urea. Nylons, urethanes, and epoxy polymers contain high nitrogen levels relative to other commodity polymers.

7.7.1 Proteins and Amino Acids

Proteins are a great source of nitrogenous feedstocks. The average elemental composition of proteins is $\sim C_5H_8O_2N_1$ or about 12 wt% N.[15] Proteins are composed of 20 common amino acids, while non-ribosomal peptides and proteins may contain many other amino acids, including *N*-methylated. Each amino acid has an alpha carbon with an amino group and a carboxy group attached to it (Figure 7.8). The side chains of the amino acids contain an abundance of structures with different elemental compositions: hydrocarbon only, oxygen, sulfur, and nitrogen (Figure 7.8). These versatile functional groups have potential uses as feedstocks for many organic reaction

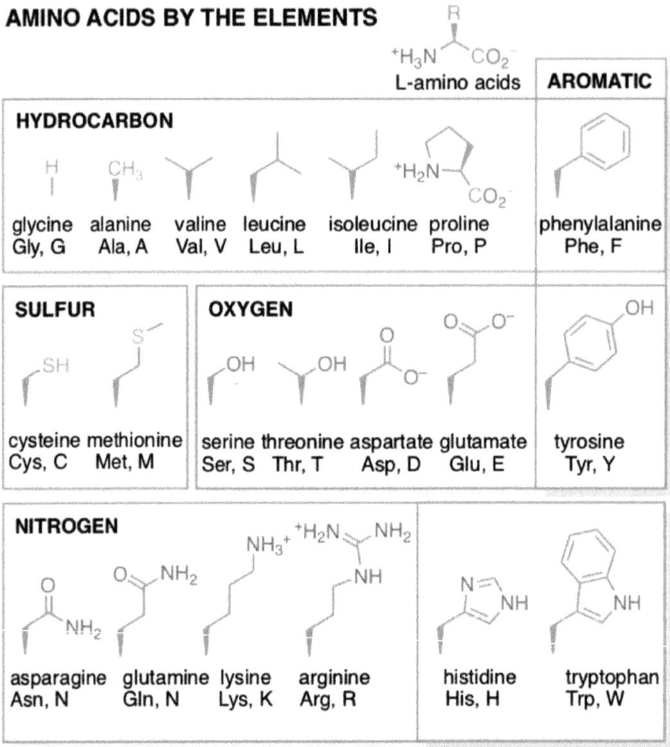

Figure 7.8 Structures of the 20 common amino acids organized by side chain elemental composition. The aromatic amino acids are grouped to the right.

processes. The asymmetric alpha carbon of the amino acids, designated "L" for the natural amino acids, and "D" for the unnatural enantiomers, also makes them useful as single-enantiomer feedstocks (Chapter 2).

7.7.2 Collagen

Collagen is the most abundant protein on earth, constituting ~30% of all vertebrate protein. Collagen is the fibrous, structural protein that makes up skin, bone, cartilage, and the sacs that contain organs. Type I and II collagens fold into a right-handed triple helix (Figure 7.9).[16] The composition of Type I collagen is rich in glycine (Gly, ~33%), proline (Pro, ~9%), and (4*R*)-hydroxyproline (Hyp, ~13%).[17] Hyp results from stereoselective post-translational enzymatic modification of the prolines in collagen by prolyl 4-hydroxylase.[16]

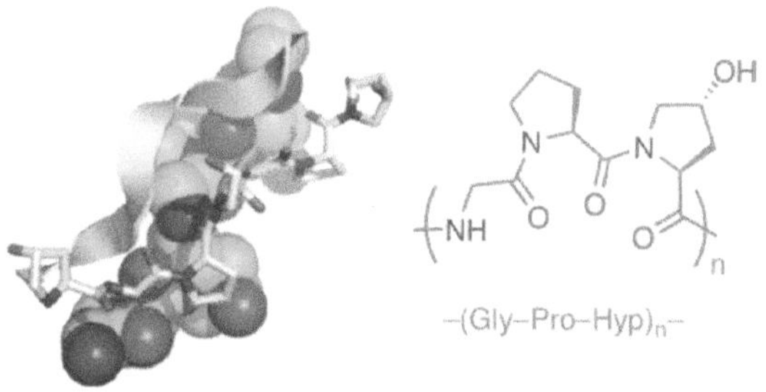

Figure 7.9 L: The triple helical structure of collagen. Each chain is shown a different way: stick, ball, and ribbon. R: Canonical sequence of the triple helical segments of collagen. Reprinted with permission of American Chemical Society from ref. 18, Copyright 2008.

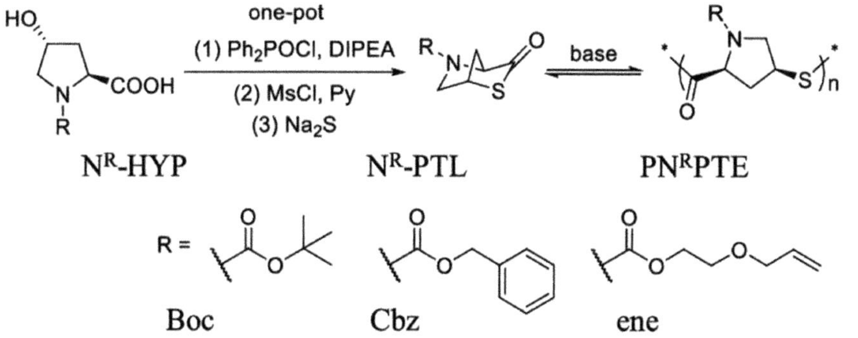

Figure 7.10 Ring-opening polymerization to make polythioesters from (4*R*)-Hyp, a biofeedstock from collagen. Reprinted with permission of American Chemical Society from ref. 19, Copyright 2019.

The canonical sequence of amino acids in the triple-helix regions of collagen is Gly–Pro–Hyp (Figure 7.9). Collagen is very strong and flexible, but not elastic. Glue and gelatin are two common consumer products made from collagen.

Yuan *et al.* designed reversible polythioesters starting from the natural feedstock (4*R*)-Hyp in two steps (Figure 7.10).[19] Ring-opening polymerization (ROP) is base-catalyzed with triethylamine, and must be stopped in a timely manner to prevent chain transfer reactions that lead to higher polydispersities.[19] The protecting groups on the amine can be removed to effect functionalization of the polymer after polymerization.[19] The polymerization is completely reversible, and the bicyclic thioester monomer is fully recoverable.[19]

7.8 Case Study: Production of Biofeedstock Dicarboxylic Acids for Nylon (Verdezyne)

Nylons are extremely strong and stable polyamides that are used wherever that stability is necessary: backpacks, tents, rain gear, even automotive engine compartment parts. Due to this stability, they do not degrade well in the environment. Nevertheless, the uses of nylon are so diverse that their synthesis from biofeedstocks is a welcome advance in green chemistry. Nylons are made from amine and carboxylic acid monomers. The monomers can be a diamine and a diacid, referred to as "nylon-*n*,*n*" where the first *n* is the number of carbons in the diamine, and the second *n* is the number of carbons in the diacid (Figure 7.11). Alternatively, a single monomer can have an amine at one end and an acid at the other end, such as nylon-6, made from 6-aminohexanoic acid. Caprolactam is the 7-membered ring feedstock for nylon-6. As of 2019, Genomatica is engineering a pathway to produce caprolactam from biofeedstocks.[14]

Adipic acid (1,6-hexanedioc acid) is the highest demand industrial dicarboxylic acid—85% of market demand is used in producing nylon-6,6.[20] To make the high purity adipic acid required for nylon-6,6 polymerization from biofeedstock, the intermediate *cis,cis*-muconic acid was made with an engineered bacterial strain of *Pseudomonas putida* KT2440 by Vardon *et al.* (Figure 7.11).[20] In this experiment, benzoate was used as a surrogate for lignin or

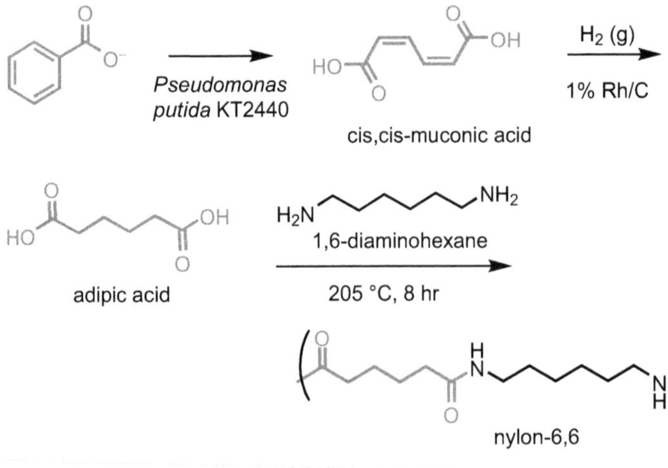

Figure 7.11 Synthesis of nylon-6,6 from bio-adipic acid produced by fermentation of benzoic acid with an engineered *P. putida* KT2440 strain.[20]

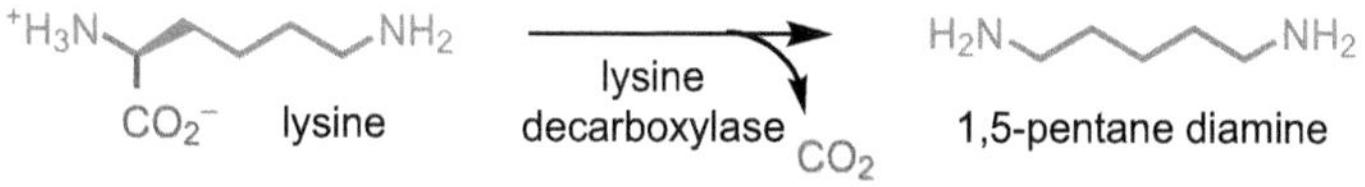

Figure 7.12 Enzymatic synthesis of 1,5-pentane diamine, a biofeedstock for nylon-5,6, from the amino acid, lysine.[21]

sugar-derived aromatic precursors to muconic acid. The production of muconate was 34.5 $g\,L^{-1}$ fermentation broth.[20] The *cis,cis*-muconic acid was then converted to adipic acid by catalytic hydrogenation over 1% Rh on activated carbon (Figure 7.11). Crystallization from EtOH produced adipic acid at the required 99.8% purity for nylon-6,6 polymerization.[20]

Verdezyne, Inc. won a PGCC Award in 2016 for the production of the dicarboxylic acids: adipic, sebacic, and dodecanedioic (DDDA). The process avoided the high temperatures, pressures, and nitric acid used in the conventional process from oxidation of a mixture of cyclohexanol and cyclohexanone to give the 6-carbon adipic acid. DDDA was produced in a yeast aerobic fermentation system from vegetable oil feedstocks. The bio-DDDA was used to make nylon-6,12 for a wide variety of products that require high chemical, moisture, or abrasion resistance: paintbrushes, hairbrushes, toothbrushes, adhesives, coatings, fragrances, automotive oils, and aviation oils.

Nylon-5,6 is a polymer of 1,5-pentanediamine and adipic acid. Toray in Japan synthesizes 1,5-pentanediamine from the amino acid lysine (Lys) by enzymatic decarboxylation for the production of bio-based Nylon-5,6 (Figure 7.12).[21] This is an efficient process because the biofeedstock already contains both nitrogens at the same oxidation state as in the final product. Only one carboxyl (CO_2) group must be removed, and that is done catalytically with the enzyme catalyst, lysine decarboxylase.

7.8.1 Nucleic Acid Bases

The side chains of nucleic acids are bases—purines: adenine, guanine, and pyrimidines: cytosine, thymine, and uracil (Figure 7.13). These bases may be useful as feedstocks for anything that includes nitrogen heterocycles, such as medicinal or polymer chemistry. Nucleic acid bases have a very high nitrogen content; for example, the RNA bases have a composition of $\sim C_{18}H_{19}O_4N_{15}$, about 40 wt% nitrogen. However, the protein content of the cell is about 3x the combined DNA and RNA content.[15] The choice of amino acid or nucleic acid as a feedstock for a nitrogen-containing product would depend on

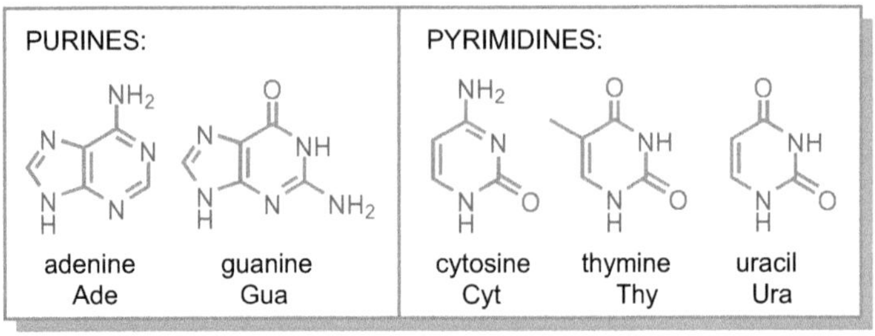

Figure 7.13 Nucleic acid bases, the purines: adenine and guanine, and the pyrimidines: cytosine, thymine, and guanidine, are potential nitrogenous feedstocks.

whether aromatic nitrogen heterocycles, as in the nucleic acid bases, or reduced amines, as in amino acids, are closer to the desired structure (Figure 7.13).

7.9 Case Study: Human Immunodeficiency Virus (HIV) Drug Carbovir from a Purine

Carbovir (Abacavir) is on the World Health Organization's Model List of Essential Medicines, both alone and in combination anti-retroviral therapy for HIV.[22] Retroviruses reproduce by inserting their DNA into the host cell DNA. The genomes of retroviruses are encoded by RNA, and they use a reverse transcriptase enzyme to generate the corresponding DNA transcript (Figure 7.14). Carbovir, and other reverse transcriptase inhibitors like azidothymidine (AZT), work by mimicking a nucleoside that results in DNA chain termination during reverse transcription because there is no 3′-hydroxyl group to attach the next nucleotide and grow the chain. Some of the most potent of these types of drugs are carbocycles, isosteres that closely mimic the ribose sugar of the nucleoside. The lack of an oxygen in the 5-membered ring makes these carbocycles less susceptible to the nuclease activity of the reverse transcriptase, ensuring greater bioavailability (Figure 7.14).

An early synthesis of carbocyclic nucleoside analogs used a bicyclic lactam, 2-azabicyclo[2.2.1]hept-5-en-3-one, as the carbocycle synthon, and 5-amino-4,6-dichloropyrimidine as the purine synthon (Figure 7.15).[23] The final step to create the adenine base was energy intensive and atom uneconomical, involving diethoxymethyl acetate to supply the last carbon of the ring, and ammonia to convert the remaining chloro into the exocyclic amine of the adenine base (Figure 7.15).[23] This synthesis was significant because of the use of

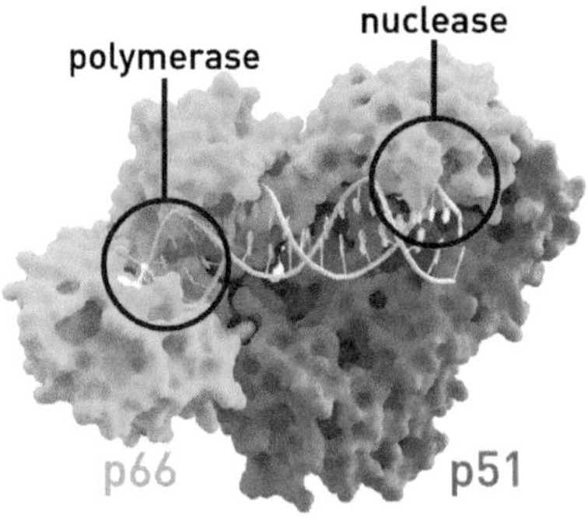

Figure 7.14 Reverse transcriptase structure from crystal structure PDB 3KLF. Reprinted from https://commons.wikimedia.org/w/index.php?curid=33598425 under the terms of the CC BY-SA 3.0 license, https://creativecommons.org/licenses/by-sa/3.0/.

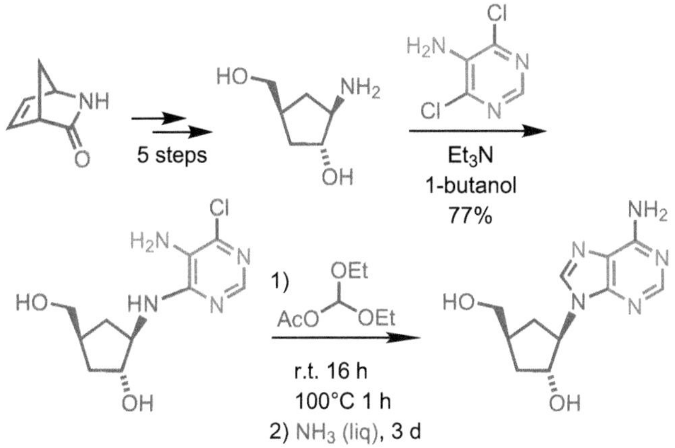

Figure 7.15 An early synthesis of carbocyclic nucleoside analogs by Daluge *et al.*[23]

the bicyclic lactam, which later became an essential component of a greener synthesis of carbocyclic nucleoside analogs.

A recent advance was the use of guanine as a biofeedstock for the production of 2-amino-6-chloropurine (Figure 7.16).[24] The Vilsmeier reagent is made from DMF and bis(trichloromethyl)carbonate, a safer substitute for phosgene because it is a solid (Figure 7.16). The Vilsmeier reagent was used as the chlorinating reagent directly on unprotected guanine (Figure 7.16).[24] Substitution of the hydroxyl group in the tautomer of guanine with chlorine from the Vilsmeier reagent probably takes place by an electrophilic aromatic substitution, which would be unusual for nitrogen heterocycles.

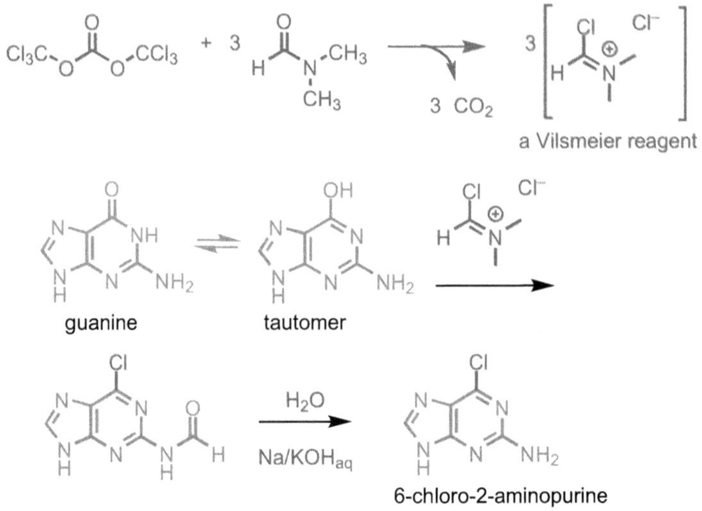

Figure 7.16 Greener synthesis of 6-chloro-2-aminopurine from guanine.[24]

The Vilsmeier reagent also adds a formyl group to the primary amine, which is removed with aqueous base (Figure 7.16).[24] The formyl group could also serve as a temporary protecting group if further modifications were necessary.

Crimmins 1996 asymmetric synthesis of carbovir began with Evan's chiral auxiliary (Chapter 2) to create the carbocyclic diacetate in five steps (Figure 7.17), including an asymmetric aldol reaction with pivaloyl 4-pentenoate, and a Ru-catalyzed Grubbs olefin ring closing metathesis (RCM).[25] Note that the use of the Evan's chiral auxiliary added two steps to the synthesis—adding the auxiliary and then removing it at the end. $Pd(PPh_3)_4$ catalyzed condensation of the single-enantiomer diacetate with 6-chloro-2-aminopurine, and substitution of Cl with cyclopropylamine, gave a reasonably good yield of single enantiomer (−)-carbovir in eight steps overall (Figure 7.17).[25] Replacement of the chloro group with cyclopropylamine in the last step probably occurs by nucleophilic aromatic substitution (S_NAr), a mechanism that is common with electron-withdrawing, nitrogen-containing heterocycles (Figure 7.17). This was an early example of using a purine, in this case guanine, as the feedstock for the synthesis of a drug, and the first asymmetric synthesis carbovir.

Taylor *et al.* (Dow Pharma) developed a whole-cell enzymatic resolution to produce the enantiopure lactam to serve as the carbocyclic precursor to carbovir (Figure 7.18).[26] The synthesis condenses the preparation of the single enantiomeric carbocycle from five steps to a

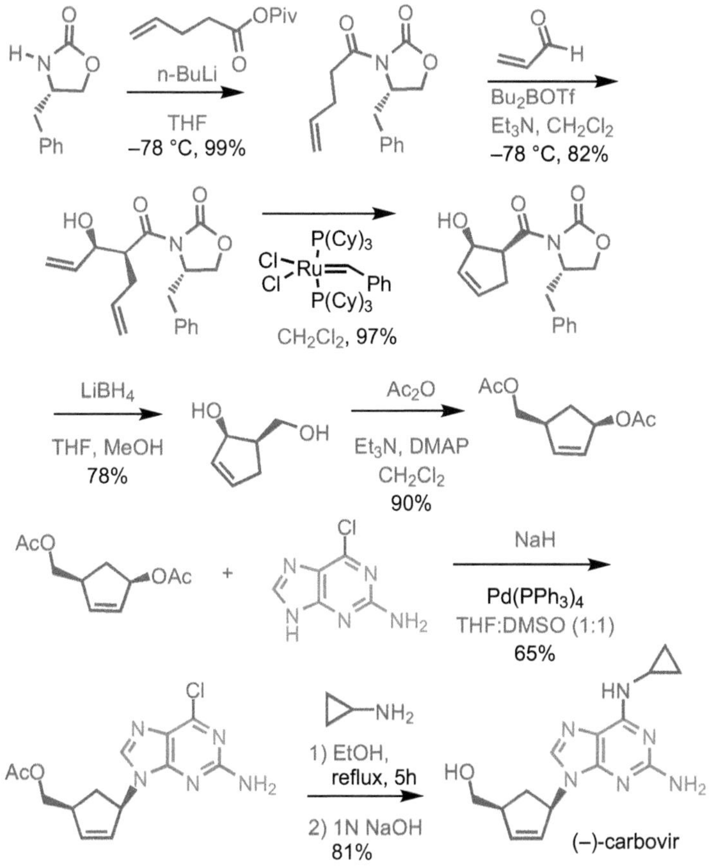

Figure 7.17 Crimmins asymmetric synthesis of single enantiomer (−)-carbovir with guanine as a feedstock. Piv = pivaloyl; Tf = trifluoromethane-sulfonate; DMAP = 4-*N*,*N*-dimethylaminopyridine.[25]

single resolution step (Figure 7.18).[26] The adenine base must then be synthesized *via* the method of Daluge *et al.* from 5-amino-4,6-dichloropyrimidine (Figure 7.15).[23]

7.10 Lipids: Fats and Oils

Lipids are triesters of glycerol, called fats if they are solid at 25 °C, and oils if they are liquid at 25 °C (Figure 7.19). Although they are an important component of cell membranes, phospholipids will not be discussed as a biofeedstock. Because lipids are mainly hydrocarbon with low oxygen content, they have a relatively high energy content per unit mass. Lipids are therefore very useful as liquid biofuels, such

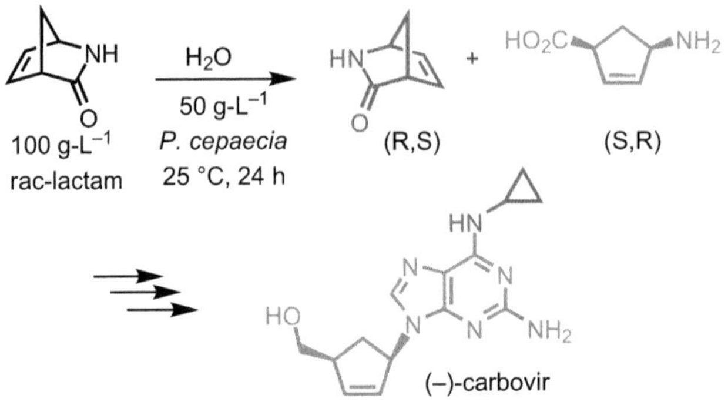

Figure 7.18 Taylor *et al.* enantioselective synthesis of carbovir by enzymatic resolution of 2-azabicyclo[2.2.1]hept-5-en-3-one lactam using the base synthesis of Daluge *et al.*[26]

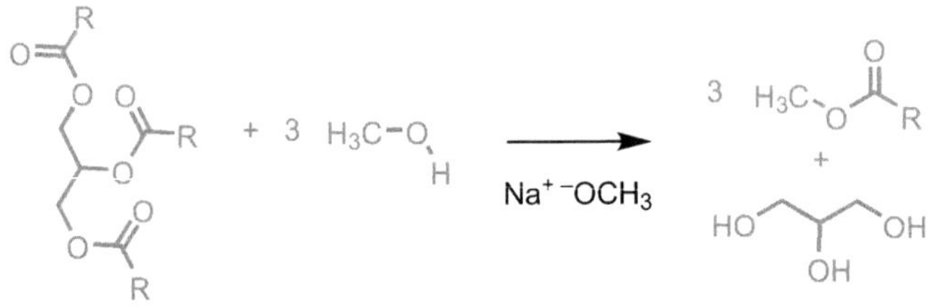

Figure 7.19 Synthesis of biodiesel from lipids.

as biodiesel (Figure 7.19). Biodiesel requires much less processing than ethanol, requiring the usual oil separation from the plant, and a single reaction with methanol, using $NaOCH_3$ as a basic catalyst (Figure 7.19). (See Appendix B and video for the biodiesel mechanism.) Ethanol comes from cellulose or starch, which must be separated from lignins, hydrolyzed to simple sugars, fermented, and dehydrated to produce ethanol. In an "all-of-the-above" strategy, biodiesel has great potential as an energy source, especially for oilseed-bearing crops that can be grown on marginal land.[27]

Lipids are much less abundant than polysaccharides in high production agricultural plants. For example, corn kernels contain about 61–78% starch, 5–6% celluloses, 6–12% protein, and 3–6% lipids by dry weight.[28] On the other hand, olives contain about 60% lipids and 28% carbohydrates plus fiber by dry weight.[29] The fatty acids of lipids, the R–CO_2H component of the ester, vary from less than five carbons (short chain) to more than 22 carbons (very long chain), but typically are C_{12} to C_{18} (Figure 7.19). Fatty acids are classified as saturated or unsaturated, referring to the hydrogen content of the side chain.

Saturated fats have alkyl groups with the maximum amount of hydrogen—they are fully reduced. Unsaturated fats have one to six *cis*-alkene double bonds, and these are non-conjugated. The allylic carbons in between the alkenes are particularly susceptible to free radical oxidation. In Chapter 5, Avoid Auxiliaries, we saw how Eastman uses lipid feedstocks for personal care products. In Chapter 9, Catalysis, we will see how Elevance uses lipid feedstocks for a wide variety of commercial products.

7.11 Natural Products

Natural products have been the source of about 49% of all new pharmaceutical entities from 1981 to 2014.[30] This includes natural product derivatives, defined botanical mixtures, and biological macromolecules (Figure 7.20). Another 21% were synthetic drugs inspired by natural products.

7.11.1 Synthetic Efficiency

If renewable feedstocks are chosen to closely match the desired product, both in terms of structure and oxidation state, synthetic

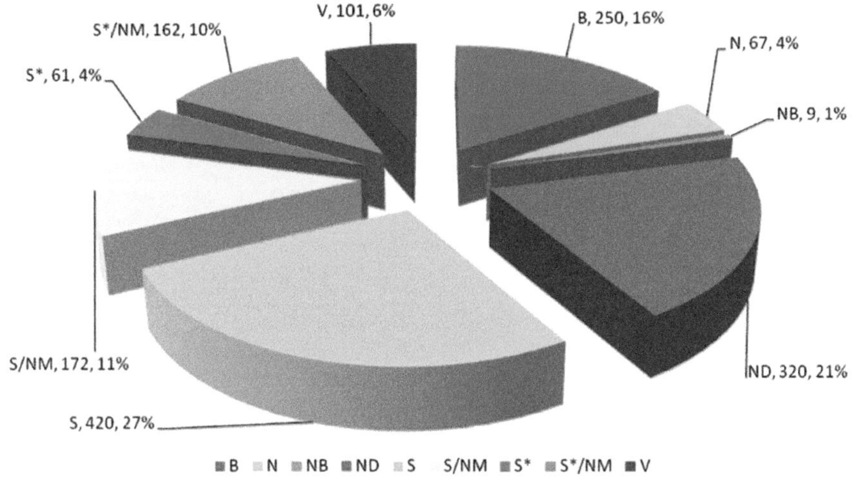

Figure 7.20 New therapeutic agents 1981–2014. Abbreviations and dates: B, Biological macromolecule, 1997; N, Unaltered natural product, 1997; NB, Botanical drug (defined mixture), 2012; ND, Natural product derivative, 1997; S, Synthetic drug, 1997; S*, Synthetic drug (NP pharmacophore), 1997; V, Vaccine, 2003; /NM, Mimic of natural product, 2003. Reprinted with permission of American Chemical Society from ref. 30, Copyright 2019.

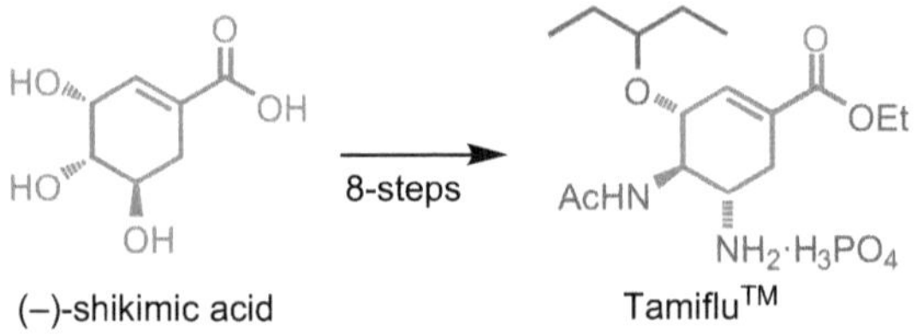

Figure 7.21 The natural product, (-)-shikimic acid, was used as the single enantiomer biofeedstock for the Nie synthesis of oseltamivir phosphate (Tamiflu™).[31]

efficiency is greatly increased. In addition, biofeedstocks are typically pure stereoisomers, which give improved atom economy if used judiciously. Two examples of this efficiency in medicinal chemistry, Tamiflu™ from (−)-shikimic acid, and Taxol™ from 10-deacetyl baccatin, are described next.

The active ingredient in Tamiflu™, oseltamivir phosphate, has been synthesized from shikimic acid, a single-enantiomer natural product that is the precursor of all aromatic amino acids (phenylalanine Phe, tyrosine Tyr, and tryptophan Trp) in plants (Figure 7.21). For the synthesis described in Chapter 2, the source of the shikimic acid was star anise, an abundant source with 1.1 kg (−)-shikimic acid per 30 kg of dried plant material.[31] Each of the stereocenters in shikimic acid was used to attain the desired stereochemistry of the product, either by retention, inversion, or double inversion. The presence of the three stereocenters in the shikimic acid feedstock provided a clear route to obtaining the desired stereochemistry of the Tamiflu™ product.

7.12 Case Study: Synthesis of Paclitaxel from Pacific Yew Tree Needles

A wonderful example of the efficient use of a biofeedstock to produce a very complex, very important anti-cancer drug in only four chemical steps, is the semi-synthesis of the anti-cancer drug, paclitaxel (Figure 7.22). The biofeedstock was 10-deacetyl baccatin III obtained from the needles of the Pacific yew tree, *T. brevifolia*. Notice the incredible complexity of the paclitaxel molecule—multiple oxidized functional groups, multiple fused rings, and many stereocenters. Yet paclitaxel is a single enantiomer (Figure 7.22). The first total synthesis reported by Holton *et al.*

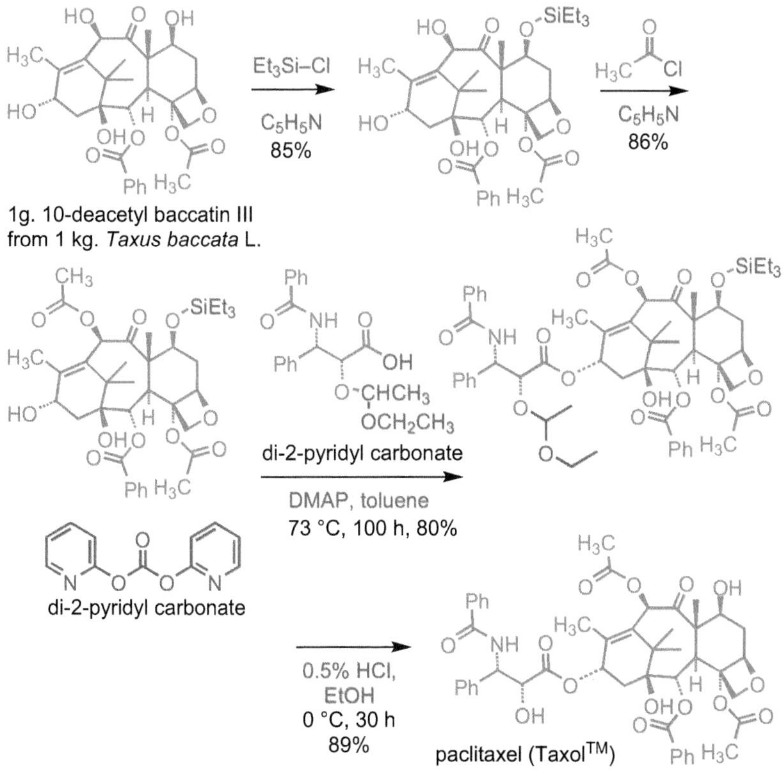

Figure 7.22 Synthesis of anti-cancer drug paclitaxel from 10-deacetyl baccatin III derived from yew tree needles. Adapted with permission of American Chemical Society from ref. 34, Copyright 1988.

in 1994, beginning with commercially available, petroleum-based reactants, comprised over 27 steps with 4–5% overall yield.[32,33] The synthetic efficiency of such a tour-de-force synthesis is almost unimaginably low.

In 1988, a semi-synthesis of paclitaxel from the natural product, 10-deacetyl baccatin III from yew tree needles was reported by the Greene and Guéritte-Voegelein groups in France (Figure 7.22).[34] The synthetic efficiency was greatly improved, including overall yield, atom economy, as well as the E-factor. The overall yield from 10-deacetyl baccatin III is extremely high compared with total synthesis. No oxidation or reduction steps are done in the 4-step synthesis, which improved the synthetic efficiency and avoided toxic reagents. The 10-deacetyl baccatin III feedstock already contains most of the correct stereocenters in the final paclitaxel product; no changes needed to be made to the stereochemistry of the complex fused-ring core. As is often the case, natural products are a great source of pure single-enantiomer feedstocks.

The synthesis still uses protecting groups, a form of derivitization (Chapter 8), in particular, one of the 2° alcohols is protected with triethylsilyl (TES) in the first step, and the TES is removed in the last step.[34] In the second step, another 2° alcohol is converted to the acetyl ester found in the product. Apparently, the 3° alcohol is unreactive, requiring no protecting group during the esterification in the key third step to add the side chain. There is also room for improvement in the use of pyridine and *N,N*-dimethylaminopyridine (DMAP) as the bases in three of the steps—less toxic alternatives could be found.

Holton and coworkers improved the semi-synthesis even further, using a highly reactive β-lactam to install the side chain in the third step (Figure 7.23).[35,36] They still used 10-deacetylbaccatin III, and the TES and Ac protecting groups were the same. TES was also used on the side chain β-lactam synthon. The key step of adding the side chain was greatly improved by deprotonation of the unprotected 2° alcohol with *n*-butyl lithium, followed by ring opening of the single enantiomer β-lactam (Figure 7.23).[35,36] The high space-time yield of the key step—0 °C for 2 h gave nearly quantitative yield—resulted in commercialization of the process for a time.[35,36]

These semi-synthetic routes to paclitaxel showcase all the benefits of a complex natural product feedstock. Just as in the use of enzymes as catalysts for organic synthesis, more companies are beginning to produce complex natural products in microbial or plant cell cultures. Commercial paclitaxel is currently biosynthesized by an efficient process in *Taxus* spp. plant cell culture by Phyton Biotech (USA) and Samyang Genex (Korea).[37] One major challenge of plant culture production is decreased productivity over time.[37] Stimulus with environmental elicitors, such as ethylene or methyl jasmonate, can lead to rejuvenation of the culture.[37]

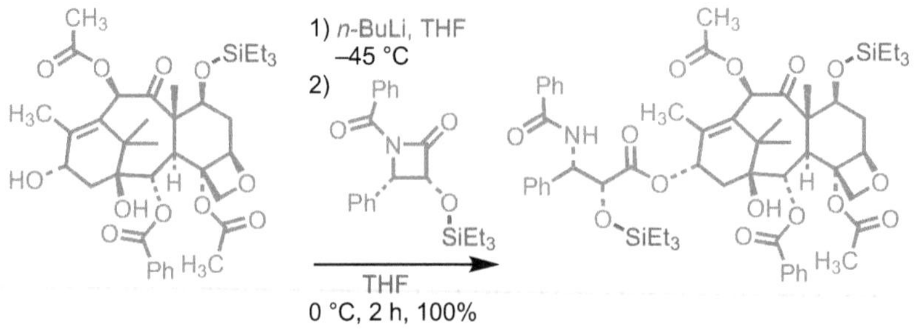

Figure 7.23 Improved semi-synthesis of paclitaxel by the Holton group.[35,36]

7.13 Summary

Because fossil feedstocks are becoming harder to extract, and the process of mining, synthesis, and disposal are all polluting, green chemistry seeks to use renewable feedstocks, typically derived from plants. Renewable feedstocks should be commercially available, best from agricultural waste products. They can come in any physical form: solid, liquid or gas. Biofeedstocks typically have the advantage of oxidation state matching—the feedstock oxidation state is chosen to match that of the desired product to avoid unnecessary steps and toxic redox reagents. Biofeedstocks can also improve synthetic efficiency if the feedstock is chosen to match the structure and oxidation state of the product. Earth-abundant elements such as Si, Al, and Fe can also be considered renewable. The most significant biofeedstocks in terms of quantity available are: cellulose, sugars, lignins, proteins, and lipids. Genomatica produces 1,4-butanediol in *E. coli* from simple sugars as feedstocks. Nucleic acid bases have been used as a source of nitrogen-containing heterocycles for medicinal chemistry. The HIV drug, carbovir, has been synthesized from both pyrimidines and purines, and an enzymatic resolution led to an efficient single-enantiomer synthesis. Tamiflu™ and Taxol™ are examples of medicines that can be produced very efficiently from complex natural product feedstocks.

7.14 Problems: Renewable Feedstocks

Problem 7.1

(a) Write the balanced reaction for the photosynthesis of glucose from CO_2.
(b) Brazil and the US have both invested heavily in bio-based ethanol as an automotive fuel. What is the biofeedstock in each country?
(c) What advantages and problems are associated with both of these biofeedstocks?

Problem 7.2

(a) How do biofeedstocks reduce greenhouse gas emissions?
(b) How do biofeedstocks improve synthetic efficiency?
(c) How do biofeedstocks reduce toxicity in synthesis?

Problem 7.3

(a) What is the most earth-abundant non-metal? ___
(b) What is the most earth-abundant transition metal? ___

Problem 7.4

(a) Write a balanced equation for the Reppe petroleum-based process for making BDO from acetylene and formaldehyde with $NaNH_2$ as catalyst and an aqueous (H_2O) work-up. (Look up the pK_a values of acetylene and ammonia in DMSO in Appendix C.)
(b) Draw an arrow-formalism mechanism for just the first step of the Reppe process to make propargyl alcohol.

Problem 7.5

(a) Name the three most significant Principles of Green Chemistry that Genomatica's processes exemplify. Explain briefly.
(b) What products are made globally from BDO?

Problem 7.6

(a) Draw the structures of the sugar feedstocks used in the Genomatica production of BDO.
(b) Draw the structure of the carboxylic acid **activating agent** used in the production of BDO.
(c) Draw only the nicotinamide ring of the **reducing agent** in *E. coli* used to make BDO. How many equivalents of hydride are required in the process? _____
(d) Draw only the nicotinamide ring of the **oxidized form** of the reducing agent. What is the driving force for the reduction?

____

Problem 7.7

(a) What four steps are required to isolate pure BDO?
(b) Which step is likely to be the most energy intensive?
(c) How could the energy use in that step be reduced?
(d) Which step is likely to be the most auxiliary intensive?

Problem 7.8

Consider the Crimmins synthesis of carbovir (Figure 7.17).

(a) How many steps are needed to create the carbocycle enantioselectively?

(b) Draw the arrow-formalism mechanism for the substitution of the 6-chloro group by cyclopropylamine on the purine ring in the last step.
(c) What is the advantage of the enzymatic resolution of the lactam (Figure 7.18)?

Problem 7.9

Paclitaxel, the common name for the important anti-cancer drug Taxol™, was synthesized in four steps from Pacific yew leaves (Figure 7.22).[34]

(a) How many stereocenters are there in paclitaxel?
(b) Calculate the overall yield for the 4-step synthesis.

Problem 7.10

Calculate the overall atom economy for the 4-step synthesis of paclitaxel in Figure 7.22. Give molecular formulas and molecular weights of each reactant in a table. Reactants are shown above the arrow, catalysts and solvents are below the arrow.

References

1. P. T. Anastas and J. C. Warner, *Green Chemistry: Theory and Practice*, Oxford University Press, New York, 1998.
2. P. J. Lemstra, *Abstracts of Papers, 241st ACS National Meeting & Exposition* 2011, CNR-4.
3. National Geographic Society, *Great Pacific Garbage Patch*, https://www.nationalgeographic.org/encyclopedia/great-pacific-garbage-patch/, accessed October 14, 2019.
4. Newlight Technologies Inc., *AirCarbon: Greenhouse Gas Transformed into High-Performance Thermoplastic*, Washington, DC, 2016, https://www.epa.gov/greenchemistry/presidential-green-chemistry-challenge-2016-designing-greener-chemicals-and-specific, accessed October 11, 2019.
5. W. Edwards, *Estimating a Value for Corn Stover*, 2014, 4, https://www.extension.iastate.edu/agdm/, accessed June 25, 2019.
6. D. Biello, *Sci. Am.*: Sustainability, 2008, https://www.scientificamerican.com/article/grass-makes-better-ethanol-than-corn/, accessed October 14, 2019.
7. B. Rice, *J. Ind. Hemp*, 2008, **13**, 145–156.
8. G. B. Kauffman, *J. Chem. Ed.*, 1993, **70**, 887.
9. M. C. Cann and T. P. Umile, *Real-world Cases in Green Chemistry*, American Chemical Society, vol. II, 2008.
10. P. Laity, *Process of Making Regenerated Cellulose Articles*, Tencel Ltd., *U. S. Pat.*, US5441689A, 1995.
11. Tencel: Sustainability, 2002, https://www.tencel.com/b2b/sustainability, accessed February 25, 2019.
12. N. Levi, A. M. Khenkin, B. Hailegnaw and R. Neumann, *ACS Sustainable Chem. Eng.*, 2016, **4**, 5799–5803.

13. M. Dani, G. Ruggiero, D. Perini and A. Bianchi, *Process for the Production of 1,4-Butanediol*, Novamont S.P.A. and Genomatica Inc., WO2015158716 A1, 2015.
14. Genomatica, *Cost-advantaged Production of Intermediate and Basic Chemicals from Renewable Feedstocks*, Washington, DC, 2011, https://www.epa.gov/greenchemistry/presidential-green-chemistry-challenge-2011-greener-synthetic-pathways-award, accessed July 25, 2019.
15. R. Milo, R. Phillips and N. Orme, *Cell Biology by the Numbers*, Garland Science, New York, NY, 2015.
16. M. D. Shoulders and R. T. Raines, *Annu. Rev. Biochem.*, 2009, **78**, 929–958.
17. P. de Paz-Lugo, J. A. Lupiáñez and E. Meléndez-Hevia, *Amino Acids*, 2018, **50**, 1357–1365.
18. N. Dai, X. J. Wang and F. A. Etzkorn, *J. Am. Chem. Soc.*, 2008, **130**, 5396–5397.
19. J. Yuan, W. Xiong, X. Zhou, Y. Zhang, D. Shi, Z. Li and H. Lu, *J. Am. Chem. Soc.*, 2019, 4928–4935.
20. D. R. Vardon, N. A. Rorrer, D. Salvachua, A. E. Settle, C. W. Johnson, M. J. Menart, N. S. Cleveland, P. N. Ciesielski, K. X. Steirer, J. R. Dorgan and G. T. Beckham, *Green Chem.*, 2016, **18**, 3397–3413.
21. Toray Industries Inc., *Toray Global: Innovation by Chemistry*, 2012, http://www.toray.com/news/rd/nr120213.html, accessed February 25, 2019.
22. World Health Organization, *WHO Model Lists of Essential Medicines*, 2017, https://www.who.int/medicines/publications/essentialmedicines/en/, accessed February 27, 2019.
23. S. Daluge and R. Vince, *J. Org. Chem.*, 1978, **43**, 2311–2320.
24. *Synthetic method for preparing high-purity 2-amino-6-chloroguanine*, *CN Pat.*, CN107312003A, 2017.
25. M. T. Crimmins and B. W. King, *J. Org. Chem.*, 1996, **61**, 4192–4193.
26. S. J. C. Taylor, R. C. Brown, P. A. Keene and I. N. Taylor, *Bioorg. Med. Chem.*, 1999, **7**, 2163–2168.
27. G. Koçar and N. Civas, *Renewable Sustainable Energy Rev.*, 2013, **28**, 900–916.
28. Y. Ai and J.-l. Jane, *Compr. Rev. Food Sci. Food Saf.*, 2016, **15**, 581–598.
29. Agricultural Research Service, *Full Report (All Nutrients): 09195, Olives, Pickled, Canned or Bottled, Green*, U.S. Department of Agriculture, 1995, https://ndb.nal.usda.gov/ndb/foods/show/2283?fgcd=&man=&lfacet=&count=&max=35&sort=&qlookup=olive&offset=&format=Full&new=&measureby=, accessed July 10, 2017.
30. D. J. Newman and G. M. Cragg, *J. Nat. Prod.*, 2016, **79**, 629–661.
31. L. D. Nie, X. X. Shi, K. H. Ko and W. D. Lu, *J. Org. Chem.*, 2009, **74**, 3970–3973.
32. R. A. Holton, H. B. Kim, C. Somoza, F. Liang, R. J. Biediger, P. D. Boatman, M. Shindo, C. C. Smith and S. Kim, *J. Am. Chem. Soc.*, 1994, **116**, 1599–1600.
33. R. A. Holton, C. Somoza, H. B. Kim, F. Liang, R. J. Biediger, P. D. Boatman, M. Shindo, C. C. Smith and S. Kim, *J. Am. Chem. Soc.*, 1994, **116**, 1597–1598.
34. J. N. Denis, A. E. Greene, D. Guenard, F. Gueritte-Voegelein, L. Mangatal and P. Potier, *J. Am. Chem. Soc.*, 1988, **110**, 5917–5919.
35. R. A. Holton, R. J. Bidediger and P. D. Boatman, in *Taxol: Science and Applications*, ed. M. Suffness, CRC Press, 1st edn, 1995, pp. 97–121.
36. R. A. Holton, *Metal Alkoxides*, Florida State University, *U. S. Pat.*, US005,229,526A, 1993.
37. M. E. Kolewe, V. Gaurav and S. C. Roberts, *Mol. Pharmaceutics*, 2008, **5**, 243–256.
38. R. Vanholme, B. Demedts, K. Morreel, J. Ralph and W. Boerjan, *Plant Physiol.*, 2010, **153**, 895–905.

8 Avoid Protecting Groups

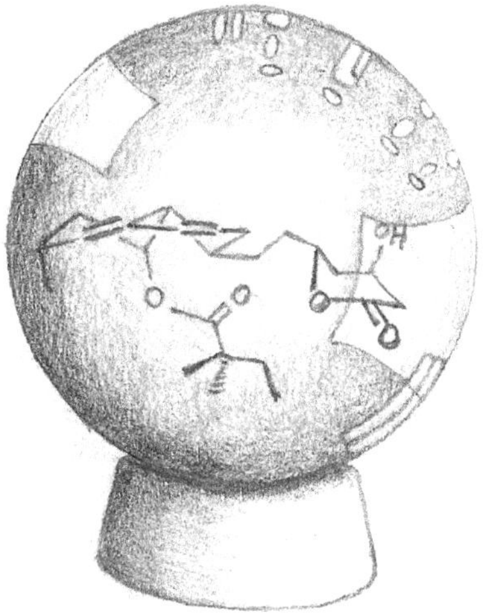

"Principle 8. Unnecessary derivatization (use of blocking groups, protection/deprotection, temporary modification of physical/chemical processes) should be minimized or avoided if possible, because such steps require additional reagents and can generate waste."[1]

8.1 Derivatives

Derivatives, for the purpose of green chemistry, are defined as temporary, covalent modifications of a molecule during a synthesis. Most frequently, derivatives are protecting groups, but chiral auxiliaries are

Green Chemistry: Principles and Case Studies
By Felicia A. Etzkorn

Published by the Royal Society of Chemistry, www.rsc.org

also used to make temporary diastereomers to prepare single enantiomers (Chapter 2).

8.1.1 Chiral Derivatives

We saw the use of chiral derivatives, such as Evan's chiral auxiliary, in Chapter 2 to isolate single enantiomer intermediates or final products. As in the use of protecting groups, syntheses using chiral derivatives include at least two extra steps: putting the chiral group on, and removing it after the asymmetric synthetic step or separation of the diastereomeric products. More efficient means of preparing single enantiomer products include chiral-pool feedstocks, asymmetric catalysis, enzymatic resolution, and crystallization of diastereomeric salts (Chapter 2).

8.1.2 Protecting Groups

To design syntheses without protecting groups, it is essential to first understand what they are and how to use them—know thy enemy. Protecting groups must be orthogonal to the reactions they undergo to survive unchanged. Orthogonal means that they are stable to the conditions of a synthesis, and they can be removed under different conditions than those used in the reactions of a synthesis, and often under different conditions from other protecting groups in the same molecule. Protecting groups can be classified by the type of conditions by which they are removed, or "labile" towards: (1) acid-labile, (2) base-labile, (3) fluoride-labile, (4) hydrogenation-labile, (5) oxidation-labile, and (6) Pd(0)-labile (Figure 8.1). Some protecting groups are harder to remove than others, even within the same class, so they require harsher conditions.

Another way to classify protecting groups is by the functional group they are meant to protect: alcohols, ketones/aldehydes, amines, *etc.* These and many other protecting groups are tabulated in a very useful way in *Greene's Protective Groups in Organic Synthesis*.[2] A very brief survey of only the most common protecting groups and standard deprotection conditions follows.

8.1.3 Acid-labile: *t*-Butyl, Acetal

The *t*-butyl group is perhaps best known for its stability as a 3° carbocation. *t*-Butyl esters and ethers can be installed using an excess of isobutylene (b.p. 19.6 °C) and an acid catalyst (Figure 8.2).

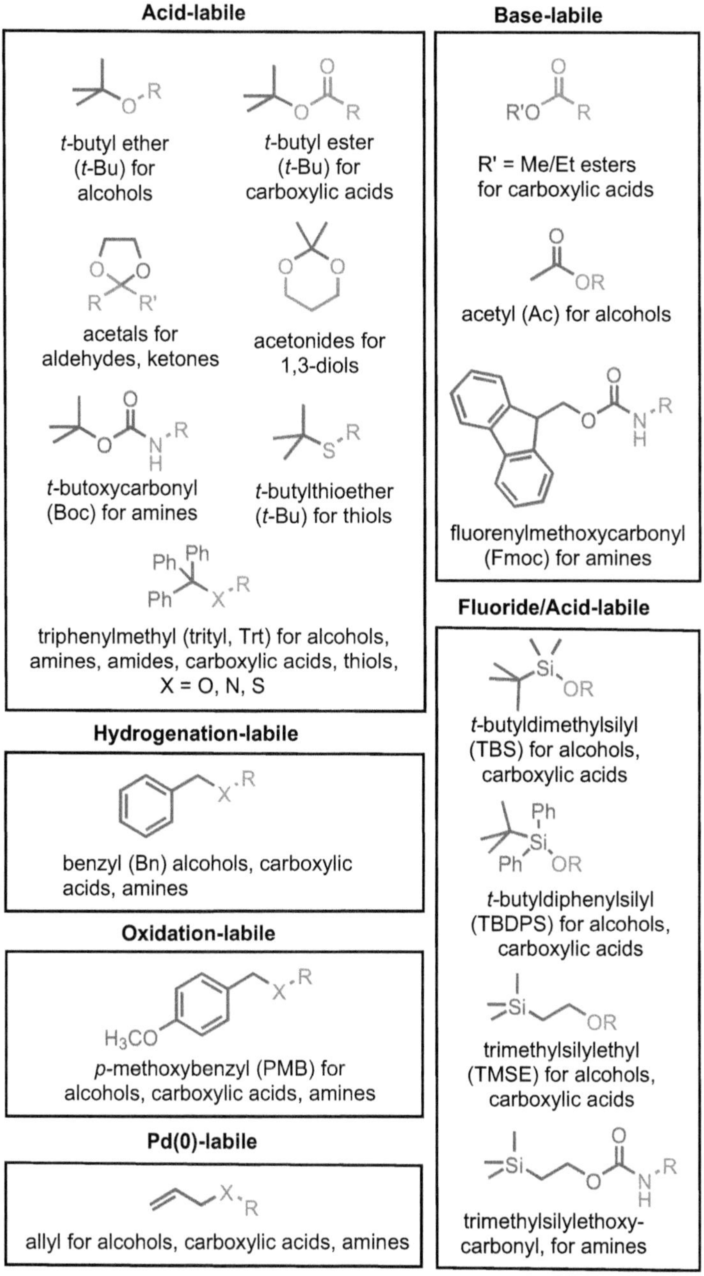

Figure 8.1 Some of the most common protecting groups (brown) are categorized by the conditions used to remove them, and the functional groups they protect (in green).

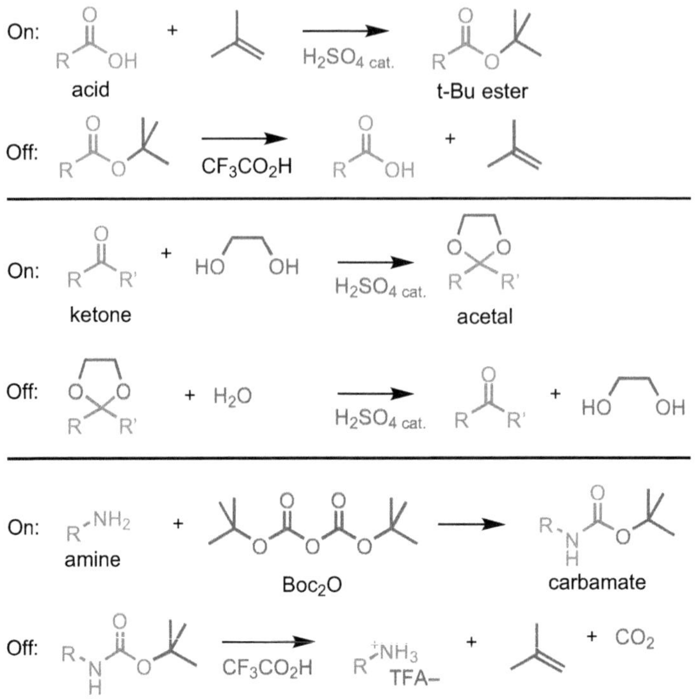

Figure 8.2 Common methods for protection ("On") and deprotection ("Off") *t*-butyl- and acetal-type protecting groups on acids, ketones, and amines.

The acid catalyst protonates the alkene to make the *t*-butyl cation, and the nucleophilic alcohol or carboxyl reacts. The *t*-butyl group is similarly removed with strong acid, like trifluoroacetic acid or HCl in an organic solvent, through an E1 elimination (Figure 8.2). This is one of the few useful E1 type eliminations, since there is no question that the correct alkene, isobutylene, will be produced, whereas E2 eliminations can be accomplished in a regio- and stereo-selective way to generate complex alkenes.

Acetals are diethers that protect ketones, or conversely ketones that protect 1,2- or 1,3-diols (Figure 8.1). Acetals are similarly removed by reversible, acid-catalyzed hydrolysis (Figure 8.2).

8.1.4 Base-labile: Ester

Esters can be used as protecting groups for carboxylic acids, typically methyl or ethyl esters, or as protecting groups for alcohols, typically acetyl esters (Figure 8.1). Simple esters like methyl or ethyl are usually made with a strong acid catalyst, such as H_2SO_4 or HCl, and

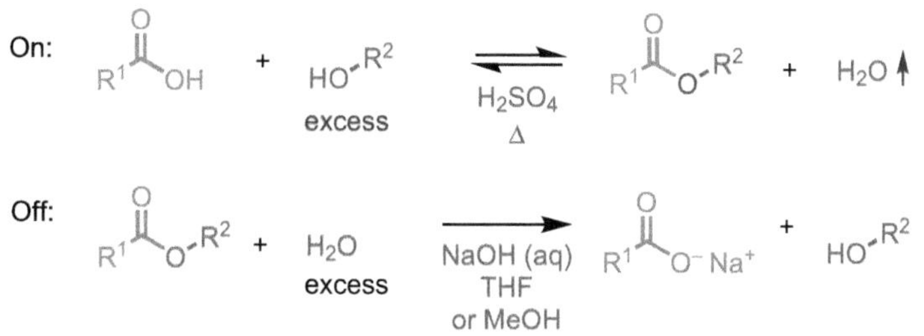

Figure 8.3 Protection and deprotection of carboxylic acids as esters.

an excess of the corresponding alcohol, methanol or ethanol, to drive the reaction (Figure 8.3). Acid-catalyzed reactions are typically reversible, so water can also be removed to drive the reaction. This is the Fischer esterification that was described in Chapter 5. Acetyl groups are usually installed with acetic anhydride or acetyl chloride.

Ester deprotection is typically base-catalyzed with an excess of water (Figure 8.3). Base-catalyzed ester hydrolysis is *not* reversible. This has to do with the relative pK_a values of the carboxylic acid (~5) and alcohol (~16) products (Appendix C). Under basic conditions, one or both products will be deprotonated. The carboxylate product is a stronger acid, and a much weaker electrophile for the reverse reaction. The alkoxide leaving group is a stronger base, and it will be protonated 10^{11}-fold more readily than the carboxylate. The resulting alcohol is a much weaker nucleophile than the alkoxide anion. Thus, both pK_a values work together to make base-catalyzed ester hydrolysis irreversible, and higher yielding.

8.1.5 Fluoride-labile: Silyl

Silyl groups are great protecting groups because their removal is so specific. The most commonly used silyl protecting group is *t*-butyldimethylsilyl (TBS, also TBDMS).[3] But the relatively high mass of the TBS protecting group makes it less atom economical. Fluoride ions remove silyl groups very selectively,[3] because silicon and fluorine form very strong bonds (~160 kcal mol^{-1}), stronger than silicon-oxygen bonds (Figure 8.4).[4] The deprotection probably proceeds by addition of fluoride through a pentavalent silicon intermediate or transition state, which silicon can accommodate with its d-orbitals, followed by elimination of the alkoxide (Figure 8.4). Since Si–Cl bonds (113 kcal mol^{-1}) are weaker than Si–O bonds (~128 kcal mol^{-1}),[4] Cl^- is used as the leaving group in silyl protecting group reagents

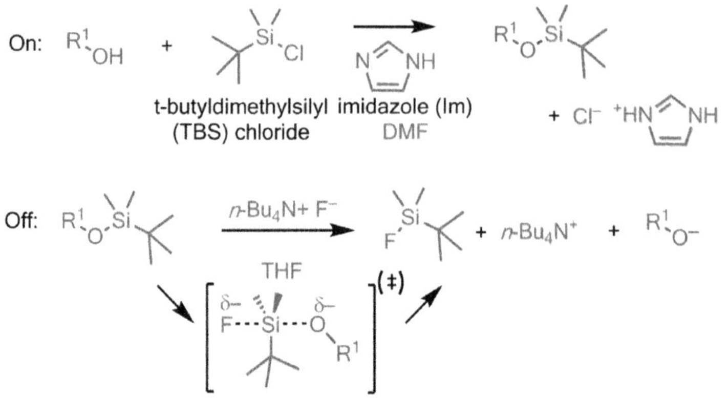

Figure 8.4 Protection and deprotection of alcohols as silyl ethers.

(Figure 8.4).[3] Because of the *t*-butyl group, the TBS group is also susceptible to strong acid E1 deprotection, which must be considered when planning a synthesis.

8.1.6 Hydrogenation-labile: Benzyl

Benzyl protecting groups are put on alcohols and amines with benzyl bromide in an S_N2 reaction (Figure 8.5). Recall that S_N2 reactions are easiest with 1° and allylic or benzylic halides. Benzyl bromide is both 1° and benzylic, and in addition, bromide is a very good leaving group. Benzyl must be put on carboxylic acids by an acid-catalyzed esterification reaction with benzyl alcohol, much as in the ester section above, because carboxylates are such poor nucleophiles (Figure 8.3).

Hydrogenation to remove benzyl groups is a very clean and atom economical reaction (Figure 8.5). However, hydrogen gas is flammable and explosive, so it is expensive and difficult to transport safely. Flammable solvents used in these reactions are very prone to catching fire when the solvent-wet catalyst is exposed to air. Hydrogenation typically uses 5% Pd on carbon (Pd/C), which is a rare and expensive metal catalyst. Some labs collect all used Pd catalysts and recycle or regenerate them.

As chemists, we often employ protecting groups to shield functional groups from reactions elsewhere in a molecule. The example often used in undergraduate organic chemistry is temporary protection of a ketone while a Grignard reagent is reacted with another group in a molecule (Figure 8.6). In this example, oxidation of the alcohol to an aldehyde must be done prior to the Grignard reaction.

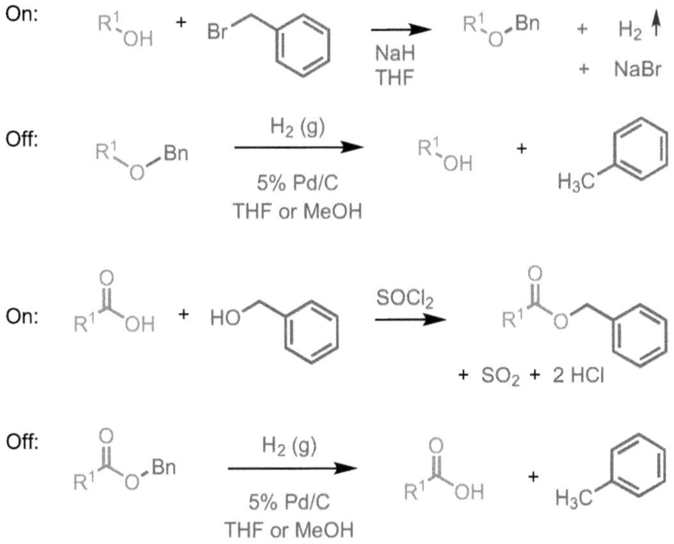

Figure 8.5 Protection and deprotection of alcohols and acids with benzyl.

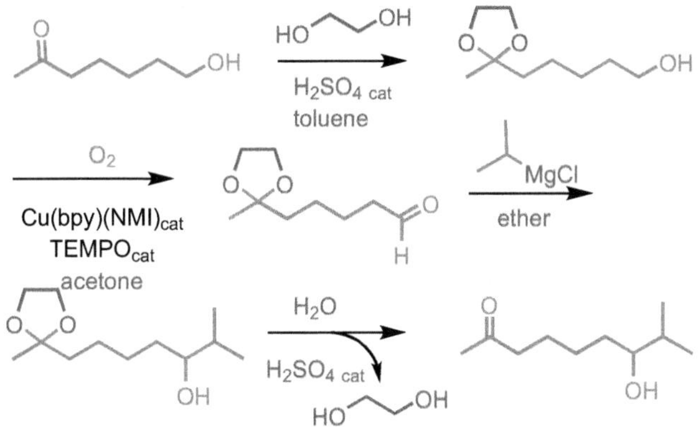

Figure 8.6 The use of a temporary protecting group for a multi-step synthesis.

If unprotected, both the ketone and the aldehyde would react with the Grignard reagent. The ketone must be protected, in this case as an acetal (cyclic 1,1-diether), in one synthetic step (Figure 8.6). Next, after oxidation of the primary alcohol to the aldehyde and the Grignard reaction, the protecting group must be removed again to unmask the original ketone. Two extra steps are required every time a protecting group is used in a synthesis! The extra steps decrease the efficiency of a synthesis, whether you measure that by atom economy, overall yield, space-time yield, E-factor, or PMI.

8.2 Renewable Feedstocks

One negative aspect of biofeedstocks is the frequent necessity of multiple protection/deprotection steps because of the number of reactive functional groups involved.

8.2.1 Peptide Synthesis

Peptide synthesis is particularly bad, with both side chain and α-amino protecting groups (Figure 8.7). Current practices use the large α-amino protecting group, fluorenylmethoxycarbonyl (Fmoc, Figure 8.1), because it is readily removed with the mild base, piperidine. This permits the use of a solid-phase synthesis resin that can be deprotected with TFA, a much less hazardous reagent than HF used in the original Merrifield peptide synthesis. Fmoc protection of the amino acids, followed by deprotection before each coupling step is very inefficient. Peptide synthesis requires one Fmoc protecting group on each α-amine, and side chain protecting groups (PG) on 12 of the 20 common amino acids (Figure 7.8), in addition to wasteful resins, coupling agents, bases, and solvents.

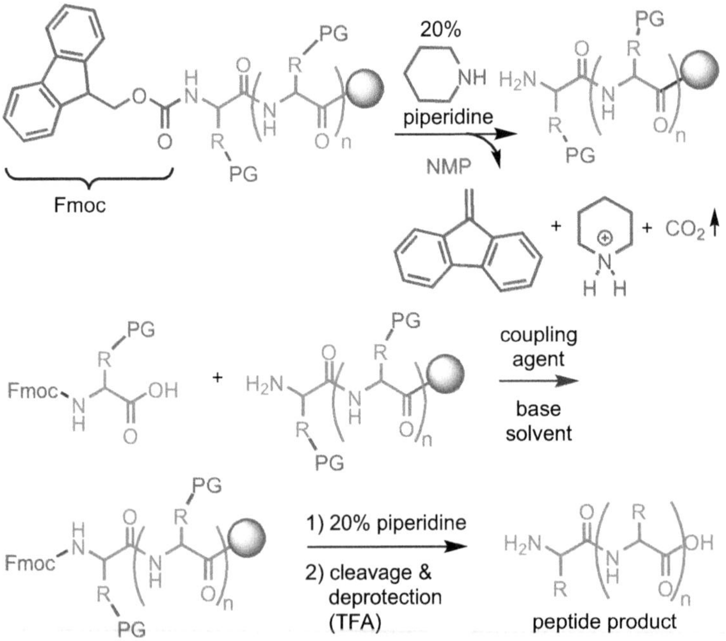

Figure 8.7 General peptide synthesis scheme, highlighting wasteful Fmoc and side chain protecting groups (PG).

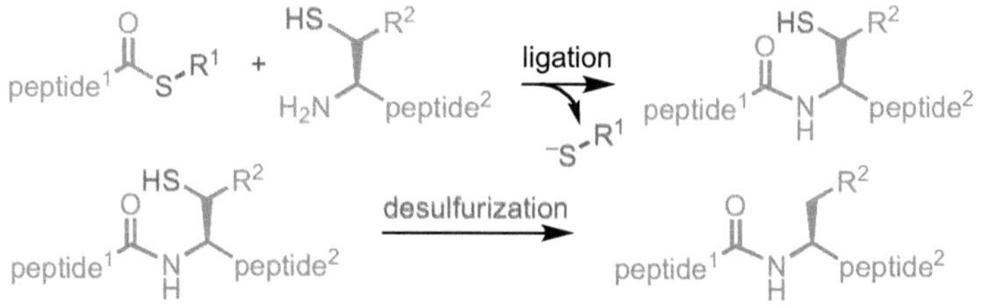

Figure 8.8 Peptide thioester ligation-desulfurization methods are used to efficiently couple unprotected peptides.

Most of the progress in eliminating protecting groups has been done with peptide ligation. Peptide ligation avoids protecting groups by creating highly reactive thioesters and α-amines with β-thiol side chains (Figure 8.8). Further advances have been made in desulfurization to leave a more common hydrocarbon side chain, such as alanine, phenylalanine, or leucine, instead of the more rare cysteine. This method has been used primarily to ligate long synthetic peptides to make full-length proteins.

Few advances in avoiding protecting groups have been successful in achieving the 99.8+% yields required for the synthesis of long peptides on solid-phase resins. Enzymatic methods reverse the reaction of peptide hydrolysis using proteases. Polysaccharide synthesis is even more challenging because all of the functional groups are alcohols, and most are secondary alcohols. Nature controls the regiochemistry of disaccharide bond formation (typically 1,4- or 1,6-) with the specificity of enzymes.

8.3 Reactive Functional Groups

8.3.1 Protic Acids: Alcohols, Carboxylic Acids, Amides

Acidic functional groups may be sensitive to basic reagents. Carboxylic acids ($R{-}CO_2H$), alcohols (R–OH), and 1° or 2° amides (R–C=ONHR′) all have a proton attached to an electron-withdrawing heteroatom. The pK_a values of many organic functional groups in water and in DMSO, which is more relevant for most organic reactions in polar aprotic solvents, can be found in Appendix C. Carboxylic acids are the strongest organic acids (pK_a in water ~ 5); alcohols ($pK_a \sim 16$), and 1° or 2° amides ($pK_a \sim 15$) are moderate acids.

If the conjugate acid of the base is a stronger acid than the acidic functional group by more than 3 pK_a units, or three orders of magnitude in concentration, then the group may not need to be protected.

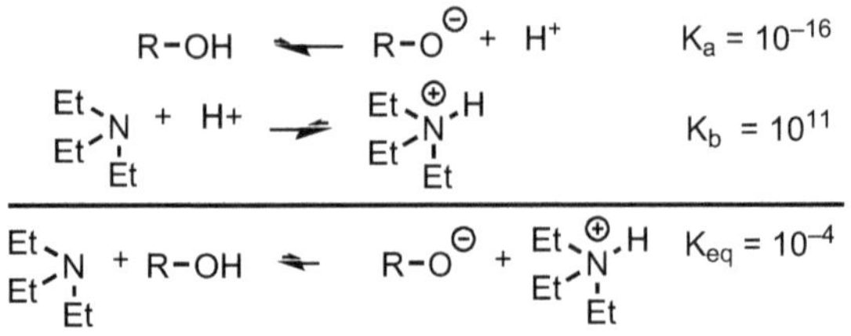

Figure 8.9 Equilibrium acidities of an alcohol and Et_3N^+H predict that Et_3N will not deprotonate an alcohol to any appreciable extent, and the equilibrium will lie to the left by four orders of magnitude.

For example, if an amine base like Et_3N ($Et_3N\text{–}H^+$ pK_a 11) is used in a reaction to deprotonate a $\text{–}CO_2H$, an alcohol OH or an amide NH may not need protection because the conjugate acid of triethylamine is four orders of magnitude more acidic than the OH or NH acid (Figure 8.9). However, if a small amount of OH or NH deprotonated by a weak base reacts further, the equilibrium will shift towards that product by Le Chatelier's Principle. Consideration of the pK_a values of functional groups and the relevant reagents of a total synthetic pathway can lead to avoiding protecting groups.

8.3.2 Acidic CH Protons Alpha to Carbonyl Groups

Because of the strongly electron-withdrawing carbonyl oxygen, carbonyl functional groups have acidic CH protons alpha to the carbonyl. The carbonyl π-bond stabilizes the conjugate base of these CH acids by resonance structures called enolates (Figure 8.10). The pK_a values of protons alpha to one carbonyl range from about 18 to 27 in DMSO (Appendix C), so any base for which its conjugate acid has a pK_a three units lower than that will not deprotonate the α-carbon to an appreciable extent (Figure 8.10). Some common strong bases of concern are the alkyl lithiums, lithium diisopropyl amide (LDA), and sodium amide ($NaNH_2$).

Most organic reactions are performed in polar aprotic solvents, like THF or DMF. A table of pK_a values in both water and DMSO is given in Appendix C. Some common weak bases (with pK_a of conjugate acid in DMSO) that should not react with the alpha CH of carbonyl compounds are triethylamine (9), piperidine (11), morpholine (9), imidazole (6.4), potassium carbonate (3.6, 6.4), and pyridine (3.4). Ketones, aldehydes, esters, and amides all have relatively moderate α-carbon pK_a values that should be considered when designing a synthesis (Appendix C).

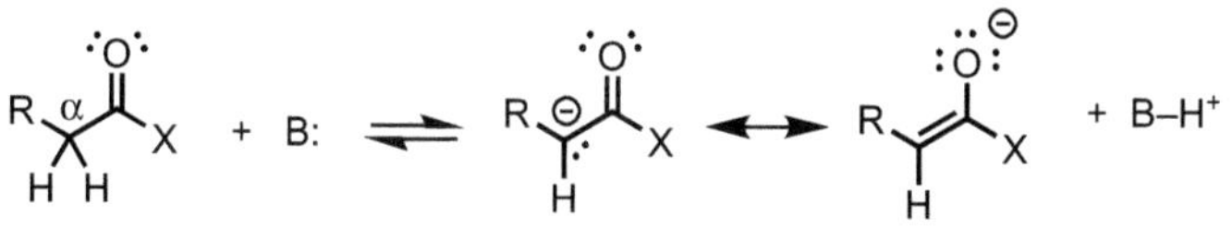

X = H, R, NRR', OR

Figure 8.10 Deprotonation alpha to a carbonyl creates a resonance-stabilized enolate.

8.3.3 Electrophilic: Ketones, Aldehydes, Esters, Amides

A second important consideration is that small bases can also be good nucleophiles. If a nucleophile is to be used in a reaction with another part of the molecule, the carbonyl electrophilicity must be taken into account. The electrophilicity of a series of carbonyl compounds is proportional to the reactivity towards aqueous hydrolysis (Figure 8.11).

Acyl chlorides and anhydrides are rarely the final target of a synthesis. They are typically made as a reactive intermediate on the way towards another carbonyl compound, such as an ester or amide. Since carboxylic acids are more likely to react as acids than as electrophiles, acyl chlorides and anhydrides are used to boost the reactivity towards nucleophiles that would deprotonate an acid. Aldehydes and ketones are typically protected as acetals (Figure 8.1). A well-designed synthesis should attempt to eliminate the protecting groups by introducing reactive functional groups only when they are needed, near the end of the synthesis if the carbonyl is in the final target. Esters and amides are difficult to protect adequately, but may not require a protecting group at all because they are less reactive as electrophiles. If there is a chance of unwanted reaction, esters and amides can be introduced near the end of a synthesis. Otherwise, esters and amides that are present in the final product can be introduced early because they are inherently unreactive under many conditions. As an example, peptide synthesis relies on the stability of the amide bond during subsequent couplings of acids with amines. Carboxylates are unreactive as electrophiles because of their delocalized negative charge.

Since most nucleophiles are also bases, carboxylic acids will react first by giving up a proton to produce an unreactive carboxylate. One situation in which carboxylic acids will react is during the synthesis of the polyamides such as nylons. The small equilibrium concentration

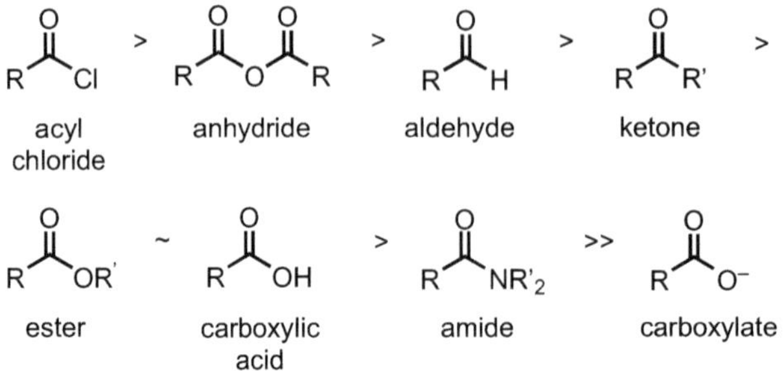

Figure 8.11 Relative electrophilicity of a series of carbonyl compounds.

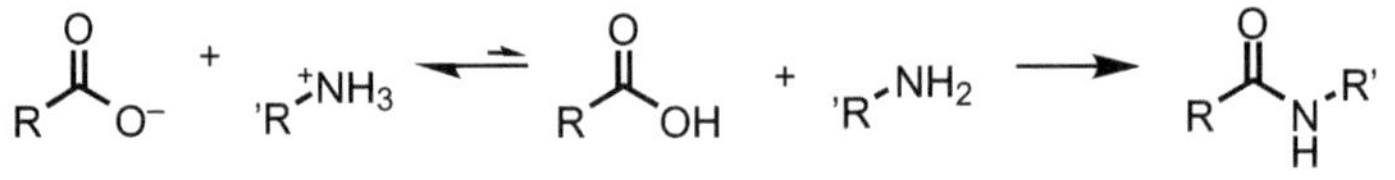

Figure 8.12 Reaction of small equilibrium levels of carboxylic acids react by Le Chatelier's Principle to polymerize nylons at high temperature.

of carboxylic acid reacts with the strongly nucleophilic amine at higher temperatures (Figure 8.12). As the acid reacts, the equilibrium shifts to produce more acid by Le Chatelier's Principle until the polymerization is complete. Once the amide bond forms, the reaction is essentially irreversible because of the low reactivity of amides (Figure 8.11). High temperatures can only be used if the reactants and products are thermally stable.

8.3.4 Basic or Nucleophilic: Amines, Thiols, Alcohols

Just as electrophilic carbonyls are potentially reactive under basic or nucleophilic conditions, nucleophilic groups are potentially reactive with electrophilic reagents or groups elsewhere in the molecule. Amines, thiols, and alcohols must usually be protected unless they are introduced late in the synthesis. Since these groups are frequently found in renewable feedstocks, protection may be necessary if the starting nucleophilic group is desired in the final product. On the other hand, an unreactive derivative of a nucleophile may be desired, so it pays to create the unreactive group immediately in a synthesis. For example, an amide desired in the final product could be installed immediately to avoid the use of a protecting group for an amine because amides are less reactive (Figure 8.13).

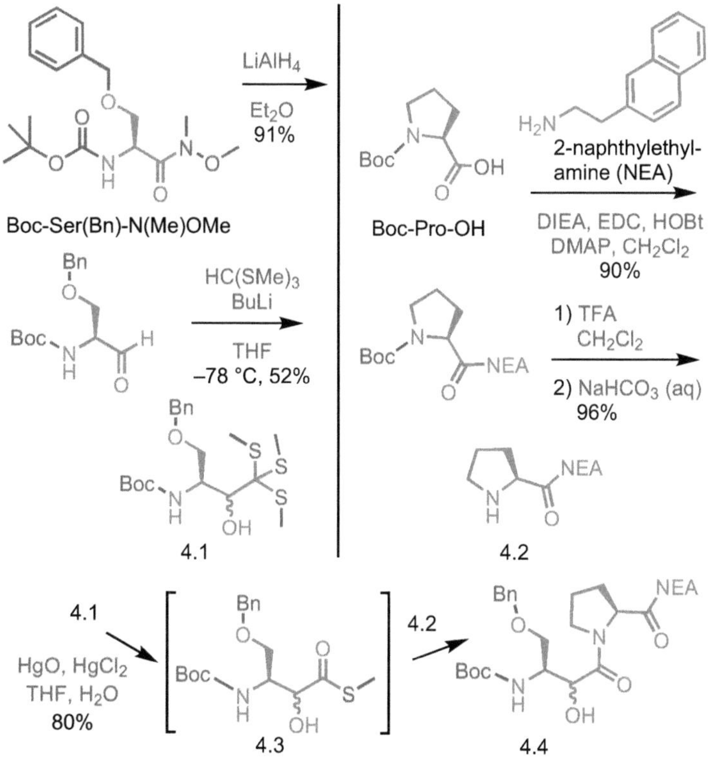

Figure 8.13 Convergent synthesis of an α-hydroxyamide avoiding a carboxylic acid protecting group for Boc-Pro-OH by making an amide at the beginning of the synthesis.[5]

8.4 Case Study: Convergent Synthesis of an α-Hydroxyamide (Etzkorn)

In the synthesis of an α-ketoamide, the natural amino acids serine (Ser) and proline (Pro) were used as starting materials (Figure 8.13).[5] This necessitated a number of protecting groups on the reactive functional groups—the benzyl ether and Boc-amine of Ser, and the Boc-amine of Pro. However, a protecting group on the Pro carboxylic acid was avoided by coupling Boc-Pro-OH directly to 2-(2-naphthyl)-ethylamine to make amide **4.2**, which served to protect the acid during the remainder of the synthesis. This synthesis is an example of a convergent synthesis, in which a similar degree of complexity in the right and left sides is developed before condensing them to make α-hydroxyamide **4.4**; intermediates **4.1** and **4.2** are of similar size and complexity (Figure 8.13). Hydrolysis of the ortho-thioester **4.1** and coupling to **4.2** were combined to eliminate a step in the synthesis of the α-ketoamide. This is possible because hydrolysis of orthothioesters with $HgO/HgCl_2$ presumably proceeds through a thioester intermediate **4.3**, in which the thiolate is an excellent leaving group.

8.5 Case Study: An Efficient Biocatalytic Process for Simvastatin Manufacture (Codexis-Tang)

8.5.1 Statin Drugs for Cardiovascular Disease

Statin drugs have been used to lower physiological cholesterol and low-density lipoproteins (LDL), and increase high-density lipoproteins (HDL). Statin drugs were designed to prevent cardiovascular disease in patients with high cholesterol. However, a recent meta-study found that the use of statins only prolongs life up to 27 days, with a median 3 to 4 days, after 5.8 years of use.[6] Side effects occur frequently. Statins inhibit the enzyme, 3-hydroxy-3-methyl-glutaryl-(HMG)-coA-reductase (HMGCR). HMGCR catalyzes the rate-limiting step in the synthesis of cholesterol, thus controlling the metabolic production of cholesterol when dietary restriction fails to lower cholesterol.

Simvastatin is the generic name (brand name Zocor®) for a synthetically modified, fungal natural product, Lovastatin (Figure 8.14). Simvastatin reduces serum cholesterol better and has less hepatotoxicity (liver toxicity) than Lovastatin. Simvastatin has one additional methyl group on the acyl side chain than Lovastatin (Figure 8.14). For the two prior syntheses described below, the overall yield was estimated to be less than 70%.

8.5.2 Prior Syntheses of Simvastatin

All of the syntheses of Simvastatin use Lovastatin as the renewable feedstock obtained from the fungal source. One commercial synthesis involves four steps: hydrolysis of the natural Lovastatin 2-methylbutanoyl group, silyl protection of the more exposed secondary hydroxyl on the δ-lactone ring, re-acylation with a 2,2-dimethylbutanoyl chloride, and finally deprotection of the silyl group (Figure 8.14). Two derivitization steps, protection and deprotection, are required for the net addition of one methyl group. The atom economy of this process is also not particularly good because the

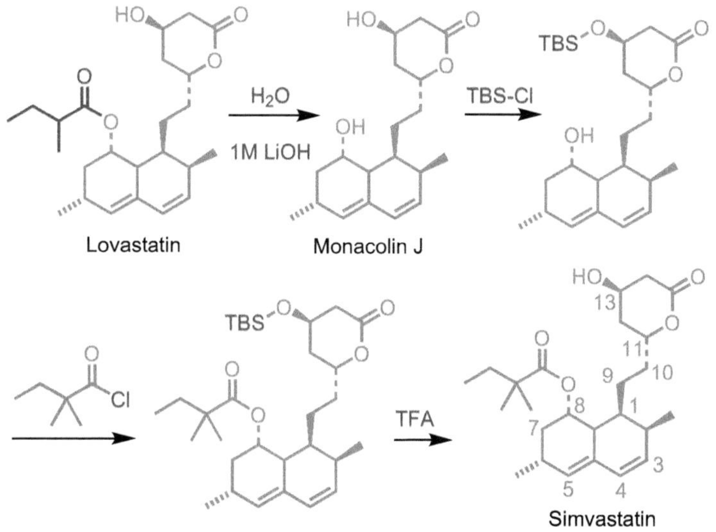

Figure 8.14 Hydrolysis/esterification route to Simvastatin uses a TBS protecting group. Atoms are green if they are retained in the product, brown if not. Adapted with permission of Codexis from ref. 7, Copyright 2012.

whole 2-methylbutanoyl group is discarded and replaced with the desired 2,2-dimethylbutanoyl group. The use of the renewable biofeedstock, Lovastatin, is the one bright spot common in these prior Simvastatin syntheses.

A second synthetic route, called "direct methylation," does not require hydrolysis of the 2-methylbutanoyl side chain because one methyl group is added to make the 2,2-dimethylbutanoyl group. Direct methylation requires two protection steps: lactone ring-opening formation of the *n*-butylamide, and *t*-butyldimethylsilyl (TBS) ether protection of the two secondary alcohols (Figure 8.15). After the key methylation step, the amide is deprotected with NaOH, and the silyl ethers with HCl, followed by ring-closure to restore the lactone (Figure 8.15). Even though the Lovastatin acyl group is not discarded in this synthesis, it is even less atom economical, with seven synthetic steps, although it appears that three of these are done in a single reactor (one pot). Many of the reagents used in this synthesis are hazardous and inefficient, even though these are very common reagents used in organic synthesis. Methyl iodide, in particular, is a potent carcinogen because it easily methylates cellular DNA, disrupting normal transcriptional regulation of genes. MeI can also cause lung, liver, kidney and central nervous system damage. The methylation step also uses pyrophoric *n*-butyl lithium.

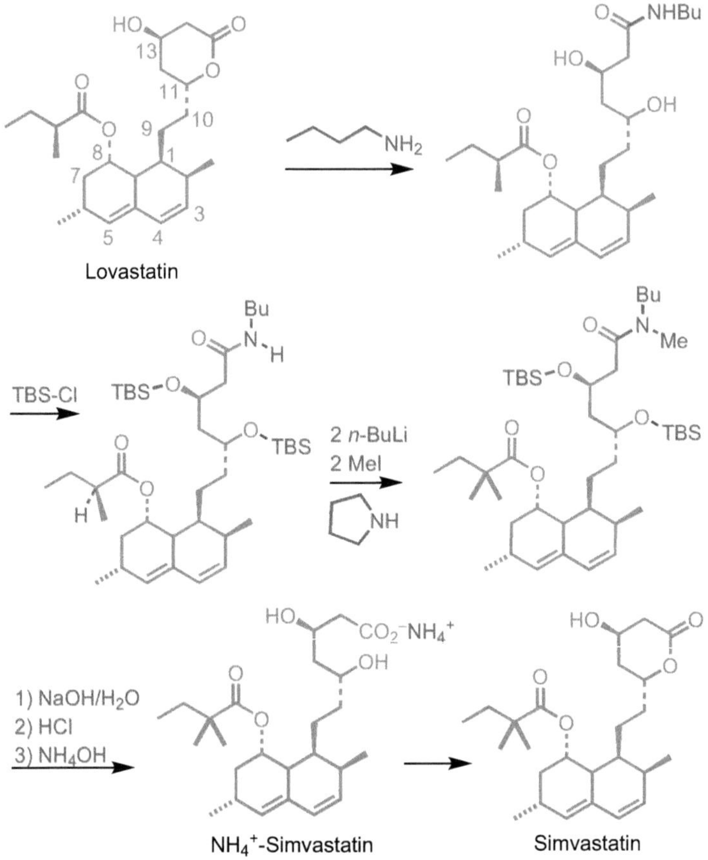

Figure 8.15 Direct-methylation route to Simvastatin uses two TBS and an *N*-butylamide as protecting groups. Same atom coloring as Figure 8.14. Adapted with permission of Codexis from ref. 7, Copyright 2012.

8.5.3 Codexis-Tang Biocatalysis of Simvastatin

Yi Tang's group (UCLA) and Codexis won a PGCC Award in 2012 for a greener synthesis of Simvastatin. They developed the key acylation step by directed evolution of the fungal enzyme, LovD acylase (Figure 8.16).[8,9] In this synthesis, the 2-methylbutanoyl group and the lactone are hydrolyzed with LiOH and NH_4OH respectively. Because the enzyme is extremely regiospecific for acylation of the desired 2° alcohol at C8, neither of the other two 2° alcohols at C11 or C13 requires a protecting group as in the previous syntheses (Figure 8.16). The hydrolysis of the lactone means that it must be restored in the last step, as in the direct methylation, although without the necessity

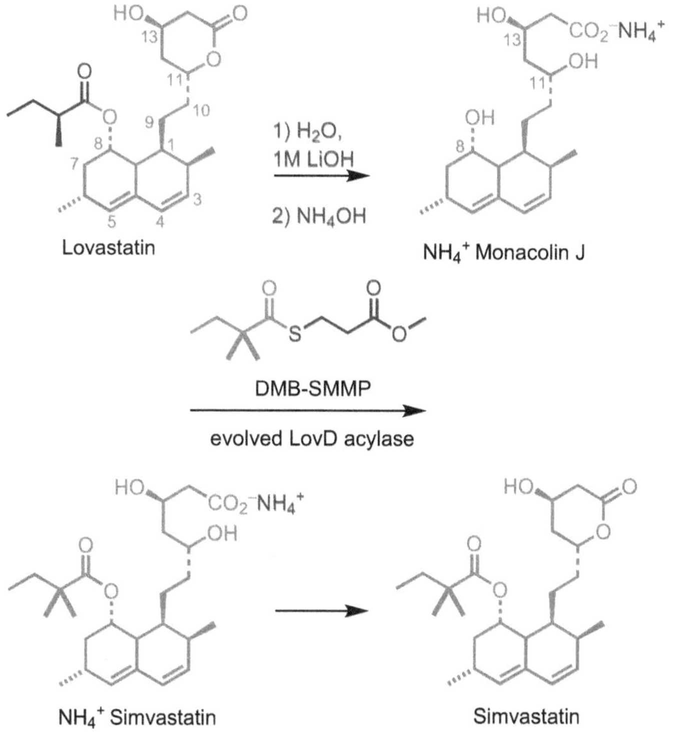

Figure 8.16 LovD acylase enzyme-catalyzed acylation of Monacolin J ammonium salt with 2,2-dimethylbutanoyl-*S*-methylmercaptopropionate (DMB-SMMP). Same atom color code as 8.14. Adapted with permission of Codexis from ref. 7, Copyright 2012.

of protection as an amide and deprotection before closing the lactone ring.

The key step in the Codexis synthesis is the acylation of NH_4^+-Monacolin J with a 2,2-dimethylbutanoyl (DMB) group (Figure 8.16). Notice the similarities and differences from the hydrolysis-esterification method (Figure 8.14). This synthesis still removes the 2-methyl-butanoyl group and replaces it with DMB, but without protection-deprotection steps. Activation of the DMB as a thioester is necessary. Codexis first optimized the thiol leaving group. In *Aspergillus terreus,* LovD transfers the 2-methylbutanoyl group from a large protein thiol, LovF (270 kD), exclusively to the desired C8 position of sodium- or ammonium-Monacolin J (Figure 8.16).[7] Since the LovF protein was not commercially available, and production of it would be inefficient, Codexis next explored the small-molecule acyl donor, DMB-*N*-acetylcysteamine (DMB-SNAC). DMB-SNAC was an inefficient acyl donor with LovD ($k_{cat} = 0.02\ min^{-1}$), but it demonstrated

Table 8.1 Codexis improvements to the LovD-catalyzed synthetic process to make Simvastatin by directed evolution of LovD and one-pot synthesis. Reproduced with permission of Codexis from ref. 7, Copyright 2012.

Parameter	Performance at initiation of evolution program	Final performance
[Monacolin J]	3 $g L^{-1}$	75 $g L^{-1}$
[Thioester]	>3 eq.	1.1 eq.
LovD loading	10 $g L^{-1}$ (natural LovD)	0.75 $g L^{-1}$ (evolved LovD)
Reaction time	18 h	36 h
Conversion	50%	97%

that the large protein, LovF, was unnecessary.[7] After trying a series of DMB donors, DMB-S-methylmercapto-propionate (DMB-SMMP) emerged as the most efficient ($k_{cat} = 0.75\ min^{-1}$), and the most cost effective, because SMMP is very cheap (Figure 8.16).[7]

Both ammonium-Monacolin J and DMB-SMMP are substrates of LovD. Sometimes when two substrates must bind to an enzyme active site before the chemical reaction can occur, the substrates can compete with each other for access to the active site. Substrate binding might need to occur in a particular order as well. In the case of DMB-SMMP, concentrations greater than 30 mM of Monocolin J did not competitively inhibit LovD, which is important for high production efficiency of Simvastatin. To make the synthesis even more efficient, the SMMP acid by-product is recyclable.

To optimize production in whole-cell *E. coli*, the host was mutated to remove a competing esterase (*bioH*), which rapidly hydrolyzed the DMB-SMMP acyl donor. The reaction was monitored by HPLC, complying with Principle 11 Real-Time Analysis. The conversion of Monacolin J at 10 $g L^{-1}$ to Simvastatin was 50% in 18 hours (Table 8.1).[7] The fermentation broth was acidified, and the product was extracted with EtOAc. Two more extractions, followed by precipitation with NH_4OH, gave 98.5% pure Simvastatin ammonium salt, which was recrystallized to >99.5% purity in 91% overall yield from Monacolin J.[7]

To further improve the efficiency, Codexis used directed evolution of LovD (Table 8.1). Directed evolution is a process of mutating DNA to produce enzymes that show optimized binding to the desired substrate and catalyze the desired reaction efficiently. Over 60 000 variants were screened to discover a mutant LovD enzyme with better stability, higher production, and good tolerance against product inhibition by Simvastatin. The patented variant LovD has

30 mutations, which is 7% different from the original enzyme.[7] Only 1.1 equivalents of DMB-SMMB were required, and much less of the evolved LovD enzyme was needed (0.75 $g L^{-1}$) than before directed evolution (Table 8.1).[7] The throughput was improved by directed evolution from 0.2 to 67 g(Simvastatin) L^{-1} day^{-1} g^{-1}(LovD).[7] The process is monitored in real-time with pH-static controlled addition of ammonia.[7]

Finally, the whole process was optimized in one pot to convert ammonium-Monacolin J, which is water soluble, to the Simvastatin ammonium salt, which precipitates. The product is collected by simple filtration. The filtration step eliminated the extraction work-up with EtOAc, minimizing the solvents used in the process. The E-factor for the biocatalytic step was 25 including water and methyl *t*-butyl ether (MTBE), but only 0.6 if the solvents were recycled.[7] Some aspect of every principle of green chemistry was used in this new process; most significantly, the use of a whole-cell biocatalytic system clearly avoids the use of inefficient protecting group derivatives. By 2010, production of Simvastatin was already over 10 million tons using the new Codexis biocatalytic process.[7]

8.6 Protecting-group-free Synthesis

The total synthesis of complex natural products has been a long-standing challenge for organic chemists, stimulating new strategies for long multi-step synthesis. E. J. Corey developed the concept of design by retrosynthetic analysis,[10] and W. S. Johnson developed the concept of biomimetic synthesis (Chapter 2).[11] Natural product synthesis has also stimulated the invention of a vast array of new reactions, reagents, and catalysts. Unfortunately, the availability of a wide variety of protecting groups has become a crutch in designing a synthesis, leading to many extra steps for protection and deprotection.

Protecting groups decrease synthetic efficiency by introducing additional reagents and reaction steps in a synthesis. P. S. Baran outlined an approach to planning a synthesis without protecting groups in his group's paper on the synthesis of a natural product called Ambiguine H.[12] Baran's approach is very different from the case studies discussed in previous chapters that use enzymes as catalysts, or natural products as advanced intermediates. He assumes that an all-chemical approach, and total synthesis from commercially available, small molecule starting materials, are both desirable.

This is often the case when it comes to complex natural products that are low in natural abundance, difficult to extract, or found only in rare or endangered species. Baran's plan includes the following eight concepts.

8.6.1 Planning a Synthesis Without Protecting Groups[12]

1. "Disconnections should be made to maximize convergency."[12]

   The idea of a retrosynthetic analysis of a target molecule using conceptual "disconnections" of bonds is attributed to E. J. Corey, who won the Nobel Prize in Chemistry in 1990.[10] Corey's convergent synthesis of prostaglandin was shown as a retrosynthetic analysis in Figure 2.25. Retrosynthetic analysis leads to the design of more efficient syntheses. Convergence is the concept that complex molecules are most efficiently synthesized by bringing together smaller pieces, or "synthons," of similar complexity to decrease the losses due to side reactions and low yields.[13] Because smaller synthons have fewer functional groups, it is easier to avoid protecting groups in a convergent synthesis. The conceptual opposite of a convergent synthesis is a linear, stepwise synthesis that builds complexity one functional group at a time. A linear synthesis normally requires a protecting group for each functional group that is added in stepwise fashion, adding two steps for each protecting group, on and off. A comparison of a highly convergent synthesis with a linear synthesis will be described in the case study of Swinholide A following Baran's efficient synthesis of Ambiguine.
2. "The percentage of C–C bond-forming steps should be maximized."[12]

   Carbon-carbon bond formations are among the most challenging reactions. Much of the art of organic synthesis revolves around the power of C–C bond formations, such as the Grignard, Wittig, Diels–Alder, and aldol reactions. The difficulty is mainly due to the specificity required when most of the atoms present in a reactant are carbons, and only one or two of these carbons must be reactive in a particular transformation. Once C–C bonds have been formed, they are usually quite difficult to undo, unlike bonds involving nitrogen, oxygen, or halogens. Complexity is built up substantially in C–C bond formations.[14] Thus, in designing a synthesis, the most efficient route will have the highest percentage of C–C bond-forming steps, and the fewest protecting groups, which are nearly all designed to prevent reactions with oxygen and nitrogen containing functional groups.

3. "Redox reactions that do not form C–C bonds should be minimized."[12]

   We saw in Chapter 3 that redox reactions often require toxic and hazardous reagents, and they are often not atom economical. Renewable feedstocks are usually at a higher oxidation state similar to the desired products, consider shikimic acid and Tamiflu™ for example (Chapter 2). Petroleum feedstocks are much more reduced, requiring more redox reactions to bring them up to the oxidation state of biologically active products. The use of renewable feedstocks is one way to minimize the number of redox reactions designed into a synthetic strategy. Each C–C bond-forming reaction gets a free pass because the bonds are hard to form, and the level of complexity typically increases in accordance with C–C bond formation.[14]

4. The overall oxidation level of intermediates should change unidirectionally.[12]

   One way to minimize redox reactions is to avoid going up and down between different oxidation states. For example, it would be inefficient to reduce a ketone to an alcohol only to oxidize it back to the ketone later in the synthesis. Baran's idea that the oxidation level of intermediates should linearly escalate in the design of a synthesis is based upon the premise that most syntheses start with highly reduced petroleum feedstocks.[12] The opposite may be true when starting with renewable feedstocks; reductions might be necessary to get to the desired product. In either case, oxidation or reduction reactions should not be counterproductive to the overall oxidation-state goal of the synthesis. Again, minimizing the number of redox reactions is key to avoiding protecting groups.

5. "Cascade (tandem) reactions should be designed and incorporated to elicit maximum structural change per step."[12]

   Sequential reactions are "...a series of reaction steps in which several bonds are formed or broken, without the isolation of any intermediates..." in a single reactor.[16] Thus, new reagents may be added to the reactor for a sequential process as necessary. Cascade reactions are a special type of sequential reaction in which a series of chemical transformations take place, each reaction unveiling or producing a new functional group that reacts in the next step without the addition of new reagents.[15] The terms, "sequential" and "cascade," as well as "tandem,"

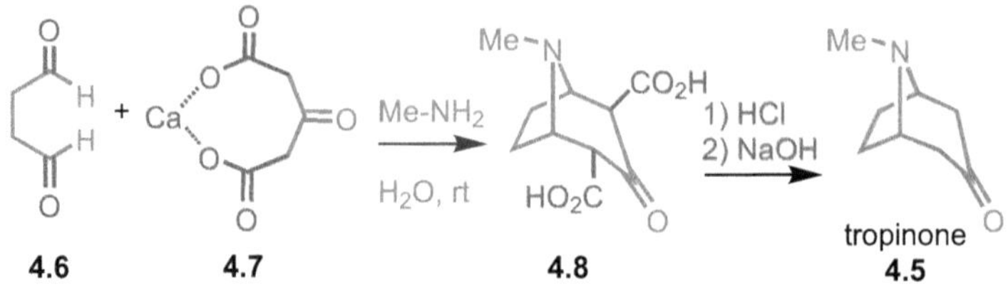

Figure 8.17 Robinson's cascade synthesis of tropinone.[15,17] Adapted with permission of John Wiley and Sons from ref. 15, Copyright 2006.

"domino," or "one-pot" reactions, are often used interchangeably. The space-time efficiency of such processes is obvious—less solvent, and fewer reagents are used. With respect to protecting groups, cascade reactions have a particularly elegant efficiency advantage; each functional group appears only as it is needed in the next reaction, so temporary protecting groups are required less often.

The earliest example of a cascade reaction, published one hundred years ago and frequently cited in the reviews, is R. Robinson's elegant synthesis of tropinone **4.5** (Figure 8.17).[15–17] The synthesis uses a double Mannich reaction of methyl amine, succinaldehyde **4.6**, and acetonedicarboxylic acid **4.7** (Figure 8.17).[17] The intermediate **4.8** is decarboxylated with acid, then neutralized. No protecting groups were used in this sequential one-pot reaction.

6. "The innate reactivity of functional groups should be exploited so as to minimize protecting groups."[12]

   "Innate reactivity" may seem an opaque term to students who are not yet initiated to multi-step synthesis in the lab. Yet this is the crux of organic synthesis, essential to designing syntheses without protecting groups.[18] Experience with the relative reactivity of functional groups, for example at the α-carbon of amides *vs.* esters as in the direct-methylation synthesis of Simvastatin (Figure 8.15), is necessary to the endeavor. The fundamental principles of organic chemistry help a lot to determine innate reactivity: pK_a values, steric and electronic environments, electronegativity, resonance structures *etc.* Beyond that, a careful review of the literature and reasoning by analogy can guide synthetic design.[18]

7. "*Invent new methodology to facilitate efficiency.*"[12]

   New reactions are constantly being invented, but those that use the Principles of Green Chemistry to guide the process are likely to be the most efficient.[19,20] An example of a highly efficient reaction is the catalytic insertion of carbon monoxide into various C–C and C–H bonds.[21] BHC used this to great effect in the protecting-group-free synthesis of ibuprofen (Figure 2.11).

8. "*Biomimetic pathways should be incorporated.*"[12]

   Nature uses no protecting groups. Enzymes evolve to catalyze specific reactions with very high chemical selectivity and stereoselectivity without protecting groups. As we saw with the Codexis process to make Simvastatin, directed evolution can be used to evolve new enzymes. What Baran means here is that biomimetic syntheses that take advantage of the innate reactivity of functional groups allows the design of efficient *chemical* syntheses without the use of protecting groups. A classic example is the biomimetic synthesis of 11α-hydroxyprogesterone by Johnson (Figure 2.26).[11]

8.7 Case Study: Synthesis of Ambiguine H Without Protecting Groups (Baran)

Baran designed an extremely efficient synthesis of a complex natural product called Ambiguine H.[12] The ambiguines are marine natural products with diverse structures, and diverse medical applications ranging from cancer to tuberculosis.[22] Baran outlines three retrosynthetic pathways to Ambiguine H: biosynthetic, standard, and no-protecting-group analyses (Figure 8.18). All three pathways are convergent, with synthons of similar size and complexity. In the standard retrosynthetic analysis, there is an emphasis on masking or "caging" oxygens and nitrogens with protecting groups that would lead to a greater number of synthetic steps (Figure 8.18).

The biosynthetic or natural synthesis was hypothesized to use a cationic cyclization of a triene and an isocyanate-substituted indole (Figure 8.18), reminiscent of Johnson's 11α-hydroxyprogesterone biomimetic synthesis (Figure 2.26).[11] The position of the isocyanate (–NC) nitrogen attached to the indole is where it would be in the amino acid tryptophan, in accord with biosynthesis. The third piece is the common isoprene with a pyrophosphate ($-OPO_3{}^{-}PO_3{}^{=}$) leaving group. In the biosynthetic pathway of Ambiguine H, protecting groups are unnecessary because of the exquisite selectivity of enzymes.

Baran's retrosynthetic analysis shares certain features with the biosynthesis, the cyclization of the indole with an isopropene-substituted cyclohexanone, and an isoprene with an X leaving group (Figure 8.18). It differs in using a ketone enolate as a

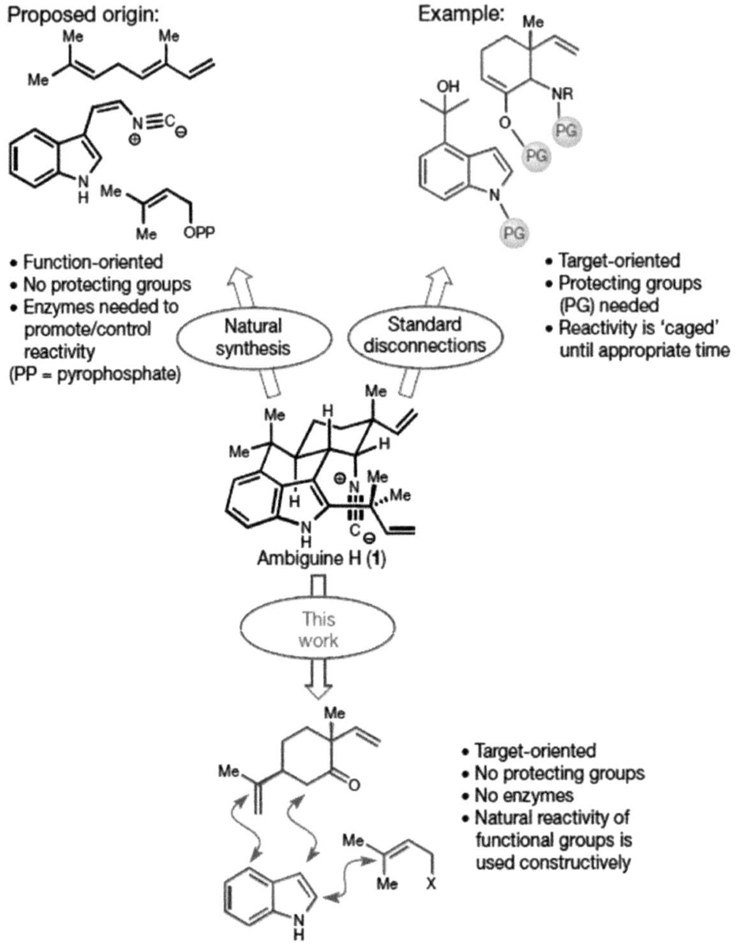

Figure 8.18 Biosynthetic compared with standard disconnection retrosynthetic analysis of Ambiguine H, and Baran's biomimetic retrosynthesis of Ambiguine H using the natural reactivity of functional groups without the use of protecting groups. Reprinted with permission of Nature Publishing Co. from ref. 12, Copyright 2007.

nucleophile to add to the indole ene-amine electrophile, and a simple indole rather than a tryptamine-like synthon.

In Baran's convergent, biomimetic synthesis, the intermediate cyclohexanone 7 was prepared in four steps from commercially available starting materials (Figure 8.19).[12] In step **a** of the forward synthesis, the ketone was deprotonated with the highly hindered, non-nucleophilic, strong base, lithium hexamethyldisilazide (LHMDS) to make the enolate (Figure 8.19). Cu(II)-catalyzed, oxidative

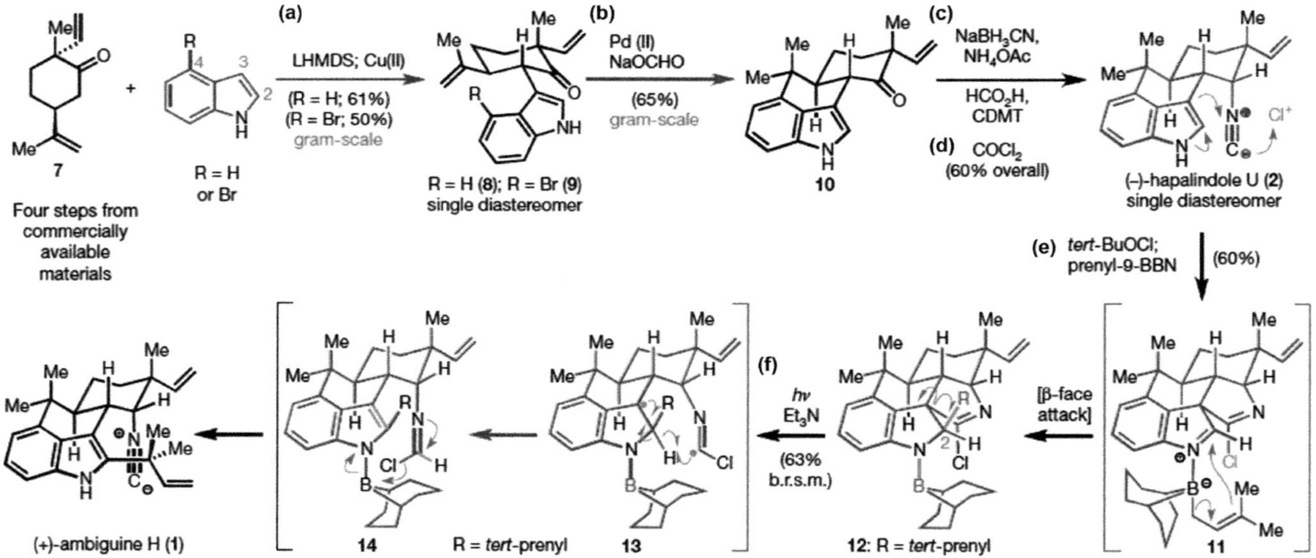

Figure 8.19 Baran's protecting-group-free, 6-step synthesis of Ambiguine H. LHMDS = lithium hexamethyldisilazide, CDMT = 2-chloro-4,6-dimethoxy-1,3,5-triazine, 9-BBN = 9-borabicyclononane. Reprinted with permission of Nature Publishing Co. from ref. 12, Copyright 2007.

addition of the enolate to the indole ring introduced the first C–C bond in the first step **a**. Pd(II)-catalyzed coupling of the indole benzene ring with the isopropene group closed the middle ring, with a second C–C bond formation in step **b**. Step **c** involved a sequential stereocontrolled (top-face) reductive amination of the ketone to make a primary amine, followed by formamide formation with the amide bond coupling reagent, 2-chloro-4,6-dimethoxy-1,3,5-triazine (CDMT). In step **d**, dehydration of the amide with phosgene produced the isonitrile group (–NC). Steps **c** and **d** were done in separate pots without isolation of the intermediate formamide.

In the fifth very challenging step **e**, instead of protecting the fragile indole NH and isonitrile, Baran exploited the natural reactivity of both groups (Figure 8.19). The 2-position of the indole was activated with a temporary bond to the isonitrile group using the oxidizing reagent, *t*-butyloxychloride. The isoprenyl-9-borabicyclononane (9-BBN), was used to introduce the final C–C bond by reducing the indole iminium ion, attaching the isoprene at the 2-position of the indole ring. Finally, in step **f**, a Norrish-type photocleavage of the chloroimidate was used to regenerate the isonitrile and cleave the N–B bond to release 9-Cl-9-BBN as a side product. Only ten synthetic steps overall were used to synthesize this incredibly complex natural product!

8.7.1 Protecting Groups as a Last Resort

Protecting groups may be necessary at times to allow the chemical synthesis of a target.[12] In some cases, no one may have been clever enough to design the synthesis of a particular target without protecting groups at all. On the other hand, nature abhors protecting groups, and enzymes are the masters of protecting-group-free synthesis. The reactivity of some functional groups is not always predictable. Natural product synthesis is rife with stories of getting most of the way through a multi-step synthesis only to discover that an unprotected tertiary alcohol reacts unexpectedly, or in the last step, one of the reactions was impossible without protecting another part of the molecule. In a linear synthesis, this often involves starting from scratch and redesigning the entire synthesis with new protecting groups. Another advantage of convergent synthesis is that the synthesis of only one of the intermediates might need to be redesigned. The challenges may involve differential protection of multiple similar functional groups, like several secondary alcohols, or protecting groups that just will not come off, necessitating redesign of a synthesis.

Sometimes there is no getting around protecting groups. In complex targets with multiple oxygen and nitrogen heteroatoms, the use of protecting groups may actually be more efficient, for example to avoid multiple redox steps. Biopolymers, such as peptides, polysaccharides, and polynucleic acids (DNA/RNA), have long and venerable histories of chemical synthesis made possible by protecting groups.[12] The monomers of these biopolymers have multiple reactive functional groups, as discussed in Chapter 7, and predictable, controlled chemical synthesis of desired sequences would be exceedingly difficult without protecting groups. Designing a synthesis using protecting groups can reduce the risk of failure and give a measure of confidence that a synthetic design will work.[12] Without the chemical synthesis of DNA primers, molecular biology would have been improbable, and the biotechnology revolution might not yet have occurred.

8.8 Convergent Synthesis

Convergent synthesis is the concept of combining small pieces of similar size and complexity to converge on the final product. Convergent synthesis is much more efficient than linear synthesis, in which the size and complexity increases linearly throughout the synthesis. For example, peptides are synthesized in a rigidly linear way with repeating steps of amine deprotection and amide bond coupling for each amino acid in the chain. Because each intermediate in a linear synthesis has an increasing number of reactive functional groups, protecting groups are required that are orthogonal to each other. Orthogonal means that the protecting groups are stable to, and removed under, different reaction conditions—some of the protecting groups must stay on, while others must be selectively removed to reveal the functional group for the next reaction. Because convergent syntheses use smaller, less complex pieces, fewer protecting groups are necessary.

8.9 Case Study: Convergent Synthesis of Swinholide A (Krische)

Swinholide A is a very complex natural product discovered in a marine sponge, *Theonella swinhoei*.[23] Swinholide A prevents the actin cytoskeleton from forming during cell division, so it is of great interest in anti-cancer research. Swinholide A is a head-to-tail homodimer macrocycle of 40 atoms. The monomer of each half is very complex with 15 stereocenters, 39 carbons, ten oxygens, three alkenes, and two pyranyl 6-membered rings (Figure 8.20).

Michael Krische and coworkers designed a highly convergent synthesis of Swinholide A that eliminated many protecting groups.[23] Krische compared the new synthesis to two previous total syntheses of Swinholide A by the following criteria: (1) number of steps in the longest linear sequence (LLS), (2) number of steps in the total synthesis (TS), (3) the number of steps to assemble the skeleton (i.e. C–C and C–O bond-forming steps), (4) the number of redox reactions, (5) the number of protection-deprotection steps, and (6) other reactions (Figure 8.20).[23] He also presented the steps as a percent of the LLS. Krische's convergent synthesis is about twice as efficient as either of the prior syntheses: the total number of steps was only 30, compared with 50 and 59 for the other two. Most impressively, his synthesis used only three redox reactions and only two protection-deprotection

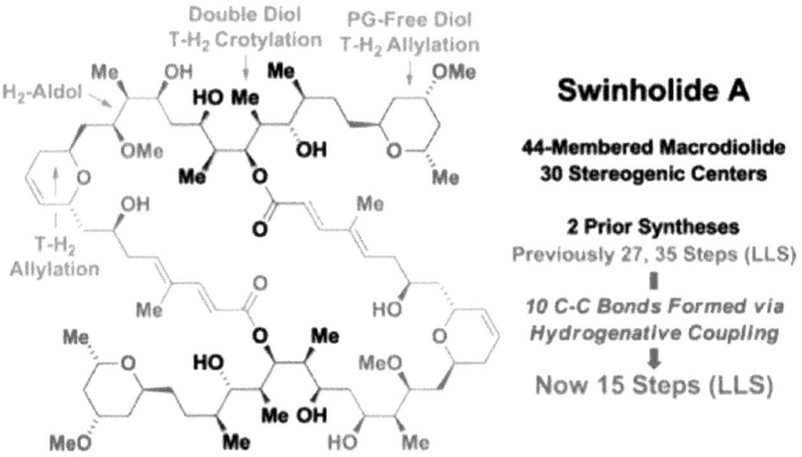

Swinholide A Total Syntheses	LLS (TS)	Skeletal Assembly	Redox Reactions	Protection-Deprotection	Other Reactions
Paterson (ref. 13)	27 (50)	10 (37%)	5 (19%)	10 (37%)	2 (7%)
Nicolaou (ref. 14)	35 (59)	9 (26%)	7 (20%)	13 (37%)	6 (17%)
Krische (*This Work*)	15 (30)	9 (60%)	3 (20%)	2 (13%)	1 (7%)

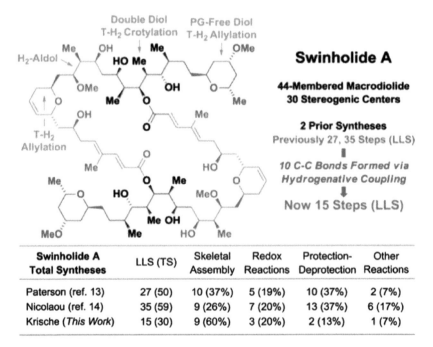

Figure 8.20 Comparison of the synthetic efficiency of linear (Paterson, Nicolau) *vs.* convergent (Krische) syntheses of Swinholide A. LLS = longest linear synthesis, TS = number of steps in the total synthesis. Reprinted with permission of American Chemical Society from ref. 23, Copyright 2016.

steps.[23] The minimal number of protecting groups for the synthesis of a molecule of such incredible complexity is spectacular.

The key to the convergent synthesis of Swinholide A was careful retrosynthetic analysis to design intermediates **A** and **B** with similar size and complexity (Figure 8.21). Also note the similar complexity of

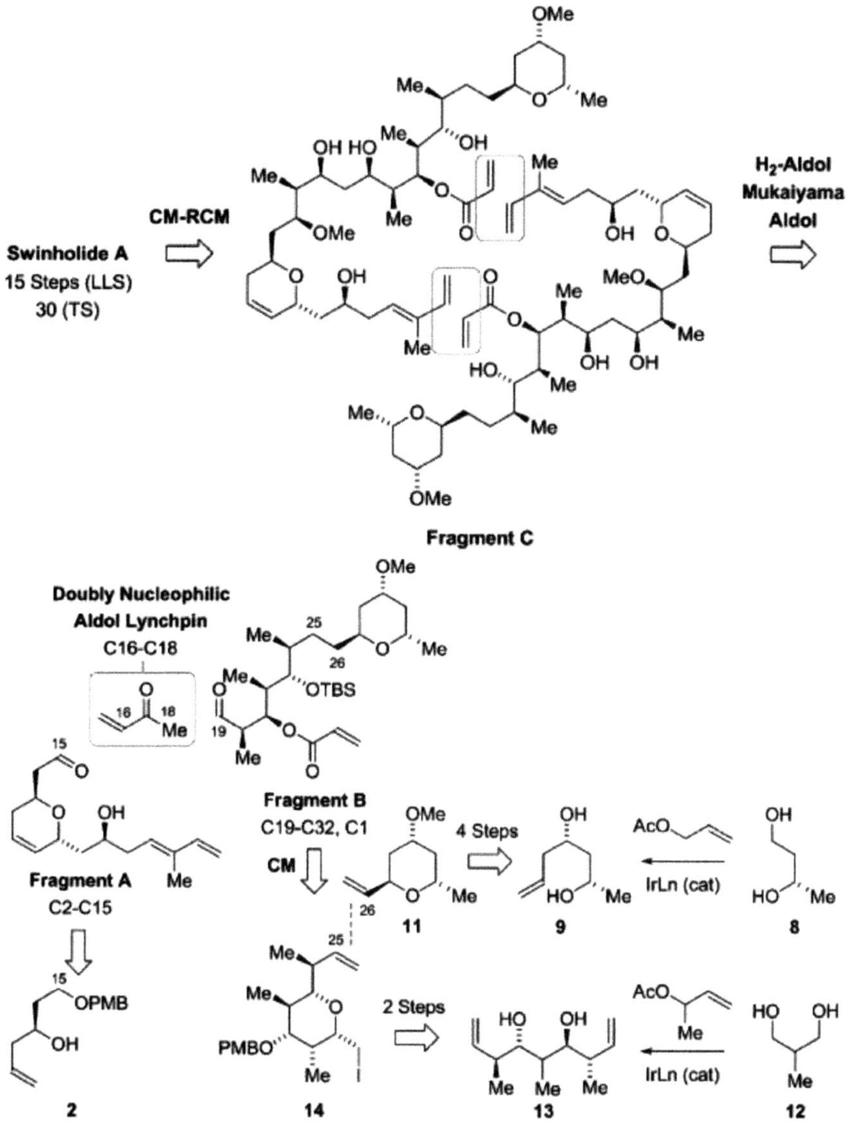

Figure 8.21 Retrosynthetic analysis of the highly convergent synthesis of Swinholide A. Reprinted with permission of American Chemical Society from ref. 23, Copyright 2016.

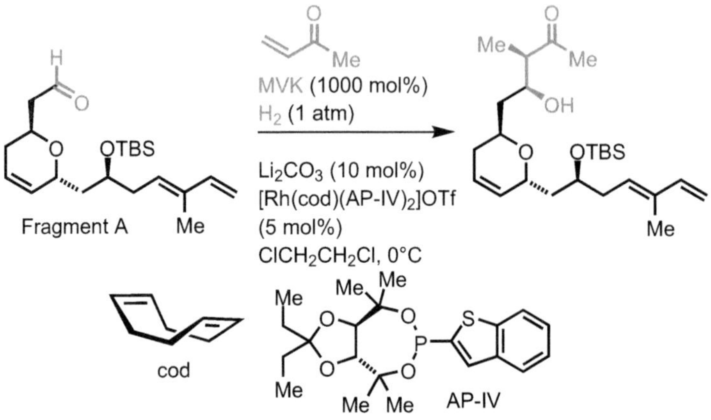

Figure 8.22 Cascade hydrogenative C–C bond coupling used by Krische to react the "lynchpin" methylvinylketone (MVK) with Fragment A in the convergent synthesis of Swinholide A. Adapted with permission of American Chemical Society from ref. 23, Copyright 2016.

starting materials **2**, **8**, **12**, and the acetoxyalkenes over the forward arrows (Figure 8.21). Homodimerization by cross-metathesis (CM)-ring-closing metathesis (RCM), which will be covered in Chapter 9 on catalysis, was used in the last, most convergent step (Figure 8.21). Unfortunately, the desired dimerization in this last step occurs in only 25% yield, while monomer cyclization occurs in 43% yield.[23]

Krische took great advantage of a new family of catalytic hydrogenative C–C bond couplings for ten steps of the convergent synthesis.[23] The Mukaiyama-aldol is a cascade hydrogenation of methyl vinylketone (MVK), followed by an aldol reaction (Figure 8.22).[24] The chemoselectivity of the hydrogenation is remarkable since there are three alkenes in Fragment A that must be retained in the product (Figure 8.22).[24] The reduced MVK probably remains attached to the asymmetric catalyst to confer stereoselectivity in the subsequent aldol. The alpha position of the 2-butanone is deprotonated by the Li_2CO_3 base, and acts as the carbon-nucleophile in the aldol reaction with the aldehyde of Fragment A (Figure 8.22).[24] These reactions tolerate a wide number of unprotected functional groups and achieve significant C–C bond-forming specificity and enantioselectivity.[24]

8.10 Summary

Protecting groups have become standard in organic synthesis, yet they contribute to synthetic inefficiency, adding two extra steps for each protecting group—putting them on and taking them off. Protecting groups can be categorized in several ways. Here, we have grouped them according to the conditions used to remove them: acid, base, fluoride, Pd, hydrogenation, or oxidation labile. Different functional groups have different reactivities that must be considered in designing a synthesis. Protic acids, α-carbon acids, electrophiles, and bases/nucleophiles each require different consideration. In the synthesis of Simvastatin, Codexis created a highly specific LovD evolved enzyme to acylate one of three 2° hydroxyls of Lovastatin. This avoided the protection and deprotection steps required in the two other commercial syntheses of Simvastatin. Baran described eight concepts to guide the design of protecting-group-free synthesis of complex natural products: convergence, maximize C–C bond formation, minimize redox reactions, linearly change oxidation state, cascade reactions, exploit innate reactivity, invent new reactions, and biomimetic synthesis. Baran incorporated these concepts into an efficient 10-step synthesis of the complex natural product Ambiguine H without any protecting groups. Krische orchestrated a highly efficient, convergent synthesis of Swinholide A with only two protecting groups.

8.11 Problems: Avoid Protecting Groups

Problem 8.1

(a) Choose the most atom economical protecting group (PG) from Figure 8.1 for the alcohol that would be stable to the strongly basic Horner–Emmons reagent in the reaction with the aldehyde to make the alkene given below.

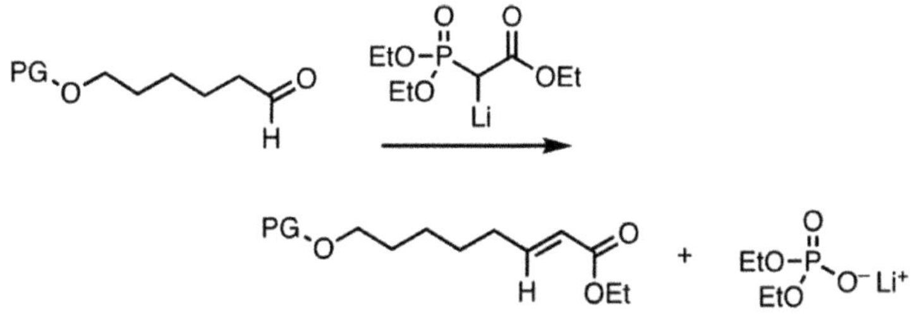

(b) What is the pK_a of a primary alcohol? _____ Estimate the pK_a of the Horner–Emmons conjugate acid (H instead of Li) based on the pK_a of a simple β-diester. _____

(c) Draw an acid-base reaction between an alcohol and a β-diester base. Which side would the equilibrium be skewed towards? By how many orders of magnitude?
(d) How could you avoid the need for a protecting group in this reaction?

Problem 8.2

(a) Draw the mechanism of a LiOH base-catalyzed ester hydrolysis making sure to show reversible steps with double arrows, and essentially irreversible steps with single arrows.
(b) For ester formation, strong acid catalysis is used. Consider your mechanism from problem 5.4 and explain why base-catalyzed instead of acid-catalyzed hydrolysis is used for ester hydrolysis. (Hint: Consider reversibility and the pK_a values of products.)

Problem 8.3

(a) How many protecting group on/off steps are in the "hydrolysis-esterification" synthesis of Simvastatin (Figure 8.14)? _____
(b) How is the silylation step selective for the one 2° alcohol over the other?
(c) What reagent was used in the last step to remove the silyl protecting group?

Problem 8.4

(a) How many protecting group on/off steps are in the "direct methylation" route to Simvastatin (Figure 8.15)? (Note that 1), 2), and 3) are separate steps over the 4th arrow.) _____
(b) What is the pK_a of an amide proton, –(C=O)–NH_2? _____
(c) What is the pK_a of the CH alpha to an amide carbonyl, –CH–(C=O)–NR_2? _____
(d) What is the pK_a of the CH alpha to an ester, –CH–(C=O)–OR'? _____
(e) How is the methylation step selective for the ester over the amide?

Problem 8.5

(a) Draw the reaction that the LovD enzyme originally catalyzes in bacteria. (Draw the 2-methylbutanoyl attached to "S-LovF" to represent the protein.)

(b) Draw and name the two small-molecule 2,2-dimethylbutanoyl (DMB) donors that were used to replace LovF.
(c) Which acyl donor was more efficient, and by how much?
(d) Is substrate inhibition a concern for the LovD enzyme-catalyzed reaction?

Problem 8.6

In a table, compare the loading of Monacolin J in $g L^{-1}$, the reaction time, the isolation and purification steps, and the % yield of the fermentations used to obtain Simvastatin before and after directed evolution of LovD and the one-pot synthesis.

Problem 8.7

List the number and name of the three most significant Principles of Green Chemistry satisfied by the Codexis synthesis of Simvastatin. Explain briefly.

Problem 8.8

(a) In what two ways should redox reactions figure into planning a protecting-group-free synthesis?
(b) How do convergent syntheses minimize the use of protecting groups?
(c) How do cascade reactions minimize the use of protecting groups?

Problem 8.9

(a) In Figure 8.19, the total synthesis of Ambiguine H, find the oxidation and reduction reactions, and list the oxidizing or reducing agent for each reaction.
(b) Which of the redox reactions are used to make C–C bonds? ____
(c) In the last step **f**, draw the structure of the ultimate waste product of the cascade reaction from compound **14**. How could it be recycled?

Problem 8.10

Refer to the hydrogenative C–C bond coupling in Figure 8.22.

(a) Calculate the atom economy of this step using only the *differences* in the atoms of reactants and products.
(b) List the catalysts used in this step.
(c) Which reagent is used most wastefully?
(d) List the desirable and undesirable properties of the solvent.
(e) After hydrogenation of MVK, an aldol reaction occurs. Draw the mechanism.

References

1. P. T. Anastas and J. C. Warner, *Green Chemistry: Theory and Practice*, Oxford University Press, New York, 1998.
2. P. G. M. Wuts and T. W. Greene, *Greene's Protective Groups in Organic Synthesis*, John Wiley & Sons, Inc., Hoboken, NJ, 5th edn, 2014.
3. E. J. Corey and A. Venkateswarlu, *J. Am. Chem. Soc.*, 1972, **94**, 6190–6191.
4. R. Walsh, *Acc. Chem. Res.*, 1981, **14**, 246–252.
5. G. G. Xu and F. A. Etzkorn, *Org. Lett.*, 2010, **12**, 696–699.
6. M. L. Kristensen, P. M. Christensen and J. Hallas, *BMJ Open*, 2015, **5**, e007118.
7. Y. Tang, X. Xie, X. Gao, A. Wojcicki, I. Pashkov, M. Sawaya, T. Yeates, R. Cacho, G. Pai, S. Collier, R. Wilson, E. L. Teo, J. Sukumaran, J. Xu, L. Gilson, W. L. Yeo and O. Alviso, *An Efficient Biocatalytic Process for Simvastatin Manufacture*, Washington, DC, 2012, https://www.epa.gov/greenchemistry/presidential-green-chemistry-challenge-2012-greener-synthetic-pathways-award, accessed July 25, 2019.
8. X. Gao, X. Xie, I. Pashkov, M. R. Sawaya, J. Laidman, W. Zhang, R. Cacho, T. O. Yeates and Y. Tang, *Chem. Biol.*, 2009, **16**, 1064–1074.
9. X. Xie, I. Pashkov, X. Gao, J. L. Guerrero, T. O. Yeates and Y. Tang, *Biotechnol. Bioeng.*, 2009, **102**, 20–28.
10. E. J. Corey and X.-M. Cheng, *The Logic of Chemical Synthesis*, Wiley-Interscience, New York, NY, 1995.
11. W. S. Johnson, R. S. Brinkmeyer, V. M. Kapoor and T. M. Yarnell, *J. Am. Chem. Soc.*, 1977, **99**, 8341–8343.
12. P. S. Baran, T. J. Maimone and J. M. Richter, *Nature*, 2007, **446**, 404–408.
13. S. H. Bertz, *J. Am. Chem. Soc.*, 1982, **104**, 5801–5803.
14. J. B. Hendrickson, *J. Am. Chem. Soc.*, 1975, **97**, 5784–5800.
15. L. F. Tietze and U. Beifuss, *Angew. Chem., Int. Ed. Engl.*, 1993, **32**, 131–163.
16. K. C. Nicolaou, D. J. Edmonds and P. G. Bulger, *Ang. Chem., Int. Ed. Engl.*, 2006, **45**, 7134–7186.
17. R. Robinson, *J. Chem. Soc., Trans.*, 1917, **111**, 762–768.
18. R. W. Hoffmann, *Synthesis-Stuttgart*, 2006, 3531–3541.
19. B. M. Trost, *Science*, 1991, **254**, 1471–1477.
20. B. M. Trost, *Acc. Chem. Res.*, 2002, **35**, 695–705.
21. R. F. Heck, in *Mechanisms of Inorganic Reactions*, American Chemical Society, 1965, vol. 49, pp. 181–219.
22. M. L. Micallef, D. Sharma, B. M. Bunn, L. Gerwick, R. Viswanathan and M. C. Moffitt, *BMC Microbiol.*, 2014, **14**, 213.
23. I. Shin, S. Hong and M. J. Krische, *J. Am. Chem. Soc.*, 2016, **138**, 14246–14249.
24. C. Bee, S. B. Han, A. Hassan, H. Iida and M. J. Krische, *J. Am. Chem. Soc.*, 2008, **130**, 2746–2747.

9 Catalysis

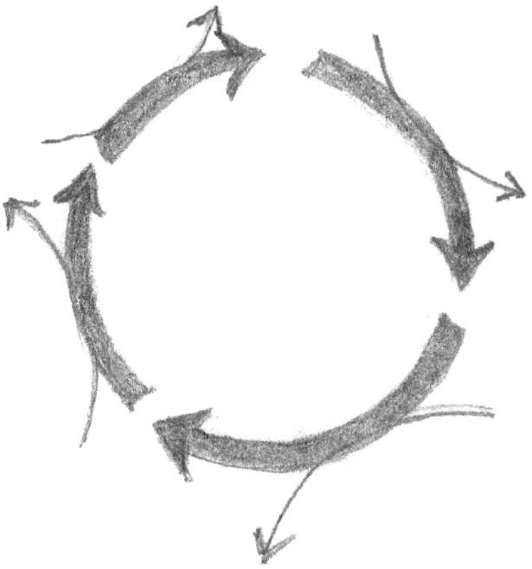

"Principle 9. Catalytic reagents (as selective as possible) are superior to stoichiometric reagents."[1]

9.1 Catalysts Accelerate Reactions

Catalysts in chemistry are broadly defined as substances that accelerate a reaction and emerge unchanged. Catalysts can be acids (Lewis or protic), bases, homogeneous metal complexes, heterogeneous metals on supports, organometallics, small organic molecules, and enzymes—the protein catalysts of biology. During the reaction, the catalyst interacts with reactants, whether forming and breaking covalent bonds, as with most metals, or forming non-covalent interactions such as hydrogen bonds and hydrophobic interactions,

Green Chemistry: Principles and Case Studies
By Felicia A. Etzkorn

Published by the Royal Society of Chemistry, www.rsc.org

as is typical of many enzymes. After the reaction, the catalyst releases products and is regenerated, ready to take on the next reactants.

Sheldon advocates for catalysts as a central solution for green chemistry.[2] Of all reagent categories, catalysts are certainly the most efficient. Perhaps that is why catalysts earned their own Principle of Green Chemistry. Typically, catalysts are used in less than 10 wt or mol% relative to the limiting reactant. Industrially, catalysts are used even more efficiently, such as 0.01 to 1 mol%. Exceptions to these common practices are found in flow systems, where the catalyst is momentarily in excess of the reactants. The efficiency is high in flow systems because the catalyst is used continuously until the reaction no longer gives high yield of the desired product. Catalysts in flow systems will be discussed in greater detail in Chapter 11 because Real-time Analysis is essential to determining when the catalyst should be regenerated or replaced.

Another significant efficiency feature of catalysts is that they can be recycled because they emerge from a reaction the same as they enter into it. In research laboratories, catalysts are infrequently recycled. In green chemistry catalyst research, the efficiency of a catalyst is often presented as the number of batch reactions that can be run before the catalyst degrades or becomes less effective.

A measure of catalytic efficiency, the turnover frequency (TOF), is an expression of the rate of the catalyzed reaction—moles per active-site time, typically in units of mol sec^{-1}. With enzymes, TOF is often as fast as mol μsec^{-1}. Higher TOF means higher efficiency, with less catalyst needed. Catalytic reactions can be so fast that the rate is diffusion limited, meaning that the rate that reactants diffuse through the solution is slower than the rate of the chemical steps that are catalyzed.

9.2 Enzymes: Nature's Catalysts Are Proteins

Enzymes are proteins that catalyze biological reactions. Because most living things do not tolerate high temperatures, enzymes evolve to operate at the optimum temperature for the organism, about 37 °C for humans. The names of enzymes usually include their primary substrate followed by the "ase" suffix. Enzyme (E)-catalyzed reactions typically have a substrate (S) binding step, the chemical step or steps, followed by product (P) release (Figure 9.1). If there are multiple substrates, the mechanism may be quite complex, with interesting names like "ordered bi-bi" and "ping-pong." Multiple books and

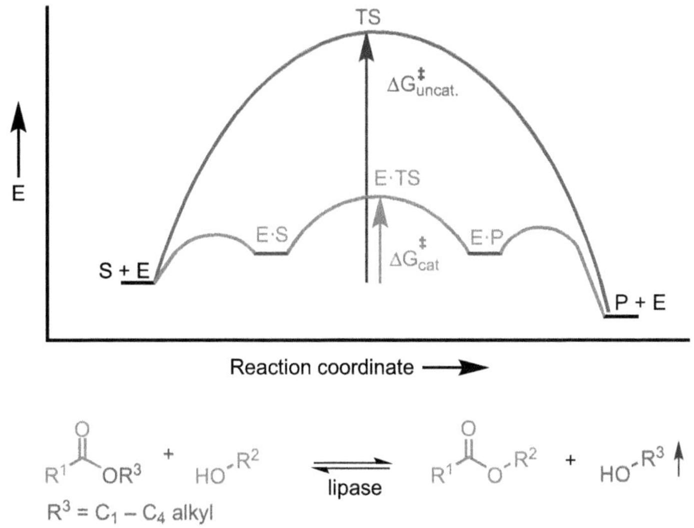

Figure 9.1 Top: Reaction coordinate diagram for an uncatalyzed and a catalyzed reaction. Bottom: The lipase reaction is shown to illustrate reactants and products of an enzyme-catalyzed reaction. The lipase mechanism is more than one catalytic step, including barriers ($\Delta G^{\ddagger}$) for each of the high-energy intermediates.

whole careers have been devoted to the determination and understanding of enzyme mechanisms. In 1946, Linus Pauling hypothesized that enzymes evolve to bind to the transition state (TS) more tightly than either reactants or products, and this is generally agreed to be the way enzymes work.[3] For isomerases, *i.e.* single-substrate enzymes, enzymes may also raise the energy of the substrate relative to the transition state. In both cases, the enzyme decreases the *difference* in energy between the substrate and the TS ($\Delta G^{\ddagger}$), thereby increasing the rate of the reaction (Figure 9.1).

In Eastman's lipase-catalyzed esterification reactions (Figure 5.5), the lipase enzyme catalyzes transesterification, the exchange of alkoxide groups R^3–O with R^2–O (Figure 9.1). The uncatalyzed reaction, or even the acid- or base-catalyzed reaction, has a much higher energy TS, and requires higher temperatures to overcome the activation barrier ($\Delta G^{\ddagger}_{uncat}$). In a catalyzed reaction, the enzyme first binds substrates to form the E · S complex, then catalyzes the reaction with much lower barriers ($\Delta G^{\ddagger}_{cat}$) for each of the mechanistic steps to give the product complex (E · P) before releasing the products (Figure 9.1). The enzyme is then free to react again. Chemical catalysts have similar energy profiles.

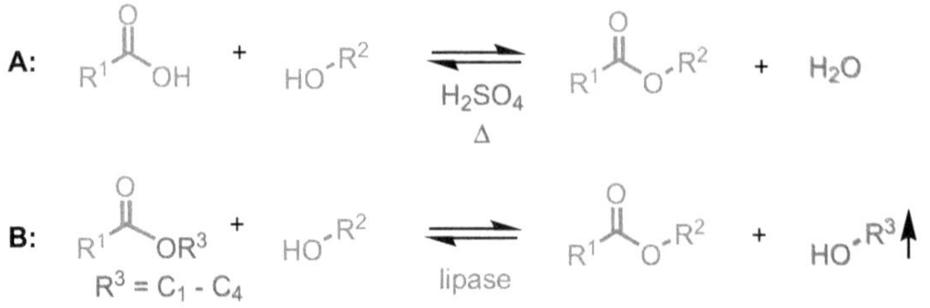

Figure 9.2 A: The Fischer esterification relies on an excess of alcohol R^2–OH to drive the reaction to completion. B: Eastman drives transesterification by removal of volatile R^3–OH product.

Note that catalysts do not change the thermodynamic equilibrium (K_{eq}) between S and P, which is dependent only on the difference in the ground-state energies of S and P (ΔG). Catalysts only change the rate of the reaction (k_{cat}), which is dependent on the difference in energy between E · S and TS ($G^{\ddagger}_{cat}$). Thus, catalysts can increase the rate in either the forward or the reverse direction—lipases can catalyze the formation of esters as well as the hydrolysis of esters. If the energy difference between substrate and product is small, the reaction can be driven in the forward or reverse direction using Le Chatelier's Principle by changing a concentration and resetting the equilibrium point. In practice, this means using an excess of one reactant, or removing one of the products. For example, the Fisher esterification, which is catalyzed by strong acid, relies on a large excess of alcohol to drive the reaction to form the ester (Figure 9.2A). Eastman's lipase transesterifications rely on removing the R^3–OH alcohol product by evaporation to drive the reaction to completion (Figure 9.2B).

9.3 Advantages of Catalysts

Catalysis gets its very own Principle number for very good reasons—catalysts can contribute to nearly every other Principle of Green Chemistry!

9.3.1 Prevent Waste and Synthetic Efficiency (Principles 1 and 2)

Decreased reaction times with catalysts lead to higher space-time yields of the desired products, preventing waste and improving synthetic efficiency. Selectivity is a key feature of catalysis that we have seen with enzymes. For example, the lipase A from *C. antarctica* stereospecifically esterifies only (3*R*)-stiripentol, not (3*S*)-stiripentol (Figure 2.21). Bacteria can be mutated to catalyze reactions other than those reactions they naturally evolved to catalyze for production of a single product. This can require directed evolution of one enzyme, such as the 30 mutations of LovD for the Codexis synthesis of Simvastatin (Chapter 8). Or several enzymes in a pathway can be mutated to avoid side products and optimize the desired product, as in the Genomatica synthesis of 1,4-butanediol (Chapter 7). In this chapter, R.H. Grubbs catalysts for olefin (alkene) metathesis are faster with less hindered and more strained alkenes, conferring selectivity for the desired product.

9.3.2 Benign Synthesis (Principle 3)

The use of catalysts can avoid toxic stoichiometric reagents, such as CrO_3, which leads to more benign synthesis, as in the case of the TEMPO/Cu catalyst system for the oxidation of alcohols with O_2 as the oxidant (Figure 3.14). Earth-abundant iron catalysts allow the use of silanes (phenylsilane f.p. 7 °C), which are less hazardous than boranes (borane-THF f.p. −17 °C), to reduce tertiary amides to amines (Figure 3.16). In 2014, Catalytic Technologies Ltd. announced a titanium catalyst to replace the toxic metal catalyst, antimony oxide, in the polymerization of polyethylene terephthalate (PET).[4] The PET has greater thermal stability, allowing less plastic to be used per bottle, greater brightness and clarity due to fewer impurities, and without increased cost to the consumer.[4] The catalyst also uses less energy to produce PET.[4]

9.3.3 Benign Products by Chemical Selectivity (Principle 4)

Elimination of by-products using catalysts that are highly specific for the desired product can also decrease toxic contaminants in the end product. For example, TAML oxidant activators use H_2O_2 as the oxidant for paper bleaching instead of Cl_2 or NaOCl, which both create highly toxic chlorinated dibenzodioxins and dibenzofurans (Figure 3.11). In 1989, a dietary supplement thought to be helpful with depression, the amino acid L-tryptophan (Trp), was found to cause eosinophilia-myalgia syndrome (EMS).[5] This was the result of a single company, Showa Denko K.K., Tokyo, making changes in the bacterial fermentation process to boost production of Trp that produced a toxic side-product contaminant, 1,1′-ethylidene-bis(tryptophan)

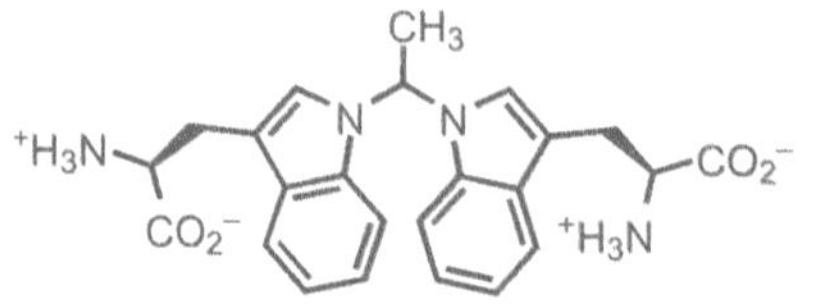

1,1′-ethylidenebis(tryptophan)

Figure 9.3 Contaminant by-product in bacterial fermentation product L-Trp.[6]

(Figure 9.3).[5] The introduction of a new bacteria strain (strain V) of *Bacillus amyloliquefaciens*, which was a new catalyst, and alteration of the purification method resulted in the contaminant.[6] Thus, having the *right* catalyst with the *right* selectivity is very important to producing benign products.

9.3.4 Avoid Auxiliaries (Principle 5)

The appropriate catalyst can decrease the amount of solvent used in a reaction. The Eastman transesterification used no solvents at all, and the enzyme was produced as a solid, which allowed it to be filtered from the liquid products and reused (Chapter 5). The higher selectivity of catalysts allows elimination of auxiliary-intensive purification steps.

9.3.5 Energy Efficiency (Principle 6)

The use of catalysts permits lower temperatures and pressures, which conserves energy. By lowering the activation barrier of reactions, catalysts directly decrease the energy needed for reactants to attain the high-energy transition state of the reaction. One consideration in a life-cycle analysis, however, is the energy needed to produce the catalyst itself assessed against the lifetime of the catalyst in the process.

9.3.6 Avoid Protecting Groups (Principle 8)

In the case study of the Codexis synthesis of Simvastatin, the use of an enzyme catalyst eliminated the need for protecting groups, and nearly eliminated the need for purification (Chapter 8). Simvastatin was produced in such high yield and purity that it precipitated from the reaction and was isolated by filtration. Filtration is about as efficient a method of purification possible.

9.3.7 Accident Prevention (Principle 12)

Reactions with lower energy requirements are also safer reactions, so accidents can be prevented by using catalysts at lower temperatures and pressures. Enzyme-catalyzed reactions are usually run in aqueous solutions, decreasing the need for flammable solvents.

9.4 Case Study: Greener Manufacturing of Sitagliptin Enabled by an Evolved Transaminase (Merck–Codexis)

Sitagliptin is an important drug for the treatment of diabetes. In process medicinal chemistry, the use of toxic heavy metal catalysts is undesirable because traces of heavy metals remaining in the final drug product can be very difficult to remove completely. These trace metals may cause adverse side effects, which is especially bad for the treatment of chronic illnesses like diabetes that require taking a drug for life.

Merck partnered with Codexis to develop an enzyme catalyst for a key step in the synthesis of Sitagliptin in their 2010 PGCC Award-winning process. To obtain the single enantiomer drug, the previous manufacturing route used an asymmetric hydrogenation catalyst. The greener route eliminated the heavy metal catalyst through the use of an artificially evolved transaminase enzyme.

Both the original Merck, and the improved Merck–Codexis routes for the synthesis of Sitagliptin ketoamide use a highly convergent-cascade reaction between three intermediates with similar complexity: 2,4,5-trifluorophenylacetic acid **9.1**, Meldrum's acid **9.2**, and a bicyclic heterocycle, 3-(trifluoromethyl)-triazolopiperazine **9.3**, to make ketoamide **9.4** (Figure 9.4). A significant portion of Meldrum's acid atoms is not incorporated into the final product resulting in a less than ideal atom economy. The efficiency of the convergent-cascade process far outweighs this negative effect.

The proposed mechanism for the 3-way condensation, determined by kinetic analysis, is shown in Figure 9.5.[7] A mixed anhydride of the trifluorophenylacetic acid with pivaloyl chloride (*t*-BuCOCl) was prepared to activate the acid for reaction with Meldrum's acid (malonate acetal). DMAP acts as a secondary activation catalyst, pumping the energy of the activated ester even higher. Deprotonated Meldrum's acid (pK_a 5) behaves as a carbon nucleophile in the reaction. Each of

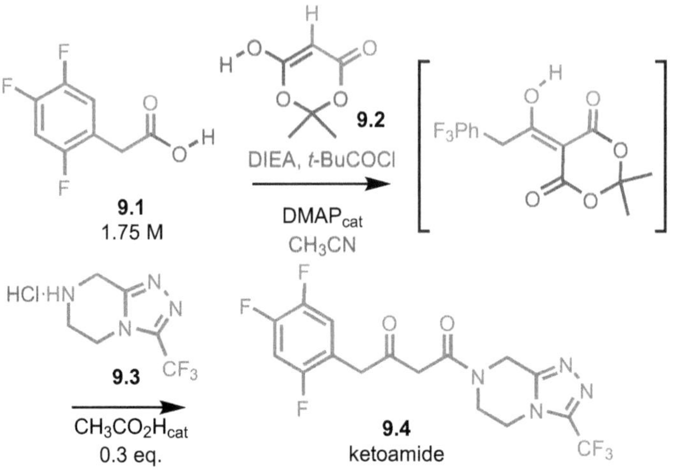

Figure 9.4 One-pot synthesis of the key intermediate β-ketoamide **9.4** for the first step of both current and greener Merck-Codexis biocatalytic manufacturing routes of Sitagliptin.[7]

these first three steps proceeds through a familiar tetrahedral intermediate. The diisopropylethylamine (DIEA, Hunig's base) salt of the product is stable and could be isolated. The optimized process was carried out in one pot by addition of 0.3 equivalents (eq.) of trifluoroacetic acid (TFA, pK_a −0.25) to rearrange the Meldrum's acid adduct intermediate, generating the highly reactive ketene ($R_2C{=}C{=}O$) intermediate. (Notice the loss of CO_2 and acetone in the third to last step, leaving only the two carbons and one oxygen of the ketene. This is not very atom economical, yet the removal of the gaseous products can be used to drive the reaction to completion.) As the DIEA salt of the amine was added to the reaction mixture, one equivalent of acid was liberated, allowing the use of less TFA. Although the intermediate β-ketoamide could be isolated and purified, the process was optimized by carrying out the next two steps in the same reactor. In the final improved process, the ketoamide was treated with dilute NaOH to isolate the crude ketoamide as a crystalline hemihydrate (0.5 H_2O) in 95% yield.

In the previous process, an enamine was prepared by reacting the β-ketoamide with NH_4OAc in 94% yield (Figure 9.6). The enamine was hydrogenated to the single enantiomer secondary amine of Sitagliptin with a homogeneous (soluble), stereoselective catalyst, Rh-*t*-Bu-Josiphos, in 97% enantiomeric excess (ee) (Figure 9.7). This catalyst had been optimized after screening Rh and Ir catalysts with eight different chiral phosphine ligands.[7] Treatment with activated

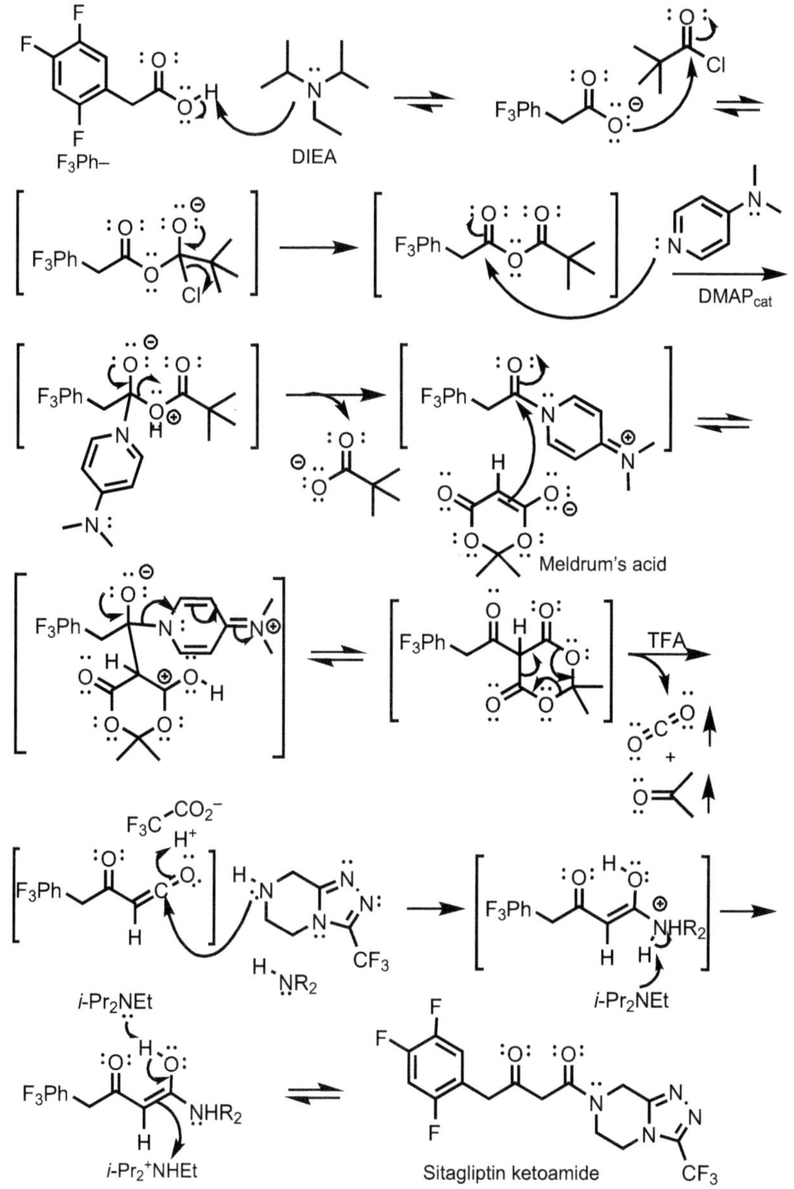

Figure 9.5 Proposed mechanism, including the ketene intermediate, for the convergent-cascade reaction to make the β-ketoamide intermediate **9.4** for Sitagliptin.[7]

carbon and crystallization from hexane/*i*-propanol produced the amine with the necessary enantiomeric purity, >99.5% ee, in 84% yield. Amine-containing drugs are typically supplied as an ammonium

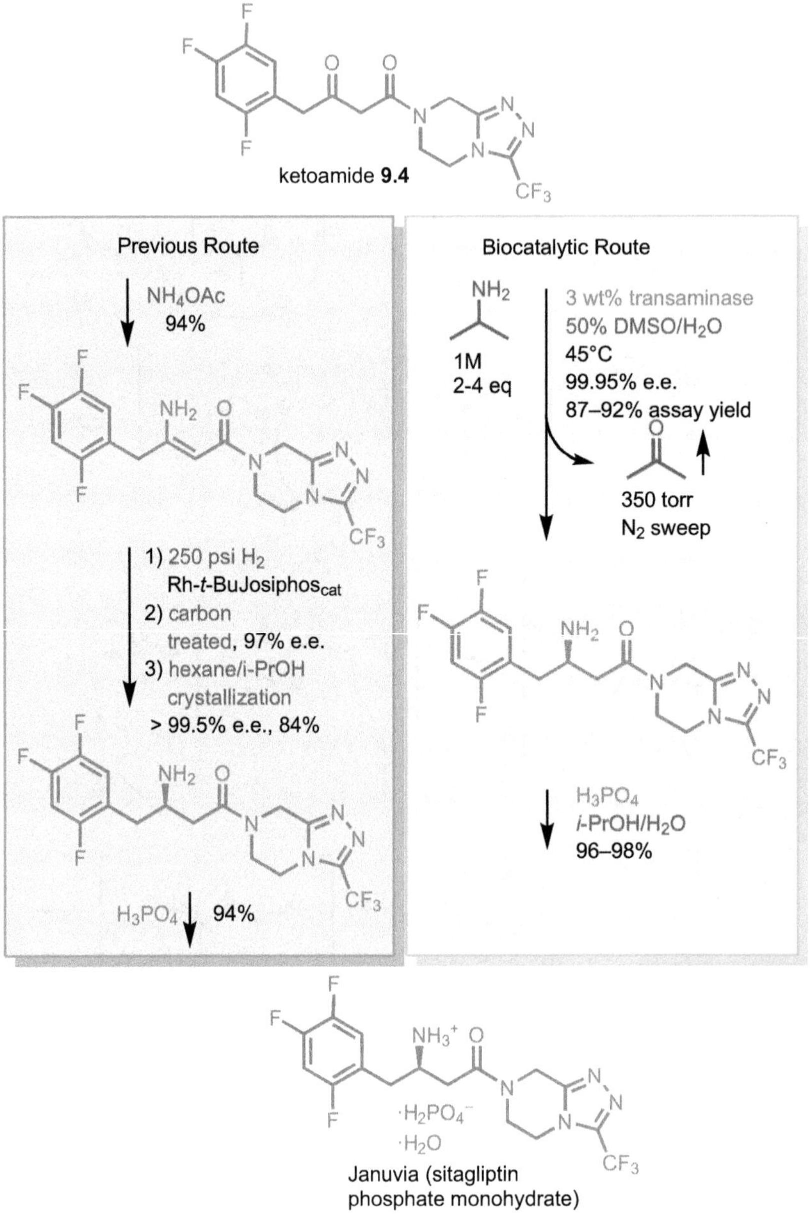

Figure 9.6 Asymmetric chemical (tan) and enzymatic (green) catalytic syntheses of the diabetes drug, Sitagliptin.[8]

salt to prevent decomposition by reaction with air and light. Treatment with phosphoric acid gave Januvia, the Sitagliptin phosphate monohydrate in 94% yield.

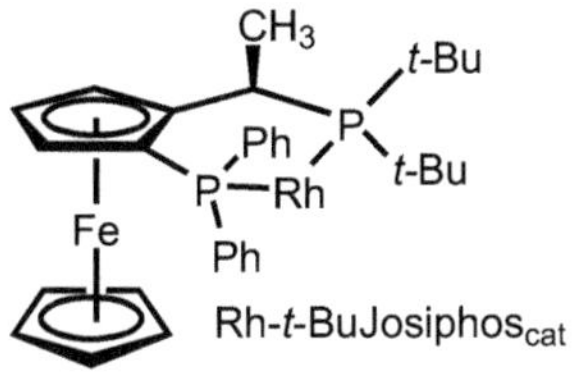

Figure 9.7 Single enantiomer Rh-*t*-Bu-Josiphos catalyst structure.

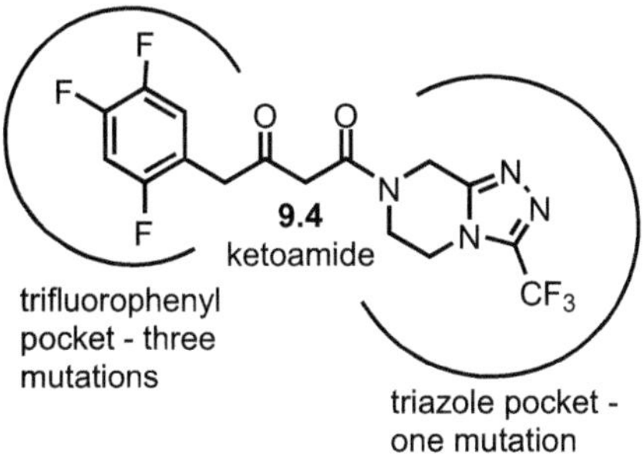

Figure 9.8 *In silico* transaminase redesign to bind β-ketoamide in the orientation to give a single enantiomer product.[8]

The highly convergent first step of this synthesis was considered very green, but there was still room for improvement. The major issues with this synthesis of Sitagliptin were the crystallization required to improve the enantiomeric purity, and the high-pressure hydrogen gas (250 psi) for the reduction, which required a large, expensive, explosion-proof vessel (Figure 9.6). Rhodium is also quite expensive ($760–$10 000 oz^{-1}) in addition to the difficulty of removing the potentially toxic homogeneous Rh catalyst from the final drug product.

The award-winning biocatalytic route developed by Merck–Codexis used a transaminase to catalyze the transfer of the amine from isopropylamine to the ketone. The transaminase was optimized by *in silico* redesign of the active site (Figure 9.8). The large pocket that bound the bicyclic triazole was mutated at one key residue, and the smaller pocket that bound the trifluorophenyl group was mutated at three residues to be more specific for these two parts of the transaminase substrate (Figure 9.8). These mutations allowed the transaminase to bind the substrate in the orientation to give the correct enantiomer of the product amine. Four rounds of directed evolution then led to 1000-fold improvement over the initial quadruple mutant. The reaction was further optimized by increasing the DMSO co-solvent from 5% to 50%. The ketoamide substrate in aqueous 50% DMSO was fed slowly over 2–4 hours to prevent two phases from forming during the reaction. The temperature was increased from 25 °C to 45 °C to increase the reaction rate. The overall improvement was greater than 25 000-fold over the first reaction detected with the quadruple mutant transaminase.

The transaminase-catalyzed reaction is in equilibrium between the ketone and isopropylamine reactants, and the amine and acetone products. In order to drive the reaction to completion, the acetone formed was removed with a mild vacuum (350 torr, 0.46 atm) and a nitrogen sweep (Figure 9.6).[8] The crude product could then be treated with phosphoric acid in *i*-PrOH/water to give Januvia in 96–98% yield after simple drying.[8] The enzyme was recovered from the reaction by precipitation at pH 2 and collection by filtration, or by extraction with *i*-PrOH/*i*-PrOAc and evaporation of the solvents.[8] The overall yield for the biocatalytic process from the β-ketoamide intermediate **9.4** was 84–87%, compared with 74% for the previous process (Figure 9.6).[8]

Life-cycle analysis confirmed that the use of hydrogen from reforming natural gas, and the mining and purification of iron and rhodium are much more energy and material intensive than the use of enzyme and the commodity chemical, *i*-$PrNH_2$.[8] The E-factor for the biocatalytic process was 31 $kg\,kg^{-1}$ Sitagliptin, while the Rh-catalyzed process was 38 $kg\,kg^{-1}$, a savings of 7 kg waste per kg of Sitagliptin produced.[8] The enzyme is produced in *E. coli* from renewable feedstocks. Further improvement with immobilized enzyme and a fully continuous flow process are envisioned.

9.5 Earth-abundant Metal Catalysts

A periodic table with the earth abundance of all the elements, developed by Todd Helmenstine for sciencenotes.org, is given as Appendix D. Iron is the most abundant heavy metal at 56 300 $mg\,kg^{-1}$ earth's crust. Remarkably, iron is the last element of the periodic table created in the core of stars before they explode, when the remaining natural elements up through uranium are created afterwards.[9] Iron is the fourth-most abundant element overall, behind oxygen (461 000 $mg\,kg^{-1}$), silicon (282 000 $mg\,kg^{-1}$), and aluminum (82 300 $mg\,kg^{-1}$). Sand is made mainly of SiO_2, which fits neatly with the popular concept of a very large number—"as many as all the grains of sand on the earth." Some of the base metals commonly used for catalysis are Fe, Cu, Ni, Co, and Zn. Unfortunately, these are not as easily molded into good catalysts as the more traditional Ru, Rh, Pd, and the Os, Ir, Pt group below them. Recently, both Fe and Cu have been widely adopted in catalysts for a wide variety of reactions, of which we will sample just a few in this section. Fe and Cu are both fairly benign—they are micronutrients in trace quantities, but they can still be toxic if too much is ingested.

9.6 Catalytic Mechanisms

9.6.1 Determination of Catalytic Mechanisms

Reaction mechanisms can never be proven without a doubt, rather we can definitively *disprove* all of the possible alternatives that we can imagine. Catalytic mechanisms are determined by similar methods to non-catalytic methods. The tools at our disposal are as diverse as our creativity. The analytical methods available will be reviewed in Chapter 11, Real-time Analysis. The first step is typically identification of all products, including the by-products that are often not drawn in organic reactions, but critically important in green chemistry to determine the efficiency of the reaction and any toxic by-products. Separation methods are important to isolate each of the products. Then any of the various forms of spectroscopy can be brought to bear to identify products. Labeling substrates with stable isotopes (^{2}H or ^{13}C) or radioactive isotopes (^{3}H or ^{14}C), followed by determination of the location of the label in the products is useful to determine where each of the atoms in the products originate, and which bonds are formed and broken in the reaction. Kinetics are used to determine the order of the reaction in each substrate, and the activation energy, E_a. These kinetic parameters are measured by the change in substrate or product concentration as a function of time at varied temperatures to give the rate of the reaction. Catalytic mechanism determination will of course include measuring rates as a function of catalyst concentration. The relative contributions of enthalpy and entropy may be obtained from thermodynamic measurements, $\Delta G = \Delta H - T\Delta S$. More advanced studies can include isolation and structure determination of high-energy intermediates, and kinetic isotope effects. Even with all of these methods applied, determining a reaction mechanism remains somewhat speculative—we should always call it a "proposed" mechanism—we say mechanism for brevity. The statistician, George Box, once said, "All models are wrong, but some are useful."[10] Think of the mechanism of a reaction as a model that is useful to improve a reaction, or to make a reaction greener.

9.6.2 Drawing Catalytic Mechanisms

When we draw the mechanism of a reaction, we are depending upon a vast literature—many papers contribute to the determination of a single reaction mechanism. The application of a particular reaction to a new substrate may result in different outcomes, and this can be due to a change in mechanism. Drawing the mechanisms of acid–base or radical catalyzed organic reactions fits neatly into the Lewis structures and curved arrows method. The mechanisms of enzyme-catalyzed reactions involve the functional groups on the side chains of amino acids in the active site of the enzyme, and they are often quite similar to organic mechanisms.

Inorganic and organometallic catalyzed mechanisms are usually not drawn by the curved arrow method, rather in a catalyst-centered cyclical mechanism to emphasize regeneration of the catalyst. During the mechanism, most transition-metal catalysts fluctuate between 16 e^- and 18 e^- species. Simplistically, these can be thought of as corresponding to filled p-level (6 e^-) and d-level (10 e^-) for a total of 16 e^-, or s- (2 e^-), p- (6 e^-), and d-levels (10 e^-) for a total of 18 e^-. To fill the shells, these electrons often come from ligands that offer pairs of electrons, such as P, N or O lone pairs. The loss of a small, loosely bound, neutral ligand, such as CO, NH_3, *i*-PrOH, or H_2O, is frequently the first step in the mechanism. The more tightly bound ligands, often donating two or more pairs of electrons, are critical in conferring chemical and stereoselectivity on the catalyst, but they are not shown repeatedly in the cyclic mechanism because they are usually complex. Because of the d-orbitals on transition metals, organic molecules can bind in a side-on orientation, for example forming two single bonds with the two carbons of an alkene.

9.6.3 Monsanto Process for Acetic Acid

A relatively simple place to start understanding organometallic mechanisms is the Monsanto process to produce the commodity chemical, acetic acid. The active catalyst is *cis*-$Rh(CO)_2I_2^-$, in which the two CO ligands are on the same side (Figure 9.9). The catalyst has the electronic configuration $[Kr]4d^8 5s^1$ for Rh, with 9 e^-, plus 1 e^- for the charge, plus 2 e^- for each CO ligand, plus 1 e^- for each iodine, a total of 16 e^-.

The first step is oxidative addition to the CH_3–I bond to give the hexa-coordinate, or octahedral, 18 e^- species on the right.[11] It is

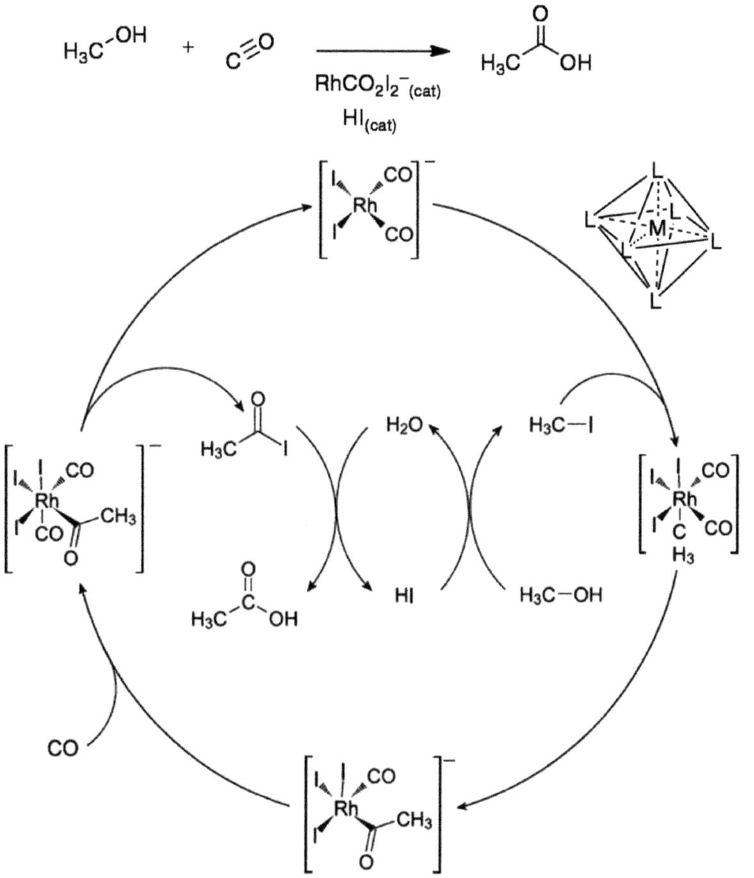

Figure 9.9 Overall reaction and the mechanism of the Monsanto process for acetic acid production. Upper right: metal in an octahedral ligand environment. Adapted from https://commons.wikimedia.org/w/index.php?curid=8945431, under a CC BY-SA 3.0 license, https://creativecommons.org/licenses/by-sa/3.0/de/.

called octahedral because the ligands are situated at the corners of an eight-sided figure composed of two square pyramids (Figure 9.9). Carbon monoxide then inserts into the Rh–CH_3 bond, which can alternatively be thought of as methyl migration to the CO ligand, to give the penta-coordinate species at the bottom. Addition of the next CO ligand then returns it to an 18 e^- species on the left, which undergoes reductive elimination of acetyl iodide and releases the original active catalyst at top.

In the middle of the cycle shown, acetyl iodide undergoes hydrolysis to give HOAc and HI (Figure 9.9). The HI then protonates the CH_3OH feedstock, and I^- substitutes for H_2O in an S_N2 reaction.

The net reaction is insertion of CO into CH_3OH to give HOAc. Water and HI can be thought of as co-catalysts because they are regenerated in each cycle. Even though the selectivity is >99%, and the conditions are milder than previous processes (150–200 °C and 30–60 atm), this is not a particularly green process because Rh is very rare and quite expensive. The process requires excess water which has to be removed from the HOAc product. HOAc for white vinegar is usually produced by fermentation.

9.6.4 Sharpless Epoxidation

Consider the Sharpless epoxidation as an example of a transition-metal catalyzed asymmetric reaction, as is the Rh-Josiphos catalyzed reduction to give Sitagliptin. The Sharpless epoxidation is catalyzed by an asymmetric titanium complex prepared *in situ* from $Ti(i\text{-PrO})_4$ and either L- or D-diethyltartrate, depending upon the desired stereochemistry of the epoxide product (Figure 9.10).[12–14] Each of the titanium atoms is in an octahedral ligand environment, with six ligand atoms each (**9.5**, Figure 9.10).[13] Six oxygens each contribute 2 e^-, and since Ti has 4 e^- in the outer shell, each Ti is a 16 e^- species. In the first step of the mechanism, one of the isopropoxy (*i*-PrO) ligands is protonated, comes off, and is replaced by the *t*-BuOOH (blue) oxidant (**9.6**, Figure 9.10).[14] A proton is transferred from the *t*-BuOOH to the *i*-PrO leaving group to maintain balance and give the neutral leaving group *i*-PrOH. Next the allylic alcohol reactant (red) displaces an axial ester C=O ligand, and the second oxygen of the *t*-BuOO ligand displaces a second *i*-PrO ligand (**9.7**, Figure 9.10). In the key step, one oxygen of the bound *t*-BuOO is transferred to the alkene to make the epoxide (**9.8**, Figure 9.10). Finally, both epoxyalcohol and *t*-BuOH products leave, and the catalyst (**9.5**) is regenerated by the binding of two *i*-PrO ligands and the axial tartrate ester C=O ligand again (Figure 9.10).

This sequential loss of ligand, binding of reactant, reaction, loss of product, binding of ligand is very typical of organometallic catalyst mechanisms. Stereoselectivity is usually the result of steric hindrance with one or two asymmetric ligands on the catalyst. In the case of the Sharpless epoxidation, the two tartaric ester ligands create the asymmetric environment. The allylic alcohol of the substrate acts as an anchor, binding the substrate to the Ti in a specific orientation. Steric interactions between the substrate R group and the catalyst ester groups force addition of the oxygen from one face of the alkene only, *si* or *re*, depending upon the tartaric acid

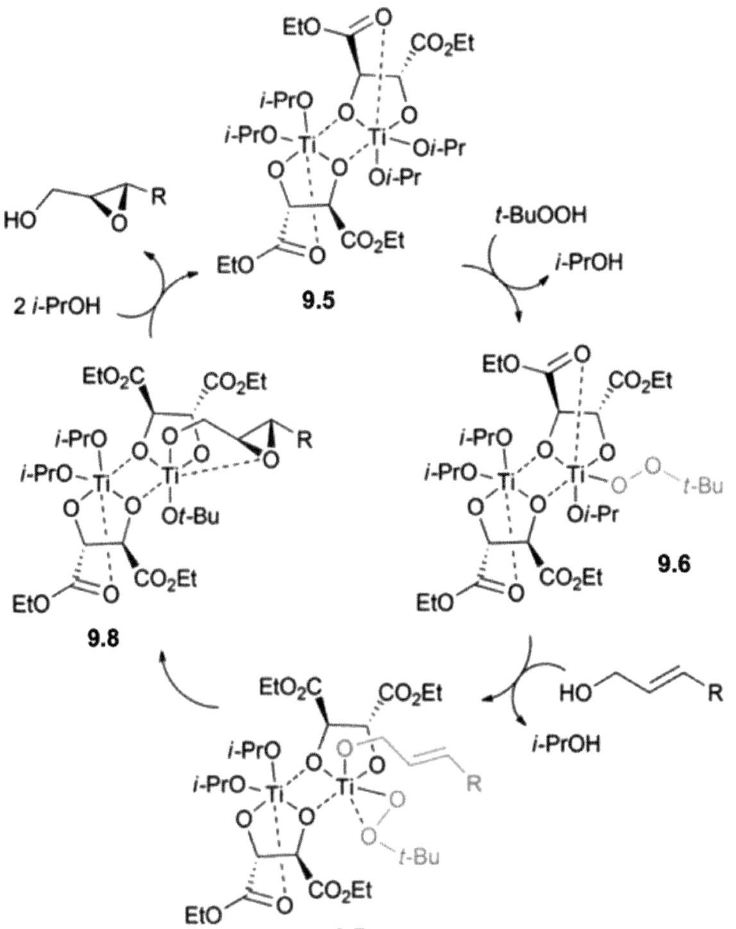

Figure 9.10 Mechanism of the Sharpless epoxidation with titanium catalyst. Adapted from ref. 14 with permission from the Royal Society of Chemistry.

enantiomer used. The Sharpless epoxidation catalyst is already fairly green, and it is easily made *in situ* for the reaction from readily available Ti(O*i*-Pr)$_4$, and the diester of natural product, tartaric acid. Ti is an earth-abundant metal at 5560 mg kg^{-1}, only about 10-fold less abundant than Fe. The catalyst is highly enantioselective using (+)- or (−)-tartrate ester. (Fun fact: Louis Pasteur was the first to separate enantiomers from a racemic mixture. He selected enantiomeric crystals of tartrate salts with tweezers.)[15] The *t*-BuOOH oxidant is flammable, less atom economical, and approximately as toxic as H_2O_2, but the steric bulk is necessary for the stereoselectivity of the Sharpless epoxidation.

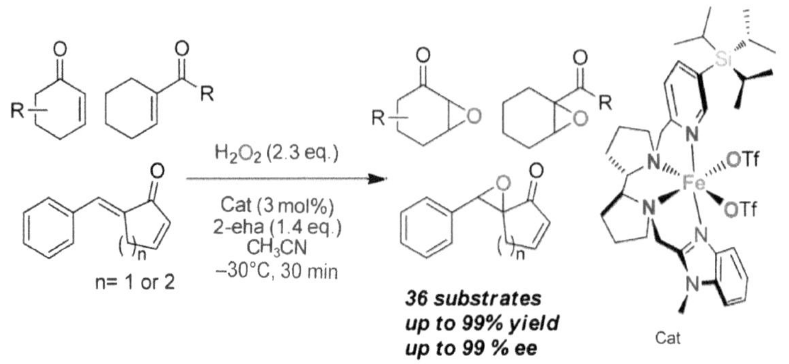

Figure 9.11 An earth-abundant iron catalyst for epoxidation of enones. 2-eha = 2-ethylhexanoic acid. Reprinted with permission of American Chemical Society from ref. 16, Copyright 2016.

9.6.5 Iron for Enone Asymmetric Epoxidation

An iron catalyst for asymmetric epoxidation of enones was created in 2016 (Figure 9.11).[16] The catalyst was very efficient with phenylcyclopentenyl alkenes, and somewhat less effective with cyclohexenones. The greater the difference in bulk of the substrate alkene's substituents, the better the yield and enantioselectivity (% ee) obtained. Mechanistic studies showed that the catalyst was an electrophilic oxidant, and that the epoxide oxygen originated only from the $H_2{}^{18}O_2$ oxidant, not water or O_2.[16] Both *t*-BuOOH and H_2O_2 gave similar % ee, so a common Fe=O intermediate was proposed.[16] H_2O_2 is cheaper and more atom economical. Iron is the least toxic, most earth-abundant transition metal available for catalysis. The complex asymmetric heterocyclic amine ligand makes the catalyst stereospecific. This method is complementary to the Sharpless epoxidation since it uses ketone substrates instead of alcohols.

9.6.6 Iron for Thermal [2 + 2] Alkene Dimerization

In Chapter 6, Energy Efficiency, we saw how Teshik Yoon uses ambient light as the energy source for the photochemically allowed [2 + 2] cycloaddition of carbonyl-substituted alkenes (Figure 6.12). By Woodward–Hoffman rules of orbital symmetry, the [2 + 2] cycloaddition of alkenes is not allowed thermally, only photochemically (Figure 6.11). Chirik's group circumvent the Woodward–Hoffman rules by causing the [2 + 2] thermal reaction to go by a different mechanism through the use of a bis(amino)pyridyl Fe or Co catalyst (Figure 9.12).[17,18]

The Fe catalyst contains a redox-active pyridine-diimine diradical ligand (Figure 9.12).[17,18] One-electron transfer between this ligand and the Fe metal is reversible, conferring catalytic activity on the Fe.[17]

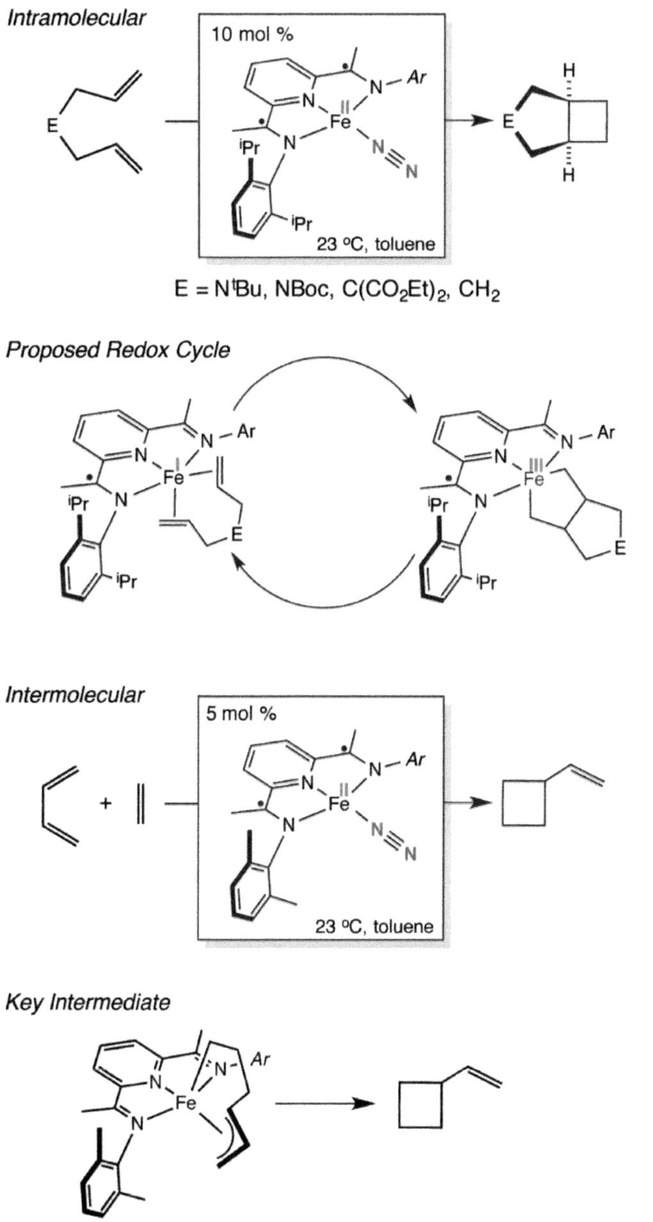

Figure 9.12 Intramolecular Fe-catalyzed [2+2] cycloaddition can occur if the "E" ring of the reactant is not too strained. Intramolecular reaction occurs through a two-step coordination and reaction of a diene with an alkene. Intermolecular cyclization of a diene and an alkene gives vinylcyclobutanes. Reprinted with permission of American Chemical Society from ref. 18, Copyright 2015.

Unactivated hydrocarbon alkenes, available from petroleum or natural sources (terpenes), react in this catalytic [2+2] cycloaddition. Intramolecular or intermolecular reactions work well with two monosubstituted alkenes. For the intramolecular case, these alkenes must be at two ends of the molecule, the α and the ω.[17] Since bicyclic cyclobutanes are created, the reactivity depends upon how easily the fused ring can be formed. Strained bicycles with 3-, 4-, or 7- to 13-membered rings fused to the cyclobutane may be more difficult to form, and 5- or 6-membered E-rings are preferred (Figure 9.12).[17,18]

In the intermolecular case, one ligand is proposed to bind in the axial position, and the other in the equatorial position of the octahedral Fe complex (Figure 9.12).[17,18] In the intermolecular case, a coordinated diene and an alkene first react to form a sigma bond, then the coordinated alkyl end and the allylic radical of the diene react to form vinylcyclobutanes (Figure 9.12). The catalytic cycle for a similar Co(I) version of the catalyst is proposed to proceed through ligand exchange, oxidative cyclization, and reductive elimination involving tetracoordinated Co(I) and Co(III) (Figure 9.13).[18]

9.7 Case Study: Using Metathesis Catalysis to Produce High-performing, Green Specialty Chemicals at Advantageous Costs (Elevance)

Elevance Renewable Sciences, Inc. won the 2012 PGCC award for olefin metathesis chemistry based on Nobel Prize-winning olefin metathesis catalysis.[19] Elevance uses renewable oil biofeedstocks to replace petrochemicals.[20] The feedstocks include palm, soy, and canola oils—many others are possible. Using oils from the local geographical area saves the energy of transport. Elevance has three global facilities with an expected combined annual production capacity of one million metric tonnes at the time of the award. A biodiesel plant in Mississippi was repurposed to operate as a biorefinery in 2013. Construction of conventional facilities costs between \$920 to \$2300 per metric tonne, while Elevance estimates costs of \$215 to \$535 per metric tonne of annual production capacity, due in part to the green chemistry.

9.7.1 Olefin Metathesis Reaction

The Nobel Prize in chemistry was awarded to Yves Chauvin, Robert H. Grubbs and Richard R. Schrock in 2005 for olefin (alkene) metathesis catalysts.[21] The second generation Grubbs catalyst is stable towards water and air, and incredibly efficient for olefin

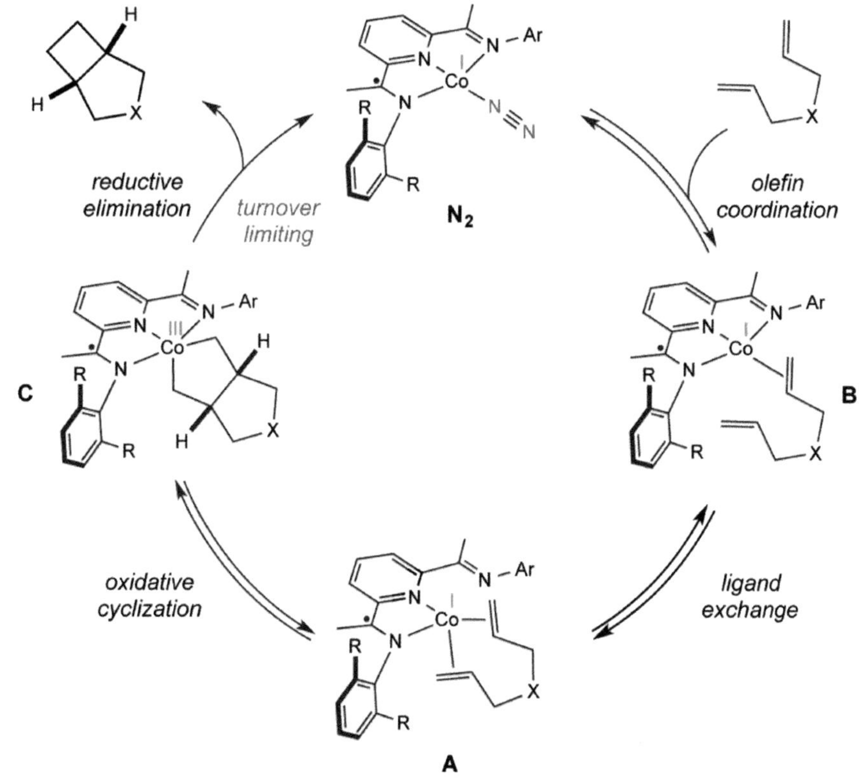

Figure 9.13 Proposed catalytic cycle for a bis(amino)pyridine-dinitrogen-Co(I) catalyzed [2+2]-cycloaddition. Reprinted with permission of American Chemical Society from ref. 18 Copyright 2015.

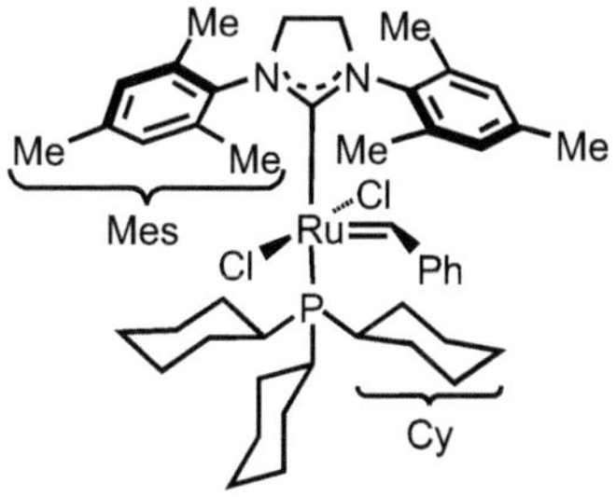

Figure 9.14 The structure of Grubbs 2nd generation olefin metathesis catalyst. Mes = mesityl; Cy = cyclohexyl.[22]

metathesis (Figure 9.14).[22] These catalysts function by having alkenes change partners, as in a square dance; both carbons of the alkene retain a new partner to make a new alkene (Figure 9.15). Vegetable oils can be converted into high-value bi-functional products—ester/alkenes and higher MW alkenes that are hard to

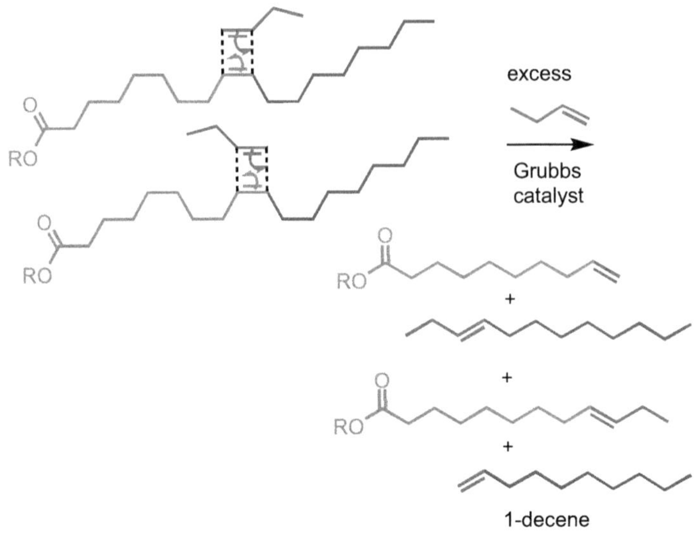

Figure 9.15 Elevance process for converting unsaturated fatty esters and low-MW alkenes into new product alkenes. The curved arrows do not show the mechanism, rather the new locations of the double bonds.[20]

obtain, such as 1-decene (Figure 9.15). Olefin metathesis is a rearrangement reaction. Recall from Chapter 2 that rearrangements are the most atom economical of the four types: elimination, substitution, addition, and rearrangement. In methathesis, both products must be used to be atom economical. The reaction will continue until all high MW alkenes have reacted with the excess of the less valuable alkene, such as butene, and then the catalyst is removed. The more stable *trans*-alkenes and terminal alkenes are the major products.

In the reaction of an oleate ester with 1-butene, the ester will lose the long alkyl tail, and gain a $=CH_2$ or $=CHCH_2CH_3$ group, while the long tail also gains either a $=CH_2$ or $=CHCH_2CH_3$ group, until a steady-state product mixture of terminal and *trans*-disubstituted alkenes is obtained (Figure 9.15). Low molecular weight alkenes—ethene, propene, and 1-butene—are gaseous reactants, derived from petroleum, that require pressure to maintain an excess during the reaction. Flammable gases under pressure are at high risk of explosion if something goes wrong during the process.

9.7.2 Elevance Products

Disubstituted alkenes formed are typically the thermodynamically more stable *trans*-alkenes. Specific feedstocks and conditions can be used to produce *cis*-alkenes. Ring-closing metathesis (RCM), ring-opening metathesis (ROM) and ring-opening metathesis polymerization (ROMP) can be used to produce other products. The driving force for ring-opening metathesis is typically the release of ring strain. Ester products from methathesis are further reacted—by transesterification, hydrolysis, and hydrogenation—to make other products. The Elevance process produces low toxicity products and by-products from renewable feedstocks with high materials efficiency.

The flexibility of the process towards multiple feedstocks reduces risk and maximizes profit. Elevance expects to consume a minimal percentage of the global supply of natural oils. The metathesis catalytic process is low-pressure, low-temperature, and catalytic. The energy efficient process is estimated to eliminate over 50% of greenhouse gas production by substituting biomass for petrochemical feedstock, and by using the Grubbs catalyst.

The Elevance process can be used to make waxes, surfactants, lubricants, additives, polymers, coatings, and thermoplastics.[20] Potential consumer products include cold-water detergents, anti-frizz and shine additives for leave-in hair-care products.[20] The estimated market sizes are $31 billion for consumer detergents and personal care products, $29 billion for lubricants and fuel additives, and $116 billion for engineered polymers and coatings.[20]

9.8 Catalyst Reuseability

A key feature of catalysts is that they are regenerated after a reaction. This should be tested in each case. During the reaction, catalyst regeneration is defined by the turnover frequency (TOF), the number of reactions per unit time ($mol\,sec^{-1}$). Here, we are concerned with the number of batch reactions that can be run using the same catalyst.

9.8.1 Recovery: Heterogeneous *vs.* Homogeneous Catalysts

In batch reactions, catalysts must be separated from products, by-products, and solvents for reuse in the next batch reaction. With heterogeneous solid catalysts, this can be a simple matter of filtering or decanting the reaction mixture. Some iron catalysts are magnetic and they can be held in place with a permanent magnet while decanting. The catalyst normally requires washing to remove all traces of other chemicals before reuse. Trace amounts of even solid catalysts can end up in products, which is particularly problematic for medicinal chemistry. The Sitagliptin case study is a good example of the advantages of switching from a heavy metal catalyst to an enzyme catalyst in the production of a drug.

Homogeneous catalysts are soluble in the reaction solvent. Extraction or precipitation can be used to recover the catalyst. Alternatively, solid products can be precipitated and collected by filtration, while the solvent can be removed from the catalyst by evaporation, or simply be reused along with the catalyst if it is clean. The Rh catalyst of the Monsanto process is recycled by evaporating the lighter components, including MeI, MeOH, and the product HOAc, and returning the aqueous solution of catalyst to the reactor.[11] Eventually all catalysts degrade to some extent, but it is usually cost efficient for precious metals, like Pd, Pt, Au, Rh, Ru, and Ir, to remanufacture recovered metals back into the original or another catalyst.

Polymer chemistry is frequently an exception to the rule that catalysts can be recycled. In polymer chemistry, the catalyst often initiates the polymerization, remaining attached at the end of the growing polymer chain. This is true for free-radical initiators, as well as heavy metal catalysts. Without catalysts, polymerizations must often be run at much higher temperatures, which is energy inefficient. Even though polymerizations may use as little as 0.01 wt% catalyst, large amounts of precious metal catalysts are incorporated into polymers, which ultimately end up in landfills or the oceans.

9.8.2 Reusability: Number of Reactions

The reusability of a catalyst depends on how easy it is to separate cleanly from the reaction mixture, and catalyst poisoning. A classic example of poisoning is the loss of catalytic activity due to adsorption of amines onto the reactive sites of Pd/C catalysts during hydrogenation. On the other hand, selective poisoning is purposely used to prepare the Lindlar catalyst, Pd/$CaCO_3$ poisoned with $Pb(OAc)_2$ and quinoline (Figure 9.16). This poisoning allows reduction of alkynes to *cis*-alkenes, and prevents over-reduction to the alkane.

9.8.3 Continuous Flow Reactors: Lifetime of Catalyst

In flow reactors, estimation of catalyst reusability is a question of time. Reactions should be monitored in real-time (Chapter 11) to

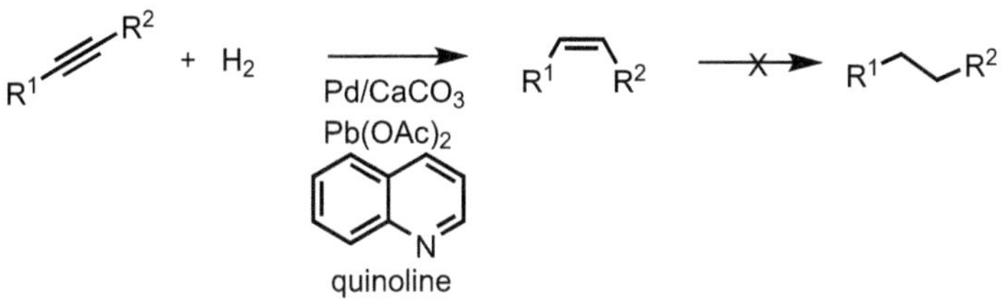

Figure 9.16 Lindlar catalyst poisoning to produce *cis*-alkenes from alkynes.

assess catalyst degradation or loss of activity during the process. Periodically, flow reactors must be paused and catalysts changed, regenerated, or remanufactured.

9.9 Summary

Catalysts contribute to nearly every Principle of Green Chemistry. Relatively low quantities of catalysts accelerate reactions at lower temperatures, giving greater structural and stereochemical selectivity for desired products with greater space-time yield. Merck and Codexis developed a transaminase enzyme catalyst for the synthesis of Sitagliptin in 84–87% yield and 99.95% ee. Determination of catalytic mechanisms requires a myriad of techniques: spectroscopic, chromatographic, kinetic, and labeling, to assist in optimizing the space-time yield and energy efficiency of a reaction. The Monsanto acetic acid process uses a highly optimized Rh catalyst. The Sharpless allylic alcohol epoxidation is catalyzed by a very green, stereospecific Ti-diethyltartrate catalyst. Earth-abundant Fe catalysts have been developed for asymmetric enone epoxidation, and thermal [2 + 2] alkene dimerization. Elevance produces a wide variety of industrial commodity chemicals and consumer products from unsaturated vegetable oils using a Grubbs olefin metathesis Ru catalyst. Catalyst recovery, reuseability, and lifetime in flow processes are parameters to be considered in assessing the green potential of a catalyst.

9.10 Problems: Catalysis

Problem 9.1

(a) How do enzyme catalysts fit into Principle 7 Renewable Feedstocks?

(b) How can catalysts assist with Principle 10 Design for Degradation?

Problem 9.2

Both syntheses of Sitagliptin use a 3-component reaction in the first step (Figure 9.4).

(a) Calculate the atom economy of the reaction. Do not include catalysts, solvents or work-up auxiliaries. Give the molecular formulas, masses and show the equation.
(b) Explain two general features of this reaction that make it efficient (see Chapter 8).

Problem 9.3

(a) Draw the structure of Rh-*t*-Bu-Josiphos used to catalyze the hydrogenation step in the chemical manufacturing of Sitagliptin.
(b) Why is it stereoselective?

Problem 9.4

(a) Draw the balanced equation for the transaminase-catalyzed reaction of pro-Sitagliptin β-ketoamide with isopropylamine.
(b) List two ways to shift the equilibrium to increase the product yield.

Problem 9.5

(a) Draw the side chains of the active-site catalytic residues of aspartate transaminase, and the pyridoxal phosphate (PLP) cofactor (search the internet) that catalyze the conversion of an amine into a ketone.
(b) Why is the evolved transaminase stereoselective?
(c) Draw the uncatalyzed transamination mechanism of a ketone with i-PrNH$_2$ to make an amine. Simplify the reactants as R^1-(C=O)-R^2 and H_2N-R_3. Combined proton transfer steps, and intramolecular transfers are ok for the sake of brevity.

Problem 9.6

(a) Draw the ligand that confers stereospecificity on the Fe-catalyzed epoxidation of enones.

(b) Draw the Fe-ligand that catalyzes one-electron redox chemistry to permit thermal [2 + 2] cyclization of alkenes. Explain whether this is an oxidation, reduction, or rearrangement.

Problem 9.7

(a) List the numbers and names of the **three** most significant Principles of Green Chemistry satisfied by Elevance processes and explain briefly.
(b) List one disadvantage of the Elevance process, relate it to a Principle, and explain briefly.
(c) Name and draw the general reactions of the three non-metathesis chemical steps Elevance uses.

Problem 9.8

(a) Draw the balanced metathesis reaction of **linoleic acid** with excess **ethene.**
(b) How many possible products can be formed? ____

Problem 9.9

Draw the mechanism of the Grubbs generation 2 catalyzed olefin metathesis reaction as a catalytic cycle. (Abbreviate the catalyst as: Ru=CH–Ph, and the *trans*-alkene reactant as $R^1CH{=}CHR^2$, and use ethene as the second alkene.)

Problem 9.10

(a) List two ways to recover a heterogeneous catalyst.
(b) List two ways to recover a homogeneous catalyst.

References

1. P. T. Anastas and J. C. Warner, *Green Chemistry: Theory and Practice*, Oxford University Press, New York, 1998.
2. R. A. Sheldon, I. Arends and U. Hanefeld, *Green Chemistry and Catalysis*, WILEY-VCH Verlag GmbH & Co. KGaA, Weinheim, Germany, 2007.
3. L. Pauling, *Chem. Eng. News Archives*, 1946, **24**, 1375–1377.
4. Catalytic Technologies LTD, *Breakthrough PET Catalyst Technology from Catalytic Technologies Ltd. Achieves Market Success*, pressreleasefinder.com, 2014, https://www.pressreleasefinder.com/Catalytic_Technologies/CATPR002/en/, accessed July 24, 2017.

5. S. Naylor, *Eosinophilia Myalgia Syndrome Contaminants: Past, Present, Future*, National Eosinophilia-Myalgia Syndrome Network, 2017, http://nemsn.org/ems-eosinophilia-myalgia-syndrome-contaminants-2017, accessed July 25, 2019.
6. A. N. Mayeno, F. Lin, C. S. Foote, D. A. Loegering, M. M. Ames, C. W. Hedberg and G. J. Gleich, *Science*, 1990, **250**, 1707–1708.
7. K. B. Hansen, Y. Hsiao, F. Xu, N. Rivera, A. Clausen, M. Kubryk, S. Krska, T. Rosner, B. Simmons, J. Balsells, N. Ikemoto, Y. Sun, F. Spindler, C. Malan, E. J. J. Grabowski and J. D. Armstrong, *J. Am. Chem. Soc.*, 2009, **131**, 8798–8804.
8. Merck & Co. and Codexis Inc., *Greener Manufacturing of Sitagliptin Enabled by an Evolved Transaminase*, Washington, DC, 2010, https://www.epa.gov/greenchemistry/presidential-green-chemistry-challenge-2010-greener-reaction-conditions-award, accessed July 25, 2019.
9. S. Charley, *How to Make an Element*, Public Broadcasting Service, 2012, http://www.pbs.org/wgbh/nova/physics/make-an-element.html, accessed July 25, 2019.
10. G. E. P. Box, in *Robustness in Statistics*, ed. R. L. Launer and G. N. Wilkinson, Academic Press, 1979, pp. 201–236.
11. J. H. Jones, *Platinum Met. Rev.*, 2000, **44**, 94–105.
12. T. Katsuki and K. B. Sharpless, *J. Am. Chem. Soc.*, 1980, **102**, 5974–5976.
13. M. G. Finn and K. B. Sharpless, *J. Am. Chem. Soc.*, 1991, **113**, 113–126.
14. Y. Minko and I. Marek, *Org. Biomol. Chem.*, 2014, **12**, 1535–1546.
15. L. Pasteur, *C. R. Acad. Sci.*, 1848, **26**, 535–538.
16. O. Cusso, M. Cianfanelli, X. Ribas, R. J. M. K. Gebbink and M. Costas, *J. Am. Chem. Soc.*, 2016, **138**, 2732–2738.
17. J. M. Hoyt, V. A. Schmidt, A. M. Tondreau and P. J. Chirik, *Science*, 2015, **349**, 960–963.
18. V. A. Schmidt, J. M. Hoyt, G. W. Margulieux and P. J. Chirik, *J. Am. Chem. Soc.*, 2015, **137**, 7903–7914.
19. R. H. Grubbs, *Angew. Chem., Int. Ed. Engl.*, 2006, **45**, 3760–3765.
20. Elevance Renewable Science, *Commercial Production and Substitution of Petrochemicals through Innovative Biorefineries producing Cost Advantaged and Higher Performing Green Specialty Chemicals based on Nobel Prize-winning Metathesis Catalysis for Markets of $176 Billion*, United States Environmental Protection Agency, Washington, D.C., 2012, https://www.epa.gov/greenchemistry/presidential-green-chemistry-challenge-2012-small-business-award, accessed July 25, 2019.
21. The Nobel Prize Organisation, *The Nobel Prize in Chemistry 2005*, 2005, https://www.nobelprize.org/prizes/chemistry/2005/summary/, accessed July 25, 2019.
22. M. Scholl, S. Ding, C. W. Lee and R. H. Grubbs, *Org. Lett.*, 1999, **1**, 953–956.

10 Degradation or Recovery

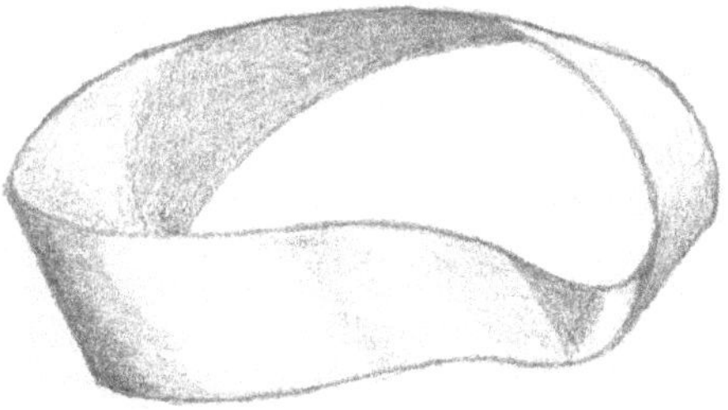

"Principle 10. Chemical products should be designed so that at the end of their function they break down into innocuous degradation products and do not persist in the environment. Products that do not degrade should be designed to fully recover the chemical materials from which they were made."[1]

10.1 Biological and Industrial Cycles

The use of renewable feedstocks is at one end of the cycle described as the "biological cycle" in *cradle-to-cradle*. At the other end is the breakdown of renewably sourced products, whether by microbes (composting), sunlight, oxygen, or water. In *cradle-to-cradle*, McDonough and Braungart also posit an "industrial cycle" that would truly recycle materials that cannot be broken down, and which should be recovered completely and reused in the same products that they came from. This could be the rare-earth elements used in a cell phone, or plastics that do not break down in the environment, but could be broken down into the component monomers to use in

Green Chemistry: Principles and Case Studies
By Felicia A. Etzkorn

Published by the Royal Society of Chemistry, www.rsc.org

resynthesis of virgin plastic. For example, the bacterial degradation of polyethylene terephthalate (PET) could be used to produce monomers for the remanufacture of virgin PET described later in this chapter. Metals, minerals, and other mined materials are an especially significant category of industrial materials that should be recovered. Products containing industrial materials should be designed to be easily dismantled to recover the raw materials—this is an engineering challenge.

10.1.1 Plastic Recycling

Our modern economic system unfortunately is still based on the assumption that the earth will absorb an infinite amount of trash. Consumer goods are designed for obsolescence rather than durability. Many different types of plastic are used in disposable goods such that confusion exists about recycling them. By the US recycling code, PETE or PET is recyclable plastic with the number 1, high-density polyethylene (HDPE) is 2, polyvinylchloride (PVC) is 3, low-density polyethylene (LDPE) is 4, polypropylene (PP) is 5, polystyrene (PS) is 6, and the "other" catch-all category is 7, including acrylics, nylons, polylactic acid (PLA), and polycarbonate (PC) made from BPA monomers (Figure 10.1). PETE, HDPE, LDPE, and PP are often recycled in curbside or community recycling systems. These mixed plastics are rarely separated and converted into monomers for resynthesis. Instead, they are used to make park benches, decking, and other building materials. This is called down-cycling, as opposed to up-cycling or true recycling, because the plastics have degraded and weakened through use or exposure to UV light. Although these are worthwhile products, they do eventually end up in the landfill. PVC, PS and most of the "other" plastics are not recycled in the US, although Europe has a broad program to recycle PVC. Because it does not contain chlorine, PP is a very good substitute for PVC in applications where durability is critical, for example in plumbing pipes. The point is that disposable products should be made from environmentally degradable materials. Such materials are usually derived from plant feedstocks, such as PLA. So, Principle 7: Renewable Feedstocks and Principle 10: Degradation or Recovery are complementary—two sides of the same coin.

The petroleum-based plastics shown in Figure 10.1 are very non-polar, and nearly completely resistant to biodegradation. The ubiquitous beverage bottle, typically made from PET, is abundant in trash and the oceans. About 6.2 billion pounds are produced

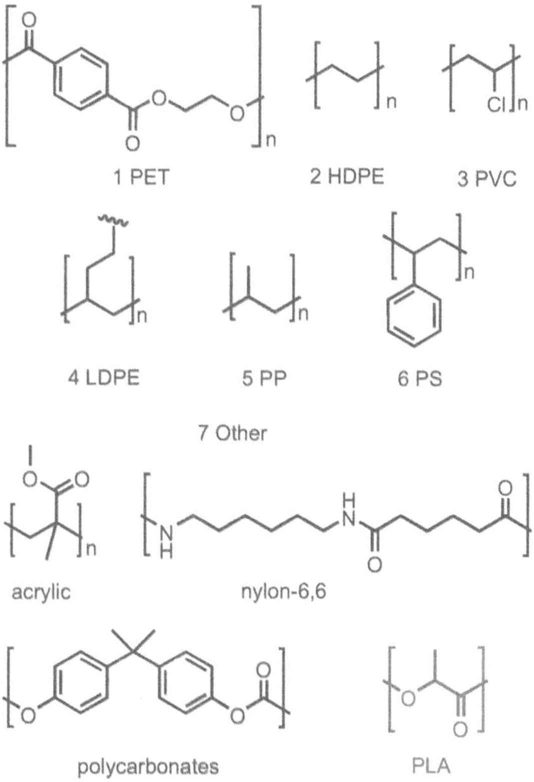

Figure 10.1 Consumer product plastics, their abbreviations, and recycling codes.

annually, yet as of 2018 only about 30% of all PET was recycled in the US.[2] HDPE, LDPE, PP, and foamed PS all float in the ocean, so they move with the wind, currents, and tides. It is unlikely that disposable plastics will go away, and even more unlikely that we will be able to keep them out of the oceans through recycling. Microscopic particles of plastics smaller than 5 mm, called microplastics, are found in nearly every marine organism, and even in organic fertilizer because it is derived from household and municipal waste.[3] One solution is to design water-degradable plastics.[4]

10.1.2 Biodegradable and Water-degradable Plastics

Biodegradable plastics have garnered the most attention from polymer chemists in recent years. Polylactic acid (PLA), polybutylene succinate (PBS), polyhydroxybutyrate (PHB),[5] and limonene-polycarbonate[6] are notable examples of biodegradable polymers.

NatureWorks® PLA is marked as compostable, but stringent conditions of microbes and aeration must be used to degrade PLA efficiently.[7] Unfortunately, in a landfill or in ocean water without microbes or much oxygen, even these biodegradable plastics can take a very long time to degrade, leaving macro- and micro-plastic pollution for aquatic organisms to ingest.

10.2 Case Study: Biodegradable Polymers from Carbon Monoxide (Coates)

10.2.1 Biodegradable Polyhydroxy Butyrate (PHB)

Geoffrey Coates of Cornell University won the 2012 Presidential Green Chemistry Challenge Award. Coates started the company Novomer Inc., which has dedicated much research effort to the design and synthesis of high-quality biodegradable plastics. One of their goals is to replace BPA epoxy coatings in food packaging. Poly-3-hydroxybutyrate (PHB) is a thermoplastic, biodegradable polyester that could serve this function. Coates devised a greener one-pot, two-step, tandem polymerization of PHB (Figure 10.2).[5]

In the first step, carbonyl insertion into (*R*)-propene oxide is catalyzed by an Al-porphyrin complex with retention of stereochemistry to give (*R*)-β-butyrolactone (BBL) (Figure 10.2). The retention of stereochemistry occurs because insertion of CO into the less sterically hindered 1° H_2C–O bond of (*R*)-propene oxide is favored over insertion into the more hindered 2° H_3CHC–O bond. The second step is ring-opening polymerization (ROP) of BBL with a Zn-diimide catalyst

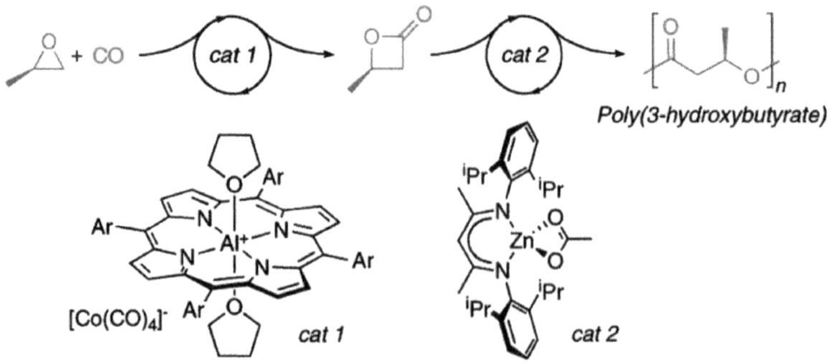

Figure 10.2 One-pot synthesis of poly-3-hydroxybutyrate (P3HB) optimized with earth-abundant aluminum and zinc catalysts. Reprinted with permission of American Chemical Society from ref. 5, Copyright 2010.

(Figure 10.2).[5] The stereochemistry is also retained in the PHB polymer, so it is a stereoregular polymer called "isotactic," which means each monomer unit of the polymer has the same stereochemistry. The opposite of isotactic is "atactic" or completely random stereochemistry. In between lies precisely alternating stereochemistry, called "syndiotactic."

The two catalysts used earth-abundant Al/Co and Zn. The best Zn catalyst was symmetric with two 2,5-di-iso-propylphenyl rings, which gave a high molecular weight (M_n) polymer of 52 kDa, and a polydispersity index (PDI) of 1.1, indicating very uniform chains.[5] PDI is the weight-average molecular mass (M_w) of the polymer divided by the number-average molecular mass (M_n): PDI $= M_w/M_n$. PDI values close to 1.0 indicate a high degree of chain-length regularity. The M_n of PHB increased linearly, and the polydispersity index remained low during the course of the reaction, indicating it is probably a living polymerization.[5]

Living polymerization is a form of chain-growth polymerization with a very low probability of chain termination or chain transfer, and a rate of chain initiation that is faster than elongation. When chain initiation is faster than elongation, all chains start at the same time, which gives a more uniform chain length, M_n. Living polymerization is most commonly observed in anionic polymerization, but the Coates PHB process is a metal-catalyzed living polymerization. Evidence suggests the Zn-catalyzed mechanism proceeds by coordination-insertion.[8] The Zn catalyst remains attached at the alkoxide living end of the polymer, while the acetate ligand caps the initiating carboxyl end.[8]

The tandem catalytic process avoids isolation and purification of the highly toxic β-lactone intermediate.[5] PHB can be produced by fermentation, but the process is energy intensive, and the polymer product must be separated from bacterial culture, including water removal.[5] Insertion of CO_2 instead of CO with different catalysts results in polycarbonates rather than polyesters.

10.2.2 Enantiopure (*R*)-Propene Oxide

The enantiopure feedstock for isotactic PHB, (*R*)-propene oxide (PO), can be made from biofeedstocks, either lactide or glycerol (Figure 10.3).[9] Both of these feedstocks have oxidized 3-carbon synthon units. We saw in Chapter 6 that 1,2-propanediol can be made from the glycerol waste from biodiesel production.[10] (*R*)-PO can be made from 1,2-propanediol by dynamic kinetic resolution of 1-*t*-butoxy-2-propanol

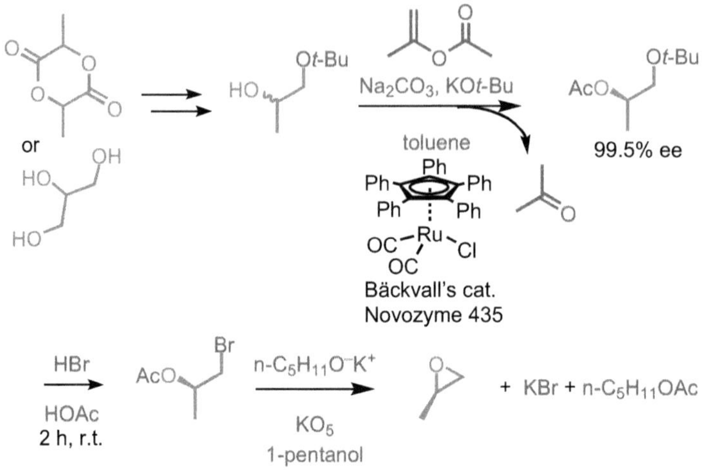

Figure 10.3 Synthesis of enantiopure (*R*)-propene oxide from renewable feedstocks, either lactide or glycerol.[9]

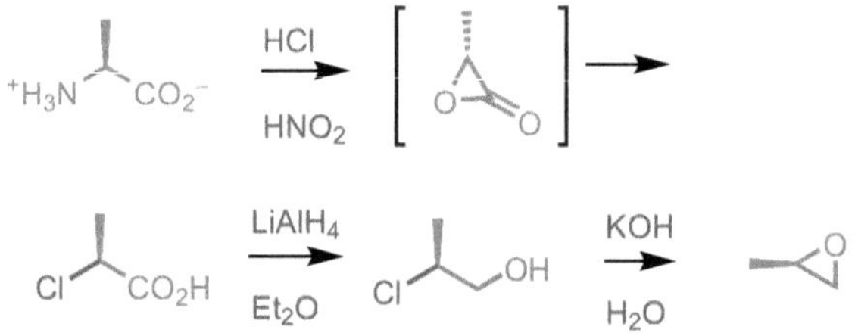

Figure 10.4 Synthesis of (*R*)-PO from the natural amino acid, L-alanine.[11]

by acetylation with Novozyme 435 and Bäckvall's Ru catalyst (Figure 10.3).[9] In this dynamic kinetic resolution, only one enantiomer of 1-*t*-butoxy-2-propanol reacts with iso-propenyl acetate in a transesterification catalyzed by the lipase, Novozyme 435 (Figure 10.3).[9] The unreacted alcohol enantiomer is then racemized by Bäckvall's catalyst and reacted with Novozyme 435 until none of the undesired enantiomer remains. The reaction can be driven by evaporation of the acetone by-product of the transesterification. To complete the synthesis, the *t*-butoxy group is converted to a bromo leaving group, followed by basic transesterification to remove the acetate, and epoxide ring closure (Figure 10.3).

Alternatively, enantiopure (*R*)-PO can be derived from the amino acid, L-alanine (Ala), also a 3-carbon synthon. This synthesis has three chemical steps with net inversion of configuration by double inversion through the high energy α-lactone intermediate, and a third inversion in the last epoxidation step (Figure 10.4).[11] In this case, a

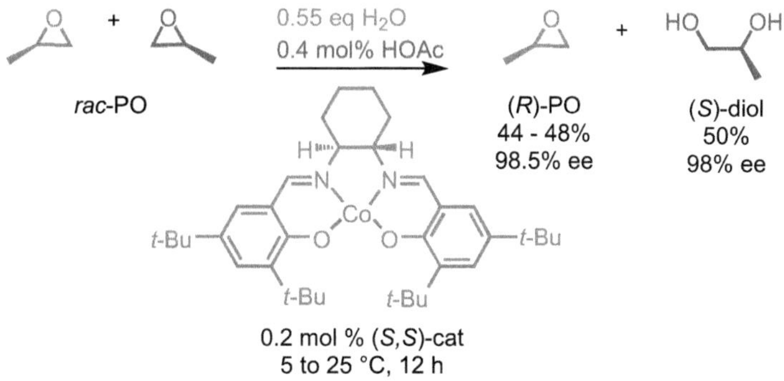

Figure 10.5 Efficient hydrolytic kinetic resolution of racemic propene oxide to produce high enantiopurity (*R*)-propene oxide (PO).[12]

chiral feedstock is used rather than introducing the stereocenter during the synthesis. This method is more atom economical; however, toxic acids, bases, flammable ether, and the reducing agent $LiAlH_4$ are required.

The most efficient method to obtain (*R*)-PO is by Jacobsen's hydrolytic kinetic resolution (HKR) with a Co-salen catalyst (Figure 10.5).[12] Racemic propene oxide could be obtained from glycerol or from petroleum feedstocks. The reaction is run in the absence of solvent and with a bare minimum (10% excess) of water as reactant (Figure 10.5).[12] The asymmetric catalyst includes fairly earth-abundant cobalt, and it is highly efficient. The catalyst showed no loss of yield or stereoselectivity after three cycles.[12] (*R*)-PO could be readily separated from the (*S*)-1,2-propane diol by fractional distillation.[12] The (*S*)-PO enantiomer may be obtained using the (*R*,*R*)-Co-salen catalyst.[12]

Industrial synthesis of propene oxide from the fossil feedstock propene, and hydrogen peroxide generates only water as a by-product, which is a much more efficient process. However, the propene oxide obtained is racemic. Although PHB can be derived from renewable feedstocks and it is biodegradable, PHB is not water soluble, and it is resistant to hydrolysis, which may be bad news for the oceans.

10.3 The Great Pacific Garbage Patch

Consumer plastic waste ends up in the oceans of the world. The "Great Pacific Garbage Patch" is an area of plastic waste floating just above or below the surface that is concentrated along the convergence zone between Japan and California.[13] Warm water from the Pacific confronts cold water from the Arctic along this convergence zone resulting in a huge whirlpool that traps ocean waste.[13] The amount of plastic entering the oceans has been conservatively estimated at 5 billion kg per year.[4] Most petroleum-based plastic never degrades, it just breaks down into ever-smaller bits that cause damage to the internal organs of wildlife. It is vital that we figure out a way to keep new plastic out of the oceans, in addition to attempting remediation.

10.4 Case Study: Water-degradable Plastics (Miller)

Steven A. Miller's (University of Florida) stated goal is to invent "...an environmentally safe, ocean-degradable alternative to petroleum plastics at a scalable volume."[4] Miller's group incorporated an acetal functional group into the backbone of PLA to design a new family of polyesteracetal (PEA) copolymers (Figure 10.6).[4] Acetals are the same group found in sugar polymers like cellulose at the anomeric position. Without the acetal group, PLA must be industrially composted—exposed to microbial enzymes, heat, and the oxygen in air, in addition to water. In 45 days, the PEA is degraded by 12%, with an estimated complete degradation in about one year (Figure 10.6). This

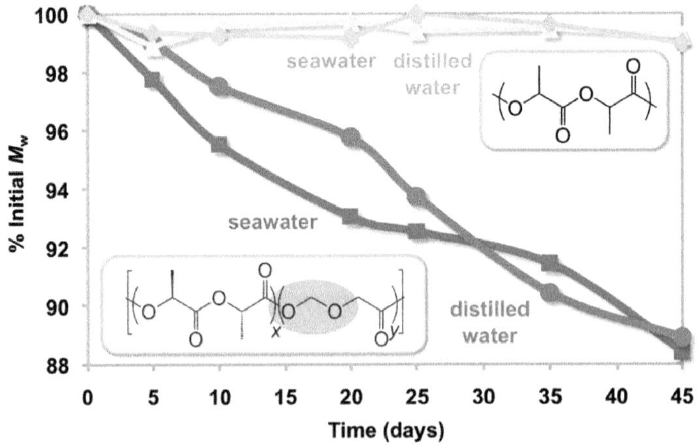

Figure 10.6 Seawater *vs.* distilled water degradation of a 0.5 mol% acetal PEA copolymer and pure PLA as measured by decreasing molecular weight (M_W). (Further analysis showed that only 11% of the polymer retained the acetal in the backbone. Formaldehyde is extruded during polymerization with heating. [S. A. Miller personal communication.]) Reprinted with permission of American Chemical Society from ref. 4, Copyright 2013.

is too fast for water bottles, which are expected to be marketable for much longer. Miller suggests that 10 years is an ideal time frame for polymers to degrade in ocean water.[4]

10.4.1 Acetal Metathesis Polymerization (AMP)

Miller has developed a general route to incorporating water-degradable acetals into polymers using acetal metathesis polymerization (AMP).[14] The diol precursor to an acetal monomer is treated with excess acetal with 1 mol% TsOH catalyst to make the bis-acetal monomer for the polymerization.

Initially, the co-monomer DMOM was used to synthesize poly-acetals. Unfortunately, the corresponding chain terminator, CH_3OH, has a higher boiling point than the DMOM co-monomer (Table 10.1). This means that the monomer would boil off faster than the chain terminator, resulting in lower molecular weights (M_n) and uneven chain lengths—a high PDI. Miller reasoned that higher molecular weights could be achieved by using the co-monomer diethoxymethane (DEOM), with a boiling point higher than the corresponding chain terminator, EtOH (Table 10.1).

In the improved process, diols could be used directly to synthesize the co-monomer *in situ* (Figure 10.7). Differing length hydrocarbon diol-ethoxy acetals, $(EtOCH_2O)_2$–R, are polymerized directly from the alkyldiol precursors (Figure 10.7).

10.4.2 1,10-Decanediol Acetal

Substitution of DEOM for DMOM worked very well with with 1,10-decanediol.[14] (Table 10.1) The AMP method uses *p*-toluenesulfonic acid (TsOH) as the catalyst.[14] The diol feedstocks can come from fats and oils, such as 1,10-decanediol derived from castor oil (Figure 10.8). Further optimization permitted the one-pot synthesis of the monomer followed by polymerization in a very efficient process (Figure 10.8).[14]

Table 10.1 Boiling points of co-monomers and chain-terminators for AMP. Higher molecular weight polymers with 1,10-decanediol were achieved with DEOM.[14]

Co-monomer	Abbrev.	b.p. (°C)	Chain terminator	b.p. (°C)	M_n
	DMOM	42°	CH_3OH	65°	17 900
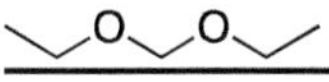	DEOM	88°	CH_3CH_2OH	78°	32 700

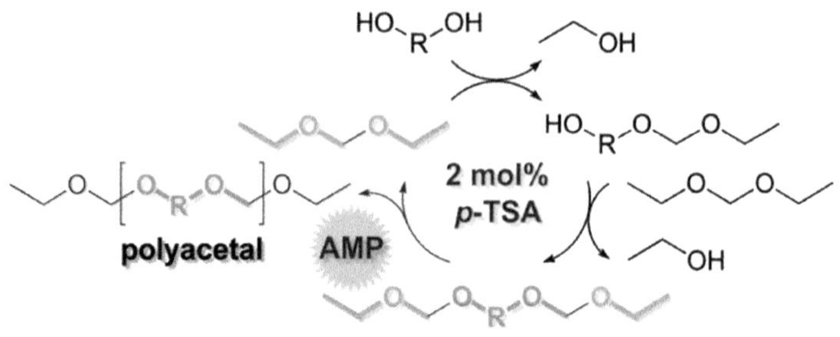

Figure 10.7 Acetal metathesis polymerization (AMP) direct route to polyacetals avoids ring-closing and ring-opening steps. Reproduced from ref. 14 with permission from the Royal Society of Chemistry.

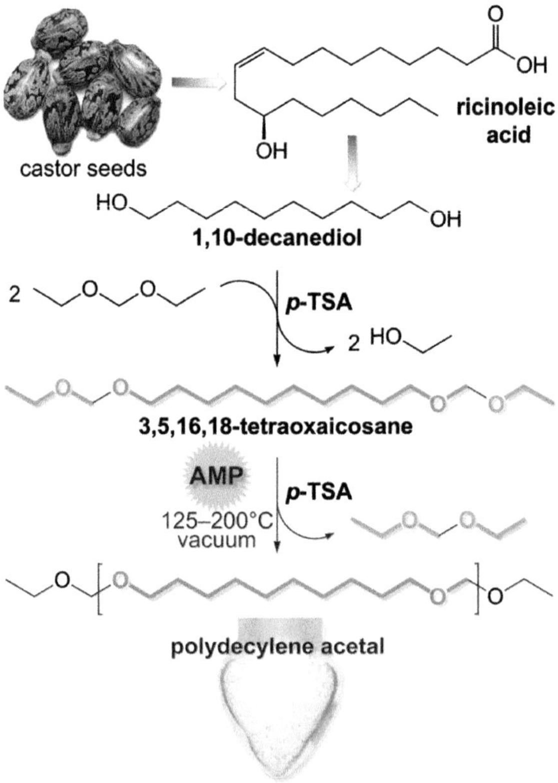

Figure 10.8 One-pot AMP route to water-degradable polydecylene acetal. 1,10-Decane diol was derived from castor oil. Reproduced from ref. 14 with permission from the Royal Society of Chemistry.

The reaction is driven by removal of the EtOH terminator by vacuum evaporation. Isolation and purification of the polymer were done by increasing the evaporation temperature to 200 °C for 12 hours.[14]

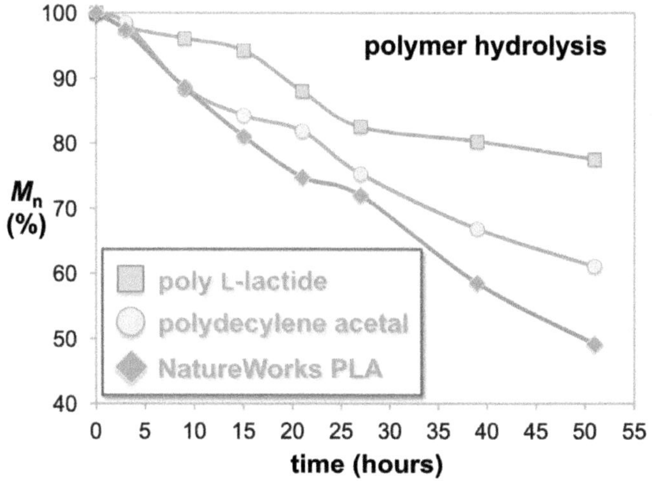

Figure 10.9 Acidic water/THF hydrolysis of polydecylene acetal compared with isotactic (all L-lactate) PLA and NatureWorks® PLA. The M_n% is relative to M_n initial. Reproduced from ref. 14 with permission from the Royal Society of Chemistry.

The mixture was cooled, quenched with basic methanol, collected by filtration and dried under vacuum to give 95% isolated yield. The M_n was 22 000 with a PDI of 1.83.[14] The DEOM removed by evaporation can be reused again in the process.

In acidic water/THF, polydecylene acetal degraded by 50% in 50 hours, slower than the rate of hydrolysis of NatureWorks® PLA-600 (Figure 10.9).[14] The acetal functional group is readily hydrolyzed under acidic conditions, even in the absence of microbes and low oxygen conditions found in landfills.

10.5 Case Study: Bacterial Degradation of Polyethylene Terephthalate (Miyamoto-Oda)

Yoshida *et al.* discovered a consortium of microbes that grow on PET by analyzing 250 soil samples contaminated with PET.[15] A bacterium, *Ideonella sakaiensis*, was isolated that contains enzymes that degrade PET, called PETase and monohydroxyethylterephthalatehydrolase (MHETase).[15] PETase is evolutionarily related to lipases.[15] When a PET film was treated with the purified PETase enzyme, the surface developed a pitted appearance; HPLC confirmed the production of MHET and TPA (Figure 10.10).[15]

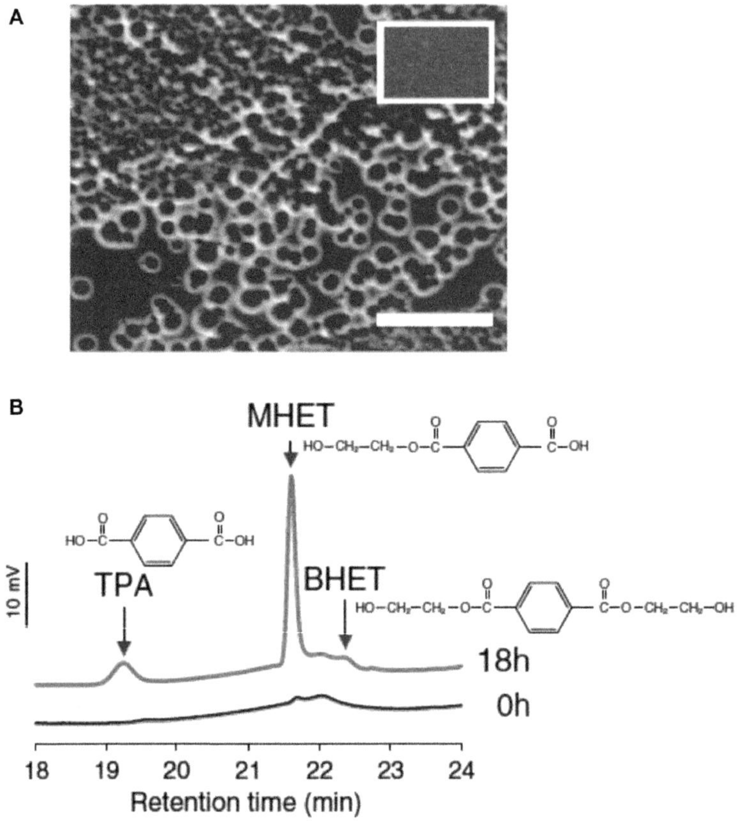

Figure 10.10 "Effects of PETase on PET film are shown in (A) and (B). PET film (diameter, 6 mm) was incubated with 50 nM PETase in pH 7.0 buffer for 18 hours at 30 °C. (A) SEM image of the treated PET film surface. The inset shows intact PET film. Scale bar, 5 μm. (B) High performance liquid chromatography chromatogram of the products released from the PET film." TPA = terephthalic acid; MHET = monohydroxyethylterephthalate; BHET = Bis-hydroxyethylterephthalate. Reprinted with permission of American Association for the Advancement of Science from ref. 15, Copyright 2016.

Enzymatic degradation of polymers depends on the flexibility of the polymer. Commercial bottle PET has a glass transition temperature of 75 °C, so it is flexible enough to be degraded by PETase and MHETase at reasonable temperatures where the enzymes are stable.[15] *I. sakaiensis* degraded smooth PET nearly completely in 6 weeks at 30 °C.[15] The bacteria are proposed to further metabolize the TPA into a 7-carbon unsaturated tricarboxylic acid, ultimately a carbon source that the bacteria can live on.[15] By manipulating the genome of *I. sakaiensis*, degradation could be stopped at the TPA and ethylene glycol monomers, which could

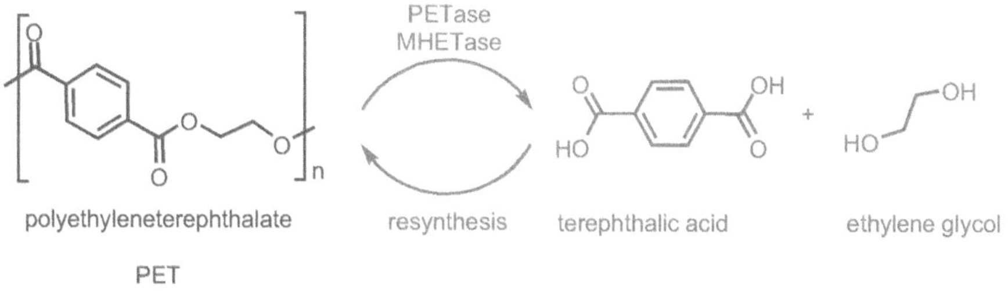

Figure 10.11 Bacterial degradation of PET into the constituent monomers. Resynthesis of PET from reclaimed monomers is proposed as an alternate source of feedstocks.[15]

then be reused in the synthesis of virgin PET (Figure 10.11).[15] This profound discovery promises potential solutions to the degradation of almost any polymer. The trick will be to keep these organisms from getting loose in the environment and degrading packaging materials before products get to the consumer. This should not be a problem because we have successfully used cellulose packaging for a very long time, and there are plenty of microorganisms that degrade cellulose.

10.6 Case Study: Biodegradable Surfactants and Sugars Replace Very Persistent Fluorinated Surfactants in Aqueous Firefighting Foams (Solberg Co.)

10.6.1 Perfluorooctane Sulfonate

In 2014, the city of Newburgh, NY was found to have perfluorooctane sulfonate (PFOS), CF_3-$(CF_2)_7$-$SO_3^-Na^+$, in their tap water at 170 parts per trillion (ppt) and people were getting sick. In May 2016, the EPA limit was lowered from 400 to 70 ppt putting Newburgh's water source level of PFOS well above federal guidelines.[16] Newburgh, an hour north of New York City, is near a military air base, Stewart National Guard Base, that used firefighting foam containing PFOS. The US EPA identified PFOS as persistent, bioaccumulative and toxic (PBT), the characteristics of a POP except for the long-range transport. PFOS was added to the Stockholm POPs Treaty with restrictions on its use (Table 4.2). PFOS is associated with cancer of the liver, thyroid, and mammary glands.[17] PFOS also has significant acute toxicity; the LD_{50} in rats was determined to be 251 mg kg^{-1}.[17]

The US manufacturing company 3M announced it would phase out production of PFOS in 2001. In 2010 and 2015, the US EPA issued a

Voluntary Stewardship Program call to eliminate long-chain fluorosurfactants from firefighting foam, thus creating a significant need for new firefighting foams. While shorter chain perfluorohexane sulfonate (C_6) is less toxic and less bioaccumulative, it still persists in the environment. Solberg Co. chose to eliminate *all* organohalogens from their firefighting foams instead of just the long-chain fluorosurfactants, because even short-chain fluorocarbons are persistent in the environment.[18] Solberg's firefighting RE-HEALING™ Foam fits the ideal of a design for degradation.[18]

10.6.2 Firefighting Non-fluorinated RE-HEALING™ Foam

Solberg won a PGCC Award in 2014 for *RE-HEALING™ Foam Concentrates–Effective Halogen-Free Firefighting* in the category of Designing Greener Chemicals.[18] Besides creating a degradable product, the function of the new product was paramount—the new firefighting foam had to be as effective or better than existing products for reasons of safety. Solberg's RE-HEALING™ Foam is now the world's most effective firefighting foam concentrate. The four parameters they considered were flame knockdown, fire control, extinguishing time, and burn-back resistance. These are important for the safety of firefighters everywhere.

Extinguishment measures whether the flame is extinguished in 3 or 5 minutes, depending on the fuel type: hydrocarbon or polar solvent. After the flame is extinguished, the foam blanket should remain in place for three times the application time—9 or 15 minutes.

The burn-back test is very specific. At the end of the extinguishment test, a one-square-foot diameter pipe is inserted into the foam blanket. The foam inside the pipe is removed, the fuel is reignited and allowed to burn for 1 minute. The pipe is removed and the whole area is allowed to burn for 5 more minutes. If less than 20% of the area covered by the foam burns, the performance of the foam is acceptable.[18]

In addition to comparable persistence in the environment, the shorter chain C_6 fluorocarbon surfactant foam was found to perform poorly relative to PFOS (C_8) in the extinguishment and burn-back tests.[18] Almost 40% more C_6 was required to match the performance of the C_8 foam, so almost 40% more fluorinated surfactant would end up in the environment than the C_8 PFOS.[18] Even with 40% more C_6, the time to extinguishment was longer, 2:00 *vs.* 1:42 hours for the C_8 surfactant.[18] Substituting C_6 for C_8 fluorocarbon surfactants is thus a poor and inadequate solution to the problem.

10.6.3 Firefighting Foam Composition

RE-HEALING™ Foam is composed of hydrocarbon surfactant(s), water, solvent, complex carbohydrates, a preservative, and a corrosion inhibitor.[18] Hydrocarbon surfactants are likely to be typical detergents, such as linear alkylbenzene sulfonates (Figure 10.12). The hydrocarbon solvents, made from renewable feedstocks, are the same as used in the health care industry.[18] The complex carbohydrates are likely to be cellulose, a β-1,4-D-glucose polymer, or starch (amylose), which is an α-1,4-D-glucose polymer with α-1,6-branching (Figure 10.12). (In the usual chair conformation, the β-1 oxygen is "up" or equatorial, and the α-1 oxygen is "down" or axial at the anomeric position.) The complex carbohydrate component gives the foam more heat-absorbing capacity than PFOS foams (Table 4.2).[18] Small amounts of a preservative were added to prevent degradation of the natural carbohydrates in storage. The corrosion inhibitor was necessary because the foam is water-based and any metal storage container could rust or corrode.

Re-HEALING™ Foams are formulated in 1%, 3%, and 6% dilutions for extinguishing Class B hydrocarbon fuels, such as gasoline, jet fuel, and diesel fuel.[18] Water-miscible solvents, like methanol, ethanol, and acetone, use foams made from the same ingredients, but with use levels modified to improve performance. The RE-HEALING™ Foam was 93% degraded in 28 days, and completely degraded in 42 days.[18]

Solberg found a way to completely eliminate fluorinated surfactants from firefighting foams to keep these persistent pollutants out of the environment. They replaced fluorosurfactants with hydrocarbon surfactants and complex carbohydrates from renewable feedstocks; these are both biodegradable. The RE-HEALING™ Foam is even more

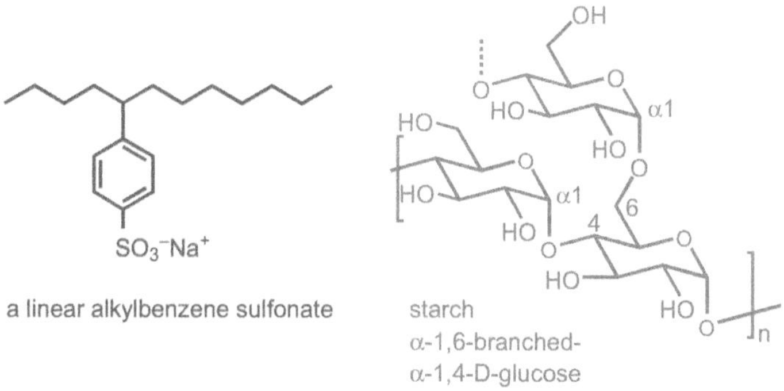

Figure 10.12 Components of Solberg RE-HEALING™ foam: sulfonate detergents and starch polysaccharide.[18]

effective than PFOS foams. The company produced demonstration videos in their environmentally safe test facility to increase acceptance by the critical firefighting customer base (https://www.youtube.com/watch?v=g6oW8J2Z5HI, accessed July 2019).

10.7 Metals Recovery

Metals and minerals are not biodegradable, so metals should be recovered at the end of the lifetime of the product, whether it is a computer, a jet plane, or a catalyst. Therefore, metal-containing products should be constructed so that they are easily dismantled, and the metals easily separated, if not physically, then chemically. The choice of which metal to use in a product or process should be based on three criteria: inherent toxicity, earth abundance, and effectiveness.

10.7.1 Toxicity

Toxicity varies greatly, as with Fe *vs.* Cr, and even different oxidation states of the same metal—Cr(III) *vs.* Cr(VI). In Chapter 12, Faraday Technology's less toxic chromium(III) plating process that won a PGCC award is described. In Chapter 4, Yttrium as a Lead Substitute in car paint primer is described. In each case, toxicity was the driving force to find a substitute for Cr and Pb. Lead pipes were a problem in ancient Rome and remain a major problem to this day. In 2016, the disastrous change in water supply in Flint, MI caused Pb to leach from the pipes into household water. There is a substantial need to replace the lead in old water pipes with something just as effective and less toxic. The alternatives under consideration are brass, stainless steel, plastic, and composites. Other familiar toxic metals are As, Cd and Hg.

10.7.2 Earth Abundance

Appendix D is a periodic table that gives the abundance of each element in the earth's crust in units of $mg\,kg^{-1}$.[19] Interestingly, oxygen is the most abundant element on earth, followed closely by Si. Al is the most abundant metal, followed by Fe as the most abundant transition metal. Most of these 4th row transition metals are abundant and cheap "base metals," like Fe, Ni, Cu, and Zn. One of the main problems with base metals is that they are reactive, especially towards oxidation—consider rust that is a mixture of iron(II/III) oxides.

In the early 4th row, Cr^0 is so stable toward oxidation it is plated on steel to protect it from oxidation. On the other hand, its next-door neighbor, V, fluctuates between so many different oxidation states it was used as a redox flow battery (Chapter 6)—in this case, it is a feature, not a bug! The typical catalyst metals: Ru, Rh, Pd, Os, Ir, and Pt, are rare precious metals, along with Au and Ag. Precious metals are well worth recovering from catalyzed reactions.

10.7.3 Separations

Chemical separation is sometimes necessary to prevent metals from going into the landfill or contaminating surface waters. These can be very cost-effective to recover, as in the case of rare-earth elements: Sc, Y plus the lanthanides. While rare-earth metals are not as rare as precious metals, they are essential in modern technologies, including electronics and magnets. Separation and recovery of metals is essential because mining is very hazardous to the environment, and the supply of many essential metals is quite limited. At one time, it was common to extract Au from played-out mines by chemical reaction with cyanide.[20]

$$4\ Au(s) + 8\ NaCN(aq) + O_2(g) + 2H_2O(l) \rightarrow 4\ Na[Au(CN)_2](aq) + 4\ NaOH(aq)$$ [20]

The gold-cyanide salts were then released in several different ways, leaving toxic cyanide salts in the abandoned mine tailings. Figuring out less toxic methods for recovering precious and rare-earth metals from consumer products at the end of their lifespan is a worthy goal of green chemistry.

10.7.4 Recovery of Heavy Metals

Heavy metals are notorious environmental pollutants, yet they are essential for computers, smart phones, and display screens. Since metals are not degradable by their nature, recovery is essential to prevent pollution. Ethylenediaminetetraacetic acid (EDTA) strongly chelates metals. D'Halluin *et al.* modified renewable cellulose as the ester with EDTA (Figure 10.13).[21] The cellulose-EDTA adsorbent works as a membrane for continuous wastewater treatment.[21] A wide variety of heavy metals, including toxic Pb, and precious Ag, can be recovered with the adsorption medium.[21] EDTA-modified cellulose gives 90–95% recovery of the heavy metals: Ag(I), Cd(II), Pb(II), Ni(II), Zn(II), Sn(II), and Cu(II) from water.[21] The removal of Cr(II) was about 76%, and the removal of ubiquitous Fe(II) was only about 66%. Pb(II)

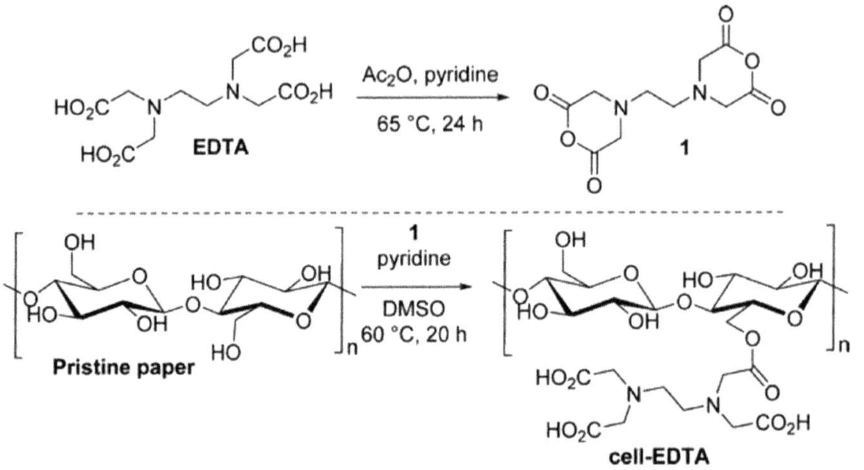

Figure 10.13 EDTA-modified cellulose for recovery of heavy metals from wastewater. Reprinted with permission of American Chemical Society from ref. 21, Copyright 2017.

absorbed from water can be almost fully desorbed and recovered, and the cellulose-EDTA can be reused for at least five cycles with consistent 93% adsorption and 96–100% desorption.[21]

10.8 Case Study: Recovery of Ecocatalysts from Plants (Grison)

Some plants have the ability to concentrate heavy metals. Claude Grison and co-workers developed a series of "EcoM" catalysts based on this ability of some water plants to hyperaccumulate transition metals.[22] These can be more efficient than single pure metal salts because EcoM catalysts have unusual or mixed oxidation states. For example, EcoNi from *Psychotria douarrei* is 7.6 times more efficient than $NiCl_2$ in a Biginelli 3-component cascade reaction to form cyclic thioureas, such as that found in the potential anti-cancer drug, monastrol (Figure 10.14).[23]

The water hyacinth (*Eichhornia crassipes*) selectively concentrates Pd in its roots from aqueous solutions.[22] Clavé *et al.* developed an EcoPd catalyst derived from the roots of *E. crassipes* for Suzuki cross-couplings with sensitive heteroaromatics like thiophenes (Figure 10.15).[22] The roots were processed by drying 6 hours at 550 °C, heating at 90 °C for 6 hours with 37% HCl, and filtering through Celite to give an active Pd(II) catalyst that is reusable as well as recyclable.[22] *E. crassipes* can be used to recover Pd from wastewater.[22] Using glycerol

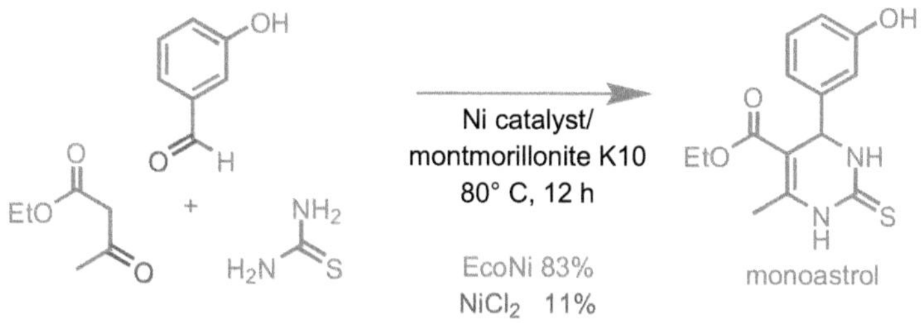

Figure 10.14 EcoNi from *P. douacrrei* catalyzed Biginelli condensation resulted in 95% conversion and 83% yield of monastrol, compared with $NiCl_2$ which gave 30% conversion and 11% yield. Adapted from ref. 23 with permission from the Royal Society of Chemistry.

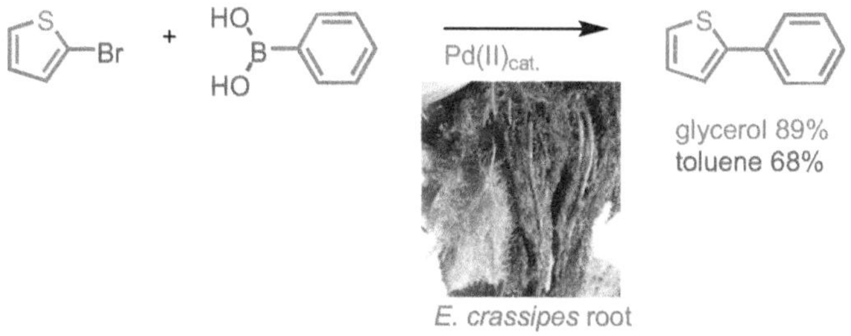

Figure 10.15 Suzuki coupling catalyzed by Pd(II) recovered from *E. crassipes* roots. Adapted from ref. 22 with permission from the Royal Society of Chemistry.

as the solvent for heteroaromatic Suzuki couplings resulted in 89% yield.[22] This yield is higher than in any traditional solvent, particularly toluene (68%), a solvent obtained from petroleum that is the standard solvent for Suzuki couplings (Figure 10.15).[22] Glycerol is a much safer solvent, with a low f.p. and low toxicity that is obtained as a by-product from the manufacture of biodiesel (Chapter 5).

10.9 Summary

To prevent pollution, environmentally degradable products of chemical processes must be designed. Plastics are of enormous concern because the most common petroleum-based plastics are not biodegradable, and have low recycling rates. Coates developed a one-pot, bi-catalytic process to make polyhydroxybutyrate more efficiently than by fermentation. The Great Pacific Garbage Patch has entered into the public consciousness. Miller has developed polyalkyleneacetals with high M_n that are water degradable inspired by the acetal linkage of polysaccharides. Yoshida *et al.* discovered enzymes that degrade PET with high efficiency. Solberg Co. developed a RE-HEALING™ firefighting foam with a biodegradable surfactant and natural polysaccharides that is more effective than toxic PFOS foams that are POPs. Heavy metals should be recovered from consumer products at the end of their lifetime. D'Halluin *et al.* made cellulose-EDTA that is very effective at recovering Pb and other heavy metals from wastewater. Grison and co-workers have developed a series of highly effective EcoM catalysts based on the ability of some plants to concentrate heavy metals.

10.10 Problems: Degradation or Recovery

Problem 10.1

(a) Draw the Zn-catalyzed mechanism of the (*R*)-BBL polymerization to give PHB.
(b) Compare the green advantages and disadvantages of the Bäckvall, L-alanine, and HKR processes to produce enantiopure (*R*)-propene oxide.

Problem 10.2

(a) What are Miller's goals for water-degradable plastics?
(b) Draw the functional group contained in cellulose that was the inspiration for Miller's biodegradable polymers.
(c) What is the environmental advantage of PEA over PLA?
(d) What is a good biological source of long-chain diols for polyacetals?

Problem 10.3

(a) Compare step-growth *vs* chain-growth polymerization.
(b) Which produces a more uniform length polymer? Why?

Problem 10.4

(a) Draw the mechanism of one step of acetal metathesis polymerization.
(b) What volatile by-product can be removed to drive the reaction?
(c) Is this step-growth or chain-growth polymerization? Explain briefly.

Problem 10.5

(a) What is Miller's ideal timeframe for water degradation of a polymer?
(b) What is the estimated time frame for petroleum-based polymer hydrolysis?

Problem 10.6

(a) What have firefighting foams been made from since the 1960s?
(b) What was suggested by the US EPA to voluntarily decrease the toxicity and bio-accumulation of firefighting foams?

Problem 10.7

Consider the persistent organic pollutants (POPs) of Chapter 4 and PFOS.

(a) What are the bond dissociation energies (BDEs) of C–F, C–C, and C–O bonds?
(b) What are the bond dissociation energies of the C–Cl bonds of ethyl chloride and phenyl chloride?
(c) How does BDE correlate with persistence in the environment?

Problem 10.8

(a) List the components of the Solberg Co. RE-HEALING™ Foam.
(b) Draw the structure of the complex carbohydrate used other than cellulose.
(c) What is the advantage of using this ingredient in fire-extinguishing foams?

Problem 10.9

(a) List the numbers and names of the **two** most significant Principles of Green Chemistry satisfied by the Solberg Co. Re-healing Foam and explain briefly.
(b) List one disadvantage of the Solberg product related to a Principle and explain briefly.

Problem 10.10

Draw the mechanism of the EcoPd-catalyzed Suzuki reaction in Figure 10.15.

References

1. P. T. Anastas and J. C. Warner, *Green Chemistry: Theory and Practice*, Oxford University Press, New York, 1998.
2. PET Resin Association (PETRA), *PET by the Numbers*, 2012, http://www.petresin.org/news_PETbythenumbers.asp, accessed April 5, 2018.
3. N. Weithmann, J. N. Möller, M. G. J. Löder, S. Piehl, C. Laforsch and R. Freitag, *Sci. Adv.*, 2018, **4**, eaap8060.
4. S. A. Miller, *ACS Macro Lett.*, 2013, **2**, 550–554.
5. E. W. Dunn and G. W. Coates, *J. Am. Chem. Soc.*, 2010, **132**, 11412–11413.
6. C. M. Byrne, S. D. Allen, E. B. Lobkovsky and G. W. Coates, *J. Am. Chem. Soc.*, 2004, **126**, 11404–11405.
7. E. T. H. Vink, K. R. Rábago, D. A. Glassner and P. R. Gruber, *Polym. Degrad. Stab.*, 2003, **80**, 403–419.
8. L. R. Rieth, D. R. Moore, E. B. Lobkovsky and G. W. Coates, *J. Am. Chem. Soc.*, 2002, **124**, 15239–15248.
9. I. A. Shuklov, N. V. Dubrovina, J. Schulze, W. Tietz and A. Börner, *Tetrahedron Lett.*, 2014, **55**, 3495–3497.
10. M. A. Dasari, P.-P. Kiatsimkul, W. R. Sutterlin and G. J. Suppes, *Appl. Catal., A*, 2005, **281**, 225–231.
11. B. Koppenhoefer, K. Lohmiller, F. V. Schurig, A. Kamal and D. R. Reddy, in *Encyclopedia of Reagents for Organic Synthesis*, ed. L. A. Paquette, John Wiley & Sons, Ltd, vol. 1, 2007.
12. M. Tokunaga, J. F. Larrow, F. Kakiuchi and E. N. Jacobsen, *Science*, 1997, **277**, 936–938.
13. National Geographic Society, *Great Pacific Garbage Patch*, http://www.nationalgeographic.org/encyclopedia/great-pacific-garbage-patch/, accessed April 6, 2018.
14. A. G. Pemba, J. A. Flores and S. A. Miller, *Green Chem.*, 2013, **15**, 325–329.
15. S. Yoshida, K. Hiraga, T. Takehana, I. Taniguchi, H. Yamaji, Y. Maeda, K. Toyohara, K. Miyamoto, Y. Kimura and K. Oda, *Science*, 2016, **351**, 1196.
16. M. Esch, *New trouble in hardscrabble town: Blood tests for chemical*, Associated Press, November 13, 2016, https://www.apnews.com/48e791e7482d44a6b2e7c6b4b0da8ac6, accessed July 25, 2019.
17. Environmental Directorate, *Co-operation on existing chemicals hazard assessment of perfluorooctane sulfonate (PFOS) and its salts*, 2002, http://www.oecd.org/chemicalsafety/risk-management/perfluorooctanesulfonatepfosandrelatedchemicalproducts.htm, accessed July 25, 2019.
18. Solberg Co., *Re-healing Foam – Compliance to the 2015 USEPA Stewardship Program and Beyond*, https://www.epa.gov/greenchemistry/presidential-green-chemistry-challenge-2014-designing-greener-chemicals-award Washington, DC, 2014.
19. T. Helmenstine, *Periodic Table of the Elements Abundance of Elements in Earth's Crust*, 2012, http://chemistry.about.com/od/periodictables/ig/Printable-Periodic-Tables/Periodic-Table—Abundance.htm, accessed July 25, 2019.
20. P. Dalley and B. Ellis, *Treatment of Ores Containing Reactive Iron Sulphides*, MultiMix Systems, https://web.archive.org/web/20091023235047/http://www.multimix.com.au/DOCUMENTS/Technical%20Bulletin1.PDF, accessed April 6, 2018.
21. M. d'Halluin, J. Ru-Barrull, G. Bretel, C. Labrugere, E. Le Grognec and F. X. Felpin, *ACS Sustainable Chem. Eng.*, 2017, **5**, 1965–1973.
22. G. Clave, F. Pelissier, S. Campidelli and C. Grison, *Green Chem.*, 2017, **19**, 4093–4103.
23. C. Grison, V. Escande, E. Petit, L. Garoux, C. Boulanger and C. Grison, *RSC Adv.*, 2013, **3**, 22340–22345.

11 Real-time Analysis

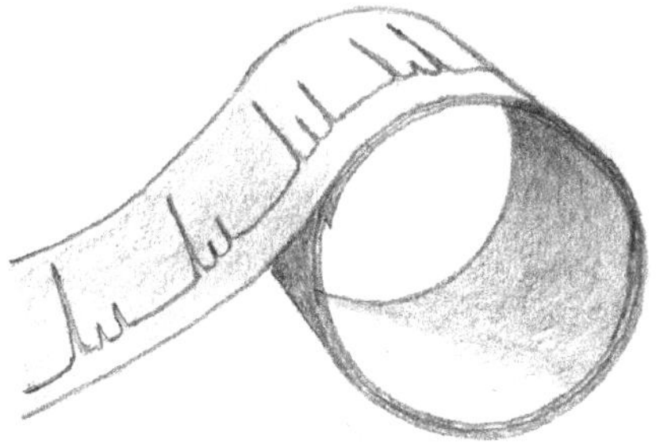

"Principle 11. Analytical methodologies need to be further developed to allow for real-time, in-process monitoring and control prior to the formation of hazardous substances."[1]

11.1 Real-time Analysis

To avoid risk, whether waste, toxic by-products, or accidents, real-time monitoring of chemical processes is essential. Monitoring can take the form of physical parameters: temperature, pressure, flow rate, heat output, pH, and light intensity, or chemical species: reactants, intermediates, products, or by-products. The focus of this chapter is on the chemical methods for real-time analysis. Monitoring the identity and the purity of the product avoids waste by allowing in-process changes to increase the yield. Minimizing by-products avoids the use of solvents and chromatography supports required to purify the product. Minimizing by-products also avoids the potential for toxic or hazardous waste. The goal of green chemistry—to design

Green Chemistry: Principles and Case Studies
By Felicia A. Etzkorn

Published by the Royal Society of Chemistry, www.rsc.org

benign, efficient chemical processes and products—depends upon analytical chemistry in real time.

11.1.1 Analysis for Each Process Step

Chemical analysis should be done in real time during every step of making a product: discovery, screening, development, process control, process optimization, and production.[2] Discovery is generally considered research, trying to find a chemical for a particular application, whether in academia or industry. Once the target function or application is known, screening is the process of looking at tens or thousands of chemicals to find the best one for the job. Development is finding the best synthetic or isolation method for producing the chemical product. Once the process is chosen, real-time analysis is used to keep the reactions under control, and to optimize the yield and purity of the product. Finally, real-time analysis is critical to production at scale to maintain the safety and efficiency of the process and the quality of the product.

11.1.2 Mechanism

Before a process can be truly optimized by real-time monitoring, it is important to study the mechanism. Temperature control is important when there is a distinct difference between kinetic and thermodynamic products (Chapter 6). Identification of by-products informs both the mechanism and the choice of methods for monitoring to avoid the by-products. Process chemists will often study the mechanism of a reaction by measuring reactant quality, kinetics, intermediates, by-products, stereochemical purity, and product ratios to determine the best way to do real-time analysis.

11.1.3 Process Analytical Chemistry

Decisions to be made for real-time analysis during production are: (1) where in the process to measure, (2) the frequency of measurement, and (3) the type of models to be used.[2] Once these choices have been made, single variable or multi-variate statistical process control can be designed.

An understanding of chemical mechanism is also helpful in deciding where in the process to take the measurements. If by-products are produced late in the reaction, perhaps as reagent concentration is depleted or as the spent reagent builds up causing side reactions, it might be best to monitor the reaction late in the process. Alternatively, if a physical parameter, such as temperature, might send a reaction down a side path early in the process, it would be better to monitor early. In a flow system, the location of real-time monitoring is an important choice as well—early in the flow process to monitor the formation of product, or late in the process to monitor complete disappearance of reactants. In all cases, the *sensitivity* of the method to the species to be monitored is critical, for example when very small amounts of a highly toxic impurity must be avoided or removed from a desired product (see tryptophan dimer, Figure 9.3).

11.2 Control Parameters

Continuous monitoring and control of these key parameters gives greater selectivity and higher yields of desired products, and greater safety to prevent accidents. Two very general parameters that can be measured for any reaction are temperature and pressure.

11.2.1 Temperature

Many reactions are run at ambient pressures, but temperature is a key variable to control the selectivity and yield of the desired products. Recall from Chapter 6 that the selectivity of a reaction can often be controlled with the temperature, especially if the kinetic and thermodynamic product mixtures are different. If the barrier to the thermodynamically more stable product is higher than the barrier to the kinetic product, a lower temperature gives more of the kinetic product because the reactants will not be able to get over the higher barrier. Catalysts are used to selectively lower the barrier to the desired product, whether kinetic or thermodynamic.

11.2.2 Pressure

Pressure monitoring is useful in a bi-molecular reaction with gases; higher pressure increases the concentration of the gaseous reactant. Examples of gaseous reactants are carbon monoxide in the BHC synthesis of ibuprofen (Chapter 2), hydrogen (H_2) in the reduction of alkenes, and ethene or 1-butene in the metathesis reactions of Elevance (Chapter 9). Decreased pressure can be used to remove a gaseous product to drive the reaction forward, as with the Sitagliptin case to remove acetone to drive the transamination (Figure 9.6).

The Eastman case used nitrogen sparging to remove low MW alcohol products to drive esterification (Chapter 5), rather than reduced pressure. Pressure is also an important parameter when using $scCO_2$ as the solvent for a reaction or chromatography.

11.2.3 Heat Output

If a reaction is strongly exothermic, it is safer to monitor and control the temperature and pressure in real time to prevent the solvent, reactants, or products from spontaneous ignition. If the reactor is a closed system, of course the reactor material must be strong enough to prevent explosions, and it would be advisable to monitor and control the pressure during the reaction. If the pressure can be reliably controlled, it may be unnecessary to overbuild the reactor to prevent explosions. However, valves can fail catastrophically (see Bhopal, Chapter 12).

11.2.4 pH

Monitoring pH in reactions in which a pH change occurs can be useful in accelerating the reaction without increasing the likelihood of side product formation. An example of this was the Codexis synthesis of Simvastatin (Chapter 8). Ammonia had to be added to prepare the water-soluble ammonium salt of the carboxylic acid, Monacolin J, keeping the pH static to prevent inactivation of the acylation enzyme, LovD.

11.2.5 Light Intensity

When light of any wavelength is used as an energy source or to catalyze a reaction, monitoring the frequency, current, or power can be important. In the case study on Yoon's photoreaction (Chapter 6), the stereochemistry of the product depended very much on the wavelength and intensity of the visible light input. Too much light and the reactants could undergo cis-trans isomerization, too little and no reaction would occur. As we begin to rely more on energy directly from the sun for chemical reactions, monitoring the light will become increasingly important.

11.3 Reaction Monitoring

In deciding what to monitor in real time during a process, knowledge of the course and the mechanism of the reaction is important. In preparation for process optimization, chemists will often determine the kinetic mechanism, temperature or pressure dependence, by-products formed, and isomeric or stereochemical purity. A great deal was made of improving the stereochemistry in the synthesis of Sitagliptin from 97% ee with Rh-*t*-BuJosiphos, to 99.95% ee with the Codexis evolved transaminase (Figure 9.6). On a large scale, this can result in many kg less waste produced and simpler purification.

11.3.1 Reactants

One way to determine whether a batch reaction has proceeded to completion is to monitor the disappearance of one or more reactant. TLC is used very commonly in the research lab for this when the properties of the desired product are unknown. The reactant can be identified by running it side-by-side with a co-spotted combination of reactant and reaction mixture on a TLC plate. Monitoring the disappearance of a reactant's spectroscopic signal in real time, as described in the next section, will be even more specific and precise.

11.3.2 Temporary Intermediates

High-energy intermediates can be extremely difficult to monitor by spectroscopic methods. However, in many cases, especially cascade reactions, a temporary, stable intermediate can be monitored in real time prior to addition of the next reagent. Consider for example, the cascade synthesis of Ambiguene H by Baran and co-workers (Figure 8.19). Once the prenylborane has added to the indole, the next step is release of the isocyanate group ($R{-}^{+}N{\equiv}C^{-}$) using Et_3N base and light (hν). Monitoring the disappearance of the isocyanate, or the formation of the chloroimine ion by IR could be a good indication of when the reaction is complete.

11.3.3 Product

Measuring formation of product is the most direct and powerful method to be certain that a reaction is proceeding as planned. However, late in the reaction, product signal is likely to overwhelm any other signal from reactants or side products. In the case where a small amount of a toxic contaminant might render the product unsuitable for market (see tryptophan dimer, Figure 9.3), measuring trace amounts of by-product formation might be more important. The dynamic range of the analytical method should be very high in order to detect very small amounts of by-product in the presence of very large amounts of product.

11.3.4 By-products

The formation of by-products has plagued organic synthesis since inception. Everything we do to make synthetic chemistry more efficient and more benign focuses on avoiding by-products. Optimizing space-time yields and stereochemical purity, eliminating purification steps, and making benign products all depend on avoiding by-products. Determination of the tryptophan dimer contaminant described in Chapter 9 was so fraught with difficulty, it was misidentified at first. The final determination was published in *Science*,[3] long after people were sickened and the product had been removed from the market. If the process had been properly monitored in real time, and the product had been thoroughly analyzed for purity, the dimer contaminant might have been avoided.

11.3.5 Identity and Purity

There are two criteria for new compounds that must be satisfied to publish in the best chemistry journals worldwide: identity and purity. Identity requires at least two forms of spectroscopy, including nuclear magnetic resonance (NMR), infrared (IR) spectroscopy, mass spectrometry (MS), and optical or ultraviolet–visible (UV–Vis) spectroscopy. Assessment of purity typically requires some form of separation, either chromatography: high performance liquid chromatography (HPLC), supercritical fluid chromatography (SFC), gas chromatography (GC), thin-layer chromatography (TLC), or electrophoresis: polyacrylamide gel electrophoresis (PAGE), agarose gel, or capillary electrophoresis (CE).

Since chromatographic methods are only capable of separating compounds, they must be coupled to spectral methods in order to identify unknown chemicals. Thus, separation methods are coupled to spectral methods such as: LC–MS, GC–MS, LC–NMR, CE–MS, *etc.* If only a separation method is to be used in real time, another way to identify peaks is by comparison with the retention time of authentic standards of known product and by-products. Determination of the identity of the unknown by-products would then have been done off-line prior to analysis by acquiring a suite of spectra as described above.

11.4 Case Study: Highly Reactive Polyisobutylene (Soltex)

Synthetic Oils and Lubricants of Texas, Inc. (Soltex) won a PGCC Award in 2015 for "A Novel High Efficiency Process for the Manufacture of Highly Reactive Polyisobutylene Using a Fixed Bed Solid-State Catalyst

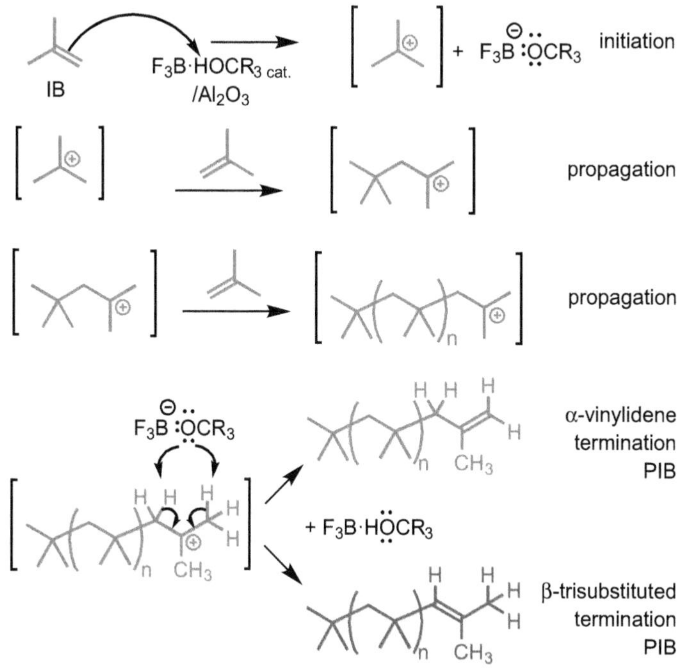

Figure 11.1 Cationic polymerization of isobutylene catalyzed by BF_3-tertiary alcohol complex on alumina.[4]

Reactor System."[4] Highly reactive polyisobutylene (PIB) contains one reactive geminal-disubstituted alkene, called an α-vinylidine, at the terminus of each polymer chain (Figure 11.1). These polymers are highly useful for attaching other molecules very specifically at the end of the polymer chain. Dispersives and surfactants are made by attaching functional end-groups to make additives for lubricants and gasoline. Undesirable low-reactive PIB contains a trisubstituted alkene at the end of each polymer chain (Figure 11.1).

11.4.1 Soltex Advantages

Prior patents describe the difficulty of PIB processes—hot spots are created, and there is insufficient contact between the liquid or slurry catalyst and the growing polymer chains in batch reactors. Prior technology required very high-purity isobutylene monomer. Workers may be exposed to hazardous liquid or gaseous BF_3 Lewis acid catalysts. High concentrations of catalysts must be used, and the catalysts are typically disposed of rather than recycled.[4] Large volumes of wastewater must be treated before release from the plant.[4]

Soltex can use any isobutylene (IB) containing feedstock.[4] The Soltex PIB product requires 50% lower catalyst to feedstock ratio, no water work-up to remove liquid catalyst, eliminates the need for neutralization, washing, and effluent treatment.[4] The capital costs to start up the Soltex 165 million lb plant was about $65 million, compared with about $100 million for plants using prior batch technology.[4] Soltex saves about $3 million per year in energy and operation costs as well.[4]

11.4.2 Cationic Polymerization

PIB is synthesized by acid-catalyzed, cationic polymerization (Figure 11.1). Initiation occurs by protonation of the IB monomer with the acid catalyst. The polymerization catalyst used previously was a Lewis acid, such as $AlCl_3$, or a liquid BF_3 complex, probably with ether or an alcohol. Soltex invented a new solid-state catalyst composed of BF_3 complexed with a tertiary alcohol (R_3C–OH).[4] Most significantly the $F_3B{\cdot}HOCR_3$ is adsorbed on a proprietary crystalline alumina (Al_2O_3) (Figure 11.1).[4] The catalyst performance depends on the crystalline form, and the pretreatment of the alumina—"not every alumina works."[4] Propagation is by addition of the next monomer IB to the cation, creating a new 3° cation (Figure 11.1). Termination by E1 type elimination can take place in two ways: (1) deprotonation of one of the terminal CH_3 groups to yield the less-stable geminal-disubstituted alkene, called an α-vinylidene, or (2) deprotonation of the last CH_2 group instead of a CH_3 group to produce the more thermodynamically stable trisubstituted alkene (Figure 11.1).

The trisubstituted alkene is more stable, and therefore much less reactive than the geminal-disubstituted alkene (α-vinylidene) due to hyperconjugation. More highly substituted alkenes are more thermodynamically stable because of hyperconjugation of the π-bond with the allylic methyl C–H, and the methylene C–C σ-bonds (Appendix B). The more allylic bonds to overlap with, the more stable the alkene. The stability order is tetrasubstituted > trisubstituted > trans-disubstituted > cis-disubstituted > geminal-disubstituted > monosubstituted (terminal) alkenes.

11.4.3 Kinetic *vs.* Thermodynamic Products

The α-vinylidene product is kinetically preferred, while the β-trisubstituted alkene is thermodynamically preferred (Chapter 6). The terminal CH_3 protons are more accessible to the sterically congested solid-phase catalyst than the internal CH_2 protons. Control over the process to maximize the α-vinylidene product is necessary to ensure the high quality necessary to produce the additives. Maintaining lower temperatures in the process ensures that the molecules do not have the energy to overcome the higher steric barrier to deprotonate the internal CH_2 group to make the trisubstituted alkene. The temperature needs to be just high enough to deprotonate the terminal CH_3 to give the α-vinylidine without deprotonating the internal CH_2. The flow system also keeps the desired α-vinylidene polymer product moving through the system, so there is less chance for reversibility to give the thermodynamic product.

11.4.4 Soltex Process Optimization

Soltex optimized their process by optimizing feedstock composition, catalyst composition, flow rates, temperature, and reactor design. Feedstock composition, temperature and flow rates are monitored and controlled continuously in real time to maintain the quality of the PIB product. Soltex created a fixed bed, solid-state flow reactor system to produce highly reactive PIB. Isobutane, C8 (octanes), C12, and C16 side products are recovered at various points in the flow system, and off-specification product is diverted after the quality control tank, and prior to collecting the PIB product.[4] The isobutene feedstock is continuously recycled. IB monomer is fed into the flow reactor at a controlled rate to maintain temperature.[4] A separate circulation loop is used to dissipate heat and because the polymerization is highly exothermic (350 BTU per lb IB).[4] Controlling the temperature for isothermal reaction is essential to controlling the molecular weight distribution of the polymer, and selectivity for the α-vinylidene product.

11.5 Spectral Methods

The choice of which analytical method to use depends on what functional group changes occur during the reaction. Each of the following methods has unique strengths in providing a way to monitor functional group changes.

11.5.1 Nuclear Magnetic Resonance (NMR)

Nuclear Magnetic Resonance (NMR) spectroscopy detects changes in the environment of magnetic nuclei. NMR is a non-destructive method, so the material analyzed can be recovered after analysis, or the flow can be returned to the flow reactor. The ^1H is the easiest magnetically susceptible nucleus to analyze, with a high natural abundance 99.98%, a strong gyromagnetic ratio, and a spin quantum number, $I=\frac{1}{2}$. Spin $\frac{1}{2}$ nuclei give splitting patterns that are simple to analyze.

For a brief review, take ethyl acetate as an example. The chemical shift is measured in parts per million from a standard. In ^1H NMR, the standard is tetramethylsilane (TMS), which is set to 0.0 ppm, because the nuclei are strongly shielded by the electrons. To the left of 0.0 ppm, the ^1H nuclei are deshielded by electron-withdrawing and aromatic groups. Figure 11.2 shows the ^1H NMR spectrum of EtOAc. The most shielded protons are the CH_3 in blue at 1.3 ppm. The green CH_3 protons at 4.2 ppm are very deshielded by the electropositive carbonyl (C=O). The CH_2 protons are slightly more shielded at 2.1 ppm because they are next to the sp^3 oxygen, which has lone pairs that shield them from the C=O. But the sp^3 oxygen is still

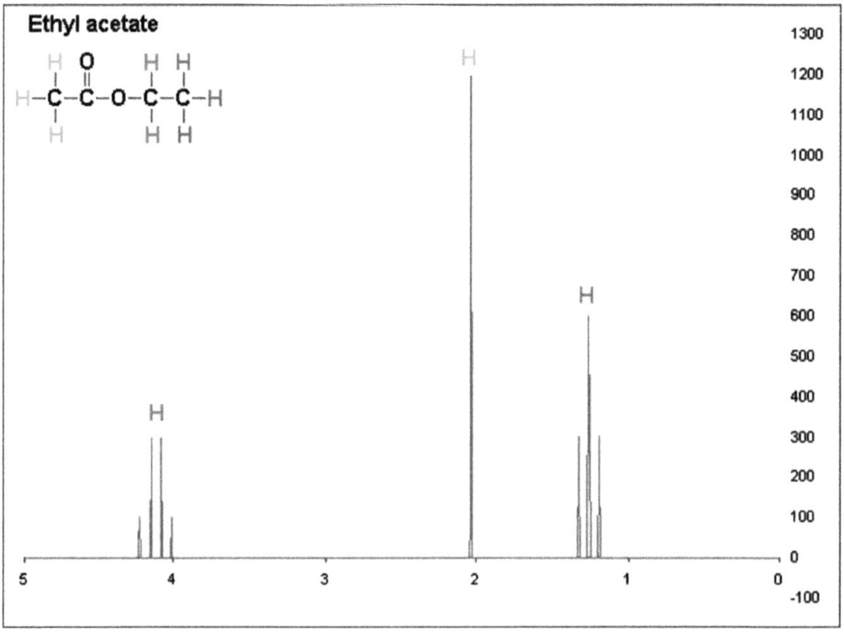

Figure 11.2 The ^1H NMR spectrum of ethyl acetate. Protons are assigned to chemical shift by color. The coupling patterns ($n+1$) show that the number of neighboring protons 3-bonds removed is n. Reproduced from https://commons.wikimedia.org/wiki/File:1H_NMR_Ethyl_Acetate_Coupling_shown_-_2.png, accessed July 8, 2019, under the terms of a CC-BY-SA 3.0 license, https://creativecommons.org/licenses/by-sa/3.0/.

electronegative, so the CH_2 protons are deshielded relative to the CH_3 protons. The splitting patterns reveal the number of nearest neighbors three bonds away.

Spin $\frac{1}{2}$ nuclei have simple spectra that give splitting patterns where the number of peaks in the pattern is equal to the number of equivalent neighbors plus one $(n+1)$. In general, the number of peaks in a splitting pattern is $2nI+1$, which simplifies to $n+1$ for $I=\frac{1}{2}$ as in ^{1}H. In the ^{1}H NMR spectrum of EtOAc, the blue CH_3 has two neighboring protons in red, so it appears as a triplet $(n+1=3)$ (Figure 11.2). The red CH_2 has three neighboring protons in blue, so it appears as a quartet $(n+1=4)$. The green CH_3 group has no neighboring protons, so it appears as a singlet. In addition to the chemical environment and the number of equivalent neighbors, the relative number of protons in each set of peaks can be determined from the area under the peaks, called the integration.

Monitoring a chemical reaction by NMR is most feasible for spin $\frac{1}{2}$ nuclei: ^{1}H, ^{19}F, and ^{31}P. Also possible are monitoring ^{13}C or ^{15}N NMR, but the natural abundances are low (^{13}C 1.1%, ^{15}N 0.37%), so they require very sensitive instrumentation and more time than ^{1}H NMR. Sitagliptin ketoamide, with fluorine in two of the reactants in the condensation step, could use ^{19}F NMR readily to monitor the coupling reaction (Figure 9.4). The ^{19}F chemical shifts for the aromatic fluorines should be quite susceptible to changes in the chemical environment upon converting from the ketone to the amine. In the enzyme catalyzed second step, we could also conceptually see that the disappearance of isopropylamine could be monitored, perhaps by MS. The isopropylamine is used in excess to drive the reaction, and the corresponding product acetone is removed (Figure 9.6).

NMR has been miniaturized, so that only μL volumes are required for analysis. In a flow system, part of the reaction volume can be routed through such an NMR for analysis.

11.5.2 Mass Spectrometry (MS)

Mass spectroscopy (MS) is extremely sensitive—modern techniques can detect picomolar (10^{-12} M) concentrations of most small molecules. Almost any reaction, except isomerization, exhibits a change in mass between reactants and products, so MS is very broadly useful. MS consists of three steps: ionization, separation of ions, and detection. Electrospray ionization (ESI) is very powerful when coupled to flow reactors or HPLC. In ESI, a mildly acidic (1% formic acid in water/acetonitrile) solution of the analyte is sprayed through a very

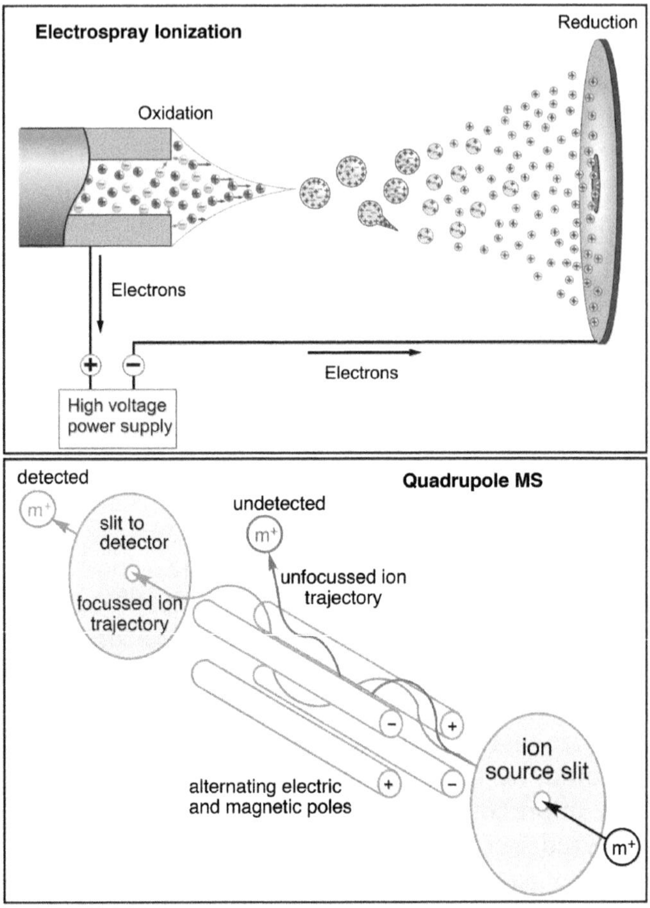

Figure 11.3 Top: Electrospray ionization. Bottom: quadrupole mass analyzer schematic showing the paths of focused and detected *vs.* unfocussed and undetected ions. Top: reproduced from https://upload.wikimedia.org/wikipedia/commons/d/d1/ESI_positive_mode_%2821589986840%29.jpg, accessed July 8, 2019; under the terms of a CC BY 2.0 license, https://creativecommons.org/licenses/by/2.0.

fine-tipped nozzle into a high vacuum, where the solvent molecules are stripped from the protonated analyte, called MH^+ (Figure 11.3). Negative ion separation and detection $(M–H)^-$ is also possible for compounds that are not readily protonated, such as carboxylates (RCO_2^-) and phosphates ($ROPO_3^=$). Separation can take place in a quadrupole, in which four alternating positively and negatively charged poles attract and repel the charged ion (Figure 11.3). There are many other separation types, including ion trap, and time-of-flight.

Magnetic sector MS is no longer used much because the newer types of instruments require less space. Detectors are typically electron multipliers that detect the charge induced or the current produced when the ion hits the surface.

MS is a destructive method; however, so little material is needed for analysis that this is typically not a concern. Inexpensive mass spectrometers are available that can be attached to a flow reactor, the headspace of a batch reactor, or samples can be extracted from a batch reactor to inject in the MS. LC–MS is probably the most common and useful combination instrument for real-time reaction monitoring.

11.5.3 Infrared (IR) Spectroscopy

Infrared (IR) spectroscopy measures the vibrational frequencies of different functional groups in molecules. IR frequencies are expressed in wavenumbers (cm^{-1}), the inverse of wavelength. The inverse relationship between frequency ν and wavelength λ ($\nu = c/\lambda$) comes directly from the equation for energy:

$$E = h\nu = hc/\lambda$$

Whereas NMR is very specific for one set of identical nuclei, IR results from the vibrations of bonds, such as the C=O double bond, the C–O single bond, and the O–H bond of a carboxylic acid. IR is highly sensitive to the environment of carbonyl (C=O) bonds of different types of compounds. The frequencies of C=O stretches are in the order: acyl chlorides > anhydrides > aldehydes > ketones > esters > amides > carboxylic acids.

Frequency changes are even clearly observed between a simple ketone C=O and an α,β-unsaturated ketone, $R_2C=C–C=O$, *i.e.* an enone. For example, compare propanal C=O at 1750 cm^{-1} to propenal C=O at 1720 cm^{-1} (Figure 11.4). Conjugation with the alkene moves the C=O to lower wavenumbers (longer wavelength λ, lower energy). Thus, any reaction involving a conversion of one type of carbonyl to another could be monitored very sensitively by IR, whereas ^1H NMR might be insensitive due to a lack of hydrogens attached to carbonyls. The other type of functional group that has good sensitivity in the IR are unsymmetrical triple bonds and cumulative bonds: terminal alkynes, nitriles (R–CN), allenes ($R_2C=C=CR'_2$), and ketenes ($R_2C=C=O$). A ketene is thought to be an intermediate in the synthesis of the Sitaglipin ketoamide (Figurc 9.5).

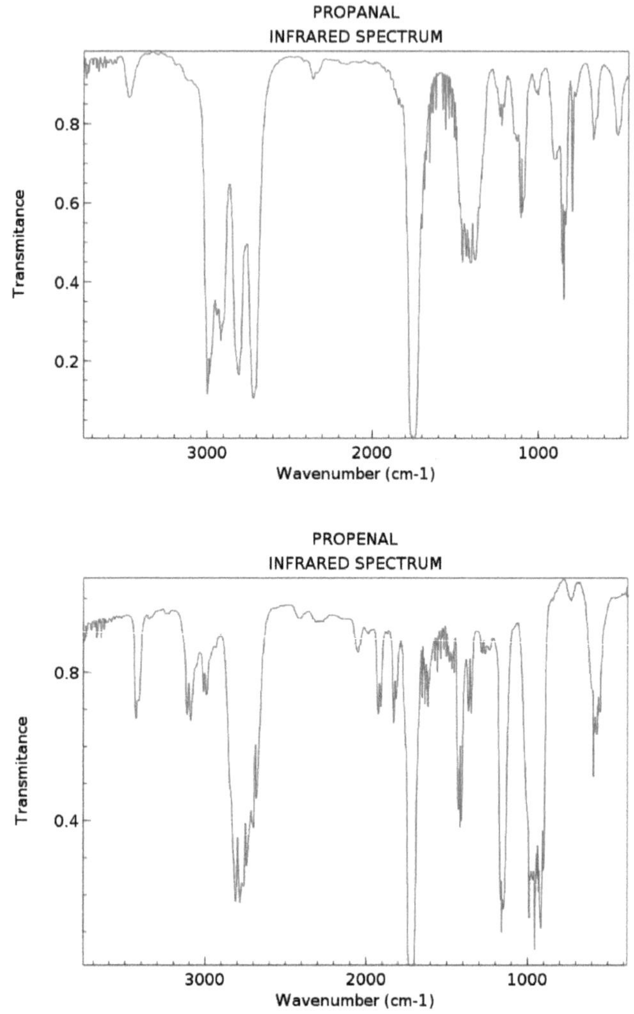

Figure 11.4 IR spectra showing C=O stretches of Top: propanal (1750 cm^{-1}), Bottom: propenal (1720 cm^{-1}). Reprinted from National Institute of Standards and Technology, http://webbook.nist.gov/chemistry, NIST Standard Reference Data (SRD); COBLENTZ SOCIETY. Used with permission. © United States of America as represented by the Secretary of Commerce.

11.5.4 Ultraviolet–Visible (UV–Vis) Spectroscopy

UV–Vis is an optical electronic spectroscopy that measures the absorption of light by excitation of outer shell electrons in the wavelength range of 200–380 nm (UV), and 380–750 nm (Vis). UV–Vis is non-destructive because the electrons relax quickly back to the ground state without encountering another reactive species.

Molecules with lone pair and π-electrons absorb light in the UV–Vis range due to the excitation of electrons, mostly from the HOMO to the LUMO, called the excited state. The signals are usually very broad due to the multiple vibrational levels within each electronic state. UV–Vis can be used to measure concentrations of a pure compound using the Beer–Lambert equation: $A = \varepsilon lc$, where A is the absorbance measured, the molecular absorptivity ε is a constant for each compound at a specific wavelength corresponding to a maximum intensity, l is the pathlength, and c is the concentration. UV–Vis is commonly used for detection when coupled to LC in a separation system.

11.5.5 Fluorescence Spectroscopy

Fluorescence is also an optical process. Outer shell electrons, usually in the HOMO, are excited by light in the UV–Vis range (200–750 nm) to unoccupied higher orbitals. As the electrons emit a photon and relax back to the ground state, they give off light, called fluorescence, at longer wavelengths and lower energy than absorbed in the process (Figure 11.5). The fluorescence detector is typically positioned at 90° from the light path in a normal UV–Vis spectrometer that measures absorption. Fluorescence can be quenched by binding to another UV-active species, or by a change in charge state due to a pH change, or by binding to metal ions.

In the Jablonski diagram of fluorescence, S_0 through S_3 are increasing energy levels of singlet states (opposite spin states of an electron pair), and v_1 through v_5 are vibrational states labeled only in the lowest singlet state (Figure 11.5). After absorption of UV–Vis light, the excited

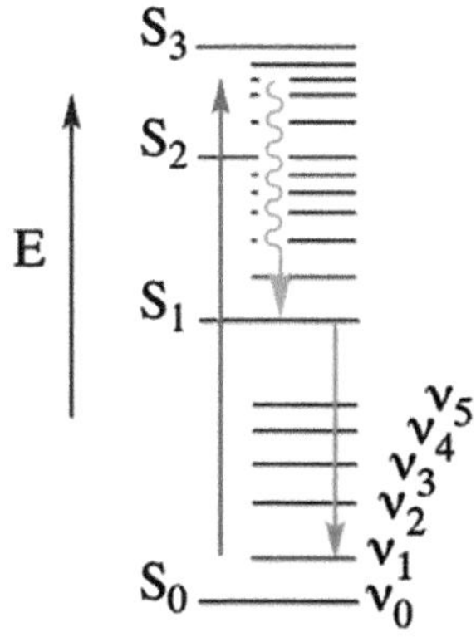

Figure 11.5 Jablonski diagram of fluorescence process showing excitation (blue arrow), internal conversion and vibrational relaxation (green wavy arrow), and fluorescence emission (red arrow).

electron undergoes loss of energy within the excited S_1–S_3 levels, called internal conversion, and vibrational relaxation (Figure 11.5). This is followed by fluorescence, the emission of light as the electron relaxes from the excited state to the ground state S_0 (Figure 11.5).

11.6 Case Study: *3D TRASAR®* Cooling System Chemistry and Control (Nalco)

Nalco Co. won a 2008 PGCC Award for real-time monitoring in water-cooling systems.[5] Water cooling is important in a large number of commercial and industrial processes. Significant quantities of water are wasted to flush out systems of scale build-up and biofilms. Scale is mostly insoluble Ca and Mg salts, primarily carbonates and phosphates. Scale corrodes metal pipes, resulting in small holes and the need to replace pipes in the system. Biofilms form from commensal bacteria and are very difficult to remove once they form. Most water-cooling systems only monitor scale, corrosion, and biofilms periodically, which is frequently too late to prevent build-up. When scale and biofilms build up too much, the entire cooling system must be flushed and refilled.

11.6.1 Scale and Corrosion Measurements

Conductivity measurements are used in most systems to measure dissolved salts, an indirect measurement of the potential for solid scale to build-up. Since water input into cooling systems can vary greatly day-to-day or even hour-to-hour, conductivity is a poor measure of scale potential. Corrosion is measured with metal "coupons," dated pieces of metal placed in the system that are removed for visual examination periodically. Dip slides, composed of nutrients in agar, are used to spot check for planktonic growth in cooling systems. Plankton include floating bacteria, archaea, algae, and protozoans. The dip slide must be incubated for about 24 hours to measure growth, *i.e.* not in real time. The dip method does not measure the plankton that inhabits biofilms, which constitutes about 95% of the bacterial population in cooling systems.

Most systems operators control for scale, corrosion, and biofilms by controlling pH, and adding scale and corrosion inhibitors. Biofilms are minimized by timer-controlled addition of bleach or other oxidizing biocides. Most of these methods are periodic, not continuous, and rely on best guesses as to the amounts of inhibitors to add and when.

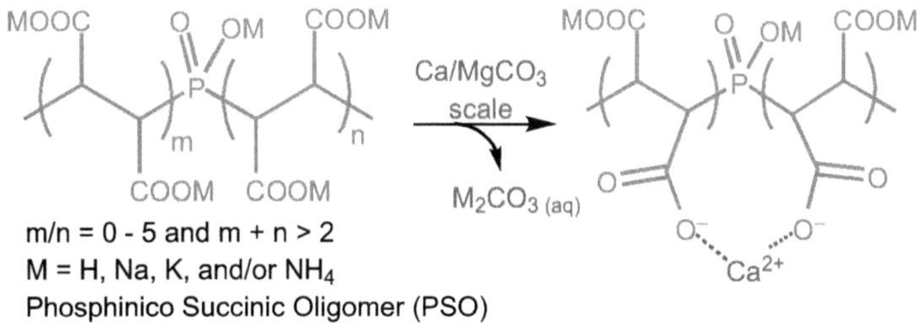

Figure 11.6 3D TRASAR® patented Phosphinico Succinic Oligomer (PSO) scale inhibitor reduces use of phosphates, water, and energy in water-cooling systems.[8]

11.6.2 Fluorescent Dye Monitoring

Nalco developed two fluorescent dyes to monitor scale build-up and biofilms simultaneously and continuously.[5,6] The fluorescence of the dyes decreases as they bind to either scale ($CaCO_3$) or to biofilms, indicating potential problems within the system that can be treated with minimal quantities of scale inhibitors or biocides. Both fluorescent dyes were chosen to be compatible with broad ranges of pH, halogen levels, and temperature.[5,6] Fluorescence is monitored in real time with an inexpensive light-emitting diode (LED) fluorometer.[7]

Nalco achieves scale inhibition with a more efficient and relatively benign scale and corrosion inhibitor that is a mixture of Phosphinico Succinic Oligomers (PSO) (Figure 11.6).[8] PSO is far more efficient and stable than previously used scale inhibitors that were high in phosphates. Phosphates pollute fresh water by an oversupply of the nutrient, causing algae blooms that choke out other aquatic life. Phosphates have been all but eliminated from detergents in the past 50 years for this reason. The 3D TRASAR® PSO oligomers are tagged with a fluorescent dye. The rate that the oligomer is consumed by scale-causing Ca/$MgCO_3$ relative to an inert tracer standard correlates directly with the build-up of scale.[5] At specific levels of PSO-dye consumption, the cooling water is either diluted or automatically purged and restored with fresh cooling solution.[5]

Corrosion of heat exchanger surfaces is the most frequent cause of catastrophic failure in water-cooling systems. The 3D TRASAR® system also uses a Nalco Corrosion Monitor (NCM100) that electrochemically monitors corrosion in the system, and automatically controls the amount of corrosion inhibitor added. PSO controls corrosion along with scale, replacing more toxic corrosion inhibitors that contain molybdate (MoO_4^{2-}) or Zn, reducing both the expense and discharge of heavy metals into the environment.[5] PSO is highly

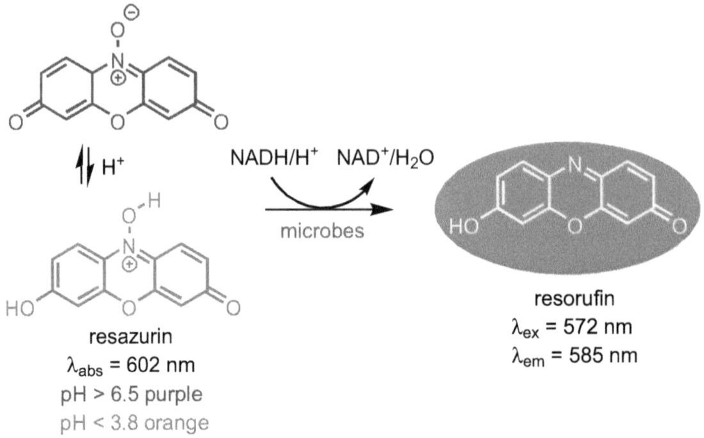

Figure 11.7 Fluorescent microbe detection by microbial reduction of resazurin to resorufin.[9] λ_{ex} = excitation wavelength; λ_{em} = emission wavelength.

resistant to halogens, it does not revert to orthophosphate, and it operates at lower phosphate levels than other phosphate systems.[5]

Biofilms have a 4.6 times higher thermal resistance than $CaCO_3$ scale.[5] Water-cooling systems have more efficient heat transfer without scale and biofilm formation, which saves energy. The dye Nalco uses to detect biofilm formation is resazurin. In the presence of microbial NADH, resazurin is reduced to resorufin (Figure 11.7).[9] Reduction converts resazurin, which is dark blue with very little intrinsic fluorescence, into resorufin, which is strongly fluorescent.[9] Typically, resorufin is excited off-maximum at about 560 nm, which does not interfere with the maximum emission wavelength at 585 nm (Figure 11.7).[9] The fluorescence of the dye in the cooling system is monitored continuously. As fluorescence increases with time, indicating bacterial growth, the system automatically adds the appropriate amount of biocide oxidant to kill the bacteria.[5] The consumption of the fluorescent dye means it must be replenished as it is consumed. The conversion of resazurin to resorufin is irreversible. The high sensitivity of the resazurin-resorufin system means that very little of the dye needs to be used, and very little resorufin builds up in the system between cooling water replacements.

Only the necessary amount of oxidant is added automatically, and addition is stopped automatically, thus minimizing the use of strong oxidants that can also be corrosive. We saw in Chapter 3 that bleach is a very cheap and relatively benign oxidizing agent, although undesired by-products such as $CHCl_3$ or even dioxins can be formed. The 3D TRASAR® system minimizes the use of bleach by only dosing the right amount at the right times to prevent the build-up of biofilms in water-cooling systems. On average, 3D TRASAR® uses 60 ± 30% less biocide to maintain the same low levels of bacteria: 1.0–1.5 gallons per day (gpd) instead of 4.0–5.0 gpd (Table 11.1).[5] When the system eventually must be flushed and cleaned, chlorinated hydrocarbons are released into the environment. Thus, minimizing the use of bleach minimizes the pollution from water-cooling system purges. Other more benign oxidants that can act as microbicides are ozone (O_3), hydrogen peroxide (H_2O_2), and oxone ($KHSO_5$) (Table 3.3).

Water savings with the 3D TRASAR® system ranges from 2.4 million gallons per year (gpy) for a hotel to 360 million gpy for a power plant, with the range of useful cycles increasing 0.5 and 3.5 times respectively (Table 11.2).[5]

Overall, the 3D TRASAR® system allows the system to be operated closer to the safety margin since it is continuously monitored and controlled.[5] This results in decreased water usage, decreased energy usage, decreased environmental pollutant discharge with less halogenated by-products and heavy metals, and less corrosion of the heat exchanger surfaces contributing to system longevity.[5] The system also includes a rapid alarm notification system that is connected to the internet and sends email automatically if the system goes out of specification limits.[5]

Table 11.1 Bio-control before and after the 3D TRASAR® system. Adapted with permission of Nalco Company from ref. 5, Copyright 2008.

Parameter	Before 3D Bio-control	3D Bio-control
Average biocide usage (gpd)	4.0–5.0	1.0–1.5
Average planktonic count (CFU-mL^{-1})	10^4–10^5	10^4–10^5
Average biofilm count (CFU-cm^{-2})	10^7	10^6–10^7
Chiller C factor	918	884

Table 11.2 "Water savings achieved by Nalco customers through the use of 3D TRASAR®."[5] Adapted with permission of Nalco Company from ref. 5, Copyright 2008.

Industry type of customer	Cycles before 3D TRASAR®	Cycles with 3D TRASAR®	Water savings (BD reduction, million gpy)
Power	7.5	11.0	360
University	3.0	4.5	44.1
Refining	4.0	8.0	32.2
Food & beverage	2.0	4.0	22.6
Manufacturing	3.0	7.0	15.0
Manufacturing	1.5	6.0	11.0
Pharmaceutical	3.5	8.0	2.8
Hotel	2.9	3.4	2.4

11.7 Chromatographic Methods

All chromatographic methods are separations methods. The name "chromatography" means writing (graphy) with color (chromato) because the first separations were done with plants, which gave visible bands of colored compounds that could be detected with the human eye. Most common organic compounds and inorganic salts are colorless and require special detection methods. Various non-specific forms of detection are used to measure the area of peaks that correspond to individual components of a mixture. Because the detection methods, such as UV–Vis for liquid chromatography, or electrical conductivity for gas chromatography, typically do not reveal the identity of the component, authentic standards must be injected or co-injected under the same separation conditions to identify each component by its retention time—the time that it elutes from the column. Modern systems can include spectrometric detectors, often mass spectrometers, that may be able to identify each component more accurately without injcction of an authentic standard.

Chromatography has already been described as differential partitioning of compounds between two physical phases in Chapter 5. Therefore, the following is a brief review.

11.7.1 Liquid Chromatography (LC)

LC uses a solid stationary phase and a liquid mobile phase. Normal phase LC uses SiO_2 as the polar stationary phase, and relatively non-polar solvents as the mobile phase, such as hexanes, ethyl acetate, THF, and isopropanol. Reverse phase LC uses a variety of surface modifications to decrease the polarity of the SiO_2, most commonly a hydrocarbon chain with 18 carbons called C-18. The mobile phase is typically a mixture of water and acetonitrile (CH_3CN) or methanol. Methanol is cheaper, but modification of MeOH with acids is more corrosive to stainless steel systems over time than is acid in CH_3CN. LC is most frequently used for separation of non-volatile compounds in pharmacology and medicinal chemistry. LC is nearly always done under high pressure (HPLC) to improve separations. The detection method is usually UV at a wavelength of 210 nm where most organic compounds absorb, with background subtraction for solvents. MS detectors for GC or LC are becoming much cheaper for viable benchtop reaction monitoring. For flow reaction monitoring, capillary HPLC columns can be connected in a loop fashion to reintroduce the reaction mixture back to the same point it was removed from.

11.7.2 Supercritical Fluid Chromatography (SFC)

SFC is a specialized form of LC that can be done with the same equipment. LC columns are typically stainless steel that readily withstand the pressures needed for $scCO_2$. An example of $scCO_2$ chromatography to purify a pharmaceutical intermediate at AstraZeneca was described in Chapter 5. In order to monitor a reaction by SFC, the interface with the SFC column could be especially challenging unless the reaction being monitored is also using $scCO_2$ as the solvent at similar temperature and pressure.

11.7.3 Gas Chromatography (GC)

GC is most useful for monitoring reactions with volatile, low MW species, particularly gases that are challenging to capture and monitor otherwise. Modern GC columns have a viscous liquid stationary phase (0.25 or 0.50 μm film thickness) coating the inside walls of a capillary column (typically 0.25 mm inside diameter).[10] The column head can be inserted into the space directly above the reaction solution to monitor gases evolved during the reaction, but a splitter must be in place to introduce the flow gas eluent, typically He.

11.7.4 Thin-layer Chromatography (TLC)

TLC is a quick and dirty way to monitor reactions in real time. Normally small, disposable silica gel plates are used. TLC cannot directly reveal the identity nor the quantity of a component of a mixture. TLC is most useful when pure standards of reactants and/or products are available to co-spot with the reaction mixture. Co-spotting means to spot a solution of a standard on the same origin spot as the reaction, while also spotting the standard and the reaction mixture separately on the same plate. If the standard elutes with the same R_f value, and appears the same under UV or chemical visualization, then the component is likely to be the same. The retention factor (R_f) value is the ratio of the distance the analyte moves to the distance the solvent front moves on the plate. Since samples must be removed during the reaction, access to the reactor is necessary. This can be challenging for large production reactors, or reactions that require water-free or oxygen-free conditions.

11.8 Reactor Design

11.8.1 Batch Processes

Batch reactors are inherently inefficient.[2] Challenges lie in transferring reactants into and removing products from large batch reactors. Heat transfer is a major problem—hot and cold spots are often found in large batch reactors. Hot spots can lead to by-products, runaway reactions, fires, or explosions. Cold spots can lead to incomplete reactions and lower yields. Inhomogeneity of reactants is also a problem. The problems of heat transfer and reactant inhomogeneity are typically solved with large mechanical stirrers; magnetic stirring is often not possible on a large scale or with solids present in the mixture. Transfer of components, heating, cooling, and stirring can all be very energy consuming (Chapter 6). Real-time analysis is also more challenging because of the inhomogeneity of the system; edges, center, top and bottom can all have different states of completion.

One solution to the scale problem of batch reactors is to run multiple reactions on a small scale in parallel. The number of reactors can be scaled-up instead of the volume of the reactor. Multiple parallel reactors can be efficient if these reactors are disposable. Single-use plastic bag reactors have become standard in producing protein drugs in mammalian cells for clinical trials.[11] The advantages of disposable reactors are smaller production plants, lower cost operation, and greener processing.[11] A caveat is that the reactors themselves become part of the waste stream, and they should be included in E-factor calculations. Another challenge with multiple reactors is that real-time monitoring is more challenging. The conditions in each reactor can vary slightly to produce undesired side products or lower yields. Robotic screening in real time from multiple reactors, while possible, is certain to be more expensive.

11.8.2 Semi-batch Processes

Semi-batch processes use a large reactor to which reactants are fed and/or products are removed, either continuously or in batches. Periodically, the reaction is stopped to thoroughly clean the reactor and start anew. The rate of addition or removal is important to be sure a catalyst is not overloaded with substrate, or that a runaway exothermic reaction does not occur. Automated liquid feeds are used to maintain control, but the rate of addition or removal will still depend on real-time analysis of reactants, intermediates, products, or by-products.

Solvents are an important consideration either to keep all reactants in solution, or to allow only the desired product to precipitate, facilitating its purification. In semi-batch reactions, temperature and pressure controls are also used to maintain control. Gases are typically the easiest to remove, but gases can cause problems in continuous-flow reactors as we will see in step 3 of the prexasertib process later in this chapter. Liquid products, whether the desired product or a by-product, can be distilled from the reaction. Removing a low-boiling product is a great strategy to drive a reaction forward by Le Chatelier's Principle, as we have seen with the removal of C1–C4 alcohols from the Eastman transesterification (Chapter 5), and acetone from the Merck-Codexis transamination to make Sitagliptin (Chapter 9).

11.8.3 Semi-batch Biodiesel

A now classic example of a semi-batch process is the production of biodiesel.[12] The reactants are vegetable oil and MeOH, which can be continuously and automatically fed into the reactor. The catalyst, NaOMe, is soluble in the vegetable oil-MeOH reaction mixture.[12] MeOH is usually used in excess to drive the reaction, and the excess can be reused.[12] Since both the oil and MeOH are liquids, no solvent is necessary. The densities of the reactants and products matter most: the density of vegetable oil is about 0.92 $g\,mL^{-1}$,[13] methyl esters of fatty acids have a low density of 0.875 $g\,mL^{-1}$,[14] while glycerol has a high density of 1.26 $g\,mL^{-1}$, and glycerol is not miscible with oil or esters.

After a batch of MeOH has been consumed, the stirring is stopped, biodiesel and MeOH are skimmed off the top of the reactor, and glycerol is drained from the bottom of the reactor. The MeOH can then be separated from biodiesel by distillation and added back into the reactor. Other by-products that have to be removed periodically are water and soap (sodium salts of fatty acids).[12] Biodiesel has also been produced by a continuous-flow process.[15]

11.8.4 Continuous Flow Processes

Continuous flow reactors are usually the greenest method, but not all are efficient enough to beat batch reactors. Continuous flow offers a great deal of flexibility to modify the system for different products. This production flexibility offers additional control to give precise product compositions. Because the actual reaction space is typically much smaller than batch reactors, continuous flow is also much safer and less likely to have runaway exothermic reactions that can lead to a fire or explosion. Heat transfer is more efficient in flow reactors, which offers better temperature control. Good temperature control is not only safer, it also avoids by-product formation to prevent waste. Continuous flow offers the possibility of faster reactions and increased throughput of product. With good process monitoring, product homogeneity is improved, which reduces the need for purification steps, and increases space-time yield. Process control prevents accidents as well as waste. Again, multiple parallel reactors are likely to be the best way to scale up to increase the rate of production, depending upon the complexity and expense of the equipment involved. A continuous flow process can more easily be shut down or started up again, so that the supply of product can be efficiently connected to the demand. Several continuous flow reaction steps in a synthetic sequence may be run in series for a highly efficient overall process.

11.9 Case Study: Kilogram-scale Prexasertib Monolactate Monohydrate Synthesis Under Continuous Flow CGMP Conditions (Ely-Lilly & Co.)

Prexasertib **11.1** is a potent anti-cancer drug candidate that inhibits Checkpoint Kinase 1 (CHK1) with an IC_{50} value less than 1 nM (Figure 11.8).[16] The mesylate salt ($MeSO_3^-$) of prexasertib was found to have better water solubility that was desirable for administration of the drug by infusion.[16] Ely-Lilly & Co. developed a method called Process Analytical Technology (PAT) to monitor the production of prexasertib under continuous-flow reaction conditions.[16] "PAT used during production included online HPLC and refractive index measurement, as well as temperature, pressure, and mass flow rates monitored by the DCS [Distributed Control System]."[16] The process condenses a 9-step batch route into a 7-step process with better bond disconnections and less toxic reagents (Figure 11.8).[16] The first three steps were still done in batch processes, and the last four steps were done in continuously monitored flow reaction systems (Figure 11.8).[16] Each of these four steps were done under Current Good Manufacturing Process (CGMP) criteria that must be met for drugs to be administered to humans.[16]

In flow reactions, precipitation of any solids can clog the tubing. Early in the process, optimization of solubility is critical. Nitrile **11.2** and aminodiazole **11.3** were soluble in THF (Figure 11.8). Flow yields in **Step 1** were optimized under 500 psi pressure at 130 °C in 0.56 mm ID stainless steel tubing and a mean residence time of 60 minutes

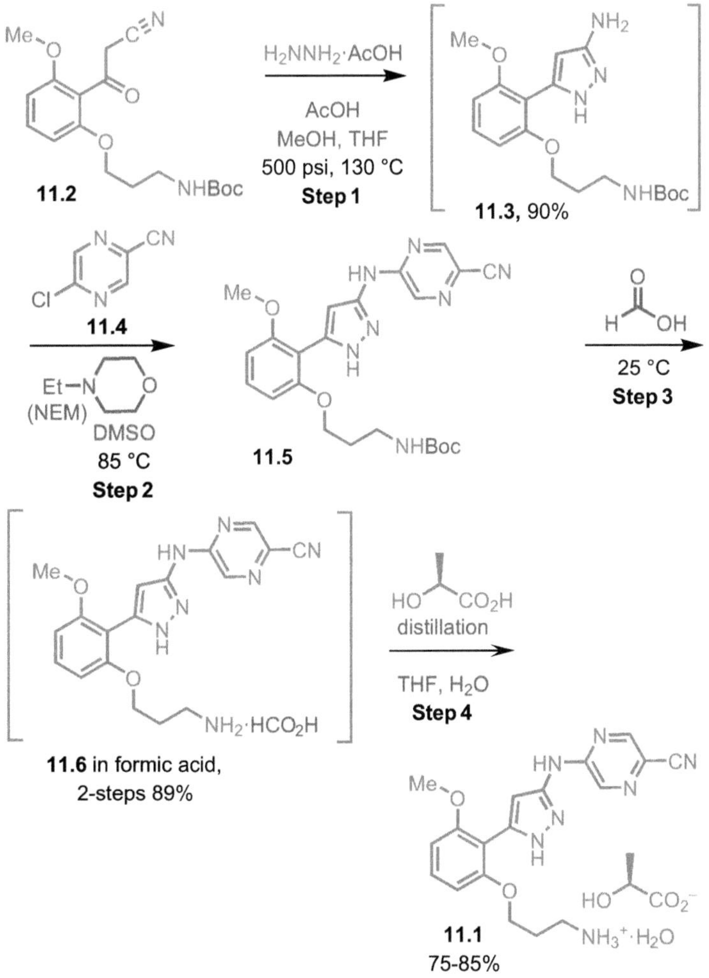

Figure 11.8 The last four steps of prexasertib·monolactate·monohydrate production were performed under continuous flow conditions. Intermediates that were not isolated are shown in brackets. Yields at production scale are given below the reaction arrows.[16]

with only a slight molar excess of hazardous hydrazine (H_2NNH_2) as the acetate salt.[16] Intermediate **11.3** was not isolated because it is a mutagen, and the overall yield was improved by skipping the isolation.[16]

In **Step 2**, nucleophilic aromatic substitution (S_NAr) of arylamine **11.3** on chloropyrazine **11.4** gave less than 5% regioisomer and other impurities by continuous HPLC monitoring (Figure 11.8).[16] The aromatic regioisomers probably arise from impurity regioisomers of the reactant chloropyrazine **11.4**. The base, *N*-ethylmorpholine (NEM) was chosen to absorb the HCl given off in the reaction because of its high solubility in the best reaction solvent, DMSO.[16] NEM gave minimal removal of the *t*-butoxycarbonyl (Boc) protecting group on the alkylamine of **11.3** and intermediate **11.5**.[16] Antisolvent crystallization of intermediate **11.5** in methanol was necessary to remove residual **11.4**, regioisomers of **11.5**, NEM·HCl, DMSO, and low levels of other process impurities.[16] Antisolvent crystallization is the crystallization of a desired product by dissolving it in a minimal amount of a solvent that it is soluble in, then adding a small amount of a solvent that the product is insoluble in. This is a fairly efficient purification method because it avoids material-intensive chromatography. Intermediate **11.5** was highly soluble in the DMSO reaction mixture, but sparingly soluble in MeOH, allowing crystallization of **11.5**, while the impurities remained in solution. The filter material and filtration process were studied in depth to optimize yields of **11.5**.[16]

Deprotection of the Boc-amine **11.5** to give free amine **11.6** in **Step 3** was accomplished with formic acid (HCO_2H) with broad acceptable temperature and reaction time ranges (Figure 11.8).[16] Formic acid was chosen rather than the more toxic, expensive, and less atom economical trifluoroacetic acid (TFA) (Figure 11.8).[16] Removal of the CO_2 and isobutylene gases formed by Boc deprotection of **11.6** reduced the amounts of by-products and improved yields significantly.[16] Major by-products found included three derivatives of **11.1**: a formamide, a *t*-butylamine, and a *t*-butylcarboxamide, which was formed by a Ritter reaction of *t*-butyl cation with the nitrile of **11.6** (Figure 11.9).[16] To avoid these by-products, the CO_2 and isobutylene were removed by bubbling nitrogen gas through the plug-flow coil reactor; the flow rates of both solution and gas were controlled by

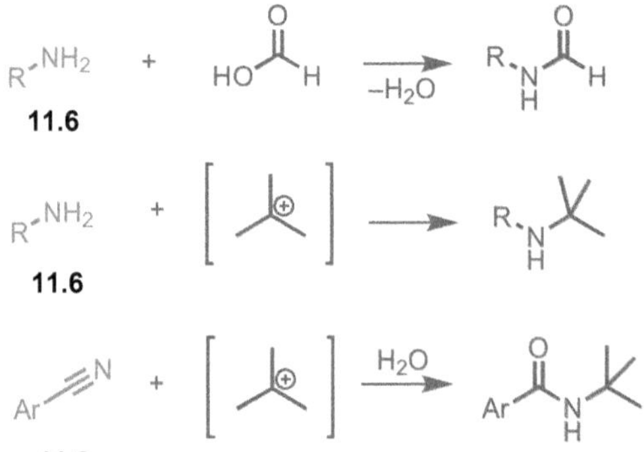

Figure 11.9 Formation of by-products in the Boc deprotection **Step 3** during manufacture of prexasertib.[16]

peristaltic pumps.[16] The combined yield of **Steps 2** and **3** was a remarkable 89% at production scale.[16]

In **Step 4**, the crude formate salt **11.6** was converted to the L-lactate salt **11.1** by addition of 8 eq. of 30 wt% lactic acid (aq.) and concentration on a 20 L rotary evaporator (Figure 11.8). Time, temperature, pressure, and mass flow parameters were monitored and controlled during the concentration step.[16] Antisolvent crystallization with aqueous THF gave the final lactate salt **11.1**.[16] By this method, 24 kg of prexasertib·lactate was produced under CGMP criteria for human clinical trials.[16] Throughout the process, real-time monitoring avoided accidents, production of by-products, and gave optimum yields of high-purity prexasertib.

11.10 Summary

Real-time analysis may be the most underappreciated Principle of Green Chemistry, yet its importance cannot be ignored. Every chemical process requires analysis, whether before, during, or after. Continuous flow processes offer opportunities for real-time analysis just about anytime and anywhere during the process. Performing analysis in real time requires consideration of what parameters are appropriate to monitor. Physical parameters include temperature, pressure, heat, pH, or light. Chemical parameters include reactants, intermediates, products, or by-products. Soltex developed a continuous flow system to produce highly reactive terminal alkene polyisobutylene (PIB) with real-time monitoring of feedstock composition, temperature, and flow rates to optimize production of the less-stable kinetic product. Analysis is necessary to determine the identity of a chemical species, as well as its purity, *i.e.* both qualitative and quantitative measures. The major types of spectral methods available for real-time analysis are NMR, MS, IR, UV–Vis, and fluorescence. Nalco developed the 3D TRASAR® method to analyze cooling water systems using two fluorescent dyes to measure biofilm, scale, and corrosion in real time. Separation of the different chemical species is an important step in determining the purity of a product or the amount of an impurity. The major chromatographic separation methods available are LC, SFC, GC, and TLC. Good reactor design includes determining where in the process to perform real-time analysis. Reactor designs include batch, semi-batch, and continuous flow. Ely-Lilly & Co. developed a continuous flow method to synthesize prexasertib, an anti-cancer drug candidate, on a kilogram-scale.

11.11 Problems: Real-time Analysis

Problem 11.1

(a) What are the two most commonly monitored physical parameters in a reaction?
(b) What are three possible chemicals to monitor during any reaction?

Problem 11.2

(a) Draw Lewis structures of the **liquid** catalysts mentioned in the Soltex PGCC application that were used to make polyisobutylene (PIB) previously.
(b) Draw the structure of the solid $BF_3 \cdot HO{-}CR_3$ on alumina (Al_2O_3) catalyst in the Soltex process.
(c) What variables were monitored in developing the Soltex process?
(d) What are the two most important industrial applications of PIB?

Problem 11.3

(a) Draw the mechanism of the high-reactive polyisobutylene (PIB) Soltex process. Begin with the correct catalyst structure.
(b) Why is the α-vinylidene more reactive than β-trisubstituted PIB? Draw simplified MO diagrams of each.
(c) Which is the kinetic product and which is the thermodynamic product?
(d) Describe three factors in the Soltex process that contribute to give α-vinylidine preferentially.

Problem 11.4

(a) Is the Soltex PIB process step-growth or chain-growth polymerization?
(b) Name a different polymerization mechanism that could be used to make PIB.

Problem 11.5

(a) List the numbers and names of the **three** most significant Principles of Green Chemistry satisfied by the Soltex process to make polyisobutylene, and explain briefly.
(b) List one disadvantage of the Soltex process related to a Principle and explain briefly.

Problem 11.6

(a) Choose an appropriate spectral method for monitoring each of the micro-flow reactions in Table 11.3 and briefly state the rationale for the method.
(b) Draw a mechanism for one reaction in Table 11.3.

Table 11.3 Micro-flow reactions.[17–23]

Reaction	Analysis method	Ref.
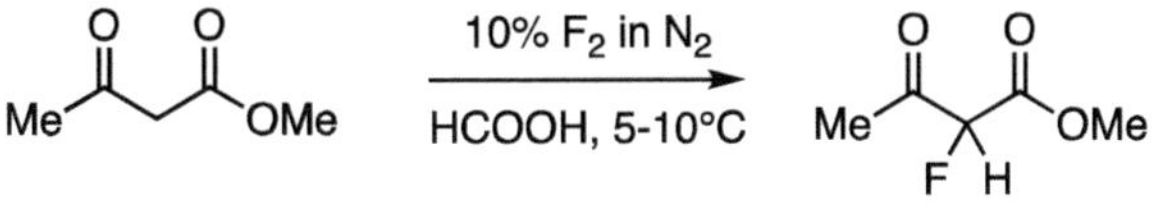		17
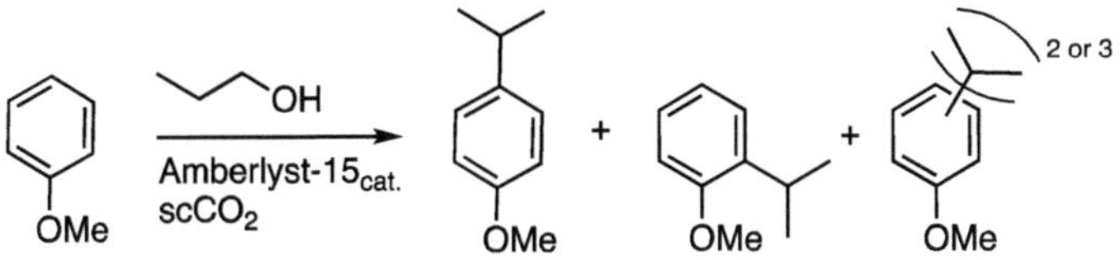		18
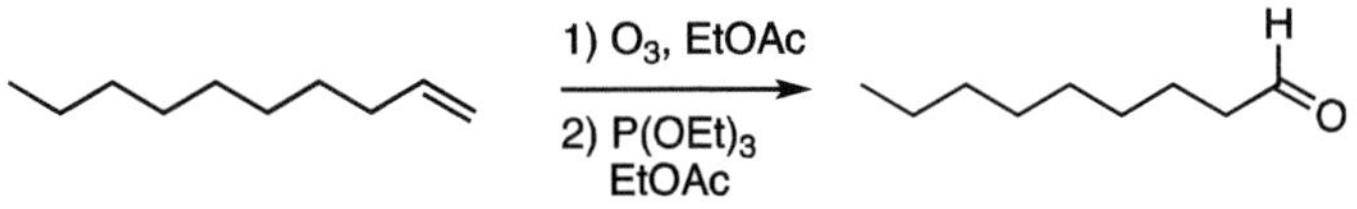		19
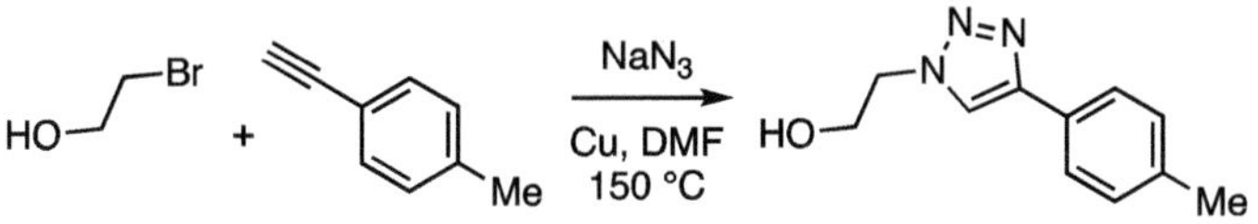		20
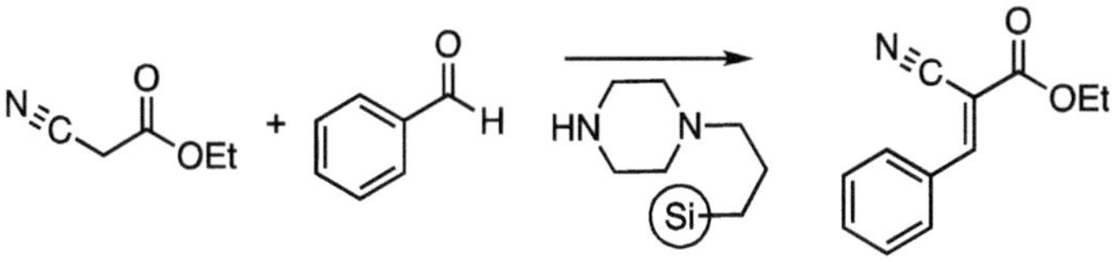		21
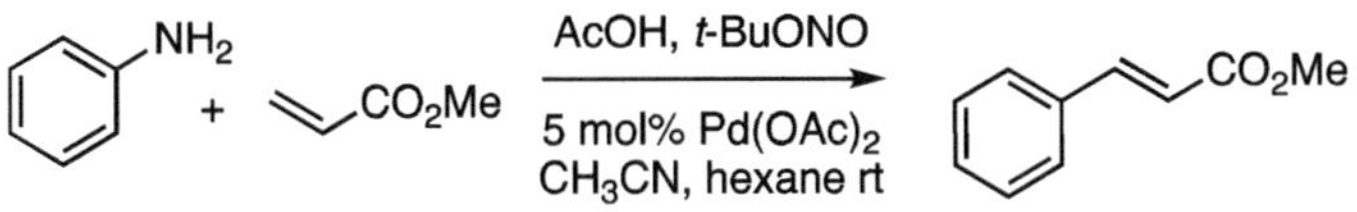		22
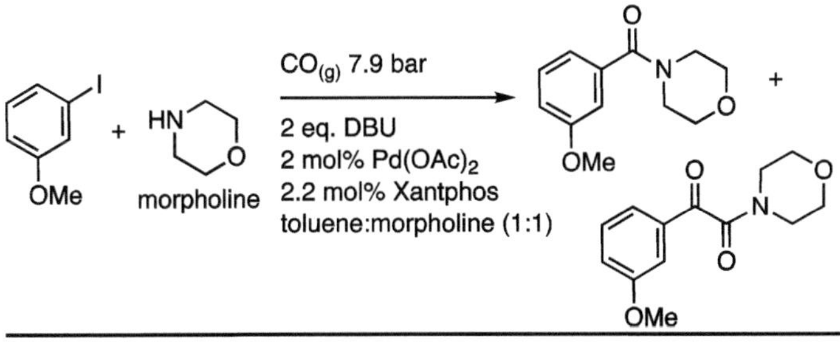		23

Problem 11.7

Consider the Nalco *3D TRASAR*® case study.

(a) Which scale compound is more soluble in water, $CaCO_3$ or $MgCO_3$? (The solubility product constants are K_{sp} $(CaCO_3) = 2.8 \times 10^{-9}$, and K_{sp} $(MgCO_3) = 3.5 \times 10^{-8}$)[24]
(b) Which is typically present at higher levels in tap water?
(c) What spectroscopic method could be used to measure concentrations of these salts in water continuously in real time? Explain briefly.

Problem 11.8

(a) Draw the balanced reaction for the synthesis of biodiesel in a semi-batch process.
(b) Describe how reactants are added and products removed.[12] (Hint: Consider the density and b.p. of each reactant and product.)

Problem 11.9

Consider **Step 2** in the synthesis of prexasertib (Figure 11.8).

(a) Draw the mechanism of the reaction of amine **11.3** with pyrazine chloride **11.4**.
(b) What is the driving force for the chloride to leave?
(c) Draw two other pyrazine products that could form in this S_NAr reaction.
(d) What methods were used for monitoring this step?

Problem 11.10

Consider **Step 3** in the synthesis of prexasertib (Figure 11.8).

(a) Draw the mechanism of the formic acid-catalyzed deprotection of the Boc-amine **11.5**.
(b) Draw the structures of the by-products described.
(c) What is the driving force for the reaction?
(d) What method was used to avoid by-product formation?

References

1. P. T. Anastas and J. C. Warner, *Green Chemistry: Theory and Practice*, Oxford University Press, New York, 1998.
2. D. Constable, *Green Chemistry Principle 11. Time to Get Real*, ACS Green Chemistry Institute, 2013, https://www.youtube.com/watch?v=nOnFv67hLFU&list=PLLlMW6nMYOaneIau3Jfm21vq7bM_rlPti&index=12, accessed July 25, 2019.
3. A. N. Mayeno, F. Lin, C. S. Foote, D. A. Loegering, M. M. Ames, C. W. Hedberg and G. J. Gleich, *Science*, 1990, **250**, 1707–1708.
4. Soltex (Synthetic Oils and Lubricants of Texas Inc.), *A Novel High Efficiency Process for the Manufacture of Highly Reactive Polyisobutylene (PIB) Using a Fixed Bed Solid State Catalyst Reactor System*, Washington, DC, 2014, https://www.epa.gov/greenchemistry/presidential-green-chemistry-challenge-2015-greener-reaction-conditions-award, accessed July 25, 2019.
5. Nalco Company, *3D TRASAR® Cooling System Chemistry and Control*, Washington, DC, 2008, https://www.epa.gov/greenchemistry/presidential-green-chemistry-challenge-2008-greener-reaction-conditions-award, accessed July 25, 2019.
6. B. Moriarty, N. M. Rao, K. Xiong, T. Y. Chen, S. Yang and M. Chattoraj, *Control of cooling water system using rate of consumption of fluorescent polymer*, Nalco Company LLC, *U. S. Pat.*, US7179384B2, 2004.
7. J. P. Rasimas, M. J. Fehr and J. E. Hoots, *Modular fluorometer*, Nalco Company LLC, *U. S. Pat.*, US6369894B1, 2000.
8. J. F. Kneller, V. Narutis, B. E. Fair and D. A. Johnson, *Oligomer containing phosphinate compositions and their method of manufacture*, Nalco Company LLC, *U. S. Pat.*, US5085794A, 1990.
9. C. Bueno, M. L. Villegas, S. G. Bertolotti, C. M. Previtali, M. G. Neumann and M. V. Encinas, *Photochem. Photobiol.*, 2002, **76**, 385–390.
10. Supelco, *How to Choose a Capillary GC Column*, Sigma-Aldrich, 2018, https://www.sigmaaldrich.com/analytical-chromatography/gas-chromatography/column-selection.html, accessed April 11, 2018.
11. R. Mullin, *Chem. Eng. News*, 2016, **94**, 21–22.
12. G. Laming, *The GL Eco-System Processor*, 2019, http://www.make-biodiesel.org/Biodiesel-Processors/the-gl-eco-system-processor/All-Pages.html, accessed July 25, 2019.
13. I. Dorfman, *Density of Cooking Oil*, 2000, https://hypertextbook.com/facts/2000/IngaDorfman.shtml, accessed July 25, 2019.
14. E. Alptekin and M. Canakci, *Renewable Energy*, 2008, **33**, 2623–2630.
15. Y.-H. Chen, Y.-H. Huang, R.-H. Lin and N.-C. Shang, *Bioresour. Technol.*, 2010, **101**, 668–673.
16. K. P. Cole, J. M. Groh, M. D. Johnson, C. L. Burcham, B. M. Campbell, W. D. Diseroad, M. R. Heller, J. R. Howell, N. J. Kallman, T. M. Koenig, S. A. May, R. D. Miller, D. Mitchell, D. P. Myers, S. S. Myers, J. L. Phillips, C. S. Polster, T. D. White, J. Cashman, D. Hurley, R. Moylan, P. Sheehan, R. D. Spencer, K. Desmond, P. Desmond and O. Gowran, *Science*, 2017, **356**, 1144–1150.
17. R. D. Chambers, M. A. Fox and G. Sandford, *Lab Chip*, 2005, **5**, 1132–1139.
18. R. Amandi, P. Licence, S. K. Ross, O. Aaltonen and M. Poliakoff, *Org. Process Res. Dev.*, 2005, **9**, 451–456.
19. Y. Wada, M. A. Schmidt and K. F. Jensen, *Ind. Eng. Chem. Res.*, 2006, **45**, 8036–8042.

20. A. R. Bogdan and N. W. Sach, *Adv. Synth. Catal.*, 2009, **351**, 849–854.
21. C. Wiles, P. Watts, S. J. Haswell and E. Pombo-Villar, *Tetrahedron*, 2005, **61**, 10757–10773.
22. B. Ahmed, D. Barrow and T. Wirth, *Adv. Synth. Catal.*, 2006, **348**, 1043–1048.
23. E. R. Murphy, J. R. Martinelli, N. Zaborenko, S. L. Buchwald and K. F. Jensen, *Angew. Chem., Int. Ed.*, 2007, **46**, 1734–1737.
24. E. Vitz, J. W. Moore, J. Shorb, X. Prat-Resina, T. Wendorff and A. Hahn, *14.10 The Solubility Product*, Chemistry LibreTexts, 2019, https://chem.libretexts.org/Bookshelves/General_Chemistry/Book%3A_ChemPRIME_(Moore_et_al.)/14Ionic_Equilibria_in_Aqueous_Solutions/14.10%3A_The_Solubility_Product, accessed July 12, 2019.

12 Prevent Accidents

"Principle 12. Substances and the form of a substance used in a chemical process should be chosen to minimize the potential for chemical accidents, including releases, explosions, and fires."[1]

12.1 Eliminate Hazards

Chemistry has long been a haven for adults who became fascinated with things that catch fire or explode when they were kids. There is a certain space cowboy machismo among chemists; telling stories about near escapes is a favorite pastime at conference dinners. Yet we are angry and sad when we hear of an explosion at a chemical plant. We worry about our families, friends and neighbors, or even total strangers on the opposite end of the world, like Bhopal. We feel to some extent responsible for chemical practices everywhere.

Green Chemistry: Principles and Case Studies
By Felicia A. Etzkorn

Published by the Royal Society of Chemistry, www.rsc.org

Chemistry is built on tradition. We are loathe to change what works; a regrettable side-effect of standing on the shoulders of giants. If a transformation, such as oxidation of an alcohol by chromate reagents, gives high yields, we use it over and again. The effect of tradition then snowballs, giving us more and more reports in the literature of the efficacy of chromate oxidations. Green chemistry asks us to pause, to step back, and to imagine a different way of doing things.

12.1.1 High Volume Hazardous Chemicals

To eliminate risk, we must eliminate hazards, not just exposure to hazards. The Clean Air Act requires the EPA to regulate chemical accident prevention for specific hazardous chemicals.[2] A list of the most common industrial toxic chemicals is given in order of amount used (Table 12.1).[3] Industries that use these chemicals in amounts above the established threshold must develop a Risk Management Plan (RMP). Ammonia and chlorine account for just over 50% of the hazardous chemicals that require an RMP (Table 12.1).[3]

Ammonia and chlorine are of greatest concern as corrosives. Ammonia is predominantly used in fertilizer, while the highest use of chlorine is to make vinyl chloride for polyvinylchloride (PVC) (Figure 12.1).[4] Alternative fertilizers and durable polymers, or

Table 12.1 "Frequency Distribution of Risk Management Plan Chemicals." Reprinted from ref. 3 with permission of Elsevier, Copyright 2001.

Chemical	Number of processes	Percentage of total
Ammonia (anhydrous)	8343	32.5
Chlorine	4682	18.3
Flammable mixtures	2830	11.0
Propane	1707	6.7
Sulfur dioxide	768	3.0
Ammonia (20% aq. or more conc.)	519	2.0
Butane	482	1.9
Formaldehyde	358	1.4
Isobutane	344	1.3
Hydrogen fluoride	315	1.2
Pentane	272	1.1
Propylene	251	1.0
Methane	220	0.9
Hydrogen	205	0.8
Isopentane	201	0.8
All Others	4139	16.1

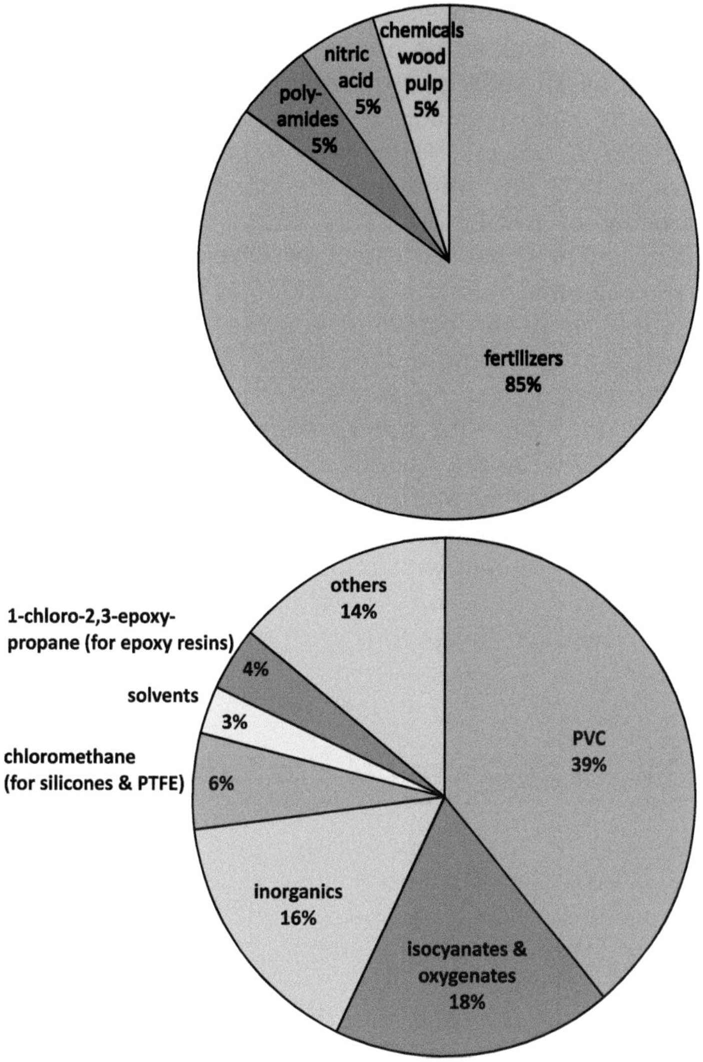

Figure 12.1 Industrial uses of the two highest abundance hazardous chemicals: top, ammonia; bottom, chlorine.[4] Adapted with permission, from Essential Chemical Industry, a web site, edited by J. N. Lazonby and D. J. Waddington, Department of Chemistry, University of York, UK.

alternative processes to make them, are needed to avoid the production and use of these highly corrosive compounds.

Ammonia represents the greatest use of a hazardous chemical (32%) because 85% of ammonia is widely used for making fertilizers (Figure 12.1).[4] Nitrogen is essential to the growth of plants to make amino acids, nucleic acids, and enzyme cofactors (vitamins).

Fritz Haber invented a process to synthesize ammonia from nitrogen and hydrogen, originally for nitrate explosives in Germany during World War I. Bosch subsequently commercialized the Haber process to make both explosives and fertilizer, and it became known as the Haber-Bosch process. Haber was awarded the Nobel Prize in Chemistry in 1920 for "improving the standards of agriculture and the well-being of mankind."[5] This award was unfortunate, since Haber subsequently made a career of developing poison gases as weapons—ammonia, chlorine, and Zyklon B, the gas used in Hitler's concentration camps during World War II.[5]

Ammonia is a toxic, hazardous gas; it is harmful if inhaled, harmful to eyes, and causes frostbite on the skin or if ingested, due to the endothermic reaction with water. Ammonia was the original refrigeration gas. Due to the hazards of ammonia upon release, it was replaced with chlorofluorocarbons (CFCs), which we now know destroy the ozone layer. Chlorine is a deadly gas because it produces HCl when it comes in contact with moist tissues in the lungs. Chlorine is used in water sterilization, in paper bleaching, and in chemical synthesis, primarily the production of poly(vinylchloride) (PVC) (Figure 12.1).[4]

12.2 Types of Chemical Hazards

Chemicals can be hazardous because they are toxic, flammable, corrosive, or explosive. Chapters 3 and 4 were a deep dive into all the different qualitative ways that chemicals can be toxic (severity), as well as the idea that concentration matters (potency). In this chapter, we will focus on chemical hazards other than toxicity. The Globally Harmonized System considers toxicity as "health hazards." Some consider flammable, corrosive, and explosive chemicals to be health hazards or toxic as well. But this can be rather indirect—corrosion of a holding tank can allow a toxin to be released in the environment for example. This chapter will be confined to flammable, corrosive, explosive, and heavy metal hazards. The hazard classifications of a specific chemical can be found on the Safety Data Sheet (SDS), available from the vendors online.

12.2.1 Flammable Liquids and Solids

Most flammable chemicals are used as either fuels or solvents. The invention of electric light did away with the hazards of kerosene or whale oil lamps, candles, and street gaslights. Natural gas is still very much a hazard in homes and businesses, and wood-burning stoves are making a comeback because wood is a renewable fuel. Of most concern to green chemistry are the flammable solvents, which were covered at length in Chapter 5, Avoid Auxiliaries. We should also be aware of the possibility of flammable reactants, by-products and products, such as low MW alkanes and alcohols. The US Occupational Health and Safety Administration (OSHA) gives the definition: "Flammable liquid is any liquid having a flashpoint at or below 199.4 °F (93 °C)."[6] Flammable mixtures are the third highest use in the EPA Risk Management Plan list, which includes many other highly flammable chemicals (Table 12.1).

Solids can also be flammable or pyrophoric; most familiar to chemists are the alkali metals: Li, Na, and K in their elemental form, and the hydrides: boranes, LiH, NaH, and $LiAlH_4$. Pyrophoric substances, such as the Et_2Zn formerly used in the production of QLEDs (Chapter 3), ignite spontaneously in air. Flammable solids include paraffin waxes and dry powders, whether natural or synthetic.

A sugar refinery explosion and fire occurred in Port Wentworth, GA, USA on February 7, 2008 that killed 14 workers and injured 36.[7] The explosion and subsequent fire were caused by a combustible mixture of sugar dust in air, and an antiquated facility owned by Imperial Sugar. A study by OSHA found that 281 dust fires killed 119 people from 1980 to 2005.[7] OSHA released guidelines for combustible dust safety in 2009.[8]

12.2.2 Corrosives: Strong Acids and Bases

Most common corrosives are strong acids: H_2CrO_4, H_2SO_4, HNO_3, HBr, HCl, HF, and HOAc, strong bases: NaOH, KOH, NH_4OH, NH_3, and oxidants: CrO_3, NaOCl (chlorine bleach). The halogen gases: F_2, Cl_2, and Br_2 are also corrosive because they react with atmospheric or mucous membrane water to produce strong acids. Corrosives burn skin, lungs, eyes—really any biological tissues on contact. The severity is a function of area of exposure, duration of exposure, and the specific characteristics of each corrosive. We have seen that HF is particularly hazardous due to deep penetration through tissue and into bone.

Mixing corrosives can be extremely hazardous. The two common household cleaning products, bleach and ammonia react to make highly toxic chloramine ($Cl{-}NH_2$) and hydrazine ($H_2N{-}NH_2$). About

30% of current water purification plants in the US purposefully add ammonia to chlorinated water because chloramine persists longer in the plumbing distribution system to kill bacteria. Unfortunately, other toxic disinfection by-products are also produced, including highly toxic nitrosamines such as *N*-nitrosodimethylamine (NDMA) (Figure 12.2). "NDMA is a probable human carcinogen with a 0.7 ng L^{-1} concentration in drinking water associated with a 10^{-6} lifetime excess cancer risk."[9]

12.2.3 Explosives: Nitrates and Peroxides

An explosion is a sudden, violent transformation of potential energy into work, the work of rapidly expanding pressure in a shock wave that causes substantial damage to physical objects. This discussion will be confined to chemical potential energy and the resulting explosion due to the rapid production of gas contained under pressure. The most familiar explosives are nitrated organic compounds, like TNT, and organic peroxides—1,3,5-Trinitrotoluene (TNT) is the most familiar explosive, used in road, tunnel, and bridge building. The strongly oxidizing nitro functional groups combined with the reduced toluene hydrocarbon fuel in the same molecule makes for an unstable situation prone to explosion in a highly exothermic reaction. In the explosion, two moles of solid TNT instantly turn into 15 to 20 moles of hot gases via two possible routes (Figure 12.3).[10,11] The rapid expansion of these gases causes intense physical damage to objects in the vicinity.

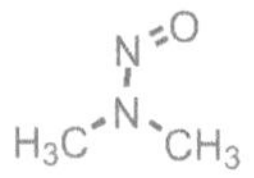

Figure 12.2 Structure of water disinfection by-product *N*-nitrosodimethylamine (NDMA).

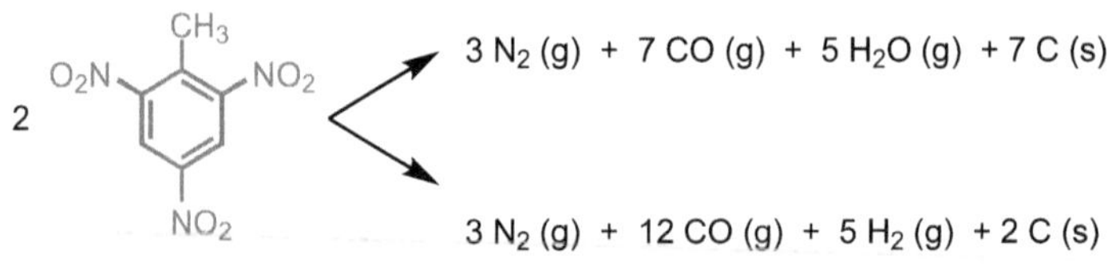

Figure 12.3 Gases produced in a TNT explosion.[10,11]

12.3 Global Harmonization System

The physical hazards definitions that follow are largely based on those of the *United Nations Dangerous Goods System* as described on Wikipedia.[12,13] Some additions and changes were necessary since the scope of the *Global Harmonization System* (GHS) includes all target audiences.[13] The hazard designations for specific chemicals can be found on their Safety Data Sheets (SDS) available from chemical vendors.

- **Explosives**, which are assigned to one of six subcategories depending on the type of hazard they present, as used in the UN Dangerous Goods System.
- **A flammable gas** is one that has a flammable range in air at 20 °C and a standard pressure of 101.3 kPa. Substances and mixtures of this hazard class are assigned to one of two hazard categories on the basis of the outcome of the test or calculation method.
- **Flammable aerosols** should be considered for classification as Category 1 or Category 2 if they contain any component, which is classified as flammable according to the GHS criteria, that is, flammable liquids, flammable gases or flammable solids.
- **Oxidizing gases** are any gas that may, generally by providing oxygen, cause or contribute to the combustion of other material more than air does. Substances and mixtures of this hazard class are assigned to a single hazard category on the basis that, generally by providing oxygen, they cause or contribute to the combustion of other material more than air does.
- **Gases under pressure** are gases contained in a receptacle at a pressure not less than 280 Pa at 20 °C or as a refrigerated liquid. This endpoint covers four types of gases or gaseous mixtures to address the effects of sudden release of pressure or freezing which may lead to serious damage to people, property, or the environment independent of other hazards the gases may pose.
- **A flammable liquid** is a liquid with a flash point of not more than 93 °C. Substances and mixtures of this hazard class are assigned to one of four hazard categories on the basis of the flash point and boiling point.
- **A flammable solid** is one that is readily combustible or may cause or contribute to fire through friction. Readily combustible solids are powdered, granular, or pasty substances which are dangerous if they can be easily ignited by brief contact with an ignition

source, such as a burning match, and if the flame spreads rapidly.

- **Self-reactive substances** are thermally unstable liquids or solids liable to undergo a strongly exothermic thermal decomposition even without participation of oxygen (air). This definition excludes materials classified under the GHS as explosive, organic peroxides or as oxidizing.
- **A pyrophoric liquid** is a liquid that, even in small quantities, is liable to ignite within five minutes after coming into contact with air. Substances and mixtures of this hazard class are assigned to a single hazard category on the basis of the outcome of the UN Test N.3.
- **A pyrophoric solid** is a solid that, even in small quantities, is liable to ignite within five minutes after coming into contact with air. Substances and mixtures of this hazard class are assigned to a single hazard category on the basis of the outcome of the UN Test N.2.
- **Self-heating substances** are solids or liquids, other than a pyrophoric substance, which, by reaction with air and without energy supply, is liable to self-heat. Substances and mixtures of this hazard class are assigned to one of two hazard categories on the basis of the outcome of the UN Test N.4.
- **Water-reactive substances** are substances that, in contact with water, emit flammable gases; or, are solids or liquids which, by interaction with water, are liable to become spontaneously flammable or to give off flammable gases in dangerous quantities. Substances and mixtures of this hazard class are assigned to one of three hazard categories on the basis of the outcome of UN Test N.5, which measures gas evolution and speed of evolution.
- **Oxidizing liquids** are liquids that, while in itself is not necessarily combustible, may, generally by yielding oxygen, cause or contribute to the combustion of other material. Substances and mixtures of this hazard class are assigned to one of three hazard categories on the basis of the outcome of UN Test O.2.
- **Oxidizing solids** are solids that, while itself is not necessarily combustible, may, generally by yielding oxygen, cause or contribute to the combustion of other material. Substances and mixtures of this hazard class are assigned to one of three hazard categories on the basis of the outcome of UN Test O.1.
- **Organic peroxides** are organic liquids or solids that contain the bivalent –O–O– structure and may be considered a derivative of

hydrogen peroxide, where one or both of the hydrogen atoms have been replaced by organic radicals. The term also includes organic peroxide formulations (mixtures). Substances and mixtures of this hazard class are assigned to one of seven 'Types', A to G, on the basis of the outcome of the UN Test Series A to H.
- **Substances corrosive to metal** are substances or mixtures that by chemical action will materially damage, or even destroy metals. These substances or mixtures are classified in a single hazard category on the basis of tests (Steel: ISO 9328 (II): 1991 – Steel type P235; Aluminum: ASTM G31-72 (1990) – non-clad types 7075-T6 or AZ5GU-T66). The GHS criteria are a corrosion rate on steel or aluminum surfaces exceeding 6.25 mm per year at a test temperature of 55 °C.

The organic peroxides category is of particular interest to chemists. Organic peroxides are often contact explosives. Some of these are useful reagents for oxidations (*e.g.* H_2O_2, *t*-BuOOH), while others are unavoidable and highly undesirable side products of solvent oxidation or reactions. Diethyl ether, THF, and MeTHF are common ether solvents that must be used up or disposed of within 6 months of receiving them to avoid explosion and fire due to the formation of organic peroxides over time (Chapter 5).

12.4 Eliminating Hazards by Design

Returning to the concept of benign by design, chemical processes should be designed to eliminate hazards and thus risk—risk of explosion, risk of fire, risk of burns, or risk of toxic leaks and spills. Explosions are the most spectacular and frightening of hazards, yet explosions are more rare than fires, burns, leaks, and spills.

12.4.1 Less Flammable Solvent or No Solvent

In green chemistry, solvents should be chosen with due consideration of the possibility of accidents. Non-flammable, non-toxic solvents are preferred. Water is great if the product can be readily isolated by precipitation and filtration to avoid energy-intensive evaporation, and if the remaining water can be readily purified for reuse or return to the environment. In practice, organic solvents must still be used for many reactions because of the low polarity of reactants or the difficulty of removing water from the product.

Most solvents in Table 5.1 are flammable, except ethylene glycol and glycerol, which are neither flammable nor combustible with flash points >93 °C. Most ionic liquids are neither flammable nor combustible, which is why they are generally considered greener solvents. Non-flammable solvents are usually high-boiling, a trade-off against Principle 6 Energy Efficiency, because solvents are usually removed from reaction mixtures by evaporation. Although chlorinated solvents are not flammable, they have toxicity issues.

12.4.2 Avoid Explosives

Chemists now understand most chemical reactions well enough to predict what reactions will produce gases and how quickly, understanding that must be used to avoid the potential for explosions. In the past, preparations for potential explosions included cooling systems and pressure relief valves. In green chemistry, we aim to avoid explosions by avoiding strongly exothermic reactions and chemicals that have the potential to produce gases rapidly. This is the difference between avoiding risk by avoiding exposure *vs.* avoiding the hazard itself.

12.4.3 Avoid Organic Nitrations

When the goal is to synthesize explosives, as is the case at the Radford Army Ammunition Plant nearby Virginia Tech in Radford, VA, organic nitrations are very tough to avoid. In the laboratory, or in process chemistry, other routes can be designed to avoid organic nitrations. The use of nitric acid as an oxidant should be avoided in any green chemistry process.

Phloroglucinol, 1,3,5-hydroxybenzene, is a useful natural product and synthetic intermediate that is synthesized in a highly inefficient process by tri-nitration of toluene (Figure 12.4).[14] This synthesis from toluene was so inefficient that it was the benchmark process that Roger Sheldon used to develop the E-factor.[15] The E-factor he calculated for the production plant did not even include the nitration of toluene. More pertinent to this chapter, TNT is an explosive intermediate in the synthesis!

12.4.4 Avoid Explosive Production of Gases

Balancing each reaction and understanding the chemical mechanism of each reaction that has been chosen for a particular process go

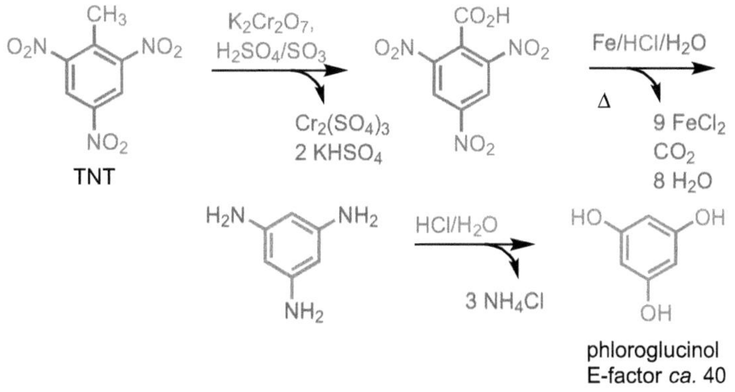

Figure 12.4 Sheldon's first E-factor analysis of phloroglucinol production. Hazardous chemicals are shown in red, atoms wasted in brown, and atoms used in green. Adapted from ref. 16 with permission from the Royal Society of Chemistry.

a long way towards avoiding the explosive production of gases. Not to say that production of *any* gas is bad—a volatile by-product can be an excellent way to avoid auxiliaries involved in purification. For example, in Figure 11.8, the deprotection of a Boc-amine in the synthesis of the anti-cancer agent, prexisertib, produced CO_2 and isobutylene gases that were removed by bubbling nitrogen through the flow system. Many similar examples have been presented in this textbook. It is the rapid, exothermic, run-away production of gases that must be avoided. The most infamous of these was the disaster in Bhopal, India in 1984.

12.5 Chemical Disaster: Explosion in Bhopal, India

A terrible explosion in Bhopal, India in 1984 at the Union Carbide India plant released the toxic gas, methyl isocyanate (MIC). MIC is used in the production of the pesticide, carbaryl (Sevin™), a carbamate insecticide (Figure 12.5). The disaster was due to multiple operating errors: old equipment, a shut-down refrigeration unit, and human error—a water line was connected to the MIC tank.[17] The explosive release of MIC caused over 2000 deaths immediately, and chronic pulmonary disease for thousands thereafter in the Bhopal disaster.[18] The Immediately Dangerous to Life or Health Concentration (IDLH) for MIC is 0.02 ppm; the LC_{50} (rat) is 6.1 ppm/6 h.[19] MIC reacts exothermically with water to produce CO_2 and

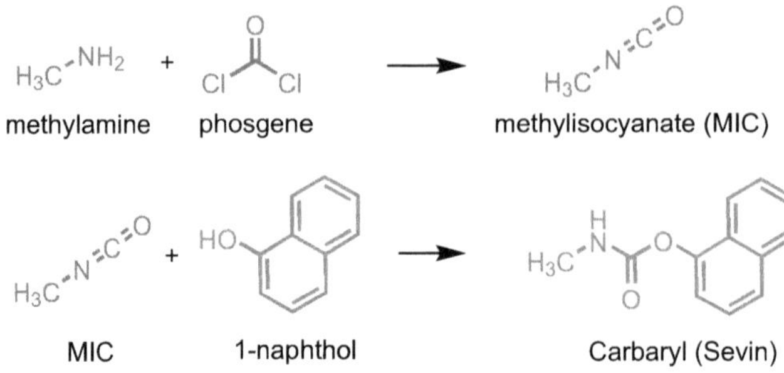

Figure 12.5 The use of MIC in the production of carbaryl at the Union Carbide India Ltd. plant in Bhopal, India. Highly hazardous chemicals are shown in red.

N,N-dimethylurea. MIC itself is toxic, reacting as a potent electrophile with nucleophiles in the body such as nucleic acid or amino acid bases.

Not only is MIC used in the production of carbaryl, MIC itself is made from extremely toxic phosgene (Figure 12.5). Exposure to phosgene gas at levels as low as 2–4 ppm causes damage to skin, eyes, nose, throat, and lungs. Low concentrations of phosgene smell of new mown hay, the sign for chemists using it to get out of the lab immediately. The IDLH value for phosgene is 0.1 ppm, and the LC_{50} (human) is 50 ppm/5 min.[19] Phosgene decomposes into HCl, CO_2 or CO, and Cl_2 gases. Methylamine is also a hazardous gas, with a f.p. of −18 °C, and an LD_{50} (oral, rat) of 698 mg kg^{-1}.[20] 1-Naphthol is less toxic, with an LD_{50} (oral, rat) of 1.87 g kg^{-1}, and relatively non-hazardous, with a f.p. of 125 °C.[21]

12.6 Replacements for Methyl Isocyanate

12.6.1 Greener Synthesis of Carbaryl

Carbaryl is an acetylcholinesterase inhibitor that is non-specific for target pest insects. Dimethyl carbonate (DMC), a non-toxic, non-volatile feedstock, has been proposed as a replacement for phosgene.[22] DMC can be made directly from CO_2 and MeOH using a Rh catalyst.[23] Two alternative syntheses of carbaryl could avoid both phosgene and the MIC gaseous intermediates (Figure 12.6). Both starting materials can be made from DMC and methylamine. Exclusive Chemistry has developed an alternative synthesis of carbaryl

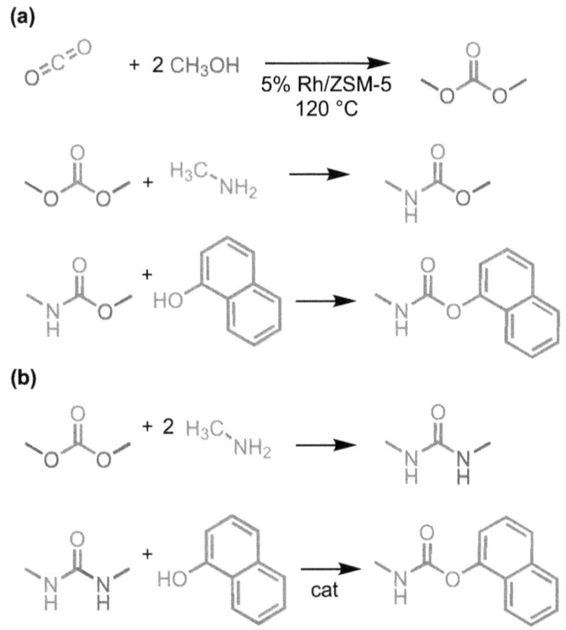

Figure 12.6 Proposed greener syntheses of carbaryl using (a) dimethylcarbonate,[23] or (b) dimethylurea.[24] Atoms used are in green, unused in brown.

that uses dimethylurea as the carbonyl starting material.[24] However, carbaryl itself is toxic to vertebrates, despite rapid metabolism. The LD_{50} (oral, rat) ranges from 302 mg kg^{-1} (male) to 312 mg kg^{-1} (female).[25]

12.7 Case Study: Hybrid Non-isocyanate Polyurethane/Green Polyurethane™ (Nanotech Industries)

Nanotech Industries, Inc. (NTI) won a PGCC Award in 2015 for a water-soluble polyurethane formulation, called Hybrid Non-isocyanate Polyurethane (HNIPU) or Green Polyurethane™. Isocyanates are particularly toxic chemicals used in the production of polyurethanes (Figure 12.7). The isocyanate functional group is a strong electrophile, facilitating rapid polymerization with diols. Previously, polyurethanes had been made with toxic isocyanates, such as diphenylmethane diisocyanate (Figure 12.7).

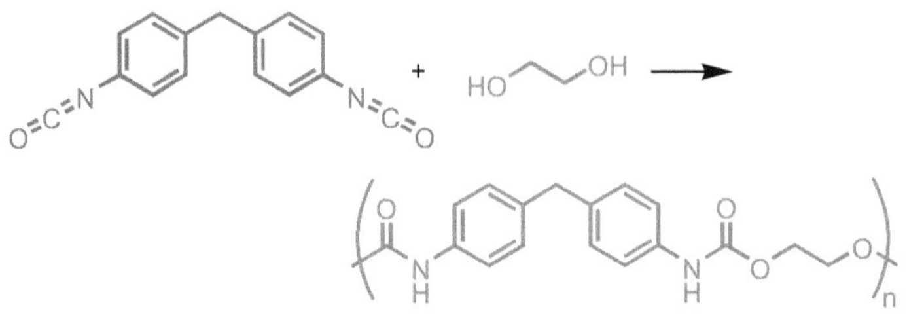

Figure 12.7 Formation of polyurethane from hazardous diphenylmethane diisocyanate and ethylene glycol.[26]

12.7.1 Hybrid Non-isocyanate Polyurethane/Green Polyurethane™ (HNIPU)

NTI replaced the isocyanates with dicarbonates, which can be synthesized from CO_2 and alcohols, to make HNIPUs (Figure 12.8). The reaction between the dicarbonates and aliphatic diamines results in a cross-linked co-polymer with mixtures of 1° and 2° alcohol β-hydroxyurethanes (Figure 12.8).[26] Aliphatic diamines can be derived from biofeedstocks, as described in Chapter 7, although this was not specified by NTI. The alcohols remaining in the polymer improve the adhesion properties of the polyurethanes (Figure 12.8).[26] Interstrand hydrogen bonds confer increased thermal stability and chemical resistance to non-polar organic solvents. The hydroxyurethanes are not susceptible to hydrolysis by moisture in the environment. These HNIPU coatings compare very favorably with conventional UV-cure floor coatings (Table 12.2).[26]

Up to 50% of the epoxidized monomers in Green Polyurethane™ were replaced with carbonated, epoxidized vegetable oils from renewable sources (Figure 12.9).[26] The trifunctional vegetable-oil-based monomers can be incorporated as crosslinkers in hydroxyurethanes and epoxides. Both carbonate and epoxides are reactive to create covalent crosslinks.

12.7.2 Hydroxy Urethane Modifiers (HUMs)

NTI also developed small molecule hydroxyurethane modifiers (HUMs) for cold-cure epoxy-amines (Figure 12.10). These are not polymers, they are small molecules that modify the properties of the Green Polyurethane™. The HUMs are synthesized by the reaction of a monocyclic carbonate and a primary monoamine instead of a diamine (Figure 12.10).[26] New HUMs were also developed on the

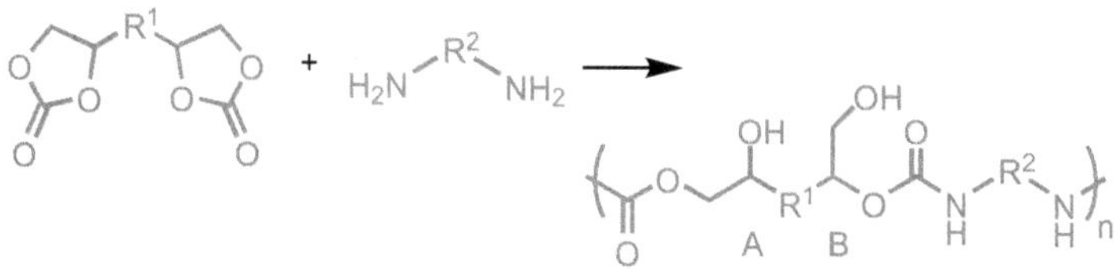

Figure 12.8 β-Hydroxyurethane moieties of non-isocyanate polyurethanes [HNIPUs]: A – with secondary hydroxyl groups; B – with primary hydroxyl groups. Adapted from ref. 26 with permission of Nanotech Industries Inc., Copyright 2014.

Table 12.2 Performance comparison of conventional and HNIPU UV-cured flooring. Reprinted from ref. 26 with permission of Nanotech Industries Inc., Copyright 2014.

Properties	Conventional UV-cured flooring	HNIPU UV-cured flooring
Adhesion	3B	5B
Pencil hardness	3H	4H
Solvent resistance	200+	200+
Gloss	84	90
Abrasion resistance, CS-17, 1000 g	150–200	100
Thickness applied	0.065–0.1 mm	0.065–1.0 mm
Primer	Required	Not required for some substrates
No. of layers	2+	1

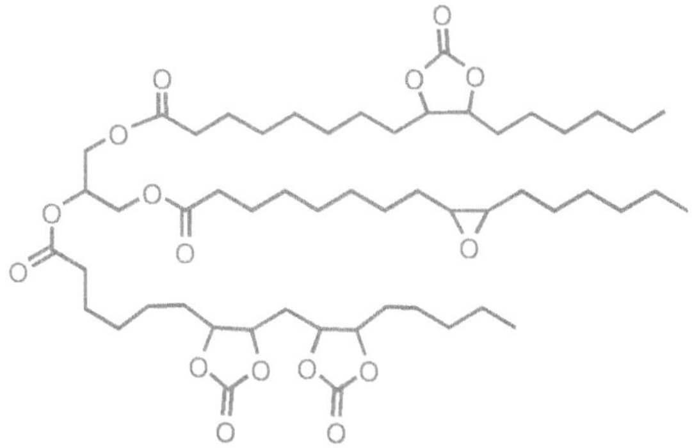

Figure 12.9 Example structure of a carbonated, epoxidized soybean oil for cross-linking urethane polymers.[26]

basis of carbonated, epoxidized soybean oil reacted with primary monoamines.[26]

HUMs can be embedded in an epoxy polymer without chemical reaction, and the extra hydrogen-bonding capacity improves the

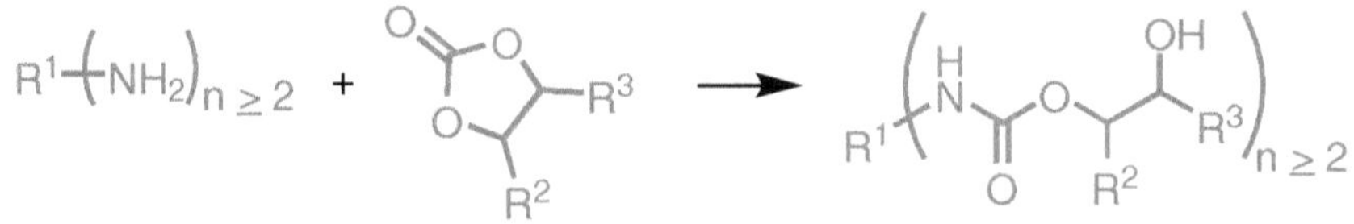

Figure 12.10 Synthesis of hydroxyurethane modifiers (HUMs).[26]

adsorption of diluents, pigments, and other additives.[26] The coating performance characteristics of the polymers are significantly improved: pot life/drying time ratios, strength-stress, and bonding to a variety of substrates (Figure 12.11).[26]

NTI used a combination of the epoxy-amine resins with a HUM to produce advanced indoor and outdoor two-component floor coatings, paint for vertical surfaces, and varnish, now sold commercially around the globe.[26] HUMs based on soybean oil are used in spray polyurethane foam (SPF), floor coatings, and UV-cured acrylic polymer coatings.[26]

12.7.3 Linear Hybrid Epoxy-amine Hydroxyurethane-grafted Polymers

NTI's cured linear hybrid epoxy-amine hydroxyurethane-grafted polymers have a controlled number of crosslinks (Figure 12.12).[26] Low molecular weight amines used as curing agents in epoxy hardeners are corrosive, toxic, and potentially carcinogenic. The low MW amines were replaced with a non-hazardous, high MW (over 1000 $g\,mol^{-1}$) amine in their hybrid epoxy-amine hydroxyurethane-grafted polymer. These new materials have a tensile strength up to 12.0 MPa and an elongation at break of 70–275%.[26]

Uncured coatings can be cleaned up during and after application with simple soap solutions. Cured product must be mechanically removed by scraping or sanding. Cured product can be disposed of in municipal landfills due to negligible toxicity. NTI is commercializing a Green Polyurethane™ do-it-yourself flooring product, and polyurethane foams, including spray polyurethane foam (SPF), and flexible foam.

In summary, NTI's HNIPU and HUM products use no isocyanates. The HUMs have no free amines and are readily biodegradable.[26] Up to 50% of the HNIPU is made with renewable vegetable oil as the feedstock. The hybrid epoxy-amine hydroxyurethane-grafted polymers use high MW amines instead of corrosive low MW amines. These coatings have chemical and mechanical resistance advantages over current polyurethane and epoxy technologies. This process also eliminates

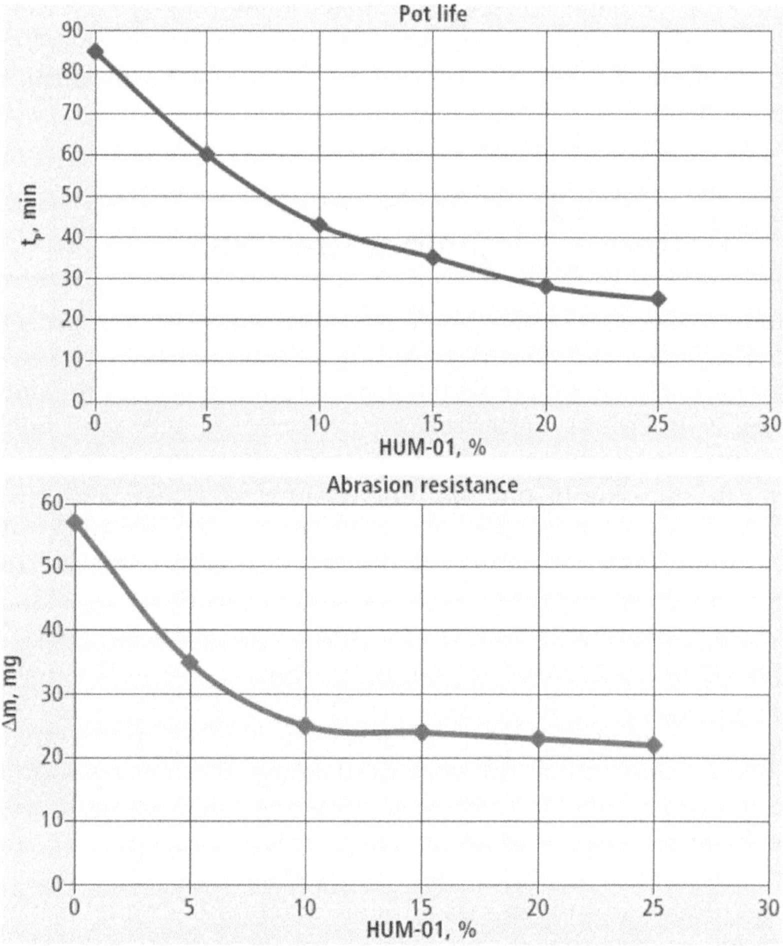

Figure 12.11 Influence of modifier HUM-01 (trimethyl-hexamethylene-diamine + propylene carbonate) on the [pot life and abrasion resistance] properties of epoxy composition based on D.E.R. 331. Reprinted from ref. 26 with permission of Nanotech Industries Inc., Copyright 2014.

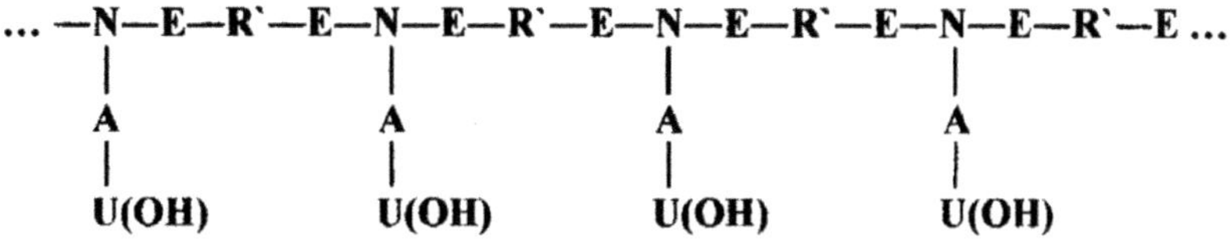

Figure 12.12 General formula of NTI linear hybrid epoxy-amine hydroxyurethane-grafted polymers, where **E–R′–E** is a residue of a diglycidyl ether; **E** is a converted epoxy group, *i.e.*, $-CH_2-CH(OH)-CH_2-O-$; **N** is a nitrogen atom; **A** is a di-primary amine; **U(OH)** is a hydroxyurethane group from a HUM. Reprinted from ref. 26 with permission of Nanotech Industries Inc., Copyright 2014.

the use of hazardous chlorine gas to make 1-chloro-2,3-epoxypropane for epoxy resins (4% of Cl_2 in use industrially) and isocyanates precursor, phosgene (18% of Cl_2) (Figure 12.2).[26] NTI's products are currently being commercialized.

12.8 Case Study: Safer Solvents for Lithium Ion Batteries (DeSimone-Balsara)

Lithium ion batteries with organic electrolytes have been in use since the 1990s. The major advantage of lithium batteries is the power-to-weight ratio. The major problem has been recalls due to flammability. The most notorious series of events involved the new Boeing Dreamliner 787 airplanes. The safety of lithium ion batteries was called into question by one fire, one incident of smoke, and several safety inspections on five different 787 planes in January 2013.[27] The most common Li-ion batteries are based on a $LiCoO_2$ cathode paired with a graphite anode for performance and safety (Figure 12.13).

The biggest safety concern is accidental shorting (electrical connection between the two terminals), which leads to thermal decomposition of the organic electrolyte, dimethyl carbonate. Hydrogen gas can be produced, leading to ignition of the organic electrolyte. Cell phones rarely catch fire because the large thermal areas on the back and front faces of a phone cool the batteries. However, larger Li-ion batteries, such as those in laptops and on the Boeing 787 have caught fire, causing great concern for their safety.[27]

The DeSimone and Balsara groups developed a series of non-flammable perfluoropolyether (PFPE) electrolytes for lithium ion batteries with greatly improved safety profiles (Figure 12.14).[29] The PFPEs readily dissolve lithium bis(trifluoromethane)sulfonimide (LiTFSI) to make the electrolyte solution. They used a $Li/Li_3NiCoMnO_6$ cell to demonstrate the feasibility of replacing the dimethylcarbonate (DMC) electrolyte solvent with PFPE-DMC in Li-ion batteries (Figure 12.14). The most promising electrolyte solvent was $PFPE_{1000}$–DMC.[29] The PFPE is chemically inert and resistant to swelling in organic solvents.[30] The electrolyte LiTFSI is corrosive and toxic.

$$\text{cathode: } CoO_2 + Li^+ + e^- \longrightarrow LiCoO_2$$

$$\text{anode: } LiC_6 \longrightarrow Li^+ + C_6 + e^-$$

Figure 12.13 Standard Li-ion battery half-reactions.[28]

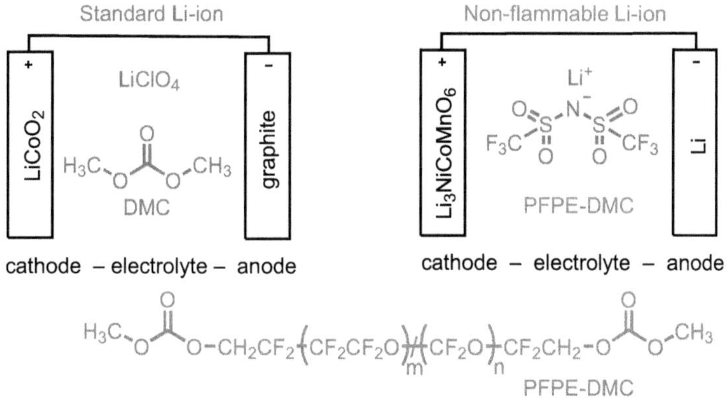

Figure 12.14 Lithium ion battery configurations, and the non-flammable perfluoropolyether (PFPE) solvent structure developed by the Balsara and DiSimone groups.[29]

Perfluorinated chemicals are persistent in the environment. For example, perfluoroalkylsulfonates (PFAS) are toxic and endocrine disrupters.[31] In 2006, the US EPA called upon manufacturers to eliminate perfluorinated chemicals voluntarily.[32] Thus, the biological activity, environmental stability and fate of PFPE solvents should be assessed prior to widespread adoption.[32] A mechanism for recovering and recycling PFPE solvents from spent Li-ion batteries would make this a much greener substitute.

12.9 Case Study: Chromium(III) Plating Process (Faraday Technology)

Accidents with toxic heavy metals can be devastating to the production workers and to the surrounding community. Chrome plating businesses are small, local and dispersed widely around the planet. Faraday Technology, Inc. won a PGCC Award in 2013 for a chrome plating process that uses trivalent Cr(III) instead of the more highly toxic hexavalent Cr(VI).[33] In the US, 6000 metric tonnes, and worldwide 135 000 metric tonnes, of hexavalent Cr are used annually. The Faraday process can eliminate the use of Cr(VI) entirely if widely implemented, a considerable reduction the risk of worker exposure to Cr(VI) globally. However, Cr(III) is not entirely benign. Cr(III) is actually more toxic once inside the cell, but it has limited bioavailability across the cell membrane. Cr(VI) has greater bioavailability, and once inside the cell it is reduced to the more toxic Cr(III) (Chapter 3).[34] The worker exposure limit for Cr(III) is an order of magnitude higher than

Cr(VI). Hexavalent chromium is classified as a hazardous air and water pollutant because it is a human carcinogen. Trivalent chromium is not an oxidizer, not known to be a carcinogen, and relatively non-toxic because of its low bioavailability. To prevent worker exposure to Cr(VI), Faraday designed an elegant new chrome plating process for functional applications of sliding wear and abrasion resistance.

Other possible replacements for hexavalent chromium are thermal-plasma spray of various metals and metal carbides, and electroplating of Co-P or Ni-based coatings. However, Faraday has invented a drop-in replacement process for electroplating trivalent chromium with no necessity for retooling. In addition, Faraday used both $CrCl_3$ and $Cr_2(SO_4)_3$ in their plating bath formulation; they found that $Cr_2(SO_4)_3$ is cheaper, and both work just as well.

12.9.1 Trivalent Chromium Electroplating

Prior attempts to plate Cr(III) were self-limiting to thin layers of about 2.5 μm suitable for decorative coatings.[33] Functional coatings for abrasion and sliding wear resistance require thicker coatings up to 375 μm; with their new process they were able to achieve this thickness easily.[33] Faraday suspected that the process of plating Cr(III) was self-limiting due to the high pH at the plating interface due to the consumption of protons and co-evolution of H_2 gas (Figure 12.15).[33]

Ordinarily in chrome electroplating, the cathodic current is turned on and held for the duration of the process. To alleviate the self-limiting deposition of Cr(III), Faraday designed a "pulse-reverse waveform" process to maintain an acidic interface and permit thick chromium plating. The pulse-reverse waveform consists of a cathodic (forward) pulse for time t_c, followed by an anodic (reverse) pulse t_a, and a relaxation period t_{off} (off-time) (Figure 12.16). Each of these times, as well as the cathodic i_c and anodic i_a peak currents, are individual process variables that are monitored in real time for process control. The pulse-reverse waveforms can be fine-tuned to control coating morphology and properties via mass transport, current distribution, and grain size.

$$Cr^{+3} + 3\,e^- \longrightarrow Cr^0\,(s)$$

$$2\,H^+ + 2\,e^- \longrightarrow H_2\,(g)$$

Figure 12.15 Self-limiting plating of Cr(III) due to high pH at the interface. Adapted with permission of Faraday Technology Inc. from ref. 33, Copyright 2013.

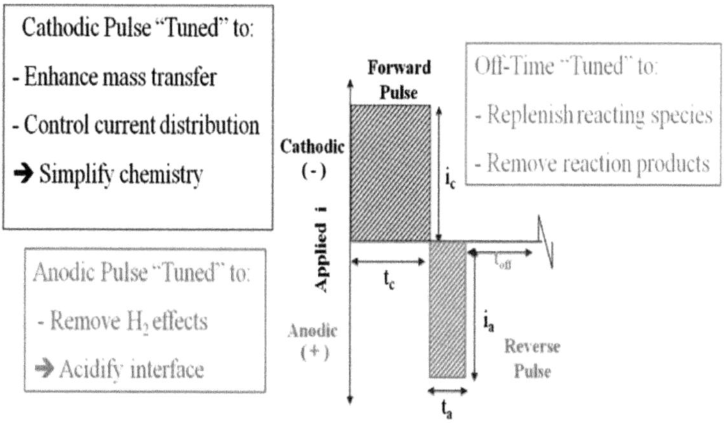

Figure 12.16 FARADAIC® TriChrome Plating pulse-reverse waveform process. Reprinted with permission of Faraday Technology Inc. from ref. 33, Copyright 2013.

12.9.2 Advantages of FARADAIC® TriChrome Plating

The plating rate for the TriChrome process was ~5 mils h^{-1}, while hexavalent chromium processes range from ~0.5 to 1.5 mils h^{-1}, up to 10 times slower. The current efficiency is defined as the ratio of the desired plating of chromium to the undesired evolution of hydrogen gas that raises the pH and causes self-limiting plating. The TriChrome process efficiency is 25 to 50%, compared to ~15% for hexavalent chromium plating. The Cr(III) is reduced to Cr(0) during the forward pulse (Figure 12.16).[33] The reverse pulse for a shorter amount of time cleans the electrode and reestablishes the acidic pH (Figure 12.15).[33] Finally, during the off-time, reacting species are replenished and the reaction products are removed before beginning another cycle (Figure 12.16).[33] After 1400 A-hr of operation, the TriChrome plating bath showed zero Cr(VI) concentration, and when the bath was spiked with Cr(VI), it was reduced to Cr(III) within 15 min.

The FARADAIC® TriChrome Plating appearance is equivalent to hexavalent chromium coatings of similar thickness (Figure 12.17). The microcracks observed are beneficial for coatings, conferring lower internal stresses, higher lubricity, better wear, and corrosion resistance.[33] The Knoop hardness of hexavalent Cr plated after baking was 887, while the hardness of trivalent Cr plated after baking was 1003.[33] For ball-on-flat and oscillating wear, TriChrome was equal or superior; axial fatigue was comparable at low-stress, and better at high-stress, to hexavalent chromium plating.[33]

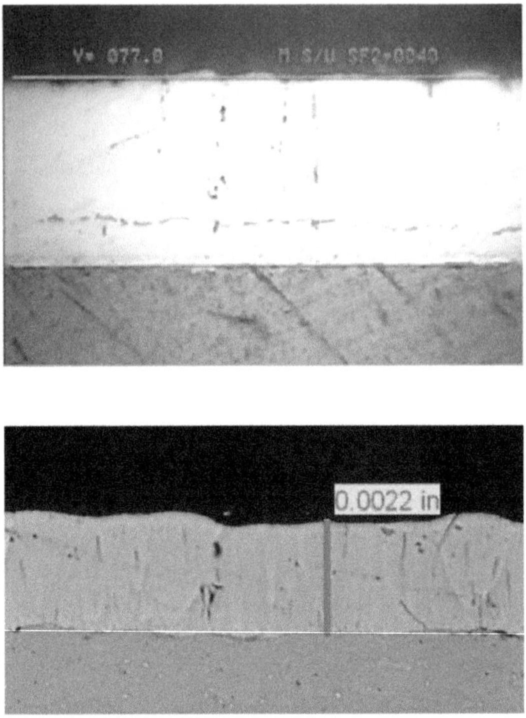

Figure 12.17 Visual comparison of FARADAIC® TriChrome Plating and hexavalent chromium plating. Top: 77 μm FARADAYIC℠ Chrome Coating; Bottom: 50 μm Hexavalent Chrome Coating. The microcracks are desirable in functional chromium platings. Reprinted with permission of Faraday Technology Inc. from ref. 33, Copyright 2013.

The global chrome plating market is estimated at ~$3 billion. Increasing government regulations of Cr(VI) make it likely that Faraday's drop-in substitute TriChrome process will succeed in the global market. Most of the funding ($999 266 out of $1 568 731 total) for developing this significant advance in chromium plating technology came from the US EPA Small Business Innovation Research (SBIR) Program.

12.10 Summary

Chemical hazards other than toxicity are classified as flammables, corrosives, and explosives. Flammables can be liquid or solid. Corrosives are typically oxidants, strong acids and bases. Explosives are numerous, with the most common being nitrates and peroxides. The Global Harmonization System categorizes chemical hazards for uniformity in commerce. Methods to eliminate hazards include avoiding flammable solvents, explosives—in particular nitrations, and explosive production of gases. The latter was exemplified by the exothermic reaction of MIC with water in Bhopal, India in 1984. Greener methods for the synthesis of carbaryl exist to displace the use of MIC. Nanotech Industries developed a Green Polyurethane™ that uses no isocyanates and soybean oil derivatives. The DeSimone and Balsara laboratories created a safer fluorinated solvent for lithium ion batteries. Faraday Technology created a drop-in chrome plating process that uses Cr(III), instead of highly toxic and corrosive Cr(VI), with superior performance and greatly decreased energy consumption.

12.11 Problems: Prevent Accidents

Problem 12.1

(a) When and where was the most infamous explosion involving an isocyanate?
(b) Draw the structure of methyl isocyanate.
(c) What are the hazards of volatile isocyanates, such as methyl isocyanate?

Problem 12.2

(a) In a chemical reaction, do you expect methyl isocyanate to behave as a nucleophile or an electrophile?
(b) Draw the mechanism of the reaction of MIC with the adenine amine.

Problem 12.3

(a) In Figure 12.5, the synthesis of carbaryl does not show the side products. Redraw and balance the reactions and consider what side products are produced in each step. Are any of these side products toxic or hazardous?
(b) Repeat the procedure for the alternative synthesis of carbaryl given in Figure 12.6 (a).

Problem 12.4

(a) List six green chemistry advantages of Green Polyurethane™ over isocyanate-based polyurethane.

(b) List the numbers and names of the **three** most significant Principles of Green Chemistry satisfied by Green Polyurethane™. Explain briefly.

Problem 12.5

(a) Draw the structure of one fatty acid chain of a **carbonated** soy oil.
(b) What purpose do HUMs serve in Green Polyurethane™ (HNIPU)?
(c) Draw the mechanism of one polymerization step to form Green Polyurethane™ (HNIPU)?

Problem 12.6

(a) Why is the perfluoropolyether-dimethylcarbonate (PFPE-DMC) solvent for Li-ion batteries non-flammable?
(b) What are the potential environmental problems for PFPE-DMC?

Problem 12.7

See Chapter 3 for Cr toxicity.

(a) Which form of Cr enters the cell most easily?
(b) Which form of Cr is the most toxic inside the cell?
(c) Inside a living cell, glutathione is prevalent reducing agent. Draw the structure of glutathione. What is the oxidized product of glutathione?
(d) What is the biological mechanism of Cr toxicity?

Problem 12.8

(a) What is the reduction potential of Cr(VI) to Cr(III)?
(b) What is the reduction potential of Cr(III) to Cr(0)?
(c) How many metric tonnes of Cr(VI) could the FARADAIC® TriChrome process eliminate per year globally?
(d) Calculate the energy **savings** per year for the drop-in TriChrome process.

Problem 12.9

(a) Draw the reaction that makes prior Cr(III) processes give only thin layers (~2.5 μm).

(b) How did Faraday prevent this self-limiting reaction? Sketch the process and label each step.
(c) What parameters are continuously monitored?
(d) What thickness was Faraday able to plate?
(e) What salts of Cr(III) were used? Which is cheaper?

Problem 12.10

(a) List the numbers and names of the **three** most significant Principles of Green Chemistry satisfied by the FARADAIC® TriChrome process to plate chromium. Explain briefly.
(b) Why are microcracks in chromium plating beneficial?

References

1. P. T. Anastas and J. C. Warner, *Green Chemistry: Theory and Practice*, Oxford University Press, New York, 1998.
2. U.S. Congress, *The Clean Air Act (Air pollution prevention and control)*, 1970, https://www.govinfo.gov/content/pkg/USCODE-2008-title42/pdf/USCODE-2008-title42-chap85.pdf, accessed July 23, 2019.
3. J. C. Belke, in *Loss Prevention and Safety Promotion in the Process Industries*, ed. H. J. Pasman, O. Fredholm and A. Jacobsson, Elsevier Science B.V., Amsterdam, 2001, pp. 1275–1314.
4. Center for Industry Education Collaboration (CIEC), *Basic Chemicals*, The University of York, 2019, http://www.essentialchemicalindustry.org/chemicals/, accessed July 11, 2019.
5. V. Smil, *Enriching the Earth: Fritz Haber, Carl Bosch, and the Transformation of World Food Production*, MIT Press, Cambridge, MA, 2004.
6. *Flammable Liquids*, https://www.osha.gov/dte/library/TrngandMatlsLib_FlammableLiquids.pdf, accessed July 26, 2019.
7. J. Johnson, *Chem. Eng. News Archive*, 2008, **86**, 5.
8. *Combustible Dust: An Explosion Hazard*, 2009, https://www.osha.gov/dsg/combustibledust/standards.html, accessed July 26, 2019.
9. T. Zeng, M. J. Plewa and W. A. Mitch, *Water Res.*, 2016, **101**, 176–186.
10. D. Furman, R. Kosloff, F. Dubnikova, S. V. Zybin, W. A. Goddard, N. Rom, B. Hirshberg and Y. Zeiri, *J. Am. Chem. Soc.*, 2014, **136**, 4192–4200.
11. *TNT*, Wikipedia, https://en.wikipedia.org/wiki/TNT#Explosive_character, accessed July 26, 2019.
12. United Nations Economic Commission for Europe, *Globally Harmonized System of Classification and Labelling of Chemicals (GHS)*, Part 3 Health Hazards, Geneva, Switzerland, 2005, 216, http://www.unece.org/trans/danger/publi/ghs/ghs_rev01/01files_e.html, accessed July 16, 2019.
13. *Globally Harmonized System of Classification and Labelling of Chemicals*, Wikipedia, https://en.wikipedia.org/wiki/Globally_Harmonized_System_of_Classification_and_Labelling_of_Chemicals, accessed April 19, 2017.
14. R. A. Sheldon and P. C. Pereira, *Chem. Soc. Rev.*, 2017, **46**, 2678–2691.
15. R. A. Sheldon, *Green Chem.*, 2007, **9**, 1273–1283.
16. R. A. Sheldon, *Chem. Commun.*, 2008, 3352–3365.
17. *Chem. Eng. News Archives*, 1985, **63**, 4–5.

18. *Chem. Eng. News Archives*, 1984, **62**, 6–7.
19. The National Institute for Occupational Safety and Health (NIOSH), *Methyl isocyanate*, Centers for Disease Control and Prevention, 2014, https://www.cdc.gov/niosh/idlh/624839.html, accessed April 19, 2017.
20. Sigma-Aldrich, *Methylamine Safety Data Sheet*, 2015, https://www.sigmaaldrich.com/MSDS/MSDS/DisplayMSDSPage.do?country=US&language=en&productNumber=8.22091&brand=MM&PageToGoToURL=https%3A%2F%2Fwww.sigmaaldrich.com%2Fcatalog%2Fproduct%2Fmm%2F822091%3Flang%3Den, accessed July 26, 2019.
21. Sigma-Aldrich, *1-Naphthol Safety Data Sheet*, 2019, https://www.sigmaaldrich.com/MSDS/MSDS/DisplayMSDSPage.do?country=US&language=en&productNumber=N1000&brand=SIAL&PageToGoToURL=https%3A%2F%2Fwww.sigmaaldrich.com%2Fcatalog%2Fproduct%2Fsial%2Fn1000%3Flang%3Den, accessed July 2, 2019.
22. P. Tundo and M. Selva, *Acc. Chem. Res.*, 2002, **35**, 706–716.
23. K. Almusaiteer, *Catal. Commun.*, 2009, **10**, 1127–1131.
24. Exclusive Chemistry, *New approach to synthesis of insecticide Sevin*, 2004–2016, http://www.exchemistry.com/sevin.html, accessed 7/26/19.
25. S. Kim, J. Chen, T. Cheng, A. Gindulyte, J. He, S. He, Q. Li, B. A. Shoemaker, P. A. Thiessen, B. Yu, L. Zaslavsky, J. Zhang and E. E. Bolton, *Nucleic Acids Res.*, 2019, **47**, D1102–D1109.
26. Nanotech Industries Inc. (NTI), *Green Polyurethane™ Coatings & Foam*, Washington, DC, 2015, https://www.epa.gov/greenchemistry/presidential-green-chemistry-challenge-2015-designing-greener-chemicals-award, accessed July 26, 2019.
27. D. Champlin, J. T. Chapin, L. Chen, S. Chi, H. Chiang, D. Grzic, E. Hjelm, K. Keighron, J. Shih, M. Tabaddor, C. Wang, E. Wang, A. Wu, D. Wu, M. Wu and J. Yen, *Multi-Level Forensic and Functional Analysis of the 787 Main/APU Lithium Ion Battery*, Northbrook, IL, 2014, http://www.ntsb.gov/investigations/AccidentReports/Documents/UL_Forensic_Report.pdf, accessed July 26, 2019.
28. *Lithium Ion Technical Handbook*, Gold Peak Industries Ltd., 2003.
29. D. H. C. Wong, J. L. Thelen, Y. Fu, D. Devaux, A. A. Pandya, V. S. Battaglia, N. P. Balsara and J. M. DeSimone, *Proc. Nat. Acad. Sci. U. S. A.*, 2014, **111**, 3327–3331.
30. J. P. Rolland, R. M. Van Dam, D. A. Schorzman, S. R. Quake and J. M. DeSimone, *J. Am. Chem. Soc.*, 2004, **126**, 2322–2323.
31. *Environmental Directorate, Co-operation on existing chemicals hazard assessment of perfluorooctane sulfonate (PFOS) and its salts*, 2002, http://www.oecd.org/chemicalsafety/risk-management/perfluorooctanesulfonatepfosandrelatedchemicalproducts.htm, accessed July 25, 2019.
32. C. Hogue, *Chem. Eng. News*, 2012, **90**, 24–25.
33. Faraday Technology Inc., *Functional chrome coatings electrodeposited from a trivalent chromium plating electrolyte (FARADAYIC®, 1 TriChrome Plating)*, Washington, DC, 2013, https://www.epa.gov/greenchemistry/presidential-green-chemistry-challenge-2013-small-business-award, accessed July 26, 2019.
34. E. J. Tokar, W. A. Boyd, J. H. Freedman and M. P. Waalkes, in *Casarett and Doull's Toxicology: The Basic Science of Poisons*, ed. C. D. Klaassen, McGraw Hill Education, New York, 8th edn, 2013, pp. 981–1030.

Appendix A Organic Functional Groups

Names, Lewis structures, and occupied (bonding) molecular orbitals.

A.1 Hybridization: sp^3

(a) Geometry: tetrahedral

A.1.1 Alkanes: R = only C and H

Linear, branched, or cyclic. Bond angles: 109.5°, Bond length: 1.54 Å

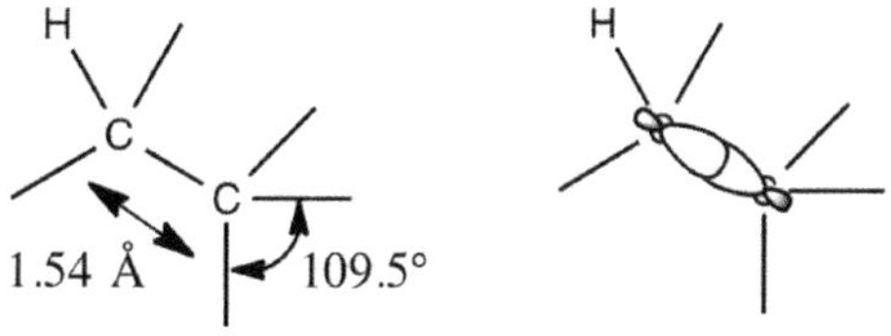

A.1.2 Haloalkanes: R–X; X = F, Cl, Br, or I

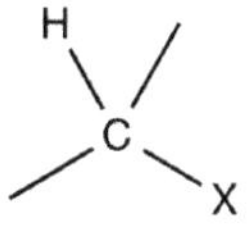

(b) Geometry: bent

Green Chemistry: Principles and Case Studies
By Felicia A. Etzkorn

Published by the Royal Society of Chemistry, www.rsc.org

A.1.3 Alcohols: R-OH; Thiols: R-SH

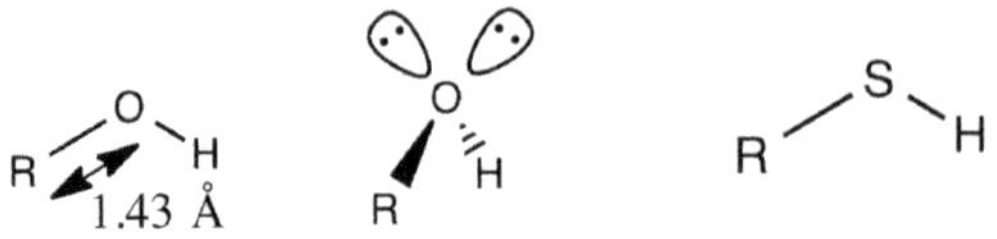

A.1.4 Ethers: R-O-R′; Thioethers: R-S-R′; R and R′ same or different.

(c) Geometry: trigonal pyramid

A.1.5 Amines: C-N; R^1, R^2, R^3 = H or alkyl; same or different

A.2 Hybridization: sp^2

(a) Geometry: trigonal planar

A.2.1 Alkenes: C=C

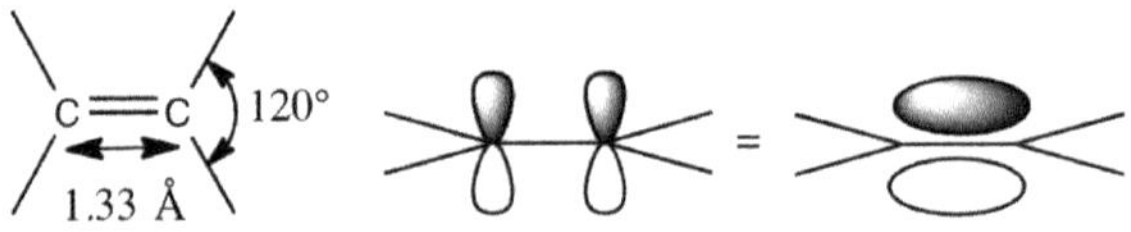

A.2.2 Aldehydes: RCHO and Ketones: (RCOR′)

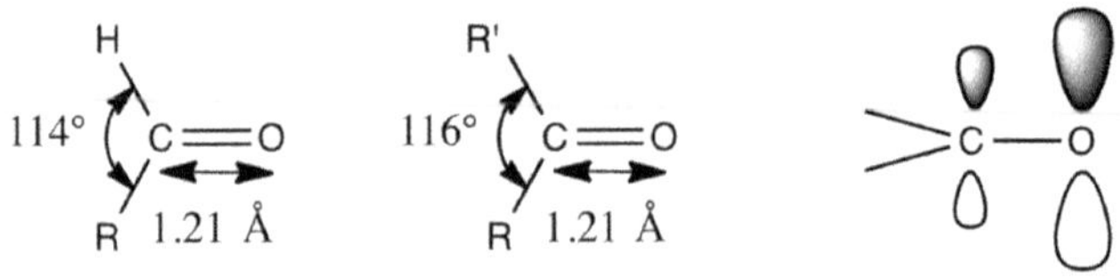

A.2.3 Esters: R^1-(C=O)-OR^2

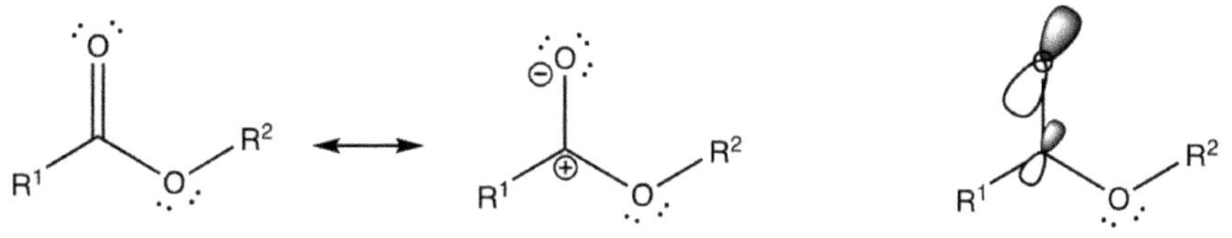

A.2.4 Amides: R^1-(C=O)-NH-R^2

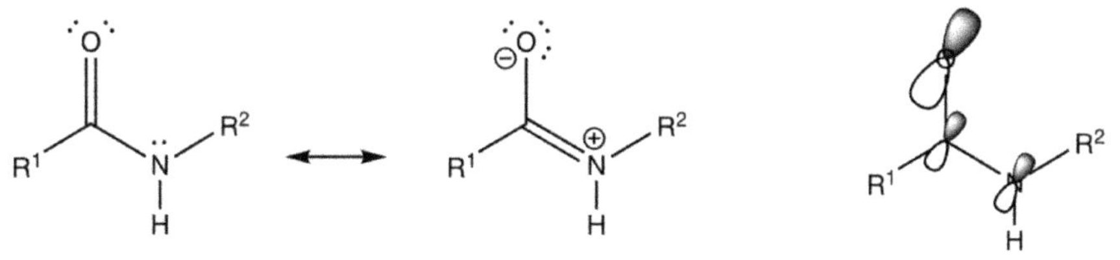

A.2.5 Carbamates: R^1-O-(C=O)-NH-R^2

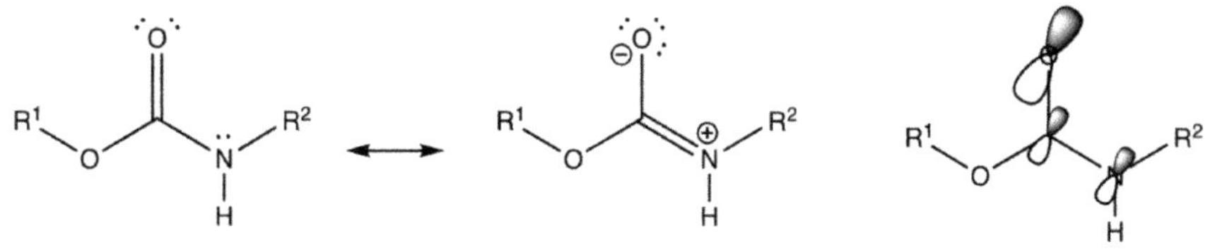

A.2.6 Imines: C=N

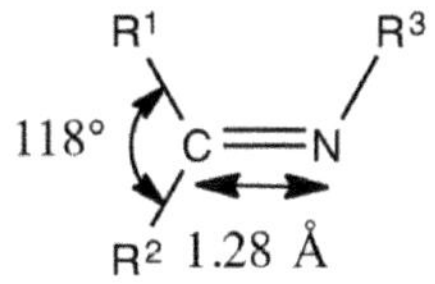

A.3 Hybridization: sp

(a) Geometry: linear

A.3.1 Allenes: C=C=C

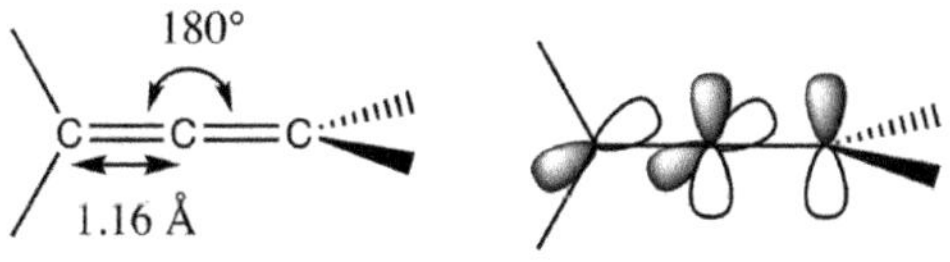

A.3.2 Carbon dioxide: O=C=O

A.3.3 Alkynes: R, R′=H or alkyl; same or different

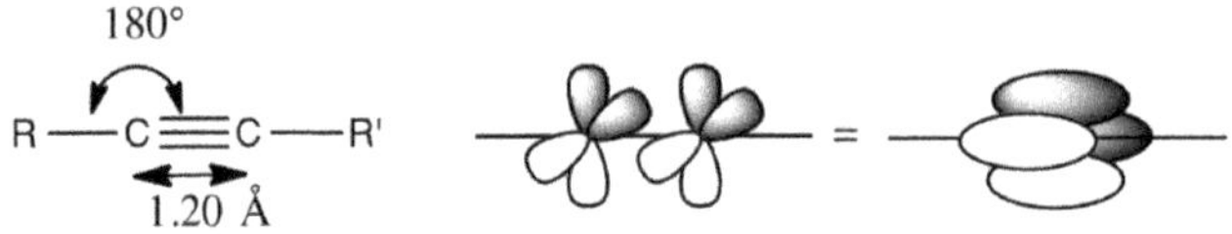

A.3.4 Nitriles: R–CN

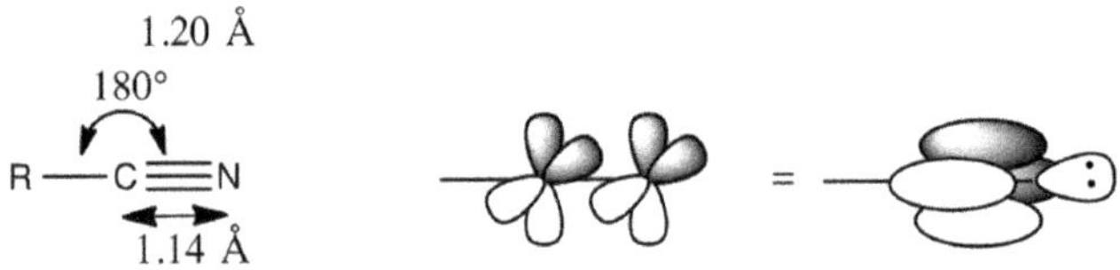

A.4 Aromatic

A.4.1 Benzene: C_6H_6

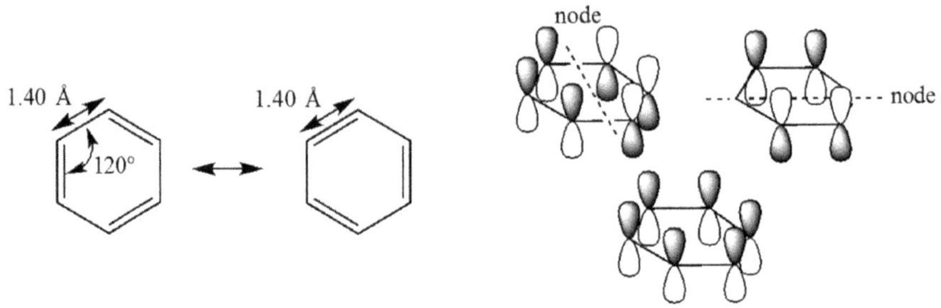

A.4.2 Pyridine: C_5H_5N

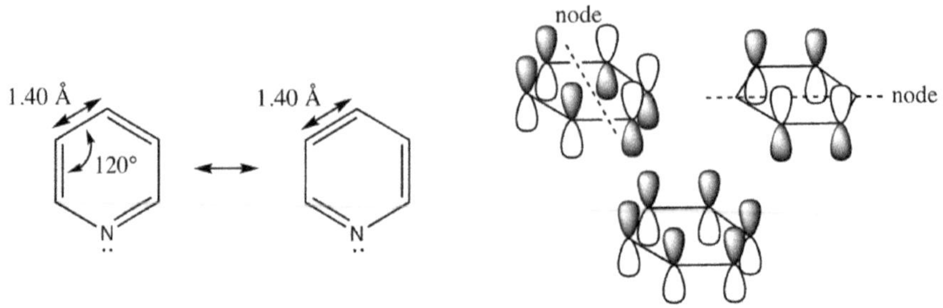

A.4.3 Imidazoles: $C_3N_2R_4$; R = H or alkyl

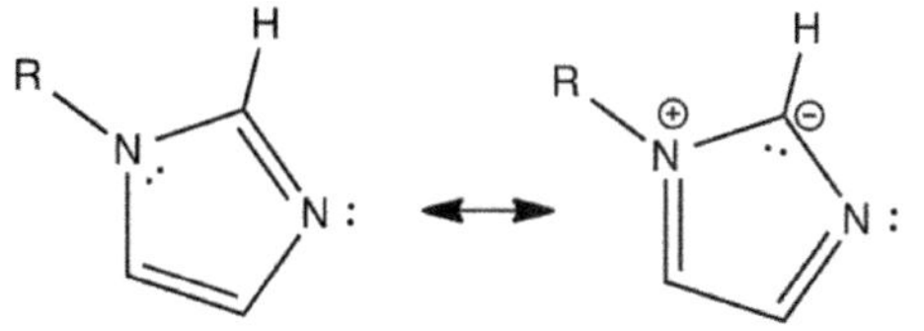

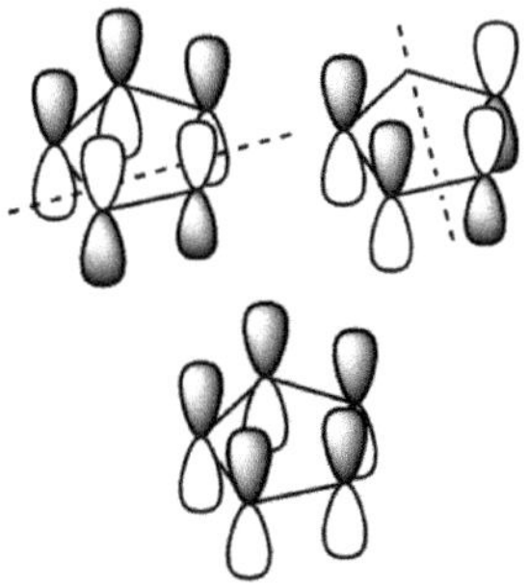

Appendix B Organic Mechanism

B.1 Motivation

In order to design greener synthetic pathways, especially with regard to synthetic efficiency, it is vital to have a working understanding of organic mechanisms. For example, in Chapter 2, HF was used as a substitute Lewis acid for $AlCl_3$ in the Friedel–Crafts reaction to synthesize ibuprofen. Thus, it is good to know what Lewis acids are so we can choose substitutes in a reaction. The mechanisms that we learned in Organic Chemistry were based on published experiments going back into the 1800s. Chemists have a pretty good sense of the molecular steps involved in most types of organic reactions; our chemical intuition can serve us well. However, every mechanism we draw is just a model to explain experimental data. We can only prove a mechanism model wrong; we can never definitively prove that a model is correct. We may say that a mechanism is best supported by all the data, and that all of the other logical mechanisms have been ruled out.

To review organic mechanisms, we will start with some general principles, then develop a strategy to approach any mechanism, and finally work through some examples relevant to green chemistry.

Green Chemistry: Principles and Case Studies
By Felicia A. Etzkorn

Published by the Royal Society of Chemistry, www.rsc.org

B.2 General Principles

B.2.1 Thermodynamics and Kinetics

Spontaneous reactions must go to a lower energy species. This is simple thermodynamics; ΔG must be negative. Not all reactions proceed to the absolute lowest energy product possible because kinetic control is also possible. By lowering the temperature, it is possible to trap products that are higher in energy than other products IF there is a lower energy barrier to get there, a kinetic barrier (Figure B.1). The lower barrier allows kinetically controlled access to the less stable product, P_1, while higher temperatures or stronger reagents allow molecules to overcome the higher barrier to access the more stable product, P_2. More detail about kinetic vs. thermodynamic control can be found in Chapter 6, Energy Efficiency.

For example, deprotonation of 2-methylcyclohexane can take place at either carbon that is α to the carbonyl (Figure B.2). The more

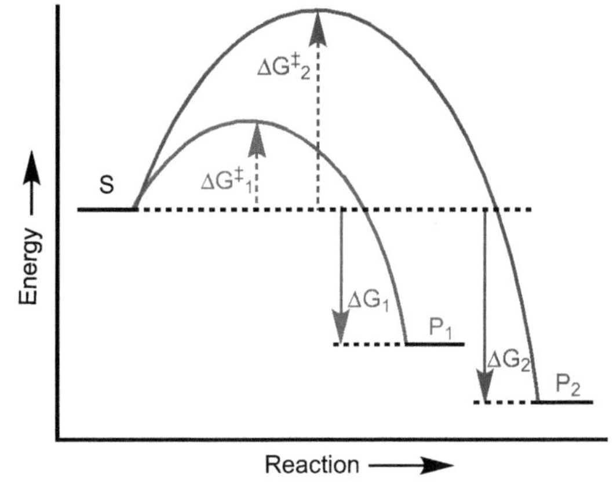

Figure B.1 Reaction coordinate diagram for thermodynamic *vs.* kinetic control of the product mixture.

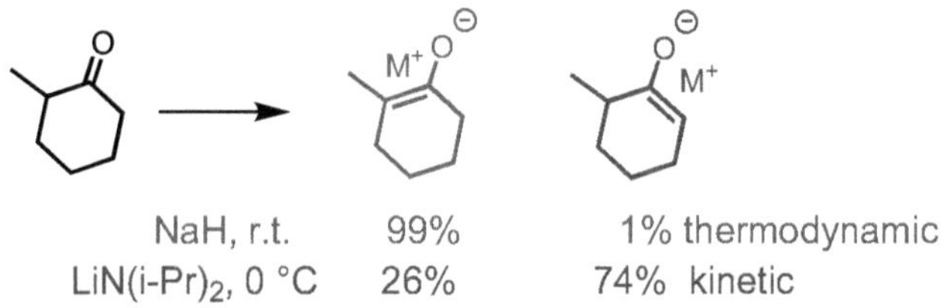

Figure B.2 Thermodynamic *vs.* kinetic deprotonation of 2-methylcyclohexanone.

substituted enolate is more stable, so the small base, NaH, will deprotonate almost exclusively at the more substituted carbon. This is thermodynamic control because the less stable enolate can deprotonate itself until the thermodynamic equilibrium mixture (99 : 1) is formed. The sterically bulky base, lithium diisopropylamide (LDA), at lower temperature gives predominantly the *less* stable enolate because sterics prevent access to the higher barrier pathway to give the more stable enolate. The reaction must be kept cool and stopped as soon as the product is formed to determine the kinetic product mixture.

B.2.2 Electrophiles React with Nucleophiles

Most polar organic reactions can be boiled down to this one general principle: electrophiles (E^+) react with nucleophiles (Nu:). Radical reactions are an exception and will be discussed separately. Acid-base reactions are simply a specific version of this concept, where the acid H^+ is an E^+ and the base B: is a Nu:. Figuring out which is which in a complex reaction is the challenge.

Starting with simple basic principles will usually allow you to determine the E^+ and the Nu: in a particular reaction. For now, ignore the vast transition metal area because we are focusing on organic mechanisms. We will consider organometallic mechanisms in Chapter 9 on catalysis.

B.2.3 Nucleophiles

Nucleophiles are found toward the right of the periodic table in the non-metals region because they are more electronegative (Figure B.3). An exception is the F^- ion, because it holds its electrons too tightly, so F^- is not nucleophilic. The nucleophilicity trend decreases from left to right of the non-metals region: $R_3C^- > R_2N^- > RO^- \gg F^-$ (R = alkyl or H). On the other hand, as you proceed down a particular column, nucleophilicity increases: $I^- > Br^- > Cl^- > F^-$ (Figure B.3). This makes sense because the inner shell electrons shield the outer shell electrons from the nucleus, so the electrons are not as tightly held. Yet iodine still prefers to have its valence shell filled with 8 electrons. In between, we have a sweet spot where RS^- and R_3P: are good nucleophiles; they are both towards the left of the non-metals, and down from the top a bit. The trend is similar for a few neutral nucleophiles: $R_2NH > ROH \gg RF$, and $R_3P > R_2S \gg RCl$. Notice that the neutral R_4C, whether R is alkyl or hydrogen, cannot be a Nu: because it has no lone pair.

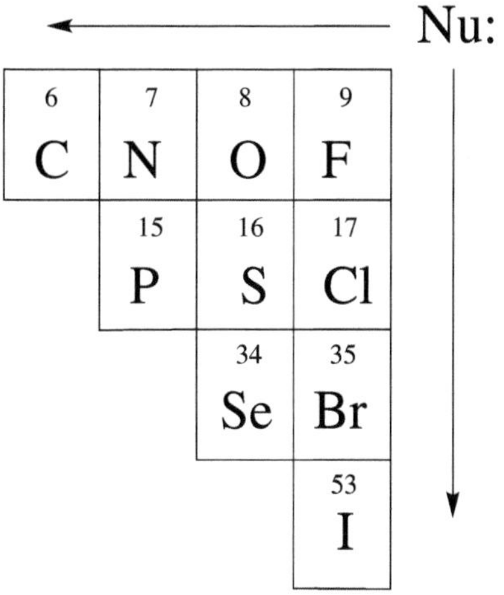

Figure B.3 Periodic trends of nucleophilicity.

There is also a solvent dependence for nucleophiles that is fairly complex. The charge on the reactant (or high-energy intermediate) in the rate-determining step will determine the effect of the solvent. If the Nu: is charged, a more polar solvent will stabilize it and slow the reaction. If the Nu: is neutral, a more polar solvent will stabilize the partially charged transition state more, and speed up the reaction. Refer to Chapter 5 on solvents for the properties of different solvents.

B.2.4 Electrophiles

Electrophiles are typically electron-poor carbons attached to electronegative elements, like oxygen, nitrogen, or chlorine: C=O, C=N, C–Cl. The exceptions are carbons with protonated heteroatoms (OH, NH), for example alcohols, ROH, acids, RCO_2H, and 1° or 2° amines, RNH_2 or R_2NH. In that case, the H^+ may act as an E^+ towards bases (nucleophiles). Similar solvent issues as described above for nucleophiles affect the E^+ of reactions.

The most common E^+ in Green Chemistry is the carbonyl carbon (C=O) because an important resonance structure for nearly all carbonyl compounds has a positive charge on the carbon, and a negative charge on the oxygen. The most common carbonyl compounds and their resonance structures are shown in Figure B.4. Aldehydes and ketones are good electrophiles because there are no additional resonance structures to stabilize the positive charge on C. Amides are the least electrophilic of those shown because nitrogen is less electronegative and tolerates the positive charge better than oxygen.

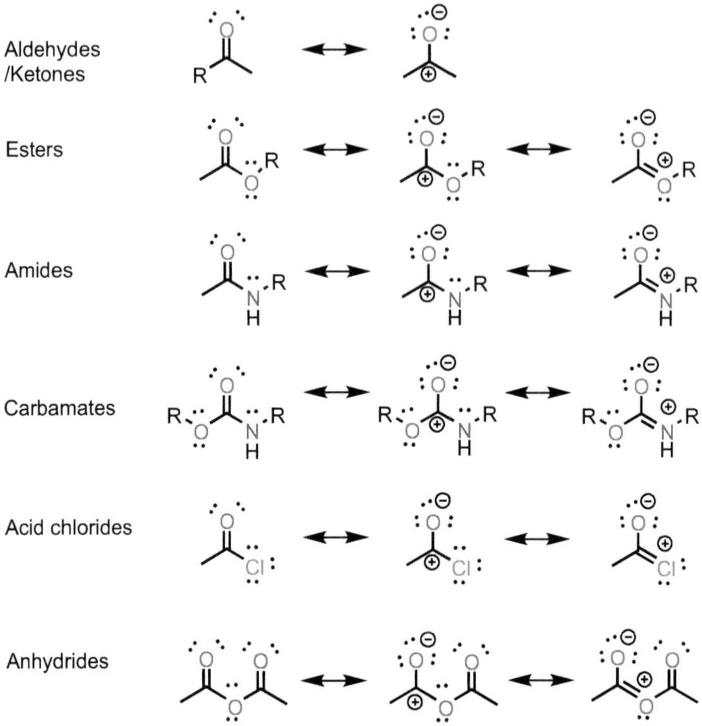

Figure B.4 Resonance structures of carbonyl compounds. Electrophilicity is based on both resonance and inductive effects. Structures for which the second resonance structure (C^+–O^-) is an important contributor are the most electrophilic.

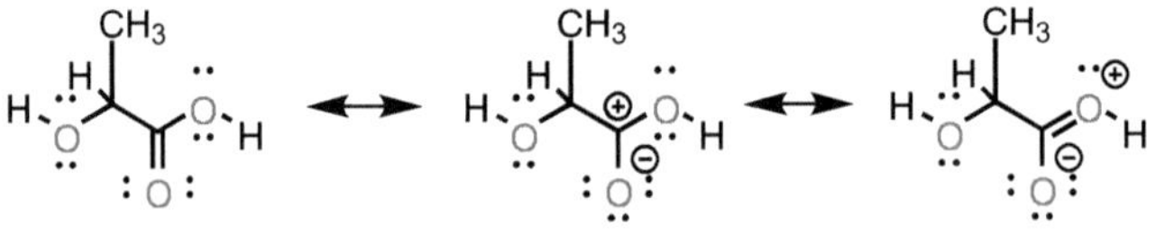

Figure B.5 Resonance structures of lactic acid.

The third resonance structure for amides contributes more to the electronic structure than the corresponding third ester resonance structure (Figure B.4). The third resonance structures for acid chlorides and anhydrides are very unimportant. Acid chlorides and anhydrides are the best electrophiles because the Cl or the additional O_2CR groups are very electron withdrawing by induction.

Carboxylic acids are not very electrophilic because deprotonation is more likely with even a weakly basic nucleophile. It is difficult for an acid, such as lactic acid, to act as an E^+ (Figure B.5).

Other electrophiles arise from having electron-withdrawing groups attached to them, so each structure must be examined carefully to determine whether it is electrophilic.

B.2.5 Catalysts

Catalysts are almost guaranteed to be involved in the first mechanistic step of a reaction—that is what they're there for, to activate a reluctant Nu: or E^+. Catalysts are frequently Brønsted acids (HCl, H_2SO_4) or bases (NaOH, KOH) that are regenerated by protonation or deprotonation at the end of each catalytic cycle.

All acids are Lewis acids, because G.N. Lewis wanted to develop the broadest possible definition of acids and bases. The non-protic acids are usually the ones called Lewis acids. Lewis acids are often used to activate carbonyl compounds by binding to the oxygen, withdrawing electron density from the carbon. For example, the Lewis acid $SnCl_2$ can be used to polymerize lactide, the cyclic dimer of lactic acid, to make polylactic acid (PLA), a green polymer (Figure B.6). Other commonly used Lewis acids in organic reactions are BF_3, $AlCl_3$, $FeCl_3$, and BCl_3. Earth abundant, base metal ions such as Ca^{2+}, Mg^{2+}, Mn^{2+}, Ni^{2+}, Co^{2+}, Zn^{2+}, Fe^{2+}, and Cu^{2+} also can serve as Lewis acid catalysts in organic reactions. A non-toxic substitute is sure to be available for any reaction requiring a Lewis acid catalyst.

All bases are Lewis bases, but Lewis base catalysts can be considered non-nucleophilic bases, such as those shown in Figure B.7. The defining feature of a non-nucleophilic base is steric hindrance around the basic site. Sometimes Lewis bases are referred to as "proton sponges." Transition metal catalysis is treated separately in Chapter 9 on catalysis.

B.3 Molecular Orbital Diagrams

To create a molecular orbital (MO) model of aromaticity, it is helpful to begin with the π-bond of ethene, and work our way up through

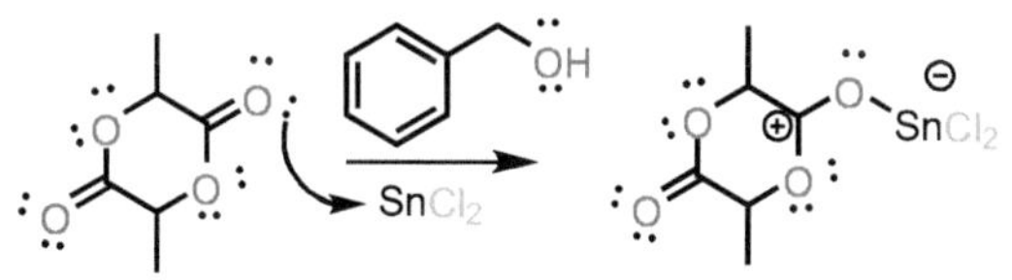

Figure B.6 Electrophilic activation of lactide with a Lewis acid, $SnCl_2$.

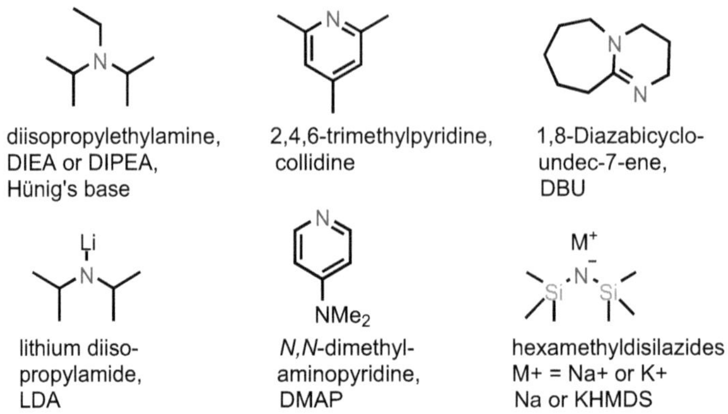

Figure B.7 Common non-nucleophilic Lewis bases.

hexatriene to benzene. By convention, the lowest energy bonding MOs are drawn near the bottom of the diagram, with increasingly higher energy anti-bonding MOs drawn towards the top. For conjugation and aromaticity, we consider only the π-bonds of the system, leaving the σ-bonds drawn as lines. Nodes are 2-dimensional mathematical planes where the electron density is zero, and the MO changes symmetry as it crosses the node. All π-bonds have a horizontal node, giving p-type and π-MOs the characteristic dumbbell shape. We indicate the change in symmetry with shaded and unshaded, or different color lobes, of the MO.

B.3.1 π-Bonds: Ethene

First, the number of MOs is determined based on the number of atoms involved in the π-bonded system. For ethene, there are 2 carbons, therefore 2 atomic orbitals (AOs), which are combined to give 2 MOs (Figure B.8). The number of MOs determines the maximum number of vertical π-nodes for the entire MO diagram, beginning at the bottom with zero π-nodes. Nodes are places where the 3D mathematical description of the orbital passes through zero. The lowest energy MO always has 0 π-nodes. For ethene then, the 2 MOs will have 0 and 1 nodes, respectively. Each of the MOs look quite different, with nodes of the same symmetry combining to give each MO a different shape. To easily draw the MOs and develop an accounting system to track the number of MOs, we simplify the process by drawing p-type AOs to represent the symmetry of the MOs. Across a π-node, the MO changes symmetry, represented by

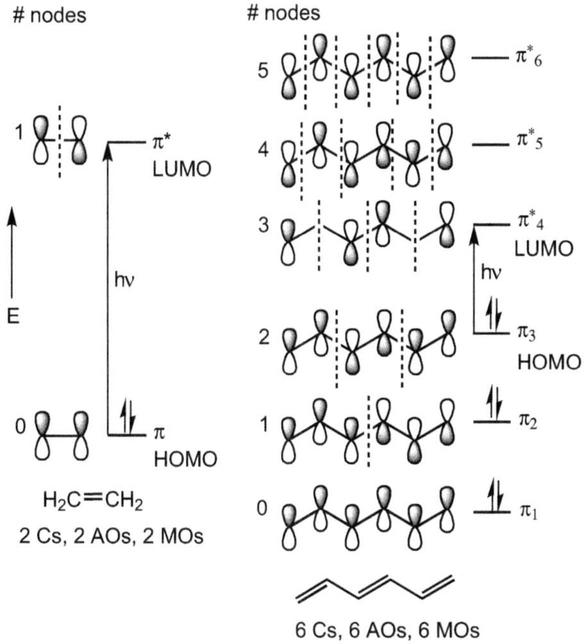

Figure B.8 Molecular orbital (MO) and energy (E) level diagrams of ethene and 1,3,5-hexatriene. Nodes (---), AO = atomic orbital, HOMO = highest occupied MO, LUMO = lowest unoccupied MO.

shading the opposite lobes. The lowest energy, most stable π-MO of ethene is thus represented with the same shading on both halves of the MO. The highest energy π*-MO has 1 node, so the symmetry and shading switches across the node. This indicates an anti-bonding relationship between the two nodes. Finally, the electrons are added in pairs from the bottom, lowest energy MO up. For ethene, the 2 electrons go into the π-bond, which is called the highest occupied MO (HOMO). The π*-MO has no electrons, and it is called the lowest unoccupied MO (LUMO).

Ethene then has 2 MOs, a π-bonding and a π*-anti-bonding MO that are far apart in energy. The energy difference, called the HOMO-LUMO gap, correlates with the absorption spectroscopy wavelength λ, that is inversely proportional to the frequency, $h\nu$ (Figure B.8).

B.3.2 Conjugation: Hexatriene

As π-bonds are added to the system, the bonding MOs get lower in energy, and the π*-MOs get higher in energy. The MO diagram for 1,3,5-hexatriene has 4 more MOs than that of ethene—6 carbons,

6 AOs, and 6 MOs (Figure B.8). With the addition of each π-bond, one MO goes down in energy, and one goes up for both bonding and anti-bonding orbitals. Starting at the lowest energy, hexatriene MOs have 0, 1, 2, 3, 4, and 5 nodes (Figure B.8). The nodes are spaced as evenly as possible across the molecule. When a node passes through a carbon nucleus, there is zero electron density at the nucleus, and no lobe is drawn there. Recall that the symmetry changes across each node, so the shading is opposite. There is a non-bonding relationship between lobes when there is a node through an atom. For the 6 MOs of the fully conjugated hexatriene, the 6 electrons are added to the 3 lowest energy bonding MOs in pairs, with the highest energy designated the HOMO. The remaining 3 anti-bonding MOs have no electrons, and the lowest energy π*-MO is designated the LUMO. The energy difference between the HOMO and LUMO, $h\nu$, is compressed relative to ethene, so the absorption wavelength is longer (Figure B.8). This is referred to as quantum compression (Chapter 3).

B.3.3 Aromaticity

Now we progress to the special stability of benzene and other aromatic compounds. The criteria for aromaticity have been developed based on a long history of experimental evidence for this extra stability. Aromatic compounds must be (1) cyclic, (2) fully conjugated, (3) planar, and (4) have $4n+2$ π electrons, where n is an integer. Benzene fulfills all four criteria with 6 π electrons ($n=1$). To build the MO diagram of benzene, we again recognize that 6 carbons gives 6 AOs, therefore 6 MOs. The lowest energy MO still has 6 lobes with the same symmetry (Figure B.9).

Here is where we depart from the MO of hexatriene. Mathematically, the nodes go all the way through the center of the ring and out the other side. There are two different ways of drawing 1 node through a ring, passing through either two bonds or two carbons (Figure B.9). These two MOs have equivalent energies, so they are called "degenerate" MOs (Figure B.9). The 1-node π-MO on the left has 4 bonding interactions between carbons (overlap with the same symmetry), and 2 anti-bonding interactions between carbons (opposite symmetry across the node), for a net of 2 bonding interactions ($4-2=2$) (Figure B.9). The 1-node π-MO on the right has 2 bonding interactions, and 2 non-bonding interactions where the node passes through 2 carbons, also for a net of 2 bonding interactions ($2-0=2$) (Figure B.9). Since these two MOs have the same degree of bonding interactions, they are degenerate, and therefore they have the same

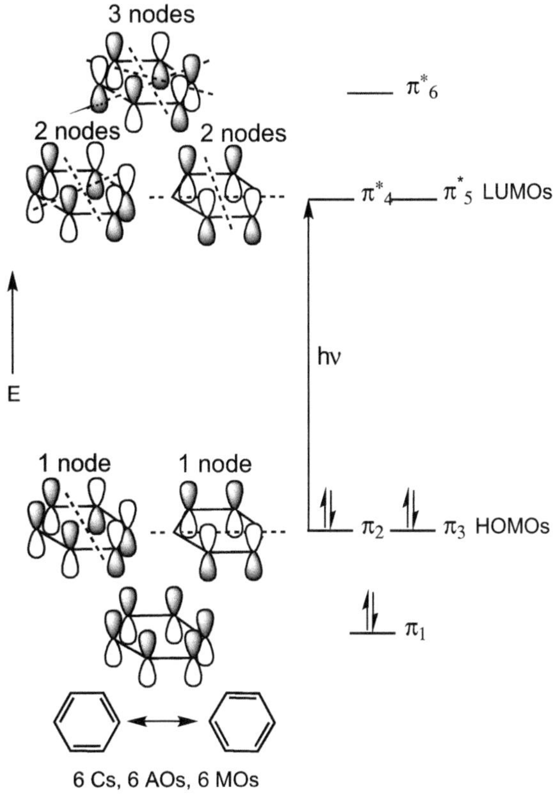

Figure B.9 MOs of the simplest aromatic compound, benzene. Abbreviations are the same as in Figure B.8.

energy. Proceeding up in energy, the 2-node π*-MO on the left has 2 bonding and 4 anti-bonding, for a net of 2 anti-bonding interactions. The 2-node π*-MO on the right has 2 anti-bonding and 2 non-bonding, again for a net of 2 anti-bonding interactions. Since these also have equivalent bonding interactions, they are also degenerate. There is only one way to draw 3 nodes through benzene, resulting in one π*-MO very high in energy with 6 anti-bonding interactions (Figure B.9). Notice that the degeneracy leads to a stabilization of the HOMO, and destabilization of the LUMO relative to hexatriene. This results in an overall more stable conjugated system with a larger HOMO-LUMO gap, and a longer wavelength absorption for benzene. In addition, the additional stability due to the degenerate MOs makes benzene highly unreactive. Before we knew that benzene was a carcinogen, it was highly favored as a solvent by organic chemists.

B.4 Hyperconjugation

Hyperconjugation describes σ-bonds that participate in π-bonding *via* conjugation. This effect can be modeled either with Lewis structures or with MOs.

B.4.1 *t*-Butyl Carbocation

As an example, take the case of a *t*-butyl cation stability. The 3° carbocation that is formed upon removal of a Boc protecting group, as in the prexisertib case study (Chapter 11), is unusually stable. We can draw Lewis structures of the *t*-butyl cation where the positive charge is delocalized onto hydrogens that are two bonds removed from the positively charged carbon (Figure B.10). Since there are 9 hydrogens attached to the 3 carbons, 9 equivalent resonance structures can be drawn.

Here again, the Lewis structures do not describe the geometry of this hyperconjugation very well. With free rotation around each of the C–C bonds, only one C–H bond at a time is properly oriented to donate e⁻ density to make the π-bond indicated by the resonance structure. The molecular orbital diagram is more informative (Figure B.11).

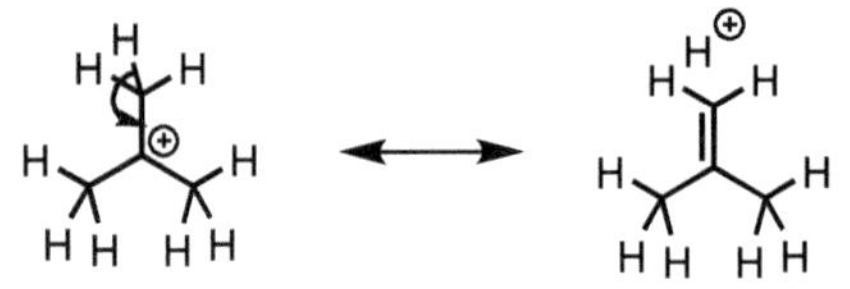

Figure B.10 The Lewis structure of *t*-butyl cation and one of the 9 hyperconjugative resonance structures.

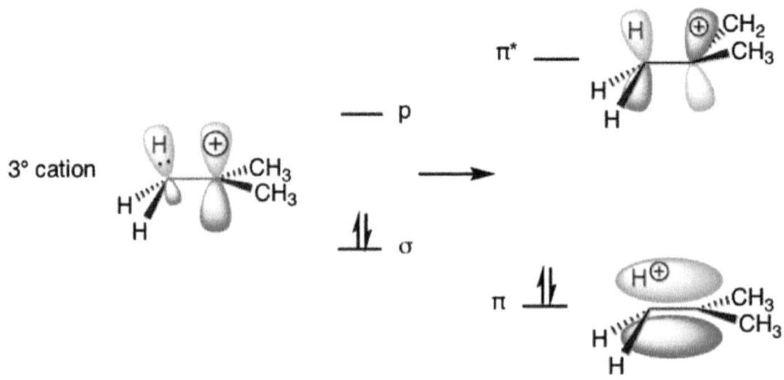

Figure B.11 MO diagrams of *t*-butyl cation from a C–H σ-bond and the empty p-orbital of the cation gives π-bonding and π*-antibonding orbitals.

The C–H σ-bond orbital is occupied with 2 e^-. When a C–H bond aligns in parallel with the unoccupied p-orbital of the carbocation, the σ-bond can donate e^- density into the p-orbital to form a quasi-π-bond. Since none of the atoms are in a different position, the π-bond is bent, and therefore less stable than a full π-bond (Figure B.11).

With enough R groups attached to the carbocation, the stabilization of the positive charge by hyperconjugative delocalization is considerable, leading to the familiar order of carbocation stability: 3° > 2° > 1° > $^+CH_3$.

B.4.2 *t*-Butyl Radical

The order of radical stability is the same, and for the same reason—hyperconjugative delocalization of the radical e^- (Figure B.12). In this case, lowering the energy of the two-electron orbital, even while destabilizing the one electron orbital, is overall stabilizing.

B.4.3 *t*-Butyl Carbanion

On the other hand, carbanions show the opposite order of stability: $^-CH_3$ > 1° > 2° > 3° because of *repulsion* between the two occupied orbitals, the C–H σ-bond and the sp^3-hybridized lone pair on the adjacent carbon. Overlap between the two occupied orbitals would create one bonding π-MO, and one anti-bonding π*-MO, so the net energy of the system would be higher than without overlap (Figure B.13).

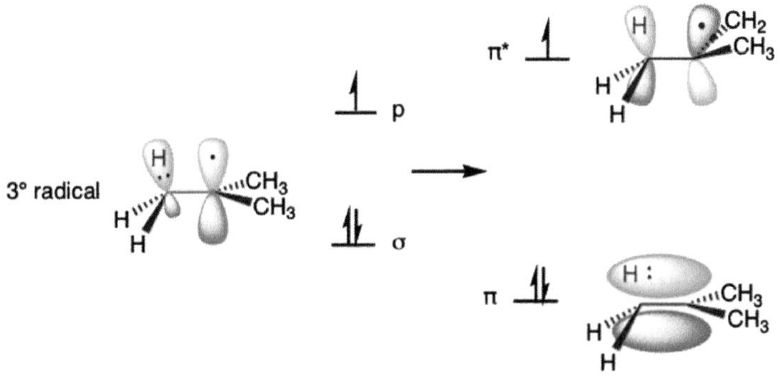

Figure B.12 MO diagrams of *t*-butyl radical from a C–H σ-bond and the singly occupied p-orbital of the radical gives π-bonding and π*-anti-bonding orbitals.

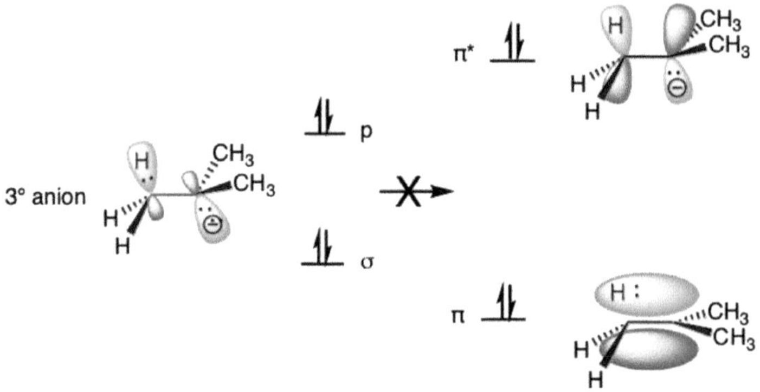

Figure B.13 MO diagrams of *t*-butyl anion from a C–H σ-bond and the empty p-orbital of the anion would give π-bonding and filled π*-anti-bonding orbitals, explaining the decreased stability of substituted carbanions.

Table B.1 Types of arrows used in organic chemistry.

Arrow	Purpose
	Reaction direction
	Proposed or hypothetical reaction
	Reversible and equilibrium reactions
	Resonance Lewis structures
	Full-headed for dipolar mechanisms
	Half-headed for radical mechanisms
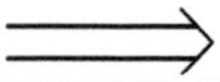	Retrosynthesis

B.5 Arrows

Different types of arrows serve different purposes in organic chemistry (Table B.1). The one-way arrow is the standard reaction arrow, often whether the reaction is reversible or not. A dashed arrow indicates that the reaction has not actually been done; it is proposed or hypothetical. Reversibility of a reaction may be indicated by forward and reverse half arrows. Double-headed arrows are used only between resonance structures. Mechanism arrows come in two flavors: full-headed curved arrows indicate the movement of two electrons toward the more electron-deficient location, and half-headed curved arrows, or fishhooks, indicate the movement of one electron as in radical reactions. Retrosynthesis arrows are used in systematic analysis to design multi-step synthesis of complex molecules, beginning with the product and working backwards towards simple, commercially available starting materials. See Chapter 2 under *Convergent Synthesis* for an example.

B.6 A Method for Mechanism

Step 1. Write the overall balanced reaction.
Step 2. Simplify the reactants and products to focus on the reacting centers.
Step 3. Determine where all the atoms in the product come from in the reactants.
Step 4. Draw the Lewis structures of reactants with all lone pairs and draw resonance structures as needed.
Step 5. Identify the catalyst for the first mechanistic step, if any.
Step 6. Identify the nucleophile (Nu:), the electrophile (E^+), and the leaving group (LG), if any.
Step 7. Draw a curved arrow that starts at the Nu: electron pair and ends at the E^+.
Step 8. Draw the intermediates with new bonds and lone pairs.

Repeat steps 6, 7 and 8 until the product is obtained.

Step 9. Regenerate the original catalyst.
Step 10. Drive the reaction to completion.

B.7 Case Study: Carbonyl Substitution by Addition-elimination

A stepwise process for proposing a mechanism will be illustrated with the synthesis of biodiesel from oils or fats.

Step 1. Write the overall balanced reaction. (Figure B.14) It is simple to see that the reaction is not balanced, but for the mechanism, methanolysis of only one ester need be considered; the other two are identical.

Step 2. Simplify the reactants and products to focus on the reacting centers. (Figure B.15) Use R groups to represent alkyl chains, Ar for aromatic groups, but retain heteroatoms like N, O and Cl, and the associated hydrogens.

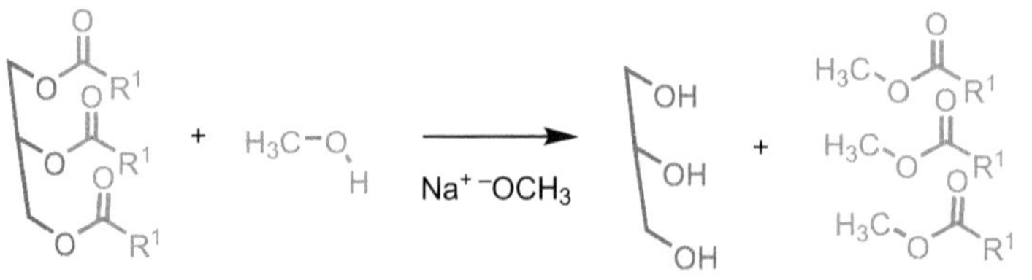

Figure B.14 Methanolysis of fat to give biodiesel. R typically represents C11 to C21 alkyl or alkenyl chains, depending upon whether the source is mammal, fish, or plant. Any alkenes present will be in the more reactive *cis* configuration, and they are spaced every 3-carbons, *i.e.* non-conjugated.

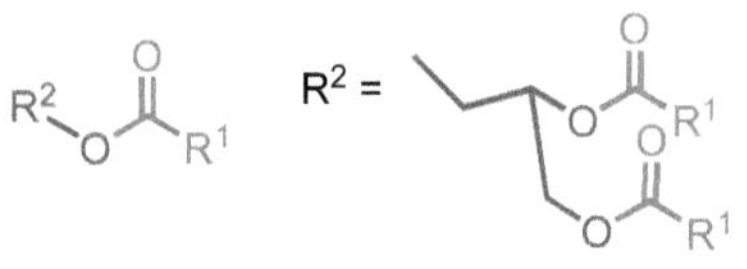

Figure B.15 Simplified structure of biodiesel to use for drawing the mechanism.

Step 3. Determine where all the atoms in the product come from in the reactants. (Figure B.16) For example, the glycerol (1,2,3-propanetriol) comes from the fat, the methyl groups come from the methanol, but where do the methoxy oxygens come from? Since O–H bonds are much easier to break than C–O bonds in alcohols, it looks like the oxygens come from the methanol. This was proven experimentally. (How could you determine the source of the oxygens in the biodiesel?) It can be helpful to color code the atoms when you first start drawing mechanisms. I use red for E^+, blue for Nu:, and green for LGs. Later you can drop this to go faster.

Step 4. Draw the Lewis structures of reactants with all lone pairs and draw resonance structures as needed. (Figure B.17) Assume that all reactants are neutral stable species with octets around all atoms, except in the case of Lewis acids and metal ions like Na^+. Count each bond as 1 e^- belonging to the atom, and each lone pair as 2 e^- belonging to the atom. Consult the periodic table to see how many electrons are needed for each atom to make it neutral. For example, C needs 4 e^-, so neutral C always has 4 bonds and no lone pairs. Neutral N has 3 bonds (3 e^-) and 1 lone pair (2 e^-) for a total of 5 e^-. O needs 6 e^-, so neutral O always has 2 bonds (2 e^-) and 2 lone pairs (4 e^-) for a total of 6 e^-.

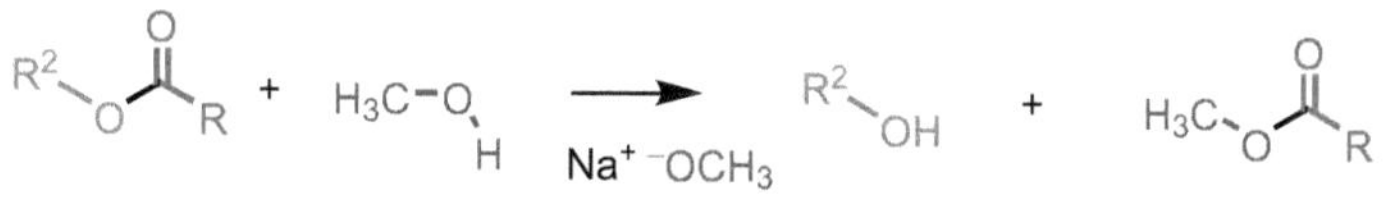

Figure B.16 Identify which atoms in the products come from which reactants. Color code used: E^+ (red), Nu: (blue), and LG (green).

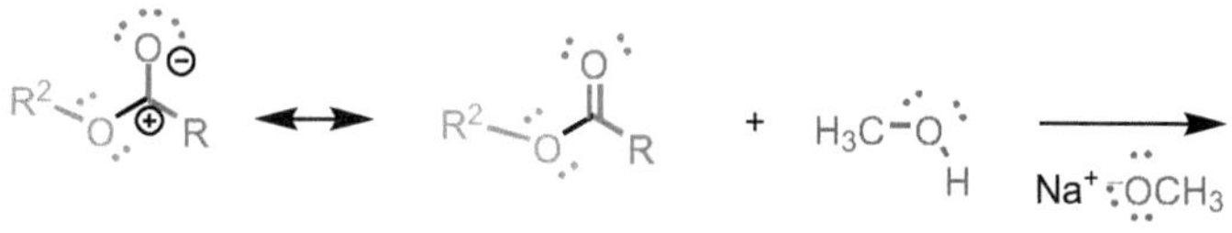

Figure B.17 Draw Lewis structures with the lone pairs, and resonance structures.

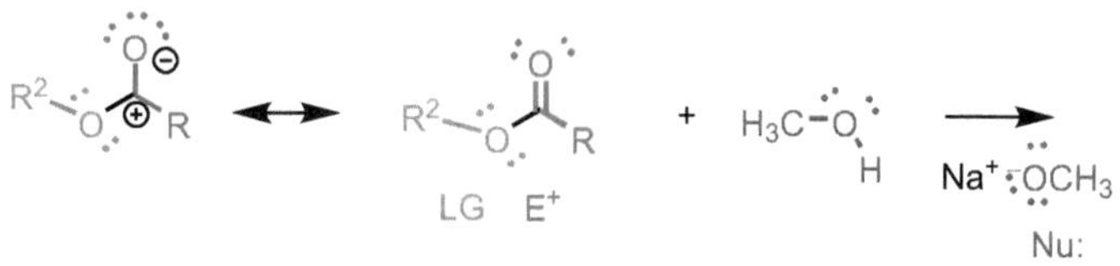

Figure B.18 Identify the Nu: (blue), E^+ (red), and LG (green).

Step 5. Identify the catalyst for the first mechanistic step. Catalysts are written under the arrow, above any solvent. Often catalysts have heavy metals in them. Is there a catalyst? What kind of catalyst? $NaOCH_3$ is a strong nucleophilic catalyst. It will clearly be involved at the beginning of the mechanism.

Step 6. Identify the nucleophile (Nu:), the electrophile (E^+), and the leaving group (LG), if any. (Figure B.18) Determine which is the most electrophilic and which is the most nucleophilic species. The catalyst is shown in the overall reaction as a charged salt, a dead giveaway that methoxide, $^-OCH_3$, is the Nu:. The E^+ is slightly more challenging, but we have already seen that carbonyl compounds, like esters, are good electrophiles because of the resonance structure shown. It is not always necessary to draw every resonance structure unless the point is that there are a lot of them to stabilize an intermediate.

Step 7. Draw a curved arrow that starts at the Nu: electron pair and ends at the E^+. (Figure B.19) If that will create an atom with too many electrons, draw a second arrow to move the π-bond e–s towards the more electronegative atom, oxygen in this case.

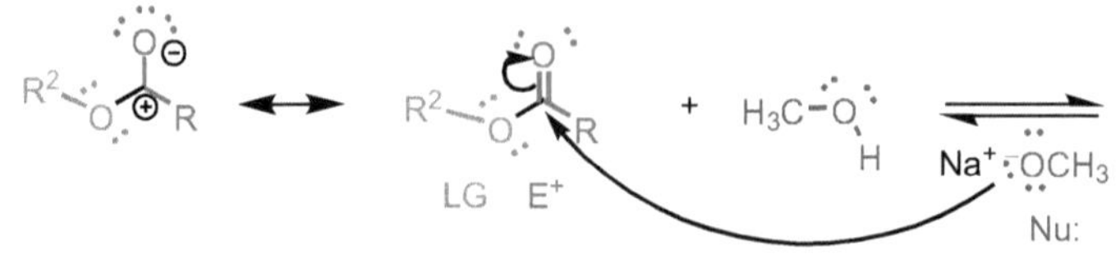

Figure B.19 Draw curved arrows from the e$^-$-rich Nu: to the e$^-$-poor E$^+$. Do not leave a structure with 5 bonds to carbon.

Step 8. Draw the intermediates with new bonds and lone pairs. (Figure B.20) The curved arrows show you exactly where to put new bonds or new lone pairs. The red oxygen now has only two lone pairs; the third was used to form the new bond. The green oxygen now has three lone pairs, one derived from the carbonyl π-bond. The Na^+ ion is a spectator; it just goes along for the ride and ends up wherever a positive charge is needed to neutralize a negative charge. Na^+ does *not* behave as an E^+. (Why?) The square brackets show that the product of this step is a high energy intermediate, in this case called a tetrahedral intermediate, describing the change of the carbonyl carbon from sp^2, or trigonal planar, to sp^3 or tetrahedral hybridized. Check the reaction: have we used all the arrows to make new bonds or lone pairs? Are all the atoms balanced? Are the charges balanced? Technically, the net charge on both sides should always be zero, but sometimes organic chemists omit the spectator ion. Don't do this because it is a good check on your work.

Repeat steps 6, 7 and 8 until the product is obtained. Identify the leaving group and the bond that needs to break to give you the product. The R^2–O^- is the leaving group. The next step is the collapse of the tetrahedral intermediate (Figure B.21). The mechanism is like the reverse of the first step, except that if $^-OCH_3$ is the leaving group, it is simply the reverse reaction, so we should draw that arrow as a reversible reaction. If R^2–O^- is the leaving group, we get to the desired biodiesel product.

Step 9. Regenerate the original catalyst. (Figure B.22) Remember that the reaction needs to happen several billion times, so the catalyst needs to be regenerated after each cycle. Also, both reactions so far are reversible. Remember the other reactant, CH_3OH. What is it there for? If there is excess methanol around, the ion, R^2–O^-, can deprotonate methanol in an acid-base reaction. Now we have regenerated the sodium methoxide catalyst, and produced both

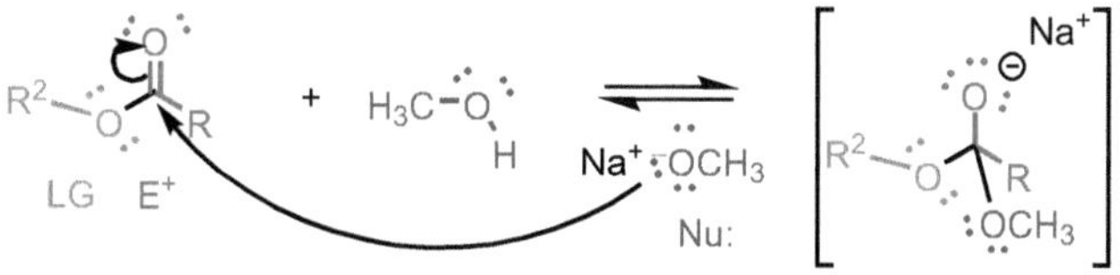

Figure B.20 Draw the tetrahedral intermediate resulting from the movement of electrons shown by curved arrows.

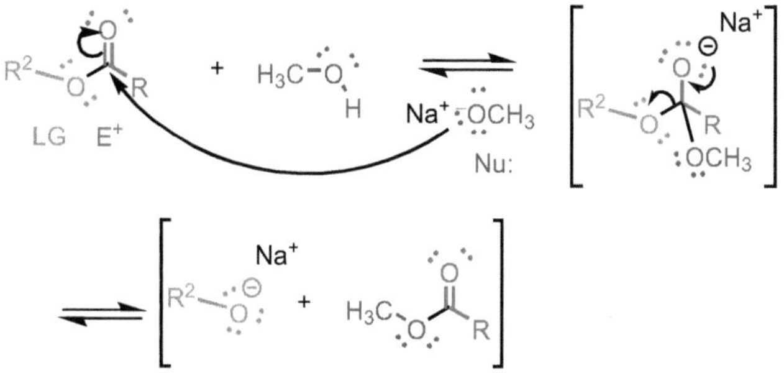

Figure B.21 Continue drawing curved arrows that lead to the product. Collapse of the tetrahedral intermediate.

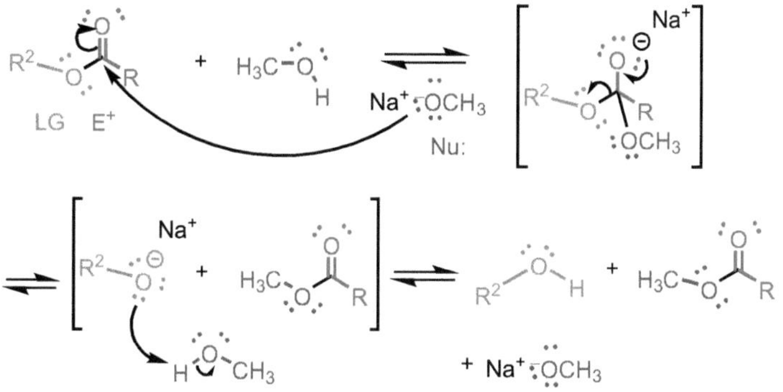

Figure B.22 Additional curved arrows show the final acid-base reaction to regenerate the catalyst.

desired products, glycerol and the fatty ester, biodiesel. Again, check the bonds and lone pairs. Check that all atoms and charges are balanced for the last step.

Step 10. Drive the reaction to completion. Whether this reaction goes to completion or not depends on the relative concentrations and pK_a values of glycerol, R^2–OH ($pK_{a1} = 13.5$), and methanol CH_3OH ($pK_a = 16$). Unfortunately, this means the equilibrium lies well towards R^2–O^- and CH_3OH, not R^2–OH and CH_3O^-. The key here is that methanol must be in excess, and by Le Chatelier's Principle, mass action will drive it toward R^2–OH. Another way to drive the reaction to completion is by removing one of the products, either glycerol or the methyl ester. Since glycerol is polar and hydrogen-bonded, it will form a denser layer and separate from the non-polar fat feedstock and methyl ester product mixture. Constant removal of glycerin and introduction of fat and methanol feedstock allow the process to be semi-batch (Chapter 6), or continuous.

B.8 Substitution Reactions

B.8.1 MOs of an S_N2 Substitution Reaction

Here is one last point that is really key to the biodiesel synthesis mechanism. For ester substitution or any other substitution at a carbonyl, one might ask, why could the mechanism not just be an S_N2 and be done with it? We need a different model that is 3D to tell us why not S_N2. Next, we will examine these reacting MOs from a geometric perspective. Why not S_N2 for carbonyl substitution? (S_N2 stands for substitution nucleophilic bi-molecular, because two molecules are required in the rate-determining step. In a standard S_N2 mechanism, the nucleophile approaches the leaving group from the backside of the electrophilic carbon (Figure B.23).)

The HOMO, the highest occupied molecular orbital, corresponds to the nucleophile, in this case the alkoxide (Figure B.23). The LUMO, the lowest unoccupied molecular orbital, corresponds to the E^+, the carbon-chlorine σ^*-MO, the anti-bonding orbital (Figure B.23). The geometries of the HOMO and the LUMO must be in alignment (Figure B.23). This gives a pentacoordinate intermediate, and inversion of configuration in the product (Figure B.23).

B.8.2 MOs of Carbonyl Substitution by Addition-elimination

In the case of nucleophilic substitution of a carbonyl, the nucleophile must approach from the backside of the leaving group. The HOMO of the nucleophile is the same, but the LUMO of the carbonyl is the C=O π^*-MO, and this is perpendicular to the nucleophile-leaving group axis. Thus, carbonyl substitution MUST occur by addition-elimination.

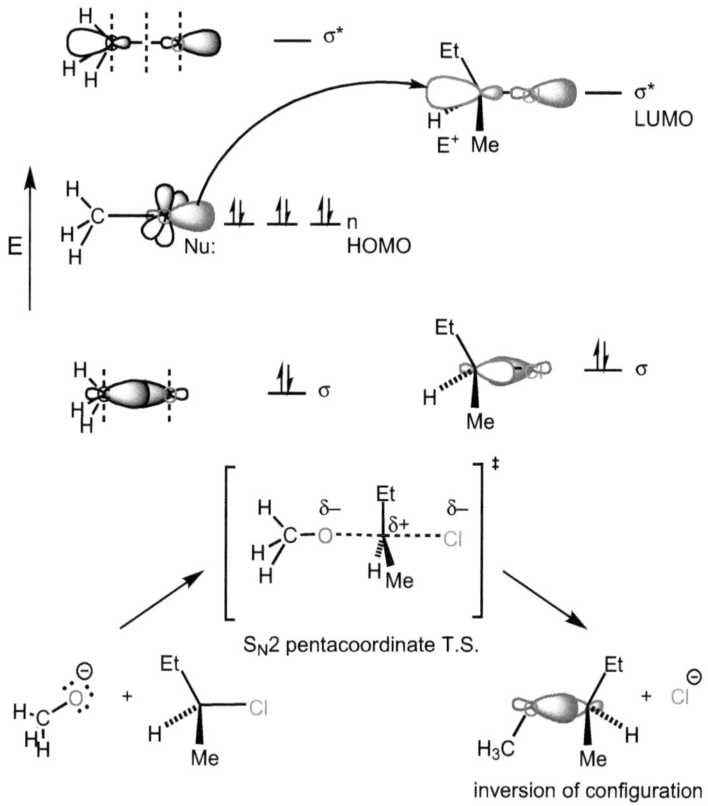

Figure B.23 MO diagram of an S_N2 substitution reaction.

For carbonyl substitution by addition-elimination, we will construct the MOs step-by-step (Figure B.24). To begin, draw the Lewis structures of the two reacting species at the bottom of a page. Above them, draw the lowest energy MOs. For the lowest energy methoxide, we will include the σ orbital of the C–O bond. We show the nodes as dashed lines. Nodes are places where the 3D mathematical description of the orbital passes through zero. There is zero probability of the electron being at a node. Then add the 3 non-bonding (n) lone pair orbitals (without nodes for clarity). The reacting MO is shown in red, but really any of the 3 n orbitals could react because they are equal in energy. We show the two σ MOs to place the n MOs on the energy scale. Non-bonding orbitals are halfway between the bonding and anti-bonding orbitals.

Next add the σ* orbital with its 3 nodes (Figure B.24). Notice that the highest energy orbitals are the ones with the most nodes. Nodes in MOs also show where the symmetry changes are. The central node between the carbon and oxygen of the σ* orbital shows the

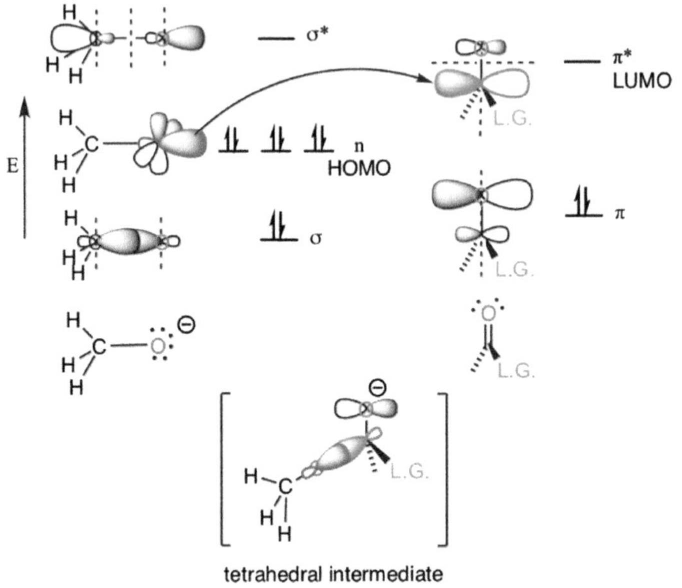

Figure B.24 MO diagram of carbonyl substitution by addition-elimination: formation of the tetrahedral intermediate.

anti-bonding arrangement of the MO, and the star indicates that is an anti-bonding orbital. Energy levels are drawn in next to each pictorial MO.

The electrons are added to each energy level as up and down arrows, indicating pairing by Hund's Rule (Figure B.24). The C–O σ-bond gets 2 electrons. The 3 n-orbitals get the 3 lone pairs on the oxide anion. There are no electrons left to put into the anti-bonding σ* orbital, which is good because that would cause the C–O bond to break. Finally, label the MOs: σ, n, and σ*. The n-orbital is the *highest occupied MO*, or HOMO. The σ* is the *lowest unoccupied MO*, or LUMO for the methoxide anion.

For the carbonyl compound, we include only the reacting π and π* orbitals of the C=O π-bond (Figure B.24). First, draw the π-bonding MO with the same symmetry of the two atomic orbitals on carbon and oxygen. Notice that these are left as AOs for the drawing, but the MO is actually overlapped where the symmetries are the same, left and right of the plane of the molecule. The lobe of the MO on oxygen, which is NOT a separate MO, is larger because oxygen is more electronegative than carbon and pulls the electron density towards it. This is reminiscent of the important resonance structure with a negative charge on oxygen. Then add the node. The second MO of the carbonyl is the π^*. Notice that the size of the two MO lobes are switched from the bonding; the larger lobe is now on the carbon. This indicates a partial positive charge on the carbon, the site of the E^+ in red. Then add the two nodes, one in the plane of the molecule for the p-type orbital symmetry, and one between the carbon and oxygen, indicating an anti-bonding relationship. Draw the energy levels in. Add the π-electrons to the bonding π-MO. Label the MOs. Since there are only two, the π is the HOMO; it is occupied with electrons. The π^* is the LUMO because it is unoccupied.

The point of all this was to show the geometric relationship between the two reacting molecule orbitals. We need a HOMO to donate electrons into a LUMO on the other molecule. The HOMO-LUMO pair that is closest in energy will react, so the methoxide HOMO donates electrons into the carbonyl LUMO. The MOs must have the same symmetry, as shown by the shaded lobes coming together to make a new oxygen–carbon sigma bond. This is the reason why the nucleophile MUST approach from the left or the right of the carbonyl, perpendicular to the plane of the molecule as shown. There is no LUMO for the Nu: to add into opposite the leaving group (L.G. in green) in the plane of the carbonyl. Thus, the tetrahedral intermediate is the only one that can arise from nucleophilic addition to a carbonyl!

B.9 Radical Mechanisms

Shaw EcoWorx carpet squares are recyclable in part because they use polyolefin, mostly polypropylene, instead of polyvinylchloride (PVC) in the backing material. The mechanism of polypropylene formation is based on radicals, or unpaired electrons. For the mechanism of radical reactions, half-headed arrows are used to show the movement of single electrons. The polymerization takes place in three steps: initiation, propagation, and termination, not always in a clean, stepwise manner.

B.9.1 Initiators

Initiation of radical reactions requires an initiator that is very prone to forming radicals. Initiators are typically highly symmetric molecules with a weak bond in the middle, such as azobisisobutyronitrile (AIBN). To see why it forms radicals so easily, we can recognize that the –N=N– in the middle of the molecule can be extruded to form extremely stable nitrogen gas, N_2 (Figure B.25). In addition, the two equivalents of radical formed are stabilized by hyperconjugation with

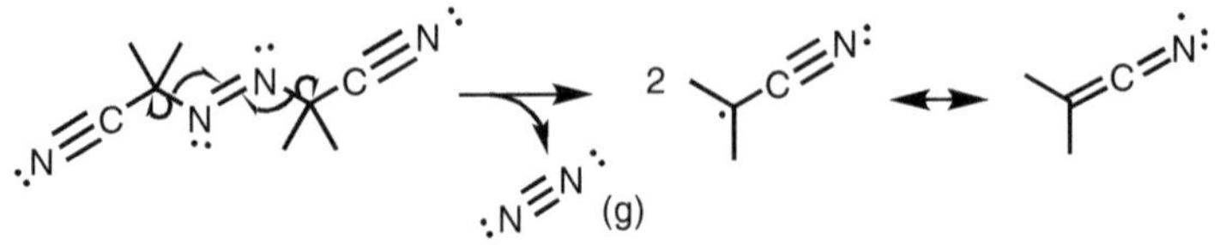

Figure B.25 Homolytic cleavage of AIBN radical initiator generates N_2 and a resonance-stabilized 3° carbon radical.

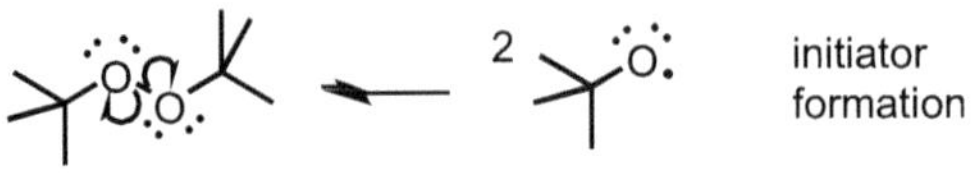

Figure B.26 Formation of the *t*-butoxy radical initiator by homolytic cleavage of di-*t*-butylperoxide.

the methyl groups, the reason 3° radicals are more stable than 2° or 1° radicals, and by conjugation with the nitrile (CN) (Figure B.25).

Notice that two half-arrows are used to show each bond breaking because there are two electrons in each bond. The radical is resonance stabilized by the nitrile group. One problem that AIBN avoids is the toxic release of CN. Even though the ends of the polymer still have the nitrile group, the absence of an alpha hydrogen prevents radical abstraction of H•, which is the first mechanistic step in the release of •CN that causes toxicity.

Other radical initiators are often used, with the one common feature of symmetry and weak central bonds. For example, di-*t*-butylperoxide is symmetric, with a weak O–O bond. The symmetry of the bond allows it to break in a *homolytic* fashion, with each oxygen taking one electron (Figure B.26), rather than a *heterolytic* bond cleavage, which gives a cation and an anion.

For the radical polymerization of polypropylene, we draw three distinct mechanistic steps after formation of the radical initiator.

B.9.2 Initiation

Regardless of which initiator is used, initiation occurs by addition of the initiator radical to the alkene. Include the hydrogens of the alkene to make sure each carbon has a full electron count at each step (Figure B.27). Note the use of half-headed arrows. The radical intermediate could reside at either of the original alkene carbons, but it is far more likely to be at the more substituted carbon both because of steric hindrance with the *t*-butoxy incoming group, and greater stability of the 2° radical formed due to hyperconjugation with the methyl group.

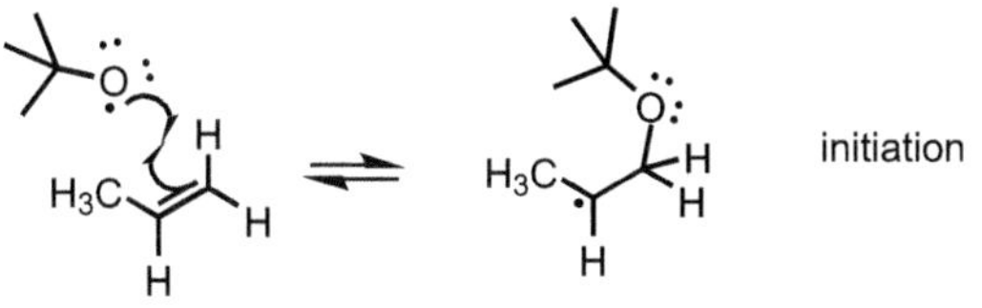

Figure B.27 Initiation by *t*-butoxy radical initiator cleavage of the propene π-bond to leave the more stable 2° radical intermediate.

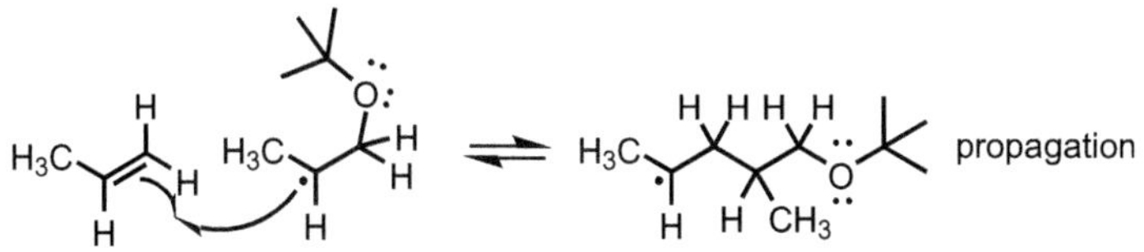

Figure B.28 Propagation step in the radical polymerization of polypropylene.

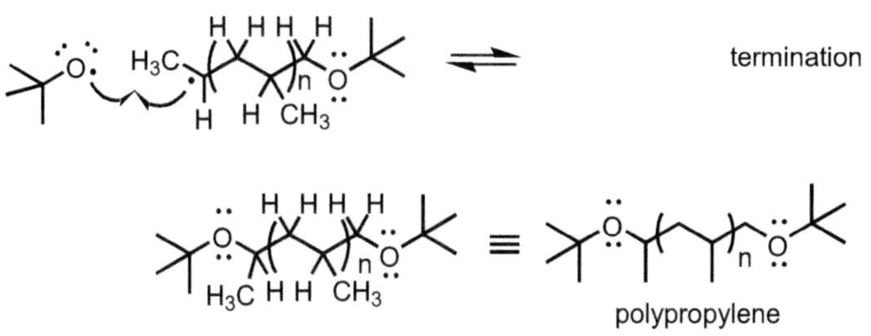

Figure B.29 Termination of radical polymerization with an initiator radical. Other termination steps are possible.

B.9.3 Propagation

The propagation step is repeated many times to create a polymer. Each step resembles the initiation in that the most stable 2° radical is created rather than a 1° radical (Figure B.28). The chain elongates in this regular way with alternating methyl groups until a termination step happens.

B.9.4 Termination

There are several possible termination steps, and termination can take place with any two radicals present in the reaction mixture (Figure B.29). This is only one possibility. Can you think of others?

The final structure of polypropylene, leaving all of the implied hydrogens off looks like this (Figure B.29). Why are the methyl groups exactly alternating, instead of randomly distributed on the backbone? What would the final polymer look like if AIBN were used instead of di-*t*-butylperoxide?

Appendix C pKₐ Tables

Reprinted with permission. Copyright 2005 D. H. Ripin and D. A. Evans.

Green Chemistry: Principles and Case Studies
By Felicia A. Etzkorn

Published by the Royal Society of Chemistry, www.rsc.org

Table C.1 pK_as of inorganic and oxo-acids.[a,b]

Substrate	pK_a H_2O	(DMSO)
Inorganic acids		
H_2O	15.7	(32)
H_3O^+	−1.7	
H_2S	7.00	
HBr	−9.00	(0.9)
HCl	−8.0	(1.8)
HF	3.17	(15)
HOCl	7.5	
$HClO_4$	−10	
HCN	9.4	(12.9)
HN_3	4.72	(7.9)
HSCN	4.00	
H_2SO_3	1.9, 7.21	
H_2SO_4	−3.0, 1.99	
H_3PO_4	2.12, 7.21, 12.32	
HNO_3	−1.3	
HNO_2	3.29	
H_2CrO_4	−0.98, 6.50	
CH_3SO_3H	−2.6	(1.6)
CF_3SO_3H	−14	(0.3)
NH_4Cl	9.24	
$B(OH)_3$	9.23	
HOOH	11.6	

Substrate	pK_a H_2O	(DMSO)
Carboxylic acids		
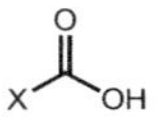		
X = CH_3	4.76	(12.3)
CH_2NO_2	1.68	
CH_2F	2.66	
CH_2Cl	2.86	
CH_2Br	2.86	
CH_2I	3.12	
$CHCl_2$	1.29	
CCl_3	0.65	
CF_3	−0.25	
H	3.77	
HO	3.6, 10.3	
C_6H_5	4.2	(11.1)
o-$O_2NC_6H_4$	2.17	
m-$O_2NC_6H_4$	2.45	
p-$O_2NC_6H_4$	3.44	
o-ClC_6H_4	2.94	
m-ClC_6H_4	3.83	
p-ClC_6H_4	3.99	
o-$(CH_3)_3N^+C_6H_4$	1.37	
p-$(CH_3)_3N^+C_6H_4$	3.43	
p-$OMeC_6H_4$	4.47	
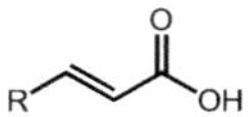		
R = H	4.25	
trans-CO_2H	3.02, 4.38	
cis-CO_2H	1.92, 6.23	

Substrate	pK_a H_2O	(DMSO)
Alcohols		
HOH	15.7	(31.2)
MeOH	15.5	(27.9)
i-PrOH	16.5	(29.3)
t-BuOH	17.0	(29.4)
c-hex_3COH	24.0	
CF_3CH_2OH	12.5	(23.5)
$(CF_3)_2CHOH$	9.3	(18.2)
C_6H_5OH	9.95	(18.0)
m-$O_2NC_6H_4OH$	8.4	
p-$O_2NC_6H_4OH$	7.1	(10.8)
p-$OMeC_6H_4OH$	10.2	(19.1)
2-Naphthol		(17.1)
Oximes & hydroxamic acids		
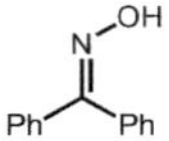	11.3	(20.1)
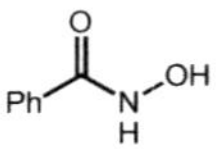	8.88 (NH)	(13.7)
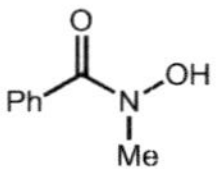		(18.5)
Peroxides		
MeOOH	11.5	
CH_3CO_3H	8.2	

Substrate	pK_a H_2O	(DMSO)
Protonated species		
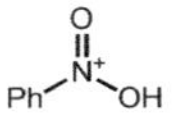	−12.4	
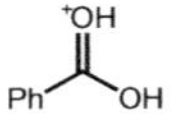	−7.8	
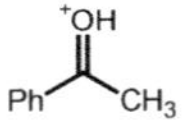	−6.2	
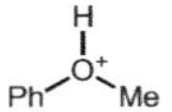	−6.5	
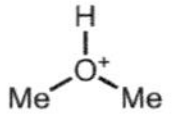	−3.8	
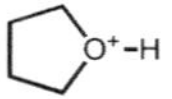	−2.05	
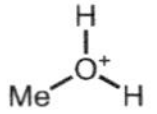	−2.2	
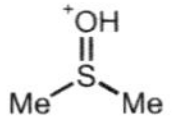	−1.8	
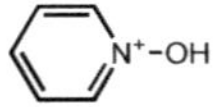	0.79	(+1.63)
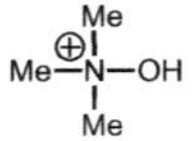		(+5.55)
Sulfinic & sulfonic acids		
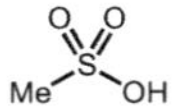	−2.6	
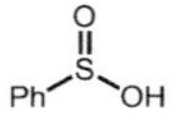	2.1	

[a] Values <0 for H_2O and DMSO, and values >14 for water and >35 for DMSO were extrapolated using various methods.
[b] For a comprehensive compilation of Bordwell pK_a data see: http://www.chem.wisc.edu/areas/reich/pkatable/index.htm.

Table C.2 pK_as of nitrogen acids.[a,b]

Substrate	pK_a	H_2O	(DMSO)
Protonated nitrogen			
N^+H_4		9.2	(10.5)
EtN^+H_3		10.6	
i-$Pr_2N^+H_2$		11.05	
Et_3N^+H		10.75	(9.00)
PhN^+H_3		4.6	(3.6)
$PhN^+(Me)_2H$		5.20	(2.50)
$Ph_2N^+H_2$		0.78	
2-Naphthol-N^+H_3		4.16	
$H_2NN^+H_3$		8.12	
HON^+H_3		5.96	
Quinuclidine 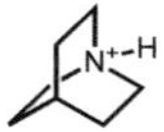		11.0	(9.80)
Morpholine 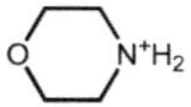		8.36	
N-Me morpholine		7.38	
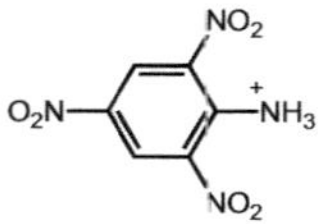		−9.3	
DABCO 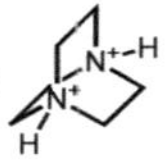		2.97, 8.82	(2.97, 8.93)
$H_3N^+CH_2CH_2NH_3^+$		6.90, 9.95	
Proton sponge 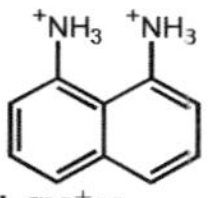		−9.0, 12.0	(—, 7.50)
$PhCN^+H$		−10	

Substrate	pK_a	H_2O	(DMSO)
Amines			
HN_3		4.7	(7.9)
NH_3		38	(41)
i-Pr_2NH		(36 THF)	
TMS_2NH		(26 THF)	(30)
$PhNH_2$	(30.6)		
Ph_2NH	(25.0)		
$NCNH_2$	(16.9)	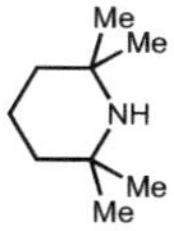	(37)
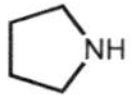	(44)		
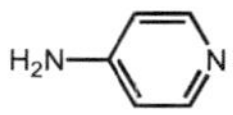		(26.5)	
Amides & carbamates			
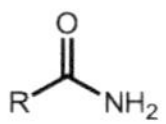	R = H		(23.5)
	CH_3	15.1	(25.5)
	Ph		(23.3)
	CF_3		(17.2)
	(urea) NH_2		(26.9)
	OEt		(24.8)
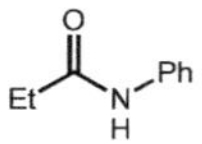	(21.6)	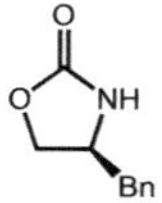 12	(20.5)
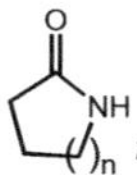	$n = 1$ (24.1) $n = 2$ (26.4)	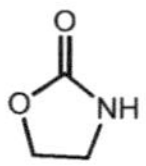	(20.8)
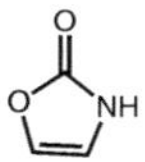	(15)	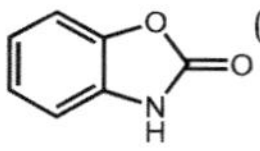	(12.1)

Substrate	pK_a	H_2O	(DMSO)
Imides			
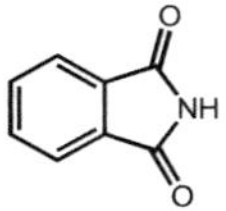	8.30	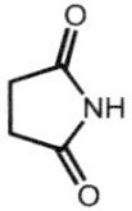	(14.7)
Ac_2NH			(17.9)
Sulfonamide			
RSO_2NH_2	R = Me		(17.5)
	Ph		(16.1)
	CF_3	6.3	(9.7)
$MeSO_2NHPh$			(12.9)
Guanidinium, hyrdazones, -ides, & -ines			
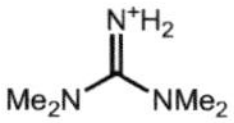	(13.6)	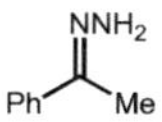	(21.6)
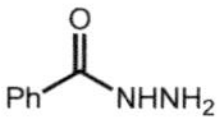	(18.9)		
$PhSO_2NHNH_2$			(17.2)
PhNHNHPh			(26.1)
Protonated heterocycles			
DBU 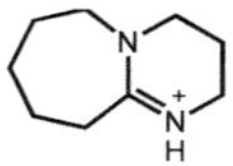		(12)	(estimate)
DMAP 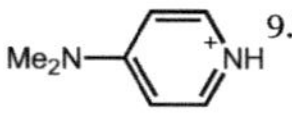	9.2	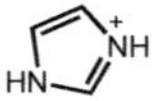	6.95
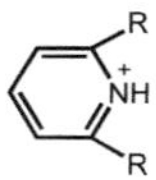	R = H (PPTS)	5.21	(3.4)
	t-Bu	4.95	(0.90)
	Me	6.75	(4.46)
	Cl, H	0.72	

Substrate	pK_a	H_2O	(DMSO)
Hydroxamic acid & amidines			
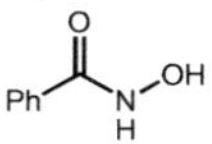		8.88 (NH)	(13.7)
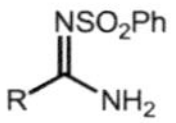	R = Me		(17.3)
	Ph		(15.0)
Heterocycles			
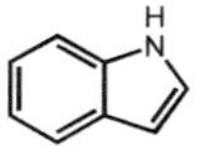	(20.95)	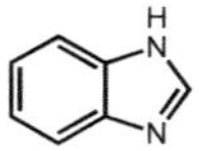	(16.4)
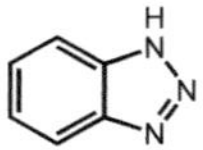	(11.9)	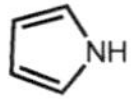	(23.0)
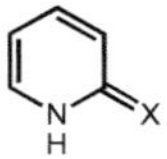	X = O (24) X = S (13.3)	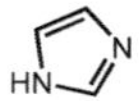	(18.6)
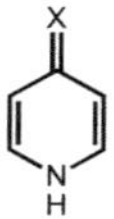	X = O (14.8) X = S (11.8)	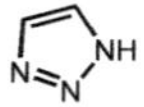	(13.9)
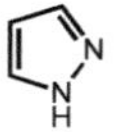	(19.8)	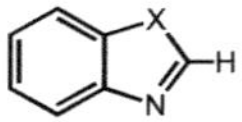	X = O (24.4) X = S (27.0)
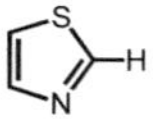	(29.4)	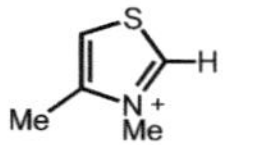	(16.5)
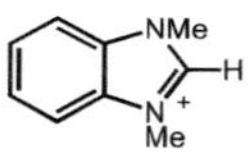	(18.4)	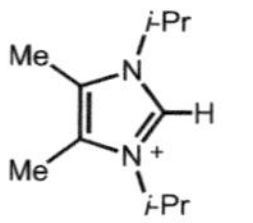	(24)

[a] Values <0 for H_2O and DMSO, and values >14 for water and >35 for DMSO were extrapolated using various methods.
[b] For a comprehensive compilation of Bordwell pK_a data see: http://www.chem.wisc.edu/areas/reich/pkatable/index.htm.

Table C.3 pK_as of CH bonds in hydrocarbons and carbonyl compounds.[a,b]

Substrate	pK_a H_2O	(DMSO)
Hydrocarbons		
$(Me)_3CH$	53	
$(Me)_2CH_2$	51	
$CH_2{=}CH_2$	50	
CH_4	48	(56)
	46	
$CH_2{=}CHCH_3$	43	(44)
PhH	43	
$PhCH_3$	41	(43)
Ph_2CH_2	33.5	(32.2)
Ph_3CH	31.5	(30.6)
HCCH	24	
PhCCH	23	(28.8)
$XC_6H_4CH_3$		
X = *p*-CN		(30.8)
p-NO_2		(20.4)
p-COPh		(26.9)
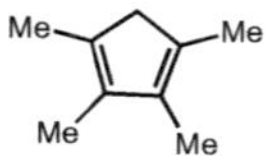		(26.1)
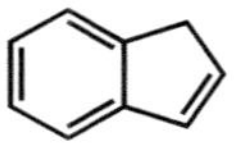	20	(20.1)
	15	(18.0)
H_2	~36	

Substrate	pK_a H_2O	(DMSO)
Esters		
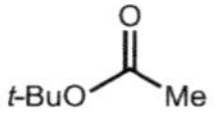	24.5	(30.3)
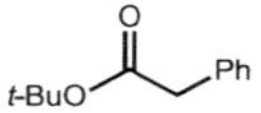		(23.6)
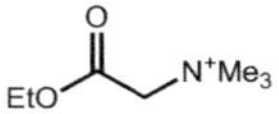		(20.0)
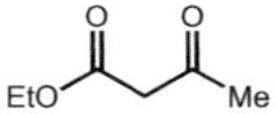	11	(14.2)
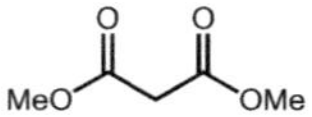	13	(15.7)
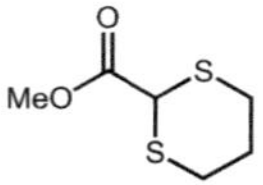		(20.9)
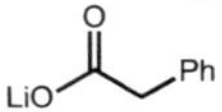	[30.2 (THF)]	
Amides		
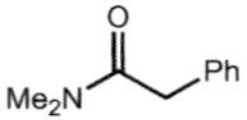		(26.6)
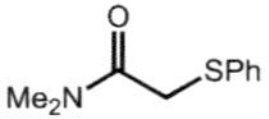		(25.9)
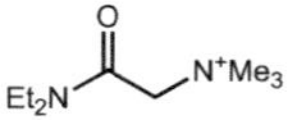		(24.9)
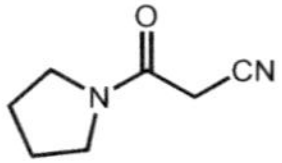		(17.2)
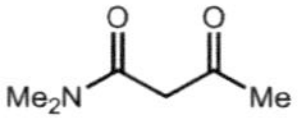		(18.2)
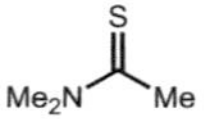		(25.7)

Substrate	pK_a H_2O	(DMSO)
Ketones		
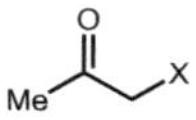		
X = H		(26.5)
Ph		(19.8)
SPh		(18.7)
$COCH_3$	9	(13.3)
SO_2Ph		(12.5)
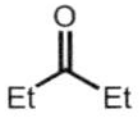	19–20	(27.1)
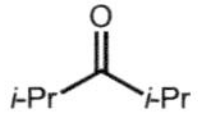		(28.3)
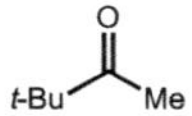		(27.7)
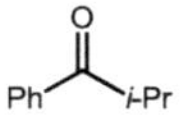		(26.3)
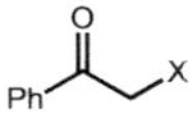		
X = H		(24.7)
CH_3		(24.4)
Ph		(17.7)
$COCH_3$		(14.2)
COPh		(13.3)
CN		(10.2)
F		(21.6)
OMe		(22.85)
OPh		(21.1)
SPh		(16.9)
SePh		(18.6)
NPh_2		(20.3)
N^+Me_3		(14.6)
NO_2		(7.7)
SO_2Ph		(11.4)

Substrate	pK_a H_2O	(DMSO)
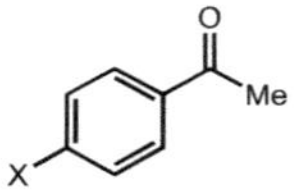		
X = H		(24.7)
OMe		(25.7)
NMe_2		(27.5)
Br		(23.8)
CN		(22.0)

n = 4		(25.1)
5		(25.8)
6		(26.4)
7		(27.7)
8		(27.4)
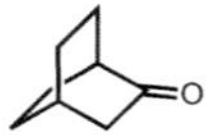		(28.1)
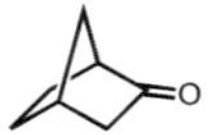		(29.0)
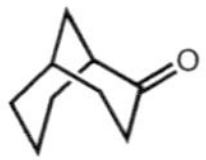		(25.5)
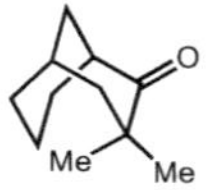		(32.4)

[a]Values <0 for H_2O and DMSO, and values >14 for water and >35 for DMSO were extrapolated using various methods.
[b]For a comprehensive compilation of Bordwell pK_a data see: http://www.chem.wisc.edu/areas/reich/pkatable/index.htm.

Table C.4 pK_as of CH bonds at nitrile, heteroaromatic, and sulfur substituted carbon.[a]

Substrate	pK_a H_2O	(DMSO)
Nitriles		
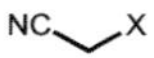		
X = H		(31.3)
CH_3		(32.5)
Ph		(21.9)
COPh		(10.2)
$CONR_2$		(17.1)
CO_2Et		(13.1)
CN	11	(11.1)
OPh		(28.1)
N^+Me_3		(20.6)
SPh		(20.8)
SO_2Ph		(12.0)
Hetero-aromatics		
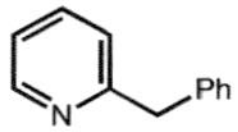		(28.2)
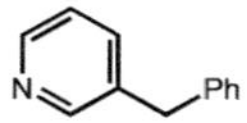		(30.1)
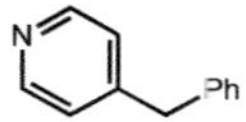		(26.7)
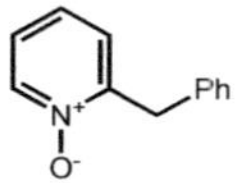		(25.2)
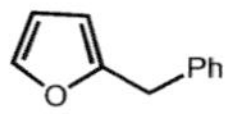		(30.2)
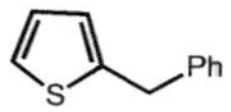		(30.0)

Substrate	pK_a H_2O	(DMSO)
Sulfides		
$PhSCH_2X$		
X = Ph		(30.8)
CN		(20.8)
$COCH_3$		(18.7)
COPh		(16.9)
NO_2		(11.8)
SPh		(30.8)
SO_2Ph		(20.5)
SO_2CF_3		(11.0)
$POPh_2$		(24.9)
$MeSCH_2SO_2Ph$		(23.4)
$PhSCHPh_2$		(26.7)
$(PhS)_3CH$		(22.8)
$(PrS)_3CH$		(31.3)
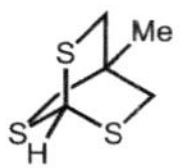		(30.5)
$(PhS)_2CHPh$		(23.0)
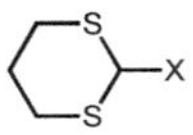		
X = Ph		(30.7)
CO_2Me		(20.8)
CN		(19.1)
$RSCH_2CH$		
R = Me		(24.3)
Et		(24.0)
i-Pr		(23.6)
t-Bu		(22.9)
$PhSCH{=}CHCH_2SPh$		(26.3)
BuSH	10–11	(17.0)
PhSH	≈7	(10.3)

Substrate	pK_a H_2O	(DMSO)
Sulfoxides		
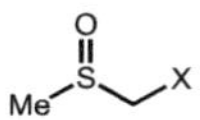		
X = H		(35.1)
Ph		(29.0)
SPh		(29.0)
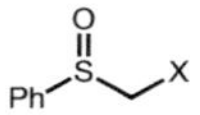		
X = H		(33)
Ph		(27.2)
SOPh		(18.2)
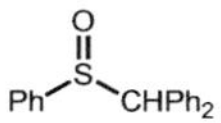		(24.5)
Sulfonium		
$Me_3S^+{=}O$		(18.2)
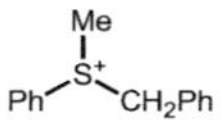		(16.3)
Sulfimides & sulfoximines		
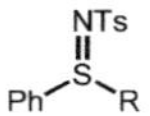		
R = Me		(27.6)
i-Pr		(30.7)
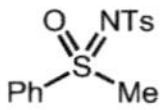		(24.5)
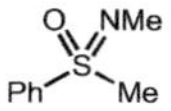		(33)
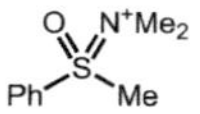		(14.4)
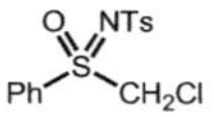		(20.7)

Substrate	pK_a H_2O	(DMSO)
Sulfones		
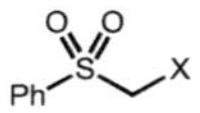		
X = H		(29.0)
CH_3		(31.0)
t-Bu		(31.2)
Ph		(23.4)
$CH{=}CH_2$		(22.5)
$CH{=}CHPh$		(20.2)
CCH		(22.1)
CCPh		(17.8)
COPh		(11.4)
COMe		(12.5)
OPh		(27.9)
N^+Me_3		(19.4)
CN		(12.0)
NO_2		(7.1)
SMe		(23.5)
SPh		(20.5)
SO_2Ph		(12.2)
PPh_2		(20.2)
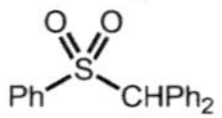		(22.3)
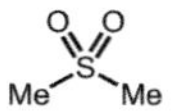		(31.1)
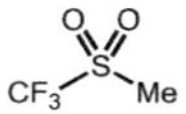		(18.8)
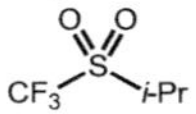		(21.8)
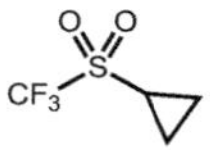		(26.6)
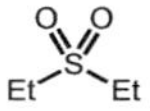		(32.8)
$(PhSO_2)_2CH_2Me$		(14.3)

[a]Values <0 for H_2O and DMSO, and values >14 for water and >35 for DMSO were extrapolated using various methods.

Table C.5 pK_as of CH bonds at heteroatom substituted carbon and references.[a,b]

Substrate	pK_a	H_2O	(DMSO)
Ethers			
CH_3OPh			(49)
$MeOCH_2SO_2Ph$			(30.7)
$PhOCH_2SO_2Ph$			(27.9)
$PhOCH_2CN$			(28.1)
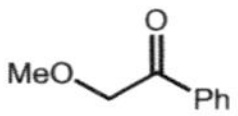			(22.85)
Selenides			
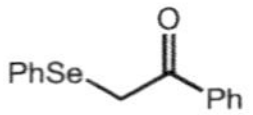			(18.6)
$PhSeCHPh_2$			(27.5)
$(PhSe)_2CH_2$			(31.3)
$PhSeCH_2Ph$			(31.0)
$PhSeCH{=}CHCH_2SePh$			(27.2)
Ammonium			
$Me_3N^+CH_2X$			
X = CN			(20.6)
SO_2Ph			(19.4)
COPh			(14.6)
CO_2Et			(20.0)
$CONEt_2$			(24.9)

Substrate	pK_a	H_2O	(DMSO)
Phosphonium			
P^+H_4			−14
MeP^+H_3			2.7
Et_3P^+H			9.1
$Ph_3P^+CH_3$			(22.4)
Ph_3P^+i-Pr			(21.2)
$Ph_3P^+CH_2COPh$			(6.2)
$Ph_3P^+CH_2CN$			(7.0)
Phosponates & phosphine oxides			
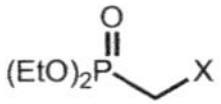			
X = Ph			(27.6)
CN			(16.4)
CO_2Et			(18.6)
Cl			(26.2)
$SiME_3$			(28.8)
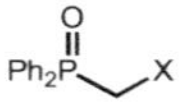			(19.4)
X = SPh			(24.9)
CN			(16.9)
Phosphines			
$Ph_2PCH_2PPh_2$			(29.9)
$Ph_2PCH_2SO_2Ph_2$			(20.2)

Substrate	pK_a	H_2O	(DMSO)
Nitro			
RNO_2			
R = CH_3		≈10	(17.2)
CH_2Me			(16.7)
$CHMe_2$			(16.9)
CH_2Ph			(12.2)
CH_2Bn			(16.2)
CH_2SPh			(11.8)
CH_2SO_2Ph			(7.1)
CH_2COPh			(7.7)
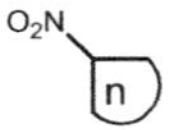			
$n = 3$			(26.9)
4			(17.8)
5			(16.0)
6			(17.9)
7			(15.8)
Imines			
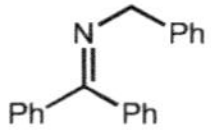			(24.3)
Oxime ethers are ~10 pK_a units less acidic than their ketone counterparts. Streitwieser, *J. Org. Chem.* 1991, **56**, 1989			

References

DMSO:

J. Am. Chem. Soc., 1975, **97**, 7007
J. Am. Chem. Soc., 1975, **97**, 7160
J. Am. Chem. Soc., 1975, **97**, 442
J. Am. Chem. Soc., 1983, **105**, 6188
J. Org. Chem., 1976, **41**, 1883
J. Org. Chem., 1976, **41**, 1885
J. Org. Chem., 1976, **41**, 2786
J. Org. Chem., 1976, **41**, 2508
J. Org. Chem., 1977, **42**, 1817
J. Org. Chem., 1977, **42**, 321
J. Org. Chem., 1977, **42**, 326
J. Org. Chem., 1978, **43**, 3113
J. Org. Chem., 1978, **43**, 3095
J. Org. Chem., 1978, **43**, 1764
J. Org. Chem., 1980, **45**, 3325
J. Org. Chem., 1980, **45**, 3305
J. Org. Chem., 1980, **45**, 3884
J. Org. Chem., 1981, **46**, 4327
J. Org. Chem., 1981, **46**, 632
J. Org. Chem., 1982, **47**, 3224
J. Org. Chem., 1982, **47**, 2504
Acc. Chem. Res., 1988, **21**, 456
F. Bordwell, unpublished results

Water:

J. March, *Advanced Organic Chemistry*, 3rd edn, 1985
W. P. Jencks, unpublished results

THF:

J. Am. Chem. Soc., 1988, **110**, 5705
See cited website below for additional data

[a]Values <0 for H_2O and DMSO, and values >14 for water and >35 for DMSO were extrapolated using various methods.
[b]For a comprehensive compilation of Brodwell pK_a data see: http://www.chem.wisc.edu/areas/reich/pkatable/index.htm.

Table C.6 DMSO acidities of common heterocycles.[a,b]

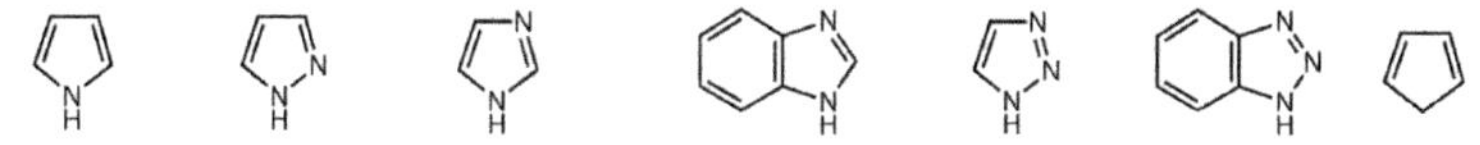

23.0	19.8	18.6	16.4	13.9	11.9	18.0

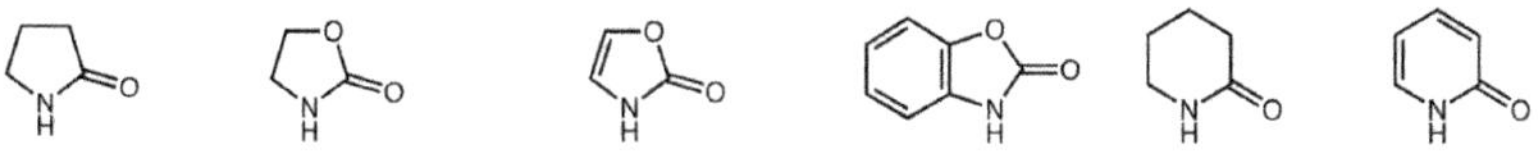

24.0	20.8	15.0	12.1	26.4	24.0

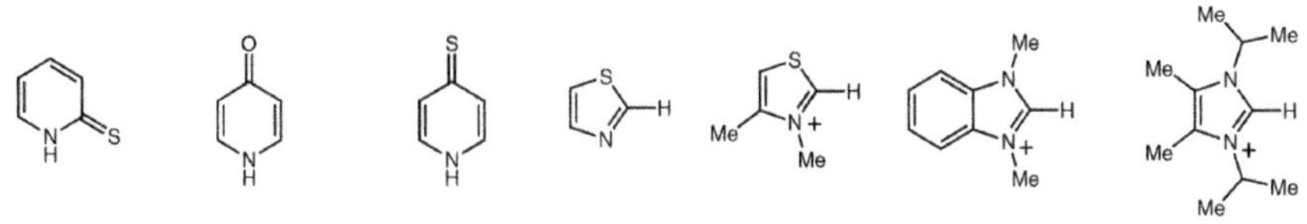

13.3	14.8	11.8	29.4	16.5	18.4	24

[a]F. G. Bordwell, *Acc. Chem. Res.*, 1988, **21**, 456.
[b]http://www.chem.wisc.edu/areas/reich/pkatable/index.htm.

Appendix D Earth Abundance Periodic Table

Green Chemistry: Principles and Case Studies
By Felicia A. Etzkorn

Published by the Royal Society of Chemistry, www.rsc.org

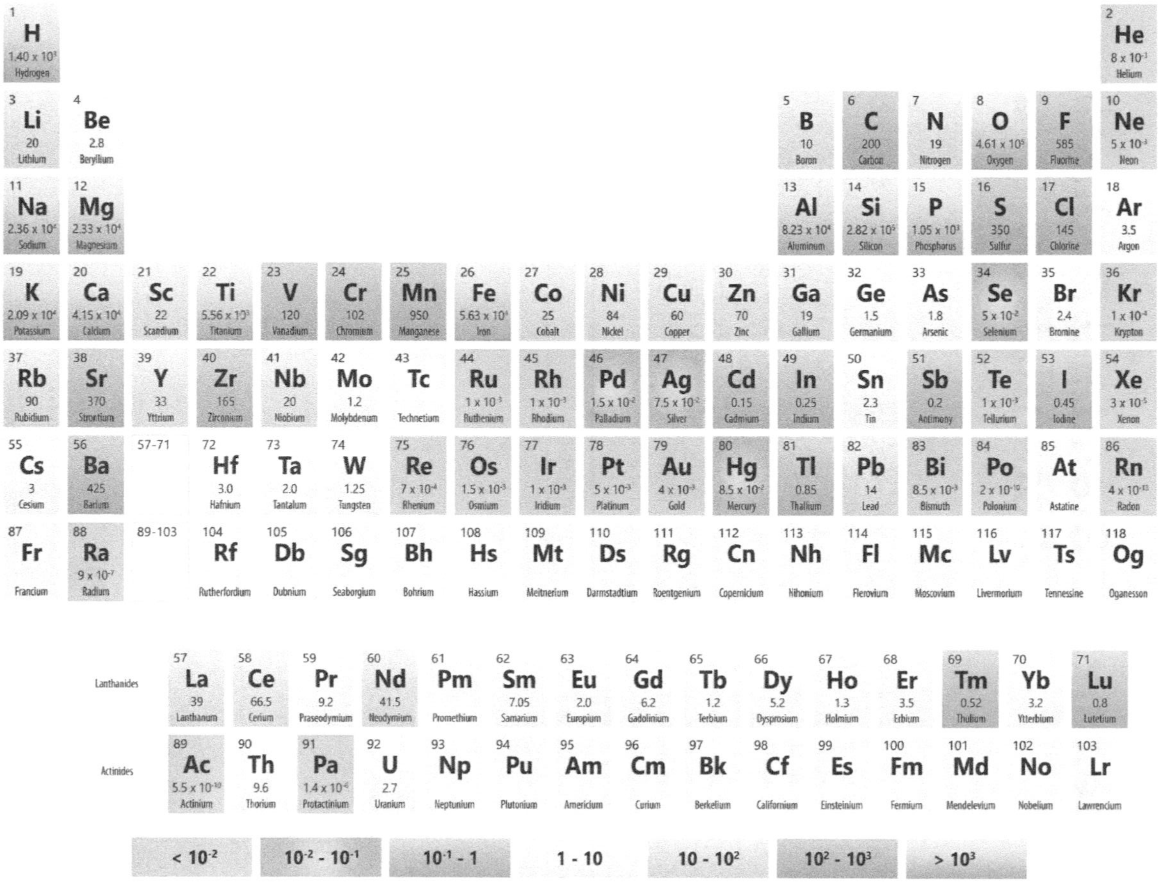

Figure D.1 Abundance of Elements in Earth's Crust. Abundance values in mg kg^{-1}. Reprinted with permission of Todd Helmenstine from sciencenotes.org, Copyright 2018.

Appendix E Standard Reduction Potentials by Value

Table E.1 E° for selected reduction reactions. Values are from the following sources: *Standard Potentials in Aqueous Solutions*, ed. A. J. Bard, B. Parsons, J. Jordon, Dekker, New York, 1985; G. Milazzo, S. Caroli and V. K. Sharma, *Tables of Standard Electrode Potentials*, Wiley, London, 1978; E. H. Swift and E. A. Butler, *Quantitative Measurements and Chemical Equilibria*, Freeman, New York, 1972. Excerpted with permission from David T. Harvey (Depauw University), P2: Standard Reduction Potentials by Value, Chemistry LibreTexts, https://chem.libretexts.org/Ancillary_Materials/Reference/Reference_Tables/Electrochemistry_Tables/P2%3A_Standard_Reduction_Potentials_by_Value.

Standard cathode (reduction) half-reaction	E° (volts)
$La^{3+} + 3e^- \rightleftharpoons La(s)$	−2.38
$Ce^{3+} + 3e^- \rightleftharpoons Ce(s)$	−2.336
$Al(OH)_4{}^- + 3e^- \rightleftharpoons Al(s) + 4OH^-$	−2.310
$AlF_6{}^{3-} + 3e^- \rightleftharpoons Al(s) + 6F^-$	−2.07
$U^{3+} + 3e^- \rightleftharpoons U(s)$	−1.66
$Al^{3+}(aq) + 3e^- \rightleftharpoons Al(s)$	−1.676
$Zn(CN)_4{}^{2-} + 2e^- \rightleftharpoons Zn(s) + 4CN$	−1.34
$Zn(OH)_4{}^{2-} + 2e^- \rightleftharpoons Zn(s) + 4OH^-$	−1.285
$Mn^{2+} + 2e^- \rightleftharpoons Mn(s)$	−1.17
$V^{2+} + 2e^- \rightleftharpoons V(s)$	−1.13
$Zn(NH_3)_4{}^{2+} + 2e^- \rightleftharpoons Zn(s) + 4NH_3$	−1.04
$O_2(aq) + e^- \rightleftharpoons O_2{}^-(aq)$	−1.0
$Cd(CN)_4{}^{2-} + 2e^- \rightleftharpoons Cd(s) + 4CN^-$	−0.943
$MoO_4{}^{2-} + 4H_2O(l) + 6e^- \rightleftharpoons Mo(s) + 8OH^-$	−0.913
$Cr^{2+} + 2e^- \rightleftharpoons Cr(s)$	−0.90
$2H_2O(l) + 2e^- \rightleftharpoons H_2(g) + 2OH^-(aq)$	−0.828
$Zn^{2+}(aq) + 2e^- \rightleftharpoons Zn(s)$	−0.7618
$Co(OH)_2(s) + 2e^- \rightleftharpoons Co(s) + 2OH^-$	−0.746

Green Chemistry: Principles and Case Studies
By Felicia A. Etzkorn

Published by the Royal Society of Chemistry, www.rsc.org

Table E.1 *(Continued)*

Standard cathode (reduction) half-reaction	E° (volts)
$Cr^{3+}(aq) + 3e^- \rightleftharpoons Cr(s)$	−0.424
$Ni(OH)_2 + 2e^- \rightleftharpoons Ni(s) + 2OH^-$	−0.72
$Ag_2S(s) + 2e^- \rightleftharpoons 2Ag(s) + S^{2-}$	−0.71
$Cd(NH_3)_4{}^{2+} + 2e^- \rightleftharpoons Cd(s) + 4NH_3$	−0.622
$Ga^{3+} + 3e^- \rightleftharpoons Ga(s)$	−0.56
$U^{4+} + e^- \rightleftharpoons U^{3+}$	−0.52
$Sb + 3H^+ + 3e^- \rightleftharpoons SbH_3(g)$	−0.510
$Ni(NH_3)_6{}^{2+} + 2e^- \rightleftharpoons Ni(s) + 6NH_3$	−0.49
$Cr^{3+} + e^- \rightleftharpoons Cr^{2+}$	−0.424
$Fe^{2+}(aq) + 2e^- \rightleftharpoons Fe(s)$	−0.44
$Cd^{2+}(aq) + 2e^- \rightleftharpoons Cd(s)$	−0.4030
$Ag(NH_3)_2{}^+ + e^- \rightleftharpoons Ag(s) + 2NH_3$	−0.373
$Ti^{3+} + e^- \rightleftharpoons Ti^{2+}$	−0.37
$PbSO_4(s) + 2e^- \rightleftharpoons Pb(s) + SO_4{}^{2-}$	−0.356
$Co^{2+}(aq) + 2e^- \rightleftharpoons Co(s)$	−0.277
$Ni^{2+}(aq) + 2e^- \rightleftharpoons Ni(s)$	−0.257
$V^{3+} + e^- \rightleftharpoons V^{2+}$	−0.255
$As + 3H^+ + 3e^- \rightleftharpoons AsH_3(g)$	−0.225
$CO_2(g) + 2H^+ + 2e^- \rightleftharpoons HCO_2H$	−0.20
$Mo^{3+} + 3e^- \rightleftharpoons Mo(s)$	−0.2
$Sn^{2+} + 2e^- \rightleftharpoons Sn(s)$ in 1 M HCl	−0.19
$Ti^{2+} + 2e^- \rightleftharpoons Ti(s)$	−0.163
$MoO_2(s) + 4H^+ + 4e^- \rightleftharpoons Mo(s) + 2H_2O(l)$	−0.152
$AgI(s) + e^- \rightleftharpoons Ag(s) + I^-$	−0.152
$Sn^{2+}(aq) + 2e^- \rightleftharpoons Sn(s)$	−0.14
$Pb^{2+}(aq) + 2e^- \rightleftharpoons Pb(s)$	−0.126
$CrO_4{}^{2-} + 4H_2O(l) + 3e^- \rightleftharpoons 2Cr(OH)_4{}^- + 4OH^-$ (in 1 M NaOH)	−0.13
$WO_2(s) + 4H^+ + 4e^- \rightleftharpoons W(s) + 2H_2O(l)$	−0.119
$CO_2(g) + 2H^+ + 2e^- \rightleftharpoons CO(g) + H_2O(l)$	−0.106
$WO_3(s) + 6H^+ + 6e^- \rightleftharpoons W(s) + 3H_2O(l)$	−0.090
$Hg_2I_2(s) + 2e^- \rightleftharpoons 2Hg(l) + 2I^-$	−0.0405
$Fe^{3+}(aq) + 3e^- \rightleftharpoons Fe(s)$	−0.037
$2H^+(aq) + 2e^- \rightleftharpoons H_2(g)$	0.00
$AgBr(s) + e^- \rightleftharpoons Ag(s) + Br^-$	0.071
$Co(NH_3)_6{}^{3+} + e^- \rightleftharpoons Co(NH_3)_6{}^{2+}$	0.1
$Ru(NH_3)_6{}^{3+} + e^- \rightleftharpoons Ru(s) + Ru(NH_3)_6{}^{2+}$	0.10
$Sn^{4+}(aq) + 2e^- \rightleftharpoons Sn^{2+}(aq)$	0.154
$Cu^{2+}(aq) + e^- \rightleftharpoons Cu^+(aq)$	0.159
$UO_2{}^{2+} + e^- \rightleftharpoons UO_2{}^+$	0.16
$Co(OH)_3(s) + e^- \rightleftharpoons Co(OH)_2(s) + OH^-$	0.17
$ClO_4{}^-(aq) + H_2O(l) + 2e^- \rightleftharpoons ClO_3{}^-(aq) + 2OH^-(aq)$	0.17
$BiCl_4{}^- + 3e^- \rightleftharpoons Bi(s) + 4Cl^-$	0.199
$SbO^+ + 2H^+ + 3e^- \rightleftharpoons Sb(s) + H_2O(l)$	0.212
$AgCl(s) + e^- \rightleftharpoons Ag(s) + Cl^-(aq)$	0.2223
$HAsO_2 + 3H^+ + 3e^- \rightleftharpoons As(s) + 2H_2O(l)$	0.240
$Ru^{3+} + e^- \rightleftharpoons Ru^{2+}$	0.249
$Hg_2Cl_2(s) + 2e^- \rightleftharpoons 2Hg(l) + 2Cl^-$	0.2682
$UO_2{}^+ + 4H^+ + e^- \rightleftharpoons U^{4+} + 2H_2O(l)$	0.27
$Bi^{3+} + 3e \rightleftharpoons Bi(s)$	0.317

Table E.1 *(Continued)*

Standard cathode (reduction) half-reaction	E° (volts)
$UO_2^{2+} + 4H^+ + 2e^- \rightleftharpoons U^{4+} + 2H_2O(l)$	0.327
$VO^{2+} + 2H^+ + e^- \rightleftharpoons V^{3+} + H_2O(l)$	0.337
$Cu^{2+}(aq) + 2e^- \rightleftharpoons Cu(s)$	0.3419
$Fe(CN)_6{}^{3-} + e^- \rightleftharpoons Fe(CN)_6{}^{4-}$	0.356
$O_2(g) + 2H_2O(l) + 4e^- \rightleftharpoons 4OH^-$	0.401
$Ag_2C_2O_4(s) + 2e^- \rightleftharpoons 2Ag(s) + C_2O_4{}^{2-}$	0.47
$Cu^+(aq) + e^- \rightleftharpoons Cu(s)$	0.52
$Cu^{2+} + Cl^- + e^- \rightleftharpoons CuCl(s)$	0.559
$H_3AsO_4 + 2H^+ + 2e^- \rightleftharpoons HAsO_2 + 2H_2O(l)$	0.560
$MnO_4{}^- + 2H_2O(l) + 3e^- \rightleftharpoons MnO_2(s) + 4OH^-$	0.60
$Sb_2O_5(s) + 6H^+ + 4e^- \rightleftharpoons 2SbO^+ + 3H_2O(l)$	0.605
$PtCl_6{}^{2-} + 2e^- \rightleftharpoons PtCl_4{}^{2-} + 2Cl^-$	0.68
$RuO_2(s) + 4H^+ + 4e^- \rightleftharpoons Ru(s) + 2H_2O(l)$	0.68
$O_2(g) + 2H^+ + 2e^- \rightleftharpoons H_2O_2$	0.695
$PtCl_4{}^{2-} + 2e^- \rightleftharpoons Pt(s) + 4Cl^-$	0.73
$Tl^{3+} + 3e^- \rightleftharpoons Tl(s)$	0.742
$Fe^{3+}(aq) + e^- \rightleftharpoons Fe^{2+}(aq)$	0.771
$Hg_2{}^{2+}(aq) + 2e^- \rightleftharpoons 2Hg(l)$	0.7960
$Ag^+(aq) + e^- \rightleftharpoons Ag(s)$	0.7996
$Hg^{2+}(aq) + 2e^- \rightleftharpoons Hg(l)$	0.8535
$Cu^{2+} + I^- + e^- \rightleftharpoons CuI(s)$	0.86
$Ru(CN)_6{}^{3-} + e^- \rightleftharpoons Ru(s) + Ru(CN)_6{}^{4-}$	0.86
$2Hg^{2+}(aq) + 2e^- \rightleftharpoons Hg_2{}^{2+}(aq)$	0.911
$HgO(s) + 2H^+ + 2e^- \rightleftharpoons Hg(l) + H_2O(l)$	0.926
$MnO_2(s) + 4H^+ + e^- \rightleftharpoons Mn^{3+}(aq) + H_2O(l)$	0.95
$VO_2{}^{2+} + 2H^+ + e^- \rightleftharpoons VO^{2+} + H_2O(l)$	1.000
$AuCl_4{}^- + 3e^- \rightleftharpoons Au(s) + 4Cl^-$	1.002
$Fe(phen)_6{}^{3+} + e^- \rightleftharpoons Fe(phen)_6{}^{2+}$	1.147
$Pt^{2+} + 2e^- \rightleftharpoons Pt(s)$	1.2
$O_2(g) + 4H^+(aq) + 4e^- \rightleftharpoons 2H_2O(l)$	1.229
$MnO_2(s) + 4H^+ + 2e^- \rightleftharpoons Mn^{2+} + 2H_2O(l)$	1.23
$Tl^{3+} + 2e^- \rightleftharpoons Tl^+$ (in 1 M $HClO_4$)	1.25
$Cr_2O_7{}^{2-}(aq) + 14H^+(aq) + 6e^- \rightleftharpoons 2Cr^{3+}(aq) + 7H_2O(l)$	1.36
$Cr_2O_7{}^{2-} + 14H^+ + 6e^- \rightleftharpoons 2Cr^{3+} + 7H_2O(l)$	1.36
$Au^{3+} + 2e^- \rightleftharpoons Au^+$	1.36
$Hg_2Br_2(s) + 2e^- \rightleftharpoons 2Hg(l) + 2Br^-$	1.392
$Ce^{4+}(aq) + e^- \rightleftharpoons Ce^{3+}(aq)$	1.44
$PbO_2(s) + 4H^+ + 2e^- \rightleftharpoons Pb^{2+}(aq) + 2H_2O(l)$	1.46
$Mn^{3+} + e^- \rightleftharpoons Mn^{2+}$	1.5
$MnO_4{}^-(aq) + 8H^+(aq) + 5e^- \rightleftharpoons Mn^{2+}(aq) + 4H_2O(l)$	1.51
$Au^{3+} + 3e^- \rightleftharpoons Au(s)$	1.52
$PbO_2(s) + 4SO_4{}^{2-} + 4H^+ + 2e^- \rightleftharpoons PbSO_4(s) + 2H_2O(l)$	1.690
$MnO_4{}^- + 4H^+ + 3e^- \rightleftharpoons MnO_2(s) + 2H_2O(l)$	1.70
$Ce^{4+} + e^- \rightleftharpoons Ce^{3+}$	1.72
$H_2O_2(aq) + 2H^+(aq) + 2e^- \rightleftharpoons 2H_2O(l)$	1.763
$Au^+ + e^- \rightleftharpoons Au(s)$	1.83
$Co^{3+}(aq) + e^- \rightleftharpoons Co^{2+}(aq)$	1.92
$O_3(g) + 2H^+(aq) + 2e^- \rightleftharpoons O_2(g) + H_2O(l)$	2.07

Appendix F Solvent Selection Guide

Table F.1 Reproduced from D. Prat, J. Hayler and A. Wells, *Green Chemistry*, 2014, **16**, 4546, with permission from the Royal Society of Chemistry.[a]

Family	Solvent	BP (°C)	FP (°C)	Worst H3xx*	H4xx	Safety score	Health score	Env. score	Ranking by default	Ranking after discussion[#]
Water	Water	100	na	none	none	1	1	1	Recommended	Recommended
Alcohols	MeOH	65	11	H301	none	4	7	5	Problematic	Recommended
	EtOH	78	13	H319	none	4	3	3	Recommended	Recommended
	i-PrOH	82	12	H319	none	4	3	3	Recommended	Recommended
	n-BuOH	118	29	H318	none	3	4	3	Recommended	Recommended
	t-BuOH[+]	82	11	H319	none	4	3	3	Recommended	Recommended
	Benzyl alcohol	206	101	H302	none	1	2	7	Problematic	Problematic
	Ethylene glycol	198	116	H302	none	1	2	5	Recommended	Recommended
Ketones	Acetone	56	-18	H319	none	5	3	5	Problematic	Recommended
	MEK	80	-6	H319	none	5	3	3	Recommended	Recommended
	MIBK	117	13	H319	none	4	2	3	Recommended	Recommended
	Cyclohexanone	156	43	H332	none	3	2	5	Recommended	Problematic
Esters	Methyl acetate	57	-10	H302	none	5	3	5	Problematic	Problematic
	Ethyl acetate	77	-4	H319	none	5	3	3	Recommended	Recommended
	i-PrOAc	89	2	H319	none	4	2	3	Recommended	Recommended
	n-BuOAc	126	22	H336	none	4	2	3	Recommended	Recommended
Ethers	Diethyl ether	34	-45	H302	none	10	3	7	Hazardous	HH
	Diisopropyl ether	69	-28	H336	none	9	3	5	Hazardous	Hazardous
	MTBE	55	-28	H315	none	8	3	5	Hazardous	Hazardous
	THF	66	-14	H351	none	6	7	5	Problematic	Problematic
	Me-THF	80	-11	H318	none	6	5	3	Problematic	Problematic
	1,4-Dioxane	101	12	H351	none	7	6	3	Problematic	Hazardous
	Anisole	154	52	none	none	4	1	5	Problematic	Recommended
	DME	85	-6	H360	none	7	10	3	Hazardous	Hazardous
Hydrocarbons	Pentane	36	-40	H304	H411	8	3	7	Hazardous	Hazardous
	Hexane	69	-22	H361	H411	8	7	7	Hazardous	Hazardous
	Heptane	98	-4	H304	H410	6	2	7	Problematic	Problematic
	Cyclohexane	81	-17	H304	H410	6	3	7	Problematic	Problematic
	Me-Cyclohexane	101	-4	H304	H411	6	2	7	Problematic	Problematic
	Benzene	80	-11	H350	none	6	10	3	Hazardous	HH
	Toluene	111	4	H351	none	5	6	3	Problematic	Problematic
	Xylenes	140	27	H312	none	4	2	5	Problematic	Problematic
Halogenated	DCM	40	na	H351	none	1	7	7	Hazardous	Hazardous
	Chloroform	61	na	H351	none	2	7	5	Problematic	HH
	CCl_4	77	na	H351	H420	2	7	10	Hazardous	HH
	DCE	84	13	H350	none	4	10	3	Hazardous	HH
	Chlorobenzene	132	29	H332	H411	3	2	7	Problematic	Problematic

Green Chemistry: Principles and Case Studies
By Felicia A. Etzkorn

Published by the Royal Society of Chemistry, www.rsc.org

Table F.1 *(Continued)*

Family	Solvent	BP (°C)	FP (°C)	Worst H3xx*	H4xx	Safety score	Health score	Env. score	Ranking by default	Ranking after discussion#
Aprotic polar	Acetonitrile	82	2	H319	none	4	3	3	Recommended	Problematic
	DMF	153	58	H360	none	3	9	5	Hazardous	Hazardous
	DMAc	166	70	H360	none	1	9	5	Hazardous	Hazardous
	NMP	202	96	H360	none	1	9	7	Hazardous	Hazardous
	DMPU	246	121	H361	none	1	6	7	Problematic	Problematic
	DMSO+	189	95	none	none	1	1	5	Recommended	Problematic
	Sulfolane+	287	177	H360	none	1	9	7	Hazardous	Hazardous
	HMPA	>200	144	H350	none	1	9	7	Hazardous	HH
	Nitromethane	101	35	H302	none	10	2	3	Hazardous	HH
Miscellaneous	Methoxy-ethanol	125	42	H360	none	3	9	3	Hazardous	Hazardous
	Carbon disulfide	46	-30	H361	H412	9	7	7	Hazardous	HH
Acids	Formic acid	101	49	H314	none	3	7	3	Problematic	Problematic
	Acetic acid	118	39	H314	none	3	7	3	Problematic	Problematic
	Ac_2O	139	49	H314	none	3	7	3	Problematic	Problematic
Amines	Pyridine	115	23	H302	none	4	2	3	Recommended	Hazardous
	TEA	89	-6	H314	none	6	7	3	Problematic	Hazardous

[a]*Only the highest scoring statements are shown. The lowest figure is given when there are more than one statement in the highest scoring category, for the sake of simplicity; #HH: highly hazardous; +Solid at 20 °C.

Appendix G Selected Bond Dissociation Energies

Table G.1 Standard bond energies.

Single bonds	$\Delta H^{\circ\, a}$	Single bonds	$\Delta H^{\circ\, a}$	Multiple bonds	$\Delta H^{\circ\, a}$
H–H	104.2	B–F	150	C=C	146
C–C	83	B–O	125	N=N	109
N–N	38.4	C–N	73	O=O	119
O–O	35	N–CO	86	C=N	147
F–F	36.6	C–O	85.5	C=O (CO_2)	192
Si–Si	52	O–CO	110	C=O (aldehyde)	177
P–P	50	C–S	65	C=O (ketone)	178
S–S	54	C–F	116	C=O (ester)	179
Cl–Cl	58	C–Cl	81	C=O (amide)	179
Br–Br	46	C–Br	68	C=O (halide)	177
I–I	36	C–I	51	C=S (CS_2)	138
H–C	99	C–B	90	N=O (HONO)	143
H–N	93	C–Si	76	P=O ($POCl_3$)	110
H–O	111	C–P	70	P=S ($PSCl_3$)	70
H–F	135	N–O	55	S=O (SO_2)	128
H–Cl	103	S–O	87	S=O (DMSO)	93
H–Br	87.5	Si–F	135	P=P	84
H–I	71	Si–Cl	90	P≡P	117
H–B	90	Si–O	110	C≡O	258
H–S	81	P–Cl	79	C≡C	200
H–Si	75	P–Br	65	N≡N	226
H–P	77	P–O	90	C≡N	213

[a]Average Bond Dissociation Enthalpies in kcal per mole (There can be considerable variability in some of these values.).

Green Chemistry: Principles and Case Studies
By Felicia A. Etzkorn

Published by the Royal Society of Chemistry, www.rsc.org

Table G.2 Bond dissociation energies.[a]

Atom or group	Methyl	Ethyl	*i*-Propyl	*t*-Butyl	Phenyl	Benzyl	Allyl	Acetyl	Vinyl
H	103	98	95	93	110	85	88	87	112
F	110	110	109		124	94		119	
Cl	85	82	81	80	95	68	70	82	90
Br	71	70	69	66	79	55	56	68	80
I	57	54	54	51	64	40	42	51	
OH	93	94	92	91	111	79	82	107	
NH_2	87	87	86	85	104	72	75	95	
CN	116	114	112		128	100			128
CH_3	88	85	84	81	101	73	75	81	98
C_2H_5	85	82	81	78	99	71	72	78	95
$(CH_3)_2CH$	84	81	79	74	97	70	71	76	93
$(CH_3)_3C$	81	78	74	68	94	67	67		89
C_6H_5	101	99	97	94	110	83	87	93	108
$C_6H_5CH_2$	73	71	70	67	83	59	59	63	81

[a]In kcal per mole.

Many of the bond energies listed here were taken from the following sources:

R. T. Sanderson, *Polar Covalence*, 1983.
R. T. Sanderson, *Chemical Bonds and Bond Energy*, 1976.

Subject Index

References to figures are given in *italic* type. References to tables are given in **bold** type.